Instant
Biochemistry

Other books by the same author

- MCQs in Clinical Biochemistry, 1999, Sci. Tech. Publ. Chennai.
- Textbook of Biochemistry—Molecular and Clinical Aspects 2nd ed., 2007, Sci Tech Publ. Chennai.
- Dictionary of Biochemistry Biotechnology and Bioinformatics 2nd ed., 2008, Sci Tech Publ. Chennai.

Instant
Biochemistry

S. Nagini

Department of Biochemistry and Biotechnology,
Faculty of Science, Annamalai University
Annamalainagar,
Tamil Nadu

Ane Books Pvt. Ltd.

New Delhi ♦ Chennai

Instant Biochemistry

S. Nagini

First Edition : 2014
Reprint : 2018, **2020, 2023**

Published by

Ane Books Pvt. Ltd.
4821, Parwana Bhawan, 1st Floor, 24 Ansari Road, Darya Ganj,
New Delhi - 110 002, Tel.: +91(011) 23276843-44, Fax: +91(011) 23276863
e-mail: kapoor@anebooks.com, Website: www.anebooks.com

Branch Office:

- Avantika Niwas, 1st Floor, 19 Doraiswamy Road, T. Nagar,
Chennai - 600 017, Tel.: +91(044) 28141554, 28141209
e-mail: anebookschennai@gmail.com, rathinam@anebooks.com

ISBN: 978-81-80521-58-4

Printed at : Repro Knowledgecast Limited, Thane

Preface

Knowledge explosion in biochemistry and molecular biology in recent years has made it a daunting task for students to keep pace with the phenomenal advances in the field especially when faced with the challenge of success in examinations. This book is intended for students who specialise in biochemistry at the undergraduate and postgraduate levels as well as medical and dental students who take up biochemistry as an ancillary subject. As the title suggests, this book will enable students to have a clear and quick grasp of the subject.

This book provides a systematic coverage of basic, clinical, molecular, and applied aspects of biochemistry organised in a concise, yet comprehensive style and can be used both as a standard textbook and as a review before an examination. The chapters are organised logically, starting from structure and functions of cells and subcellular organelles, and progressing to metabolic pathways, hormonal regulation, homeostatic mechanisms before progressing to clinical aspects. This is followed by molecular biology, genetic engineering, analytical techniques and ending with the 'omics' disciplines that have revolutionised biology. The book adopts a logically sequenced question-and-answer format with the answer for one question leading to the next question. Ample illustrations have been provided to aid understanding. I hope that students would find this book a useful companion to standard textbooks and as a source of concise information.

I am grateful to the authorities of Annamalai University for support. I thank the entire team at Ane Books Pvt. Ltd. for the support and co-operation in bringing out this book.

S. Nagini

Contents

Section 5: Homeostasis

Section 6: Function Tests and Plasma Proteins

Section 7: Molecular Biology and Genetic Engineering

Section 8: Analytical Techniques

Section 9: Special Topics

Section 1

Cell Structure and Biomolecules

- The Cell and Subcellular Organelles
- Structures and Functions of Biomolecules

Chapter 1

The Cell and Subcellular Organelles

❑ What are prokaryotes?

- Prokaryotes are organisms that lack a nuclear envelope.
- The chromosome in prokaryotes is confined to a nuclear zone called the *nucleoid*, which is in direct contact with the protoplasm.
- The prokaryotes are subdivided into *Eubacteria* (bacteria and blue-green algae) and the *Archae* (*Archaebacteria*).

❑ What are eukaryotes?

Eukaryotes are organisms that have a true nucleus bounded by a nuclear membrane. These include fungi, plants, and animals.

❑ Define autotrophs and heterotrophs.

- ***Autotrophs*** are organisms that obtain energy from photosynthesis, a process in which radiation from the sunlight is used to transform carbon dioxide and water into glucose and other complex carbohydrates. Green plants are autotrophic.
- ***Heterotrophs*** are organisms that obtain energy from carbohydrates, fats, and proteins synthesized by the autotrophs. Animals are heterotrophs.

❑ Elucidate the structural components of a bacterial cell.

Figure 1.1 shows the structure of a bacterial cell.

The components of a bacterial cell include

- ***Cell wall***
  - is the outer rigid layer surrounding the bacterial cell
  - provides mechanical protection to the cell
  - is made up of polysaccharides, lipids, and proteins
  - the most abundant polypeptide in the cell wall is *porin*, which forms channels that facilitate diffusion of solutes.
- ***Plasma membrane***
  - is the inner layer surrounding the bacterial cell
  - is composed of lipids and proteins
  - functions as a barrier for the transport of ions and molecules across the cell

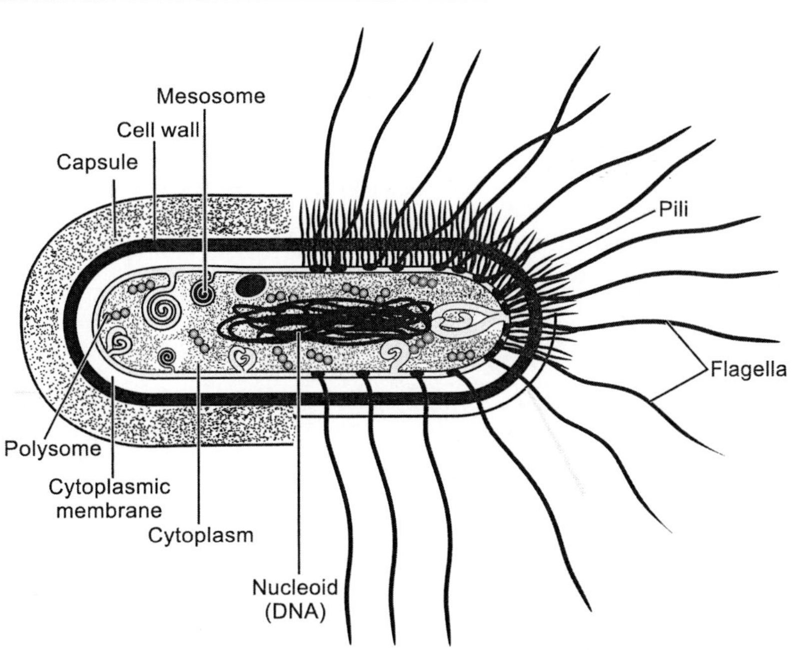

Fig. 1.1 Structure of a bacterial cell

- maintains the internal milieu of the protoplasm
- contains enzymes involved in oxidative and photophosphorylation.

Capsule

- surrounds some bacterial cell walls
- is made up of polysaccharides or polypeptides
- protects the bacterial cell from predators
- facilitates attachment of the cell to a substratum and to another cell.

Chromosome

- is found in the nucleoid
- in general, *is* a single, circular, double-stranded DNA molecule. The *E.coli* DNA is ~1 *mm* in length (10^6 nm) and contains 4.6 million bp
- some bacteria contain multiple chromosomes—linear, single-stranded DNA or RNA.

Plasmids

- are extrachromosomal, circular, double-stranded DNA capable of autonomous replication
- contain genes encoding 20–100 proteins most of which facilitate transfer of genetic information to other cells, promote the metabolism of unusual compounds, or confer resistance to chemicals, heavy metals, or antibiotics.

Ribosomes

- are ribonucleoprotein particles surrounding the DNA in the dark region of the protoplasm
- there are ~ 20,000–30,000 ribosomes present as polysomes
- they are the sites of protein synthesis
- the prokaryotic ribosomes have a sedimentation velocity of 70S, where S is the rate of sedimentation in Svedberg units
- each 70S ribosome contains two subunits—a larger 50S and a smaller 30S subunit.

- ***Flagella***
  hair like appendages in the plasma membrane that facilitate motility in some bacteria.
- ***Pili***
  - long hollow tubes that contribute to adhesiveness in some bacteria
  - the sex pili are involved in DNA transfer from a donor cell to a recipient cell during conjugation.

❑ Describe the structural components of a eukaryotic cell.

The eukaryotic cell consists of three main compartments: the cell (plasma) membrane, nucleus, and the cytoplasm (Fig. 1.2).

Plasma membrane

- separates the cellular contents from the external environment
- is a thin film of 6–10 nm thickness, made up of a phospholipid bilayer with proteins attached to the surface or intercalated into the bilayer

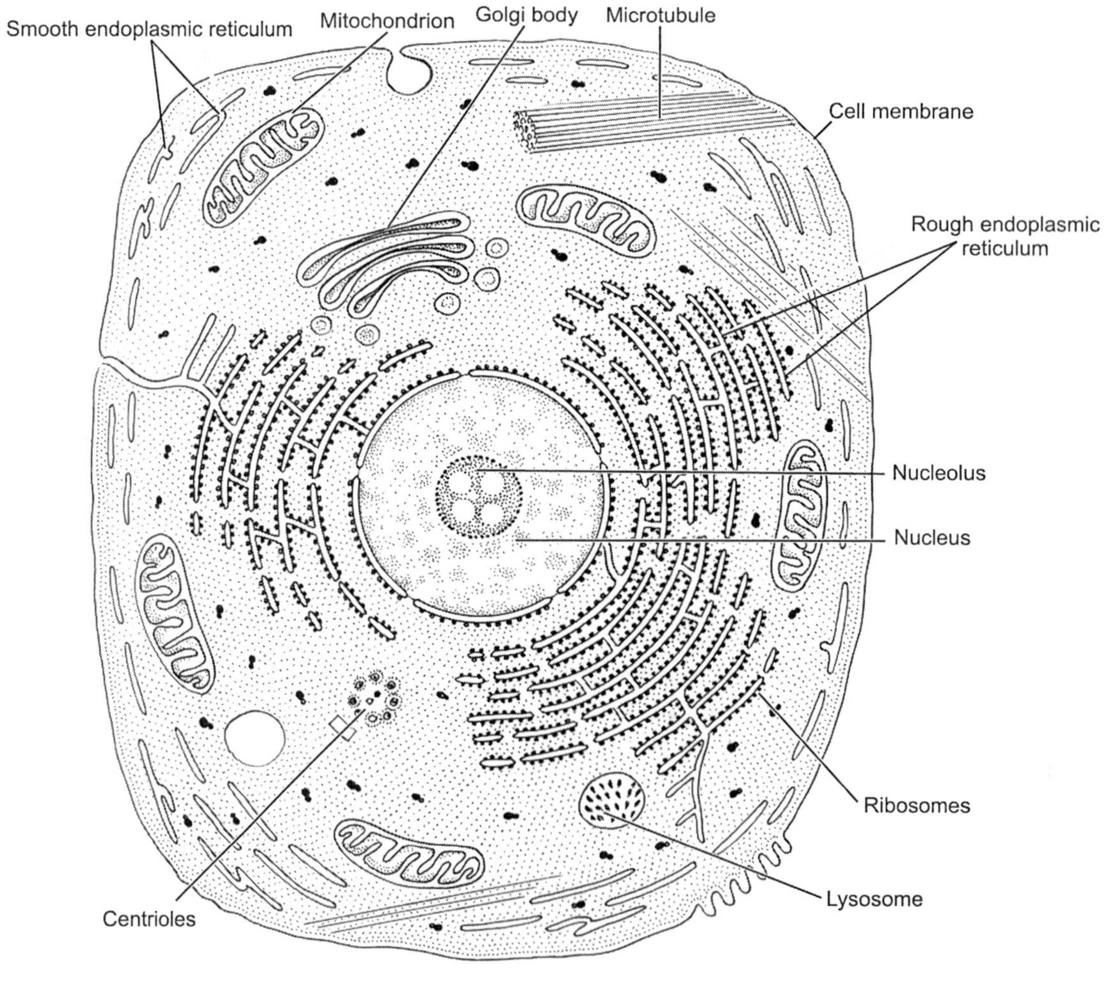

Fig. 1.2 Structure of a eukaryotic cell

- in plant cells, it is covered by a cell wall made up of cellulose fibers. In some plant cells, lignin is incorporated into the cellulose fibers. The cell wall is responsible for rigidity and offers protection
- in animal cells, it is covered by a cell coat made of glycoproteins, glycolipids, and polysaccharides. The cell coat provides protection and is involved in intercellular communication.
- ***Functions***
  - selectively permeable membrane barrier that regulates the entrance (endocytosis) and exit (exocytosis) of ions and molecules.
  - cell – cell signaling and cell adhesion.

The nucleus

In the nondividing phase of the cell cycle (interphase), the following structures can be identified in the nucleus:

- A double membrane nuclear envelope that separates the nucleus from the cytoplasm. The nuclear envelope is perforated by nuclear pores at intervals.
- The nucleoplasm that fills the nuclear space. This material represents uncondensed chromatin that corresponds to *euchromatin*.
- The chromocenters represent the condensed regions of the chromosomes that correspond to the *heterochromatin*. These are found near the nucleus and are attached to the nucleolus.
- The nucleoli, which may be single or multiple. The nucleolus contains high concentrations of RNA and protein. It is the site of rRNA synthesis and ribosome assembly.

  During cell division the nucleus undergoes the following changes:
  - the nucleolus and the nuclear envelope disappear
  - the chromatin becomes condensed to form dark staining bodies called *chromosomes*. Chromosomes contain the nucleoprotein chromatin, made of DNA and basic proteins called *histones*.

Cytoplasm

Refers to the intracellular contents between the nucleus and the plasma membrane. It can be divided into the endomembrane system and the cytoplasmic matrix.

- ***The endomembrane system*** which is involved in sequestering, packaging, and transport of substances synthesized and used within the cell or transported to the external medium.

  Components include
  - *the nuclear envelope*, which is a double membrane structure made of flattened sacs or cisternae. The two membranes merge at regular intervals to form pores through which substances move from the nucleus to the cytoplasm.
  - *the endoplasmic reticulum* (ER), which is made up of tubules and flattened sacs. It is of two types:
    i. *rough endoplasmic reticulum* (*RER*) studded with ribosomes, which are the sites of protein synthesis.
    ii. *smooth endoplasmic reticulum* (*SER*), which is agranular, is involved in the synthesis of lipids and detoxification of hydrophobic compounds.
  - *the Golgi complex* made up of stacks of flattened sacs and vesicles called *dictyosomes*. This complex is involved in processing and packaging of secretory proteins, lysosomal proteins, and membrane proteins synthesized on RER. These proteins are then released by exocytosis from the cell.

- ***The cytoplasmic matrix***, constitutes the cytoskeleton that contains the structures involved in cell shape, motility, and muscle contraction. It supports the cell membranes, helps in organization of organelles, and participates in cell movement. It comprises:
  - *the microtubules,* predominant components in cell organelles such as the asters, spindle, and centrioles, which are involved in mitosis, and the cilia and flagella which play a role in locomotion.
  - *the microfilaments* composed of the protein actin, which together with myosin is involved in muscle contraction.
  - *the intermediate filaments,* so called because of their size (10 nm) which is in between that of microtubules (25 nm) and microfilaments (4 – 6 nm). These are made up of fibrous proteins and serve a mechanical function.
  - *membrane organelles.*

The cytoplasm contains several membrane-bound organelles such as the mitochondria, chloroplasts, lysosomes, and peroxisomes.

❑ What is the difference between the cytoplasm and the cytosol?

The cytoplasm includes organelles, whereas the cytosol refers to the fluid portion of the cytoplasm outside the organelles.

❑ What is meant by karyotype?

- Karyotype refers to the characteristics of a complete chromosomal set that includes the number of chromosomes, relative size, position of the centromere, length of the arms, and satellites.
- The karyotype is characteristic of an individual, species, genus, or large groups.

❑ What are satellites?

Satellites are rounded bodies seen in some chromosomes. They are separated from the rest of the chromosomes by a secondary constriction.

❑ Elucidate the structure and functions of the mitochondrion.

- ***Structure***
  - The mitochondria are ovoid or rod shaped organelles surrounded by a double membrane (Fig. 1.3).
  - The inner membrane contains the components of the electron transport chain and is less permeable compared to the outer membrane.
  - The inner membrane is thrown into folds or *cristae,* which protrude into a central amorphous matrix.
  - The inner surface of the inner membrane contains F_0F_1-ATPase.
  - The mitochondria constitute a semiautonomous genetic system and contain their own DNA, RNA, and ribosomes. The ribosomes are of the same size as those in bacteria, supporting the theory that mitochondria have evolved from prokaryotes.
- ***Functions***
  - The mitochondria are the sites of energy production and generate energy in the form of ATP. Mitochondria are therefore regarded as the '*power houses*' of the cell.
  - Mitochondria also play a key role in programmed cell death or apoptosis.

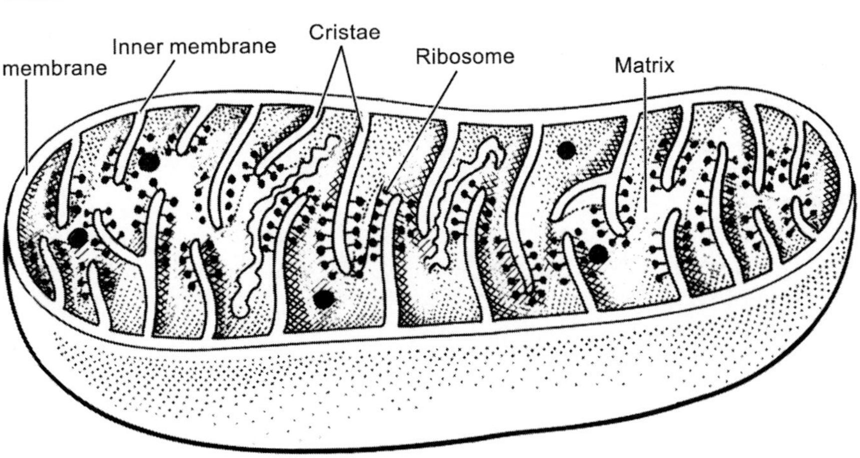

Fig. 1.3 Structure of a mitochondrion

❑ What are chloroplasts?

- Chloroplasts are subcellular organelles surrounded by a double membrane.
- They are responsible for generating energy in the form of ATP by oxidation of glucose and fatty acids.
- They constitute a semiautonomous genetic system and contain their own DNA, RNA, and ribosomes. The ribosomes are of the same size as those in bacteria, supporting the theory that these organelles have evolved from prokaryotes.

❑ Give an account of lysosomes.

- The lysosomes are subcellular organelles found in eukaryotic cells.
- They have an acidic lumen containing enzymes that digest substances internalized by the cell as well as worn out cellular membranes and organelles.
- The lysosomal enzymes themselves are resistant to acid denaturation and are inactivated at neutral pH such as that prevailing in the cytosol.
- About 50 lysosomal hydrolases have been identified, which act on different substrates. These include:
  a. nucleases, which degrade nucleic acids to mononucleotides
  b. proteases, which degrade proteins and peptides
  c. phosphatases, which remove phosphate groups from mononucleotides, phospholipids, etc.
  d. glycosidases, which degrade complex polysaccharides and glycolipids.

❑ What are peroxisomes?

- Peroxisomes are single membrane-bound organelles present in many plant cells and all animal cells.
- They are rich in the oxidative enzymes peroxidase, catalase, D-amino acid oxidase, and urate oxidase.
- ***Functions***
  - formation and degradation of hydrogen peroxide

- oxidation of fatty acids that generates acetyl groups which are transported to the cytosol and used for cholesterol biosynthesis
- detoxification of toxic substances in the liver and kidney.

❑ Discuss the structure and types of endoplasmic reticulum.

◘ *Structure*

The endoplasmic reticulum (ER) is an extensive network of membrane-bound flattened sacs called *cisternae*. It is involved in the synthesis of lipids, membrane proteins, and secreted proteins. The ER is of two types namely, rough endoplasmic reticulum (RER) studded with ribosomes, and smooth endoplasmic reticulum (SER), which is agranular.

◘ *RER*

- is granular in appearance due to the presence of ribosomes, which are the sites of protein synthesis.
- is found in most eukaryotic cells and abundant in specialized cells that make large amounts of a specific protein such as the β-cells of the pancreatic islets that synthesize insulin.
- The sedimentation velocity is 70S for prokaryotic ribosomes and 80S for eukaryotic ribosomes (S is the rate of sedimentation in Svedberg units).
- Each ribosome has a large and a small subunit.

◘ *SER*

- lacks ribosomes.
- is abundant in hepatocytes.
- contains
  a. enzymes for the synthesis of triglycerides, phospholipids, and cholesterol
  b. the electron transport systems NADH, cytochrome *c* reductase, and NADH cytochrome *b*5 reductase
  c. the hemoproteins cytochrome P450 and *b*5
  d. peptidases and glycosyl transferases
  e. enzymes for the detoxification of hydrophobic chemicals such as pesticides and carcinogens. These enzymes convert xenobiotics into water-soluble conjugated products that can be excreted from the body.

❑ Elucidate the structure and functions of the Golgi complex.

◘ *Structure*

- The Golgi complex is a differentiated portion of the endomembrane system found in both plant and animal cells.
- It consists of dictyosomes, which are collections of membrane-bound flattened vesicles or sacs (cisternae). The cisternae lack ribosomes.
- Each dictyosome contains a proximal surface termed the *forming face* and a distal surface called the *maturation face*.

◘ *Functions*

- Processing and modification of secretion products synthesized in the ER.
- Addition of carbohydrate residues to proteins and lipids. The sugar residues are added sequentially by glycosyltransferases.
- Synthesis of lipoproteins.
- Secretion of proteins that are to be exported and of enzymes present in lysosomes and peroxisomes.
- Formation of lysosomes.

Chapter 2

Structures and Functions of Biomolecules

❑ What are carbohydrates?

Carbohydrates are polyhydroxy aldehydes or ketones, or sugars that generate these compounds on hydrolysis.

❑ How are carbohydrates classified?

Carbohydrates are classified into three major classes:

Monosaccharides

- These are simple sugars that contain a single polyhydroxy aldehyde or ketone unit.
- They are referred to as *reducing sugars* because of their capacity to reduce ferric or cupric ions.
- They are subclassified into *trioses, tetroses, pentoses,* and *hexoses* based on the number of carbon atoms, and as *aldoses* and *ketoses* based on whether they possess an aldehyde or a ketone group.
- Examples of common monosaccharides are *glucose, fructose* and *galactose.* Their structures are shown in Figure 2.1.

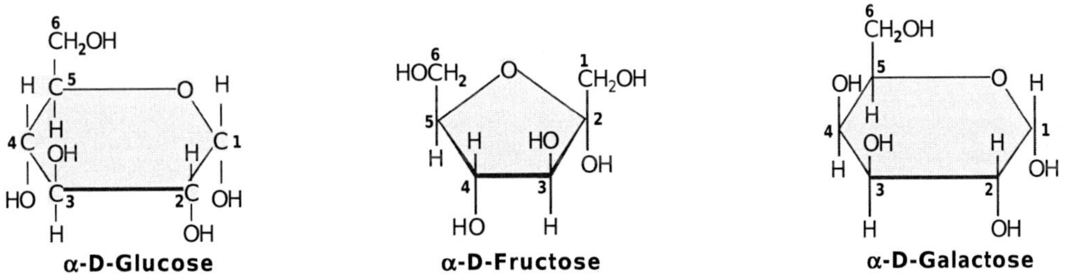

Fig. 2.1 Structures of monosaccharides

Oligosaccharides

- These contain monosaccharide units linked together by a *glycosidic bond.*
- *Disaccharides* containing two monosaccharide units linked by a glycosidic bond are the most abundant oligosaccharides in nature. They may be reducing or non-reducing sugars. Examples of common disaccharides are:

  a. *maltose,* which contains 2 glucose units

b. *lactose* (milk sugar), which contains glucose and galactose units
c. *sucrose* (cane sugar), which contains one unit each of glucose and fructose

Maltose and lactose are reducing sugars, whereas sucrose is non-reducing. Their structures are shown in Figure 2.2.

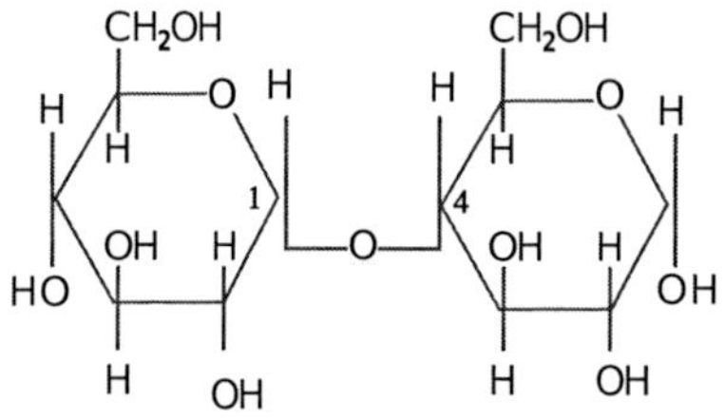

Maltose (Glc α1 → 4 Glc)

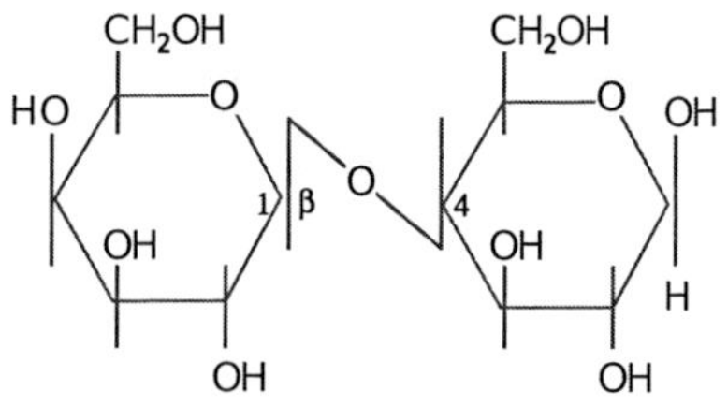

Lactose (Gal β1 → 4 Glc)

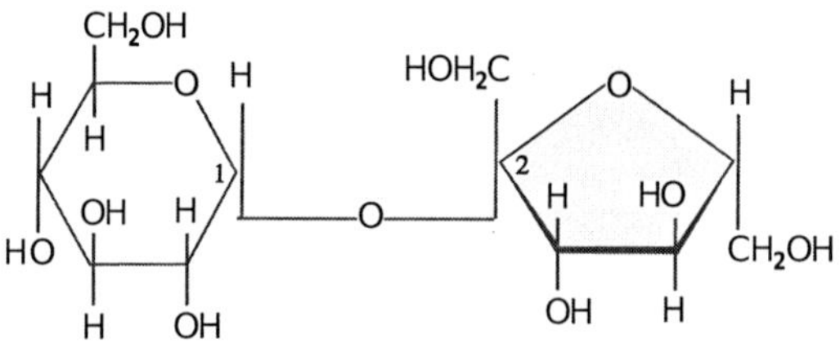

Sucrose (Glc α1 → β2 Fru)

Fig. 2.2 Structures of disaccharides

- ***Polysaccharides***

  These contain many monosaccharide units joined by glycosidic bonds. Polysaccharides are subclassified into

  - *Homopolysaccharides*: These contain only a single type of monosaccharide unit. Examples include starch and glycogen.
  - *Heteropolysaccharides* (*glycosaminoglycans*): These contain two or more types of monosaccharide units. Examples include chondroitin sulphate and heparin.

❑ What is a glycoside?

- A *glycoside* is the product of condensation between the hydroxyl group of the anomeric carbon of a monosaccharide and another compound that may or may not be a monosaccharide. If it is not a monosaccharide, then it is termed *aglycone*. Methanol, phenol, and adenine molecules are some common examples of aglycones.
- If the reaction occurs between a hemiacetal group and a hydroxyl group, the *O*-acetyl group is termed an *acetal* link.
- The resulting compound is referred to as a *glucoside*, if the hemiacetal group is glucose and a *galactoside*, if it is galactose.
- Cardiac glycosides contain steroids as the aglycone moiety and include *digitalis*, and *ouabain*, an inhibitor of Na^+K^+-*ATPase*.
- The antibiotic *streptomycin* is also a glycoside.

❑ List the physiologically important pentoses.

The physiologically important pentoses include

- *Ribose*, a structural component of nucleic acids and coenzymes such as ATP and NAD.
- *Arabinose* and *xylose* that are constituents of glycoproteins.
- *Ribulose*, an intermediate in the pentose phosphate pathway.

❑ List the physiologically important hexoses.

Physiologically important hexoses include

- *Glucose*, the most abundant monosaccharide in nature that is obtained by the hydrolysis of starch, cane sugar, maltose, and lactose. It is the principal sugar in the blood and is utilized by tissues for energy.
- *Fructose*, which occurs in high concentrations in fruit juices and honey.
- *Galactose*, a constituent of glycolipids and glycoproteins that is obtained by hydrolysis of lactose.

❑ Define epimers and anomers.

- *Epimers* are sugars that differ only in the configuration around one carbon atom. For example, D-glucose and D-galactose differ in their spatial arrangement only at C-4.
- *Anomers* are the isomeric forms of monosaccharides that differ only in the configuration around the hemiacetal or hemiketal carbon atoms.

❑ What are deoxy sugars?

Deoxy sugars are those sugars in which the hydroxyl group is replaced by hydrogen. For example, in deoxyribose, a constituent of DNA, the hydroxyl group in position 2 is replaced by a hydrogen (Fig. 2.3).

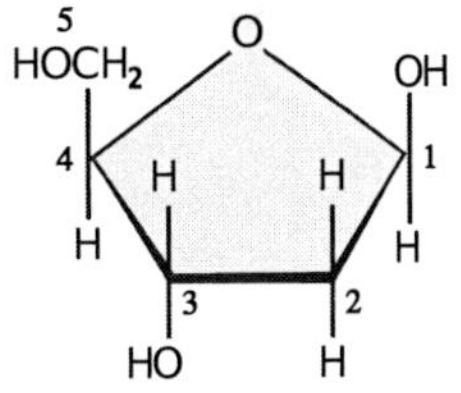

Fig. 2.3 Deoxyribose

❑ What are amino sugars?

Amino sugars are constituents of glycoproteins, gangliosides, and glycosaminoglycans. For example, hyaluronic acid contains D-glucosamine, and chondroitin contains D-galactosamine as amino sugars. Antibiotics such as erythromycin also contain amino sugars. The structure of glucosamine is shown in Figure 2.4.

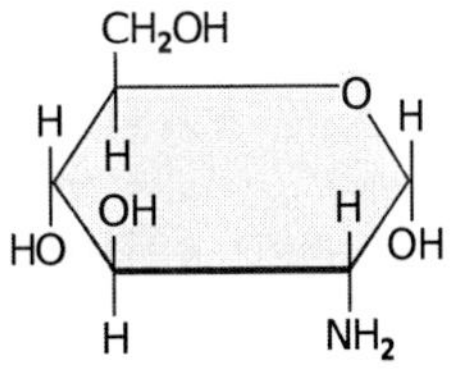

Fig. 2.4 α-D-Glucosamine

❑ Give an account of homopolysaccharides.

Homopolysaccharides are polysaccharides that contain a single type of monosaccharide unit.

Examples include

- *Starch*, a homopolymer of glucose (*glucosan*) that functions as a storage form of carbohydrate in plants. It contains two components:
  a. *Amylose*, a nonbranching helical structure in which the glucose units are linked by $\alpha 1 \rightarrow 4$ glycosidic bonds.
  b. *Amylopectin*, containing branched chains with the glucose residues joined by $\alpha 1 \rightarrow 4$ glycosidic linkages in the straight chains, and by $\alpha 1 \rightarrow 6$ linkages at the branch points (Fig. 2.5).

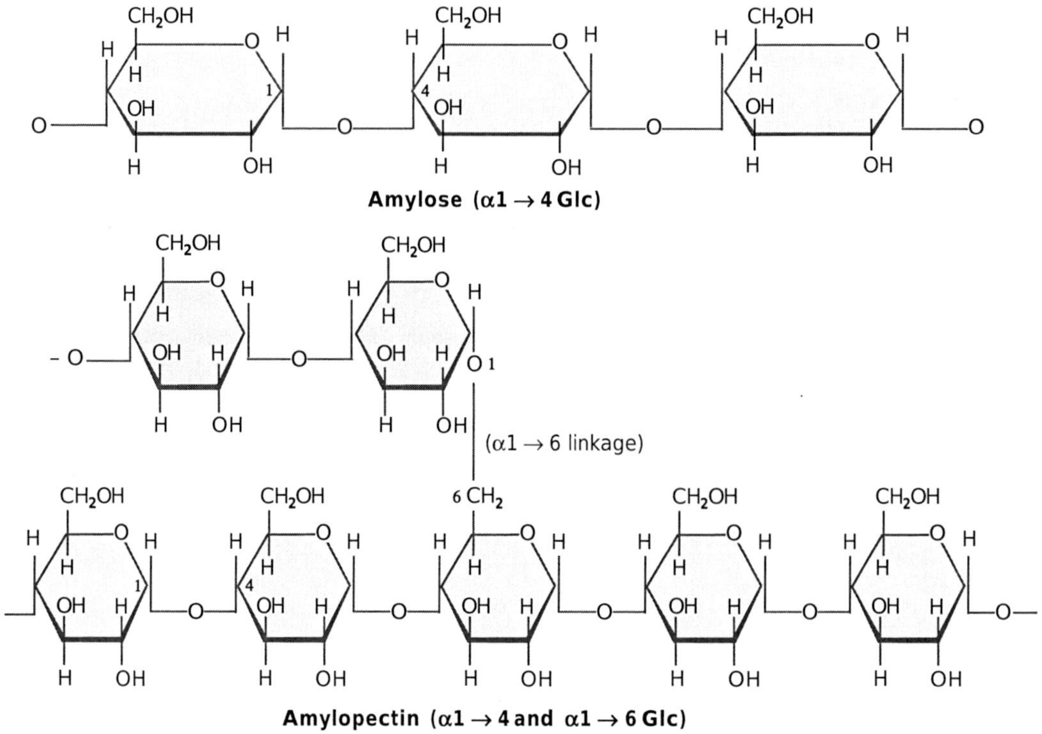

Fig. 2.5 Structures of amylose and amylopectin

- *Glycogen*, a homopolymer of glucose (*glucosan*) that functions as a storage form of carbohydrate in animals. The liver and skeletal muscle are rich in glycogen. It contains both α(1 → 4) and α(1 → 6) linkages similar to starch but is more highly branched than amylopectin.
- *Inulin*, a homopolymer of fructose (*fructosan*) that is found in tubers, and roots of dahlias, and artichokes. Inulin clearance is used to measure the glomerular filtration rate.
- *Cellulose*, a structural polysaccharide in plants is made up of β-D-glucopyranose units linked by β(1 → 4) bonds. Mammals cannot digest cellulose due to the absence of an enzyme that cleaves the β linkage. Bacterial metabolism of cellulose in the gut of ruminants hydrolyses the β linkage. The structure of cellulose is shown in Figure 2.6.
- *Chitin*, a structural polysaccharide in the exoskeleton of crustaceans and insects that contains repeating units of N-acetyl-D-glucosamine joined by β(1 → 4) linkage. The structure of chitin is shown in Figure 2.6.

❑ Write notes on glycosaminoglycans.

- *Glycosaminoglycans*, also known as *mucopolysaccharides*, are extracellular heteropolysaccharides that contain amino sugars and uronic acids.
- These polymers have a high density of negative charge, and lend viscosity, adhesiveness, and tensile strength to the extracellular matrix.
- When the glycosaminoglycan chains are attached to a protein molecule, the product is named a *proteoglycan*. Proteoglycans form the base substance of connective tissues.

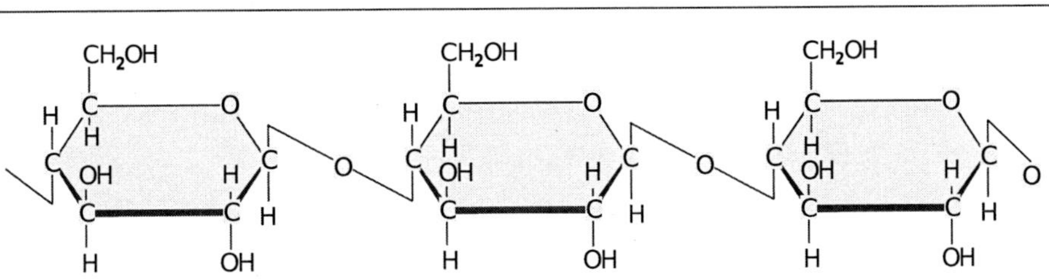

Cellulose

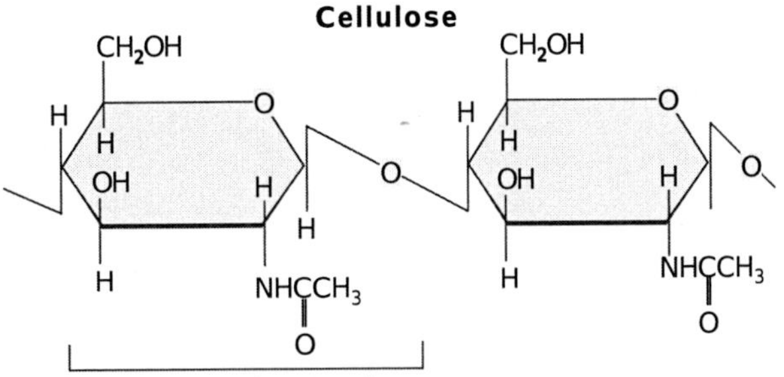

N-acetyl-D-glucosamine

Chitin

Fig. 2.6 Structures of cellulose and chitin

- The glycosaminoglycans include the following
  a. *Chondroitin sulfate.* It contains repeating units of β-glucuronic acid and sulfated N-acetylgalactosamine joined by α(1 → 3) linkage. It gives tensile strength to the cartilage, tendons, ligaments, and walls of the aorta.
  b. *Heparin.* It is an anticoagulant, made up of sulfated glucosamine and sulfated iduronic acid.
  c. *Hyaluronic acid.* It contains repeating units of β-glucuronic acid and N-acetylglucosamine. It functions as a lubricant in synovial fluid of joints, and gives tensile strength and elasticity to cartilage and tendons.

The structures of chondroitin sulfate, heparin and hyaluronic acid are shown in Figure 2.7.

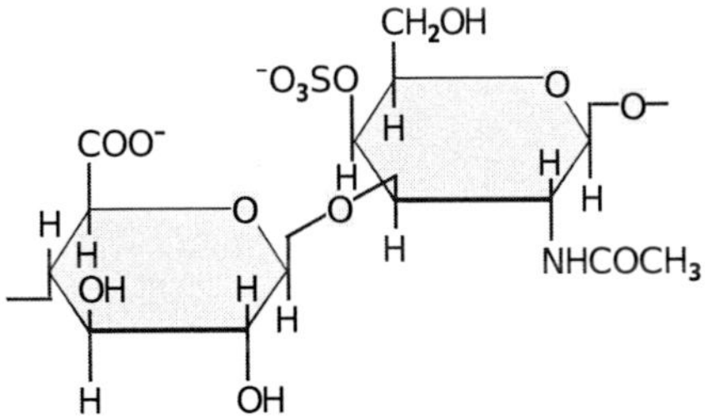

Chondroitin sulphate

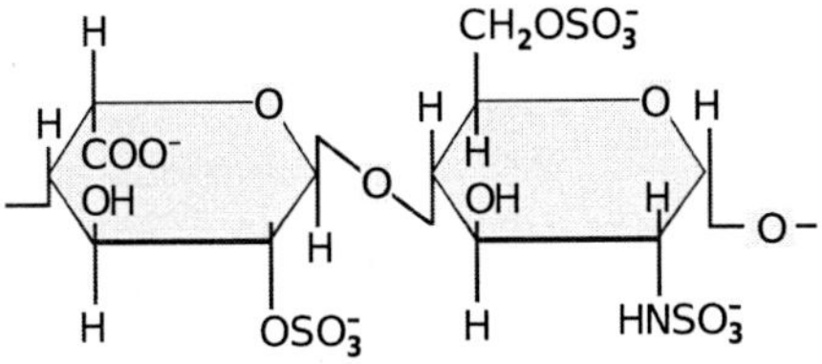

Heparin

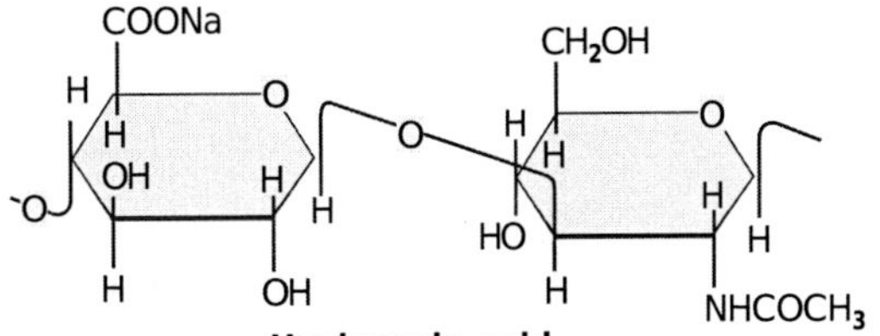

Hyaluronic acid

Fig. 2.7 Structures of glycosaminoglycans

❑ Differentiate between glycoproteins and proteoglycans.

- A *glycoprotein* is a protein containing branched or unbranched oligosaccharide chains.
  The carbohydrates found in glycoproteins include
  a. hexoses such as mannose and galactose
  b. acetyl hexosamines such as N-acetylglucosamine and N-acetylgalactosamine
  c. sialic acids, which are N- or O-acyl derivatives of neuraminic acid, a 9-carbon sugar derived from mannosamine and pyruvate.

  Figure 2.8 shows the structures of N-acetylglucosamine and sialic acid.

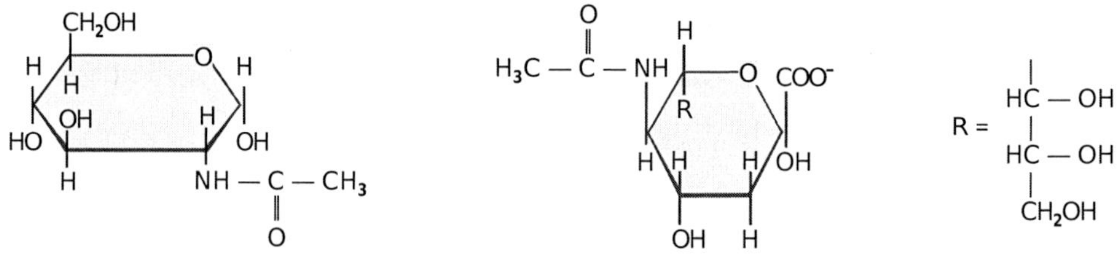

N-acetylglucosamine **Sialic acid (N-acetylneuraminic acid)**

Fig. 2.8 Structures of N-acetyl glucosamine and sialic acid

- A *proteoglycan* is a glycosaminoglycan attached to a protein molecule. Proteoglycans form the base substance of connective tissues.

❑ What are glycolipids?

Glycolipids are membrane lipids in which the hydrophilic head groups are made up of oligosaccharides, which function as recognition sites for carbohydrate-binding proteins.

❑ What are lectins?

- Lectins are proteins that contain specific carbohydrate-binding domains found on the outer surface of cells.
- These play an important role in cell – cell communication.
- They also determine the rate of degradation of some peptide hormones, blood cells and blood proteins.
- Lectins, found within the cells, mediate intracellular protein targeting.

❑ What are amino acids?

- Amino acids are the building blocks of proteins.
- Twenty L-amino acids are commonly found in proteins.
- All amino acids contain a carboxyl group (– COOH), an amino group (– NH_2), and a side chain (R group), attached to the same α-carbon atom (Fig. 2.9).

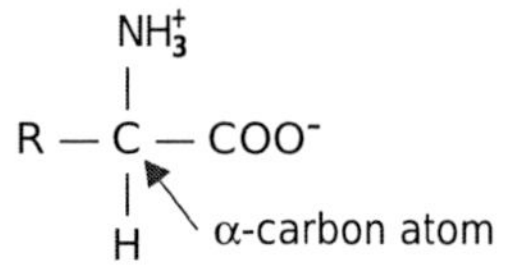

Fig. 2.9 General structural formula of an amino acid

❑ How are amino acids classified?

- Based on the chemical nature of the side chains, amino acids are categorized into seven major classes (Fig. 2.10).

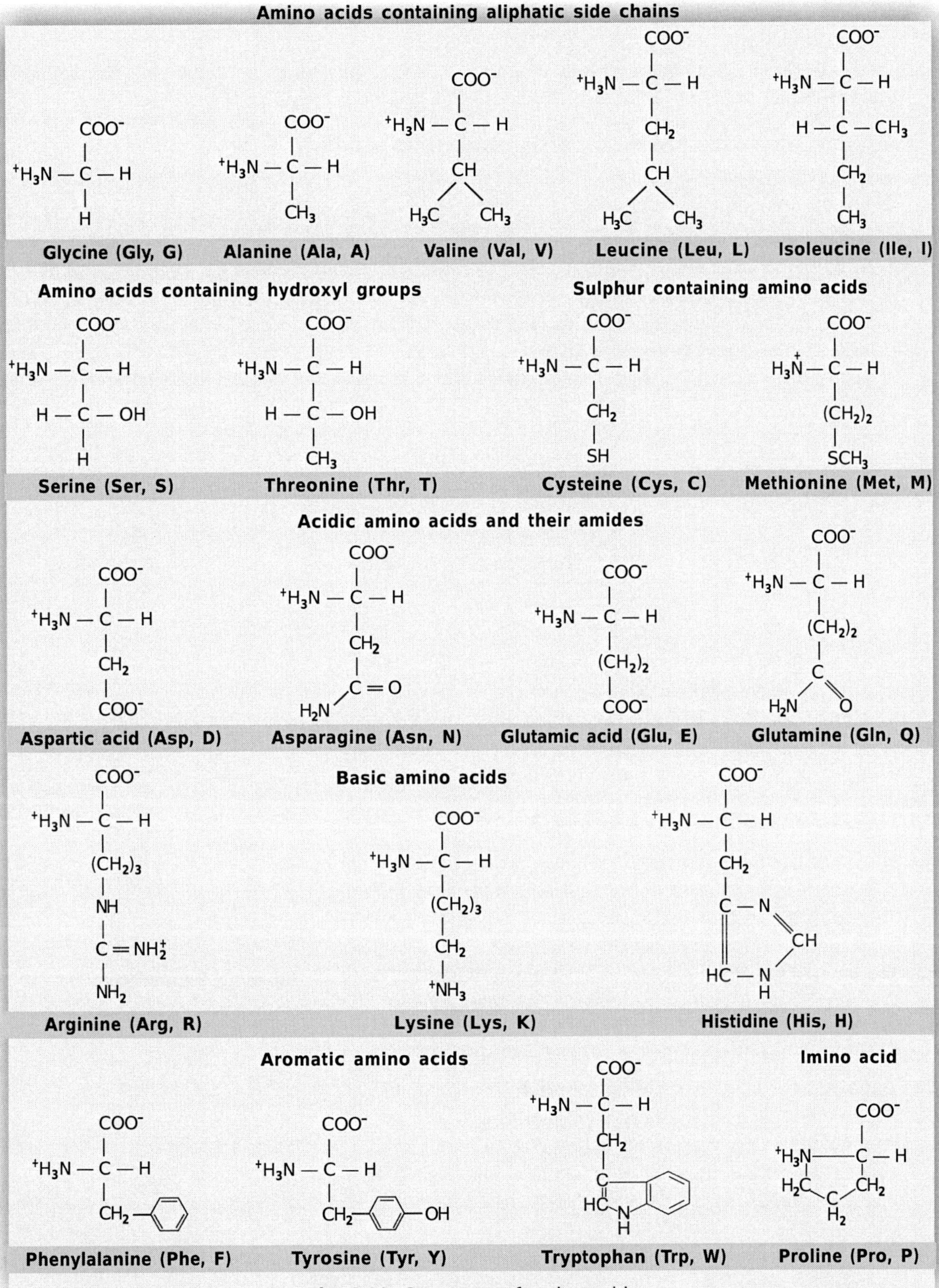

Fig. 2.10 Structures of amino acids

- Based on the nature of R group, amino acids are classified into polar (hydrophilic) and non-polar (hydrophobic) amino acids.
  - The polar amino acids include *arg, asp, asn, cys, glu, gln, gly, his, lys, ser, thr,* and *tyr.*
  - The non-polar amino acids include *ala, ile, leu, met, phe, pro, trp, and val.*

Explain the acid-base properties of amino acids.

- Amino acids are *amphoteric* in nature, *i.e.*, they can function as acids as well as bases.
- At low pH both the $-NH_2$ and the $-COOH$ groups are protonated, and the amino acid is in the cationic form.
- At higher pH, the $-COOH$ group loses a proton and the amino acid exists in the anionic form.
- The pH *at which* an amino acid bears no net charge is termed the *isoelectric point (pI) or the isoelectric pH.* At this pH, the molecule does not move in an electric field. These neutral molecular species are termed *zwitterions* (Fig. 2.11).
- The pI differs for the different amino acids and is based on the number and charge of the ionizable groups.

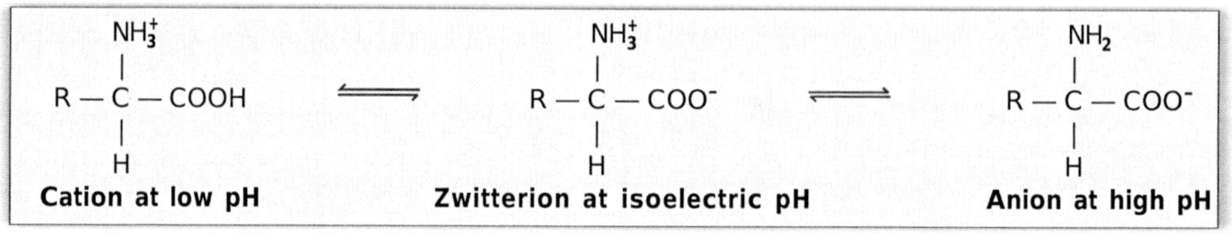

Fig. 2.11 The acid-base properties of amino acids

What are peptides?

- A peptide contains two or more amino acid residues joined together by a *peptide bond.*
- A *dipeptide* is made up of two amino acids, linked together by one peptide bond.
- An *oligopeptide* contains up to 25 amino acid residues.
- A *polypeptide* is made up of more than 25 amino acids.

How is a peptide bond formed?

- A peptide bond is a covalent bond formed between the α-carboxyl group of one amino acid and the α-amino group of another amino acid by removal of a water molecule (Fig. 2.12).
- The peptide bond is very stable and can be hydrolyzed only by the action of proteolytic enzymes.

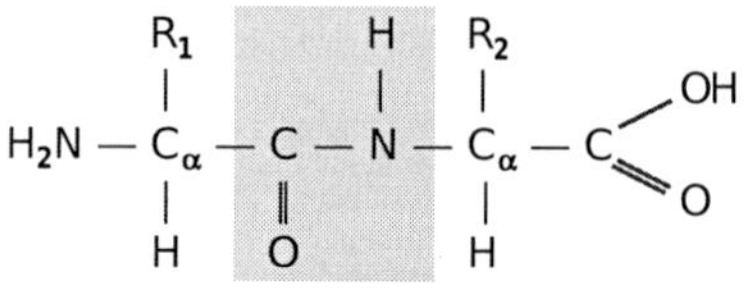

Fig. 2.12 Formation of peptide bond

Name some biologically important peptides.

Examples of peptides of biological importance include

- *hormones* such as glucagon and insulin
- *antioxidants* such as the tripeptide glutathione (γ-glutamylcysteinylglycine)
- *endogenous opiates* such as enkephalin and endorphin
- *smooth muscle hypotensive agents* such as bradykinin and kallikrein.

What are proteins?

- Proteins are made up of amino acid residues that form the building blocks.

- A protein molecule contains one or more polypeptides and can be classified into two types based on the constituents.
  1. *Simple proteins.* These contain only polypeptides.
  2. *Complex proteins.* These contain, in addition to polypeptides, a non-amino acid component called *prosthetic group*, such as:
     - lipid in lipoproteins (*e.g.*, LDL)
     - haem in hemoproteins (*e.g.*, hemoglobin)
     - metal in metalloproteins (*e.g.*, transferrin)
     - phosphate group in phosphoproteins (*e.g.*, casein).

How are proteins classified?

- Based on the shape, proteins are classified into
  - *globular proteins.* These are soluble in aqueous solutions and tightly coiled 3-dimensionally (*e.g.*, hemoglobin).
  - *fibrous proteins.* These are insoluble in aqueous solutions and found as extended linear fibres (*e.g.*, collagen).
- Based on their biological functions, proteins may be classified into:
  - *contractile proteins* (*e.g.*, actin, myosin, etc.)
  - *defence proteins* (*e.g.*, immunoglobulins)
  - *enzymes* (*e.g.*, ligases, isomerases)
  - *regulatory proteins* (*e.g.*, hormones)
  - *storage proteins* (*e.g.*, ferritin)
  - *structural proteins* (*e.g.*, collagen)
  - *transport proteins* (*e.g.*, hemoglobin).

What are the different orders of protein structure?

The orders of protein structure include

- *Primary structure*, which refers to the order in which amino acids are linked together in the polypeptide chain.
- *Secondary structure*, which refers to the conformation of the polypeptide chain.
- *Supersecondary structures*, which indicate the folding patterns in proteins.
- *Tertiary structure*, which refers to the 3-dimensional conformation of proteins generated by interactions between the amino acid residues.
- *Quaternary structure*, which refers to the spatial arrangement of subunits that are joined by non-covalent interactions. This is seen only in proteins with two or more polypeptide chains.

What is meant by primary structure of a protein?

- Primary structure refers to the order in which amino acids are joined together in the polypeptide chain. For example,

  ···· Tyr — Cys — Glu — Ile — Leu — Arg — Cys — Phe — His ····

- The amino acid sequence is written from left to right, starting with the N-terminal amino acid and ending with the C-terminal amino acid. Peptide bonds and disulphide bonds are responsible for the stability of the primary structure.

❑ Give a brief account of secondary and supersecondary structures in proteins.

Secondary structure

- Refers to the 3-dimensional conformation of the polypeptide chain and the relative positions of each of the constituent atoms in the molecule.
- Types: α-helix and β-pleated sheet (Fig. 2.13).
- ***α-helix***
  - Each peptide bond in the polypeptide chain participates in intrachain hydrogen bonding.
  - Hydrogen bonding occurs between the carbonyl of a peptide bond and the – NH of another peptide bond four amino acid residues further down the chain.
  - There are 3.6 amino acid residues/turn of the helix with the R groups protruding outward; amino acids with bulky R groups and proline disrupt helix formation.
  - Proteins in hair and wool are examples of this type.
- ***β-pleated sheet***
  - Hydrogen bonding occurs between two polypeptide chains or between two regions of a single chain.
  - The two polypeptide chains may run in the same direction (parallel) or in opposite directions (antiparallel).

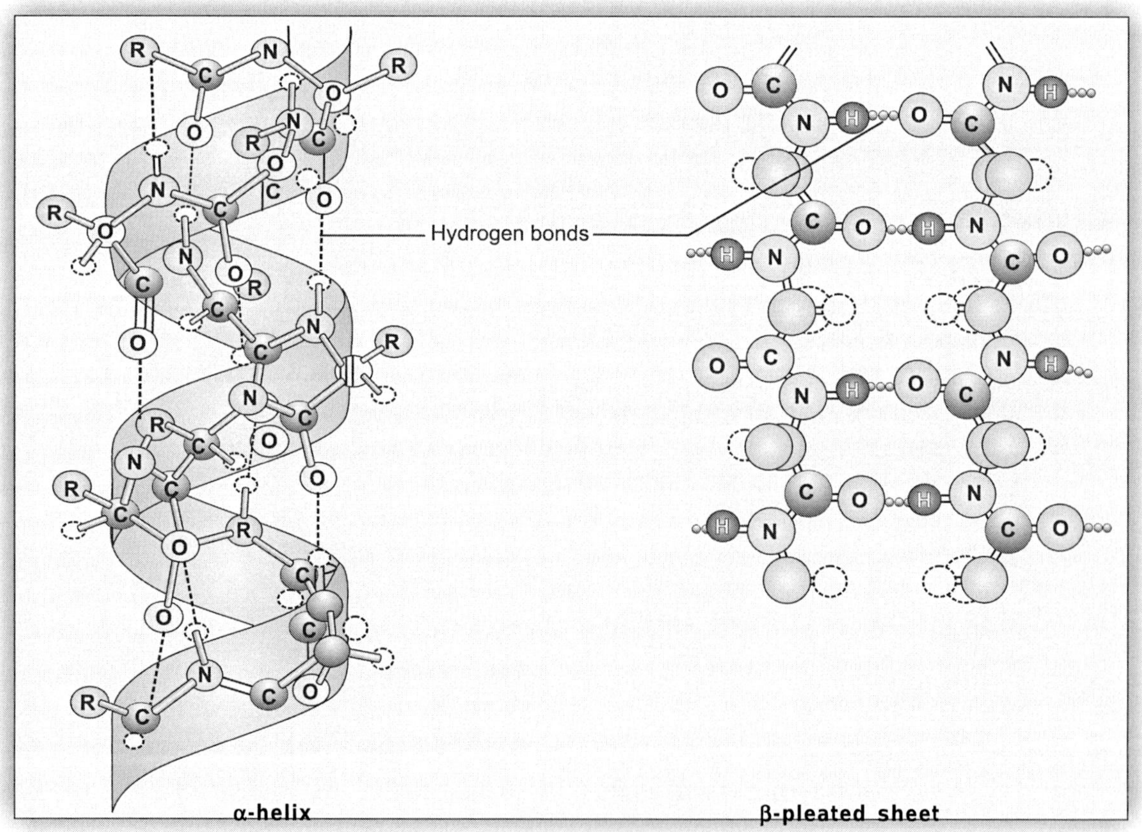

Fig. 2.13 Secondary structures of proteins

Supersecondary structures

- Refer to the folding patterns involving α-helices and β-pleated sheets.
- β-α-β_2: In this structure, an α-helix connects two strands of β-sheets.
- *β-hairpin*: This consists of antiparallel β-sheets joined by short loops.

❑ Explain the tertiary and quaternary structures of proteins.

Tertiary structure

Refers to the 3-dimensional conformation of proteins generated by interactions between amino acid residues, such as

- hydrogen bonds
- hydrophobic interactions
- disulphide bonds
- electrostatic interactions.

Quaternary structure

- Refers to the spatial arrangement of subunits linked by non-covalent interactions such as
  - hydrogen bonds
  - hydrophobic and electrostatic interactions.
- Occurs only in proteins with two or more polypeptide chains.
- Hemoglobin, which contains 2α and 2β chains, possesses a quaternary structure. In this, R group contacts are present between similar chains and there is very little contact between dissimilar chains.

❑ What are lipids?

Lipids are a group of organic compounds that are soluble in nonpolar solvents (*e.g.*, chloroform, ether) and insoluble in water.

❑ Enumerate the functions of lipids.

Lipids serve the following functions:

- membrane composition
- tissue immunity
- thermal insulation.
- cell recognition
- sources of high energy

❑ How are lipids classified?

Lipids are classified into three categories:

- ***Simple lipids.*** These are esters of fatty acids with alcohols and include
  - fats, which are esters of fatty acids with glycerol
  - waxes, which are esters of fatty acids with alcohols of higher molecular weight.
- ***Compound lipids.*** These are esters of fatty acids containing fatty acid, alcohol and some additional group, and include
  - phospholipids
  - glycolipids
  - amino lipids
  - sphingolipids
  - sulpholipids
  - lipoproteins
- ***Derived lipids.*** These include
  - steroids
  - ketone bodies
  - fat-soluble vitamins
  - fatty acids
  - hormones

❑ What are fatty acids? Give examples.

Fatty acids are aliphatic carboxylic acids with a long hydrocarbon chain and may be classified into two types:

1. *Saturated fatty acids* that contain no double bonds. Examples include lauric, palmitic and stearic acids.
2. *Unsaturated fatty acids* that contain one or more double bonds. These may be
   - monounsaturated that contain one double bond (*e.g.*, oleic acid, palmitoleic acid)
   - polyunsaturated that contain two or more double bonds (*e.g.*, arachidonic acid).

The fatty acids may occur free (*free fatty acids, FFA*) or esterified with glycerol in the form of triglycerides.

$CH_3-(CH_2)_{14}-COOH$

Palmitic acid

$CH_3-(CH_2)_5-CH=CH-(CH_2)_7-COOH$

Palmitoleic acid

Fig. 2.14 Structures of palmitic acid and palmitoleic acid

❑ What are essential fatty acids?

Fatty acids that are required in the diet and cannot be biosynthesized are termed *essential fatty acids.* These include linoleic, linolenic, and arachidonic acids (Fig. 2.15).

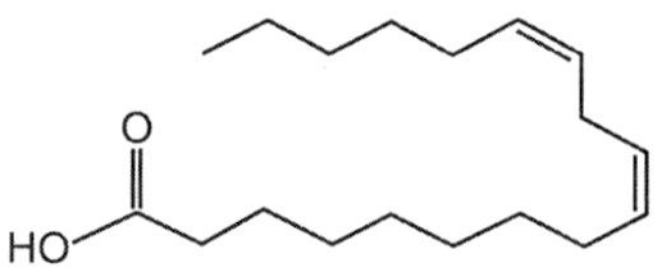

Linoleic acid

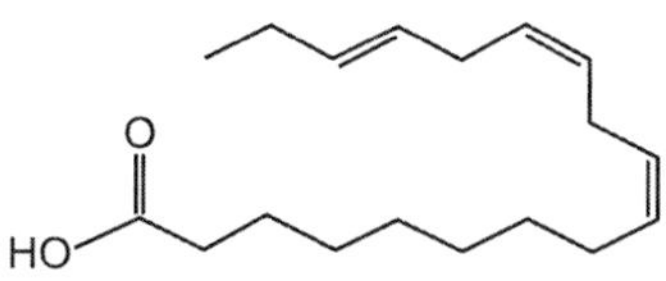

Linolenic acid

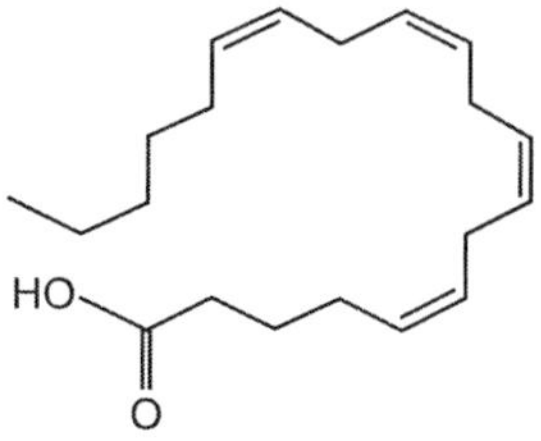

Arachidonic acid

Fig. 2.15 Essential fatty acids

❑ What are eicosanoids?

Eicosanoids are compounds containing 20 carbon atoms, derived from essential fatty acids (Fig. 2.16). These include

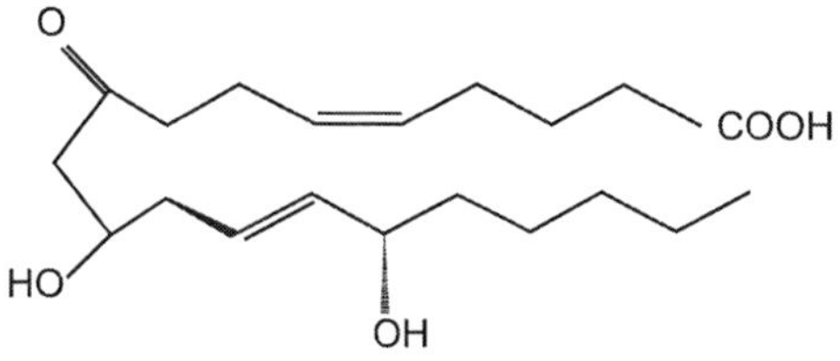

Prostaglandin E2

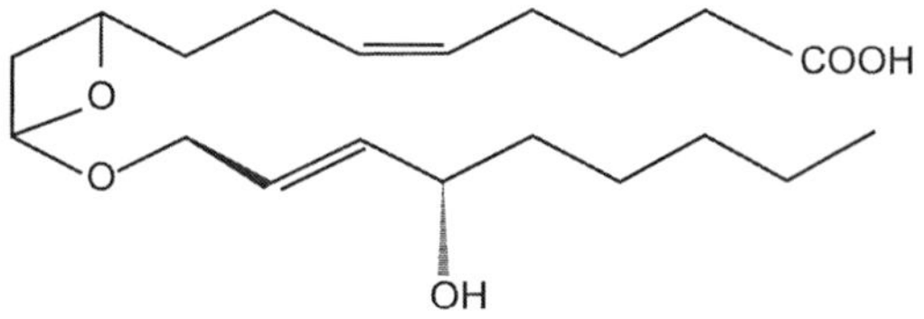

Thromboxane A2

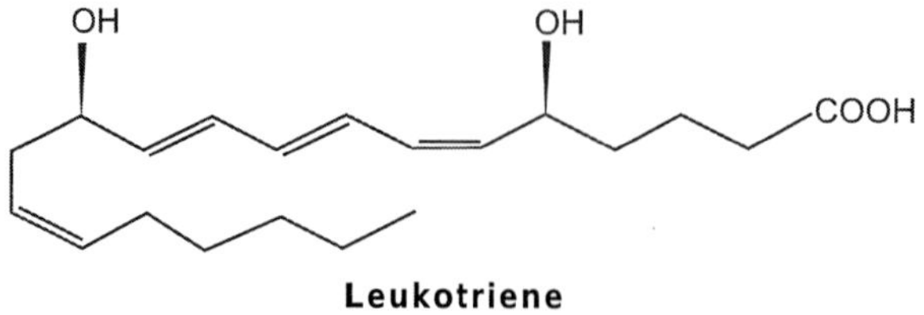

Leukotriene

Fig. 2.16 Structures of eicosanoids

- *Prostaglandins (PG),* local hormones that regulate platelet aggregation, stimulate inflammation, and modulate synaptic transmission.
- *Thromboxanes,* local hormones that act near their sites of synthesis.
- *Leukotrienes* that play a role in muscle contraction, chemotaxis, allergy, and inflammation.

❑ What are triglycerides?

Triglycerides (or triacylglycerols) (Fig. 2.17) are

- fatty acid esters of glycerol
- most abundant lipids
- major components of storage lipids in the adipose tissue
- made up of more than one type of fatty acid
- transported from the liver and intestine as lipoproteins.

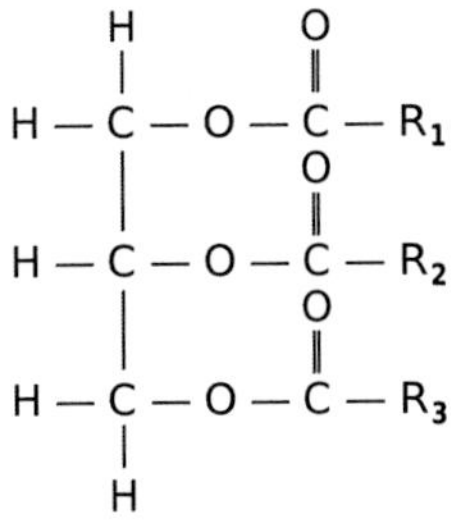

Fig. 2.17 Structure of a triglyceride

❑ Briefly explain the structure and functions of phospholipids.

Structure

- Phospholipids (or phosphoglycerides) contain glycerol, one or more fatty acids, phosphoric acid, and a nitrogenous base (Fig. 2.18).
- They have a polar head and a nonpolar tail and are *amphipathic.*
- Examples of phospholipids include *phosphatidylethanolamine* (*cephalin*), *phosphatidylcholine* (*lecithin*), *phosphatidylinositol*, *phosphatidylserine*, *cardiolipin*, *lysophospholipids* and *plasmalogens.*

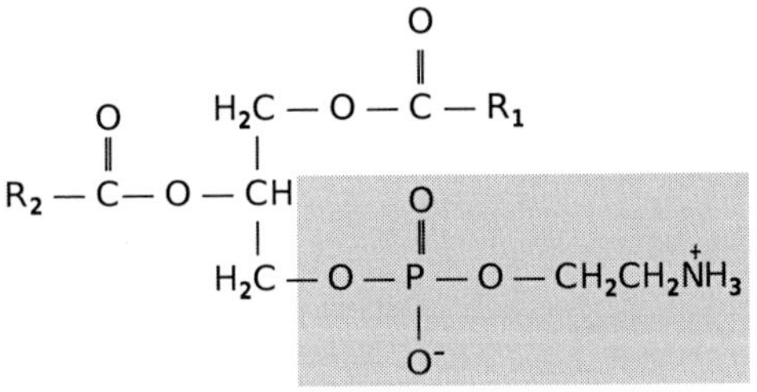

Phosphatidylethanolamine

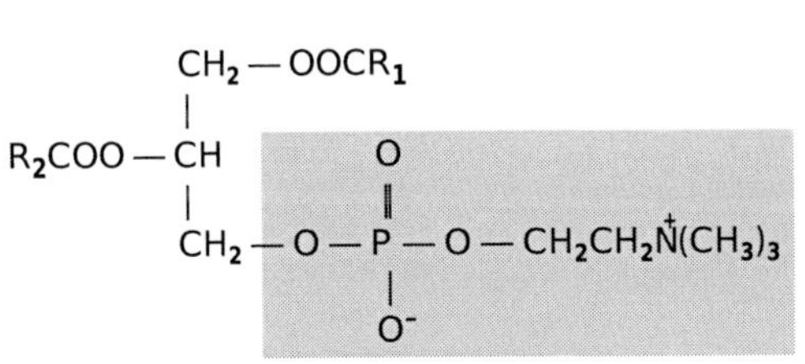

Phosphatidylcholine

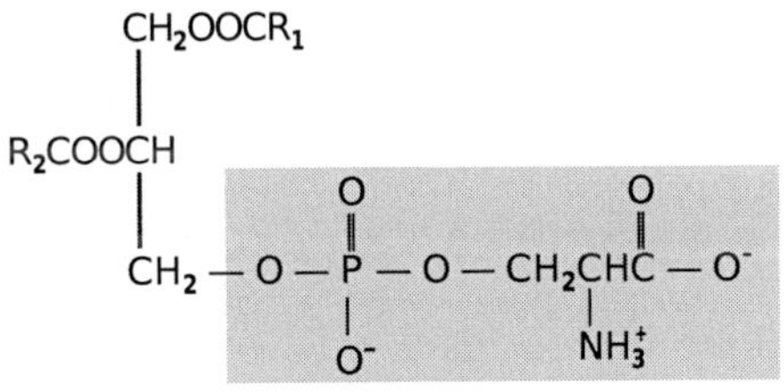

Phosphatidylserine

Fig. 2.18 Structures of phospholipids

Functions

- Major components of cell membrane
- Maintenance of cell membrane integrity
- Lowering of surface tension
- Lipid absorption in the intestine
- Lowering of blood cholesterol

- Blood coagulation
- As lung surfactant: *Dipalmitoyl lecithin,* a component of lung surfactant prevents adherence of the inner surfaces of the lungs.
- As precursor of second messengers: *Phosphatidyl inositol* is a precursor of diacylglycerol and inositol triphosphate that function as second messengers in signal transduction.

❑ What are sphingolipids?

Sphingolipids are made up of fatty acids, a long chain amino alcohol called *sphingosine* and a polar head group (Fig. 2.19). They are present in high concentrations in brain and nerve tissue and include

- *Phosphosphingolipids* (*sphingomyelins*), which contain a fatty acid, phosphate, choline, and sphingosine
- *Glycosphingolipids,* which contain the sphingosine-fatty acid *ceramide* with one or more sugar residues.

The sphingolipids abundant in nature are sphingomyelin, cerebrosides, and gangliosides.

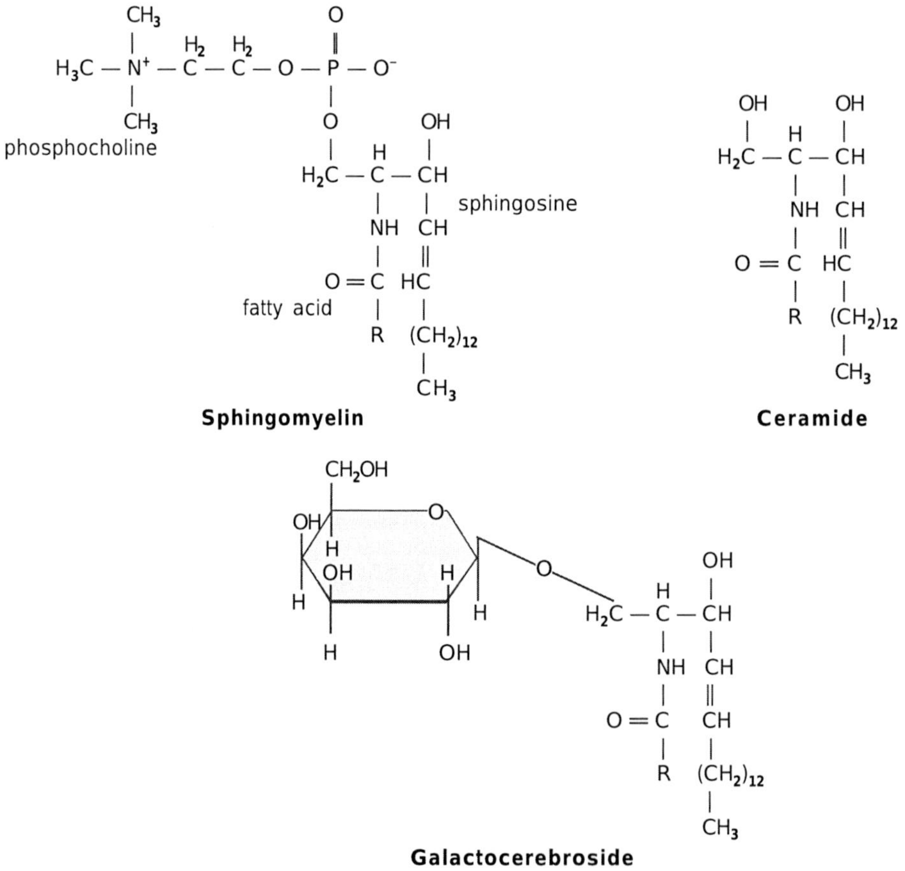

Fig. 2.19 Chemical structures of sphingolipids

❑ Write a note on cholesterol.

- Cholesterol is a steroid that contains the *cyclopentanoperhydrophenanthrene* ring system (Fig. 2.20).

- It is found in large amounts in nervous tissues, plasma membrane and in plasma lipoproteins.
- About two-thirds of plasma cholesterol is esterified with fatty acids to form cholesterol esters.
- *Functions*
  Precursor of many biologically important compounds including bile acids, steroid hormones, cardiac glycosides, sitosterols, and vitamin D.

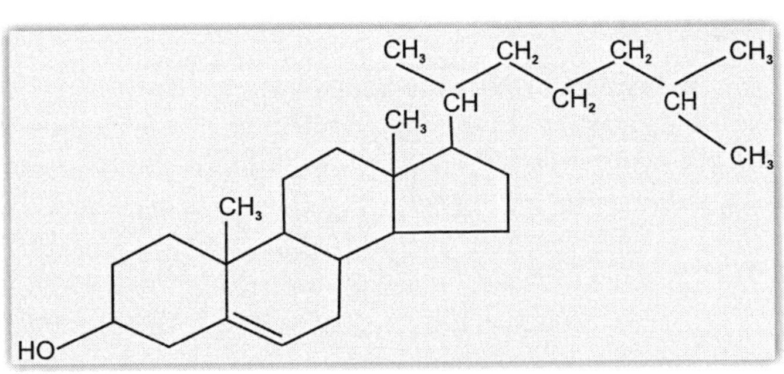

Fig. 2.20 Cholesterol

❑ What are lipoproteins?

- Lipoproteins are complexes of lipids and proteins that transport lipids in circulation.
- These are made up of an insoluble nonpolar core of triglycerides and cholesterol esters surrounded by a polar surface coating of free cholesterol, phospholipids, and proteins.
- The polar groups are oriented outwards.
- The protein part of the lipoprotein is referred to as *apolipoprotein.*

❑ How are lipoproteins classified?

Lipoproteins are classified based on their hydrated density and electrophoretic mobility (Fig. 2.21). Based on hydrated density, lipoproteins are classified into five fractions:

1. *Chylomicrons* that transport dietary triglycerides from the intestine to all cells.
2. *Very low density lipoproteins* (*VLDL*) that export triacylglycerols from the liver to cells.
3. *Intermediate density lipoproteins* (*IDL*) that contain both cholesterol and triacylglycerols. They are not detectable in normal plasma.
4. *Low density lipoproteins* (*LDL*) which are rich in cholesterol and transport cholesterol to cells. Three LDL subtypes LDL-I, -II, and -III have been identified.
5. *High density lipoproteins* (*HDL*) that transport cholesterol from the cells to the liver. Two fractions of HDL can be separated by ultracentrifugation; these are HDL_2 and HDL_3. HDL_2 contains 70% lipid and 30% protein, while HDL_3 is made up of 45% lipid and 55% protein.

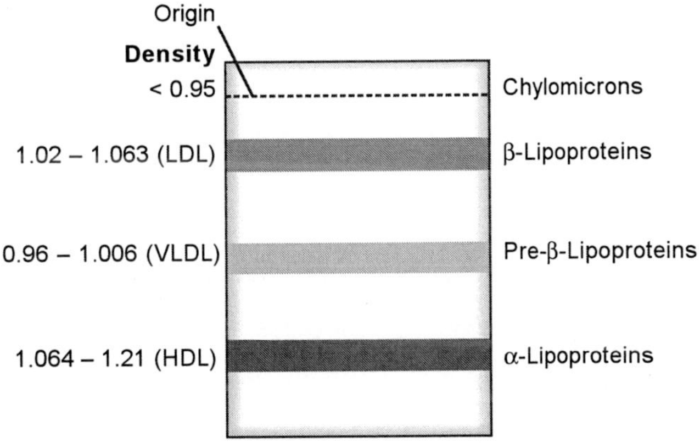

Fig. 2.21 Classification of lipoproteins

Lipoproteins can be separated according to their electrophoretic mobility into

a. *Chylomicrons*, which remain at the origin
b. *β-lipoproteins* (*LDL*)
c. *Pre-β-lipoproteins* (*VLDL*)
d. *α-lipoproteins* (*HDL*), which move fastest.

❑ What are the minor classes of lipoproteins?

In addition to the major lipoproteins, several minor lipoproteins, normally present in low concentrations in plasma, have also been identified. These include the following:

1. *Lipoprotein(a)* or *LP(a)* with pre-β mobility (also known as sinking pre-β) that is similar to LDL in composition but with higher protein content. Plasma LP(a) is found in large amount in patients with coronary heart disease.
2. *Lipoprotein-X*, which is made up of 66% phospholipids, 22% free cholesterol, 3% ester cholesterol, 6% protein, 3% triglycerides, and 2–3% bile acids. This is detected in the plasma of patients with obstructive jaundice and lecithin cholesterol acyltransferase (LCAT) deficiency.
3. HDL_c (HDL_1) that contain cholesterol and apoE, and found in the plasma of diet-induced hypercholesterolaemic animals.
4. HDL_2 and HDL_3 that are formed during HDL metabolism.
5. Pre-β-HDL, which is a component of very high density lipoprotein (VHDL).

❑ Write a note on apolipoproteins.

The protein part of lipoprotein is termed *apolipoprotein*. There are five major apolipoprotein families:

1. *Apo* A-I, A-II, and A-IV, the major apolipoproteins of HDL.
2. *Apo B* found in LDL, VLDL, and chylomicrons. Chylomicrons contain apo B-48, whereas LDL and VLDL contain apo B-100.
3. *Apo C-I, C-II* and *C-III* which are small apoproteins that freely exchange between different lipoproteins.
4. *Apo D* found in HDL.
5. *Apo E* found in VLDL and HDL, is arginine-rich.

Some apolipoproteins contain carbohydrate moieties such as glucose, galactose, mannose, fucose, glucosamine, and sialic acid (*e.g.*, apo B).

Functions:

a. Lipid transport
b. Cofactors of enzymes (*e.g.*, C-II for *lipoprotein lipase*).
c. Ligands for lipoprotein receptors (*e.g.*, apo A-I for HDL receptor).

❑ What are purines and pyrimidines?

- Purines and pyrimidines are nitrogenous bases, which are planar, aromatic, heterocyclic compounds containing a variety of substituents. They are constituents of nucleotides, which are the building blocks of DNA and RNA. The purine and pyrimidine skeletons are shown in Figure 2.22.

Purine skeleton **Pyrimidine skeleton**

Fig. 2.22 Structure of purine and pyrimidine skeleton

- The major purine bases found in nucleic acids are *adenine* and *guanine*.
- The major pyrimidine bases found in nucleic acids are *cytosine, thymine* and *uracil*.

The structures of purines and pyrimidine bases are shown in Figure 2.23.

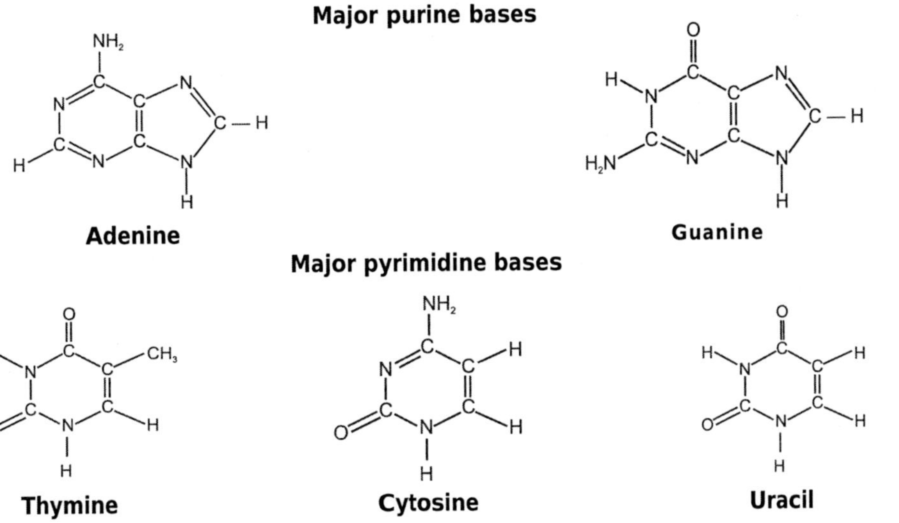

Fig. 2.23 Structures of the major bases

- Cytosine is found in both DNA and RNA, whereas thymine is present only in DNA and uracil only in RNA.

❑ What are minor bases?

- In addition to the major bases, certain unusual bases are present in small amounts in both prokaryotes and eukaryotes.
- These include 5-methylcytosine, 5-hydroxymethylcytosine, N6-methyladenine, N6, N6-dimethyladenine, N7-methylguanine, and pseudouridine. The structures of 5-methylcytosine and pseudouridine are shown in Figure 2.24.

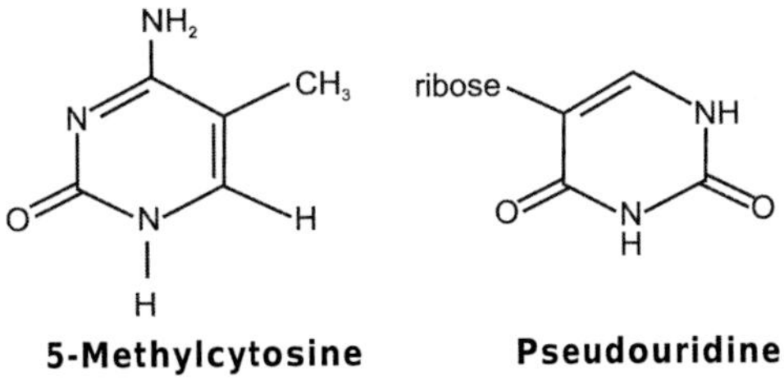

Fig. 2.24 Some examples of minor bases

❑ What is a nucleoside?

- A *nucleoside* refers to a purine or pyrimidine base covalently attached to a pentose sugar by a glycosidic bond to N-9 of purine or N-1 of pyrimidine (Fig. 2.25).
- A *ribonucleoside* is one in which the sugar residue is ribose (as in RNA).
- A *deoxyribonucleoside* is one in which the sugar residue is deoxyribose (as in DNA).
- The base positions are numbered 1, 2, etc., whereas atoms in the sugar are numbered 1' (prime), 2', etc.
- The C1' position of the ribose or deoxyribose is linked to N-1 of pyrimidine or N-9 of purine.

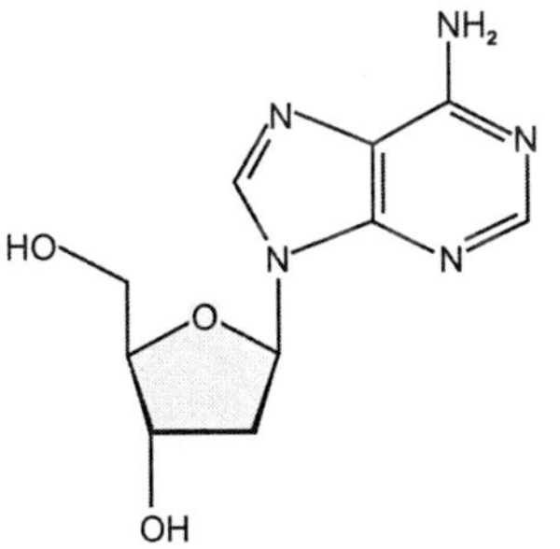

Fig. 2.25 Deoxyadenosine

❑ What is a nucleotide?

- Nucleotides are the building blocks of nucleic acids.

- Each *nucleotide* in a ribonucleic acid (RNA) or deoxyribonucleic acid (DNA) molecule consists of three components:
  - a nitrogenous base (purine or pyrimidine)
  - a pentose sugar (ribose or deoxyribose)
  - a phosphate residue.
- A nucleotide is, in other words, a nucleoside to which one or more phosphate groups are covalently attached to one or more of the –OH groups of ribose or deoxyribose (Fig. 2.26).

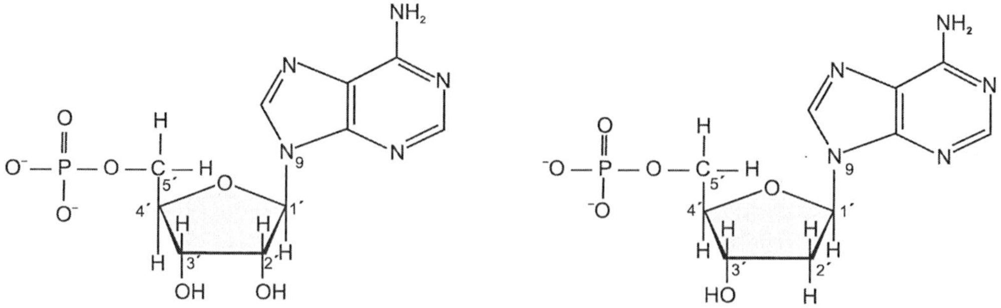

Adenylate (AMP) **Deoxyadenylate**

Fig. 2.26 Structures of nucleotides

- Based on the number of phosphate groups in the 5' position, nucleotides may be:
  - *nucleoside 5'-monophosphates (NMPs)*
  - *nucleoside 5'-diphosphates (NDPs)*
  - *nucleoside 5'-triphosphates (NTPs).*

  The NTPs are the building blocks of RNA while deoxyribonucleoside 5' triphosphates (dNTPs) are the building blocks of DNA.

Give an account of biologically important nucleotides.

A large number of nucleotides have biological importance (Fig. 2.27). These include

- ***Adenosine derivatives***
  - ADP and ATP, which participate in oxidative phosphorylation. ATP is the source of high-energy phosphate for biosynthetic reactions.
  - cAMP (3', 5'-AMP), synthesized from ATP by the action of *adenylate cyclase,* and degraded by *phosphodiesterase* to AMP. c-AMP functions as a second messenger via regulation of cAMP-dependent protein kinases and plays an important role in various cellular activities and metabolic pathways.
  - Adenosine 3'-phosphate 5'-phosphosulphate (PAPS) or "*active sulphate*" which functions as a sulphate group donor for sulfated proteoglycans or urinary metabolites of drugs excreted as sulphate conjugates.
  - S-adenosylmethionine (SAM) that functions as a methyl group donor for methylation reactions.
- ***Guanosine derivatives***
  - GTP
    1. is a high-energy phosphate compound
    2. is essential for activation of guanylate cyclase by some hormones
    3. functions as an allosteric regulator and energy source for protein synthesis.
  - cGMP that functions as a second messenger and is formed from GTP by *guanylate cyclase.*

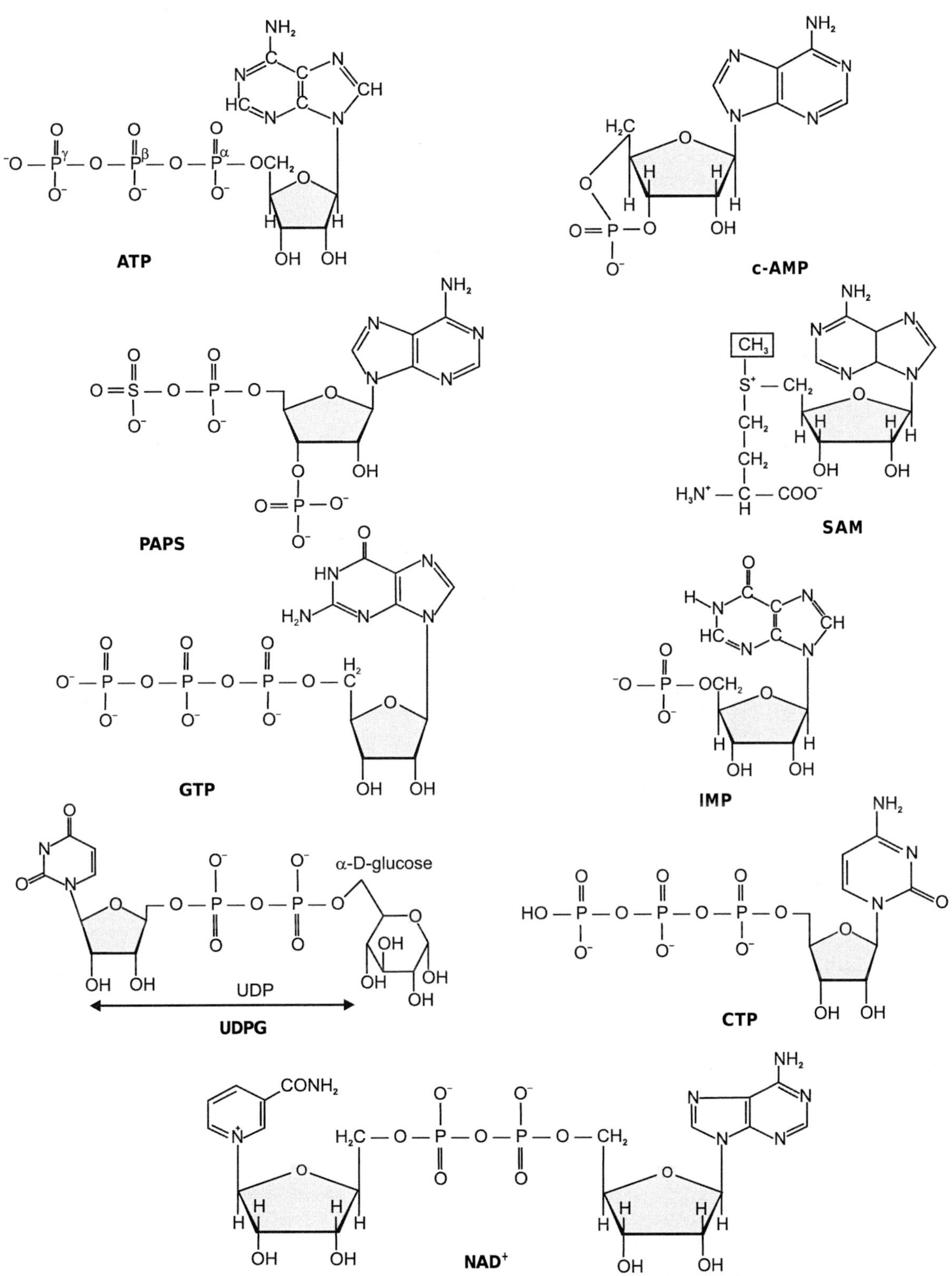

Fig. 2.27 Structures of biologically important nucleotides

- ***Hypoxanthine derivatives***
  Hypoxanthine ribonucleotide (IMP) or inosinic acid (inosinate) formed by deamination of AMP is a precursor of purine ribonucleotides.
- ***Uracil derivatives***
  UDP-sugars
  - interconvert glucose 1-phosphate to galactose 1-phosphate.
  - function as sugar donors for biosynthesis of oligosaccharides of glycoproteins and proteoglycans.
  - UDP-glucose is a donor of glucose residues during glycogen biosynthesis.
  - UDP-glucuronic acid functions in the conjugation of bilirubin and drugs.
- ***Cytosine derivatives***
  - CTP is a high-energy phosphate compound necessary for phosphoglyceride biosynthesis.
- ***Vitamin nucleotides***
  - Vitamin coenzymes such as NAD^+, and FAD incorporate nucleotides.

❑ Write a note on tautomerism in purine and pyrimidine bases.

- Alternative structures that differ only in the location of hydrogen atoms are called *tautomers*.
- The amino and oxo groups of purines and pyrimidines are in tautomeric equilibrium.
- The amino groups form an amino/imino pair and the oxo groups, a keto/enol pair of tautomers (Fig. 2.28).
- The keto tautomer is referred to as a *lactam* ring and the enol tautomer as the *lactim* ring.
- At physiological pH, the purine and pyrimidine bases are predominantly present in the amino and keto tautomeric forms.

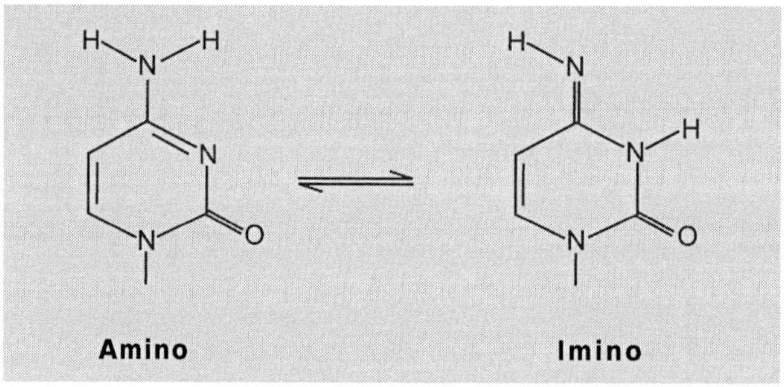

Fig. 2.28 Tautomeric forms of cytosine

❑ How is the shape of the nucleotide determined?

The shape of a nucleotide is based on

- *Sugar pucker*: The furanose ring is puckered and assumes $C^{2'}$ endo and $C^{3'}$ endo conformations that are in equilibrium at room temperature.
- *Syn-anti conformation of the glycosidic bond*:
  a. the *anti* conformation is generally favored and is essential for proper positioning of the bases in DNA.
  b. the *syn* conformation is common in left-handed helices such as Z-DNA.
- $C^{4'}$– $C^{5'}$ orientation
- Shape of the phosphate ester bonds.

Sugar pucker and syn-anti conformation of nucleotides are shown in Figure 2.29.

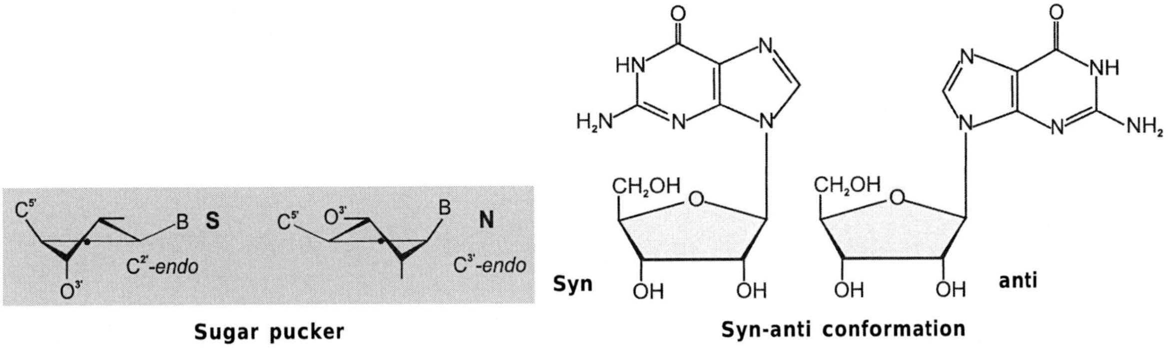

Fig. 2.29 Sugar pucker and syn-anti conformation of nucleotides

❑ Enumerate the therapeutic uses of the nitrogenous bases, nucleosides, and nucleotides.

- *Cancer chemotherapy*: 5-fluorouracil (5-FU), thymine analogs, 6-thioguanine, 6-mercaptopurine, 5- or 6-azauridine, and cytabrine (araC).
- *Gout*: allopurinol and 4-hydroxypyrazolopyrimidine.
- *Immunosuppression during transplantation*: Azathioprine.

❑ Discuss the Watson-Crick model of DNA structure.

The structure of DNA as deduced by Watson and Crick (Fig. 2.30) has the following features:

- The DNA molecule is a double helix with the two strands wound around each other in a right-handed manner.
- The two strands are *antiparallel, i.e.*, they run in opposite directions. While one strand runs in the 5' → 3' direction, the other strand runs in the 3' → 5' direction.
- The two strands are *complementary* to each other with reference to the base composition. Adenine (A) on one strand of DNA base pairs with thymine (T) on the opposite strand, while guanine (G) base pairs with cytosine (C). This is known as *Watson-Crick (W-C) base pairing.*
- Each DNA strand is a polynucleotide with the adjacent nucleotides linked together by phosphodiester bonds between the 3'–OH of one nucleotide and 5'-phosphate of the adjacent nucleotide.
- The covalent backbone is made up of alternating sugar-phosphate residues present on the outside. The phosphate groups are negatively charged.

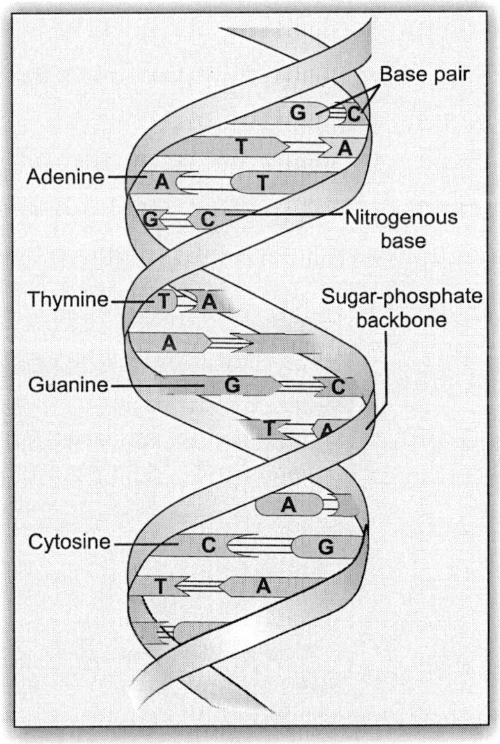

Fig. 2.30 Structure of DNA

- The hydrophobic purine and pyrimidine bases are stacked in the interior of the molecule. The distance between adjacent base pairs is 0.34 nm (3.4Å).
- There are 10.5 residues/turn of the helix accounting for the secondary repeat distance (*pitch)* of 3.4 nm (3.4Å). The diameter of the helix is 2.0 nm (20Å).
- There are two grooves on the surface of the double helix: a *major (deep) groove* and a *minor (shallow) groove* that contain many water molecules. DNA-protein interactions occur within these grooves.
- The double helix is stabilized by:
  - H-bonding between the complementary bases. There are two H-bonds between A and T and three between G and C
  - hydrophobic stacking interactions between adjacent bases
  - hydration of the grooves.

Explain Chargaff's rules.

According to *Chargaff's rules*

- The sum of purines is equal to the sum of pyrimidines.

  i.e., A + G = T + C

- Adenine and thymine are found in equimolar amounts, and cytosine and guanine are also present in equimolar amounts.

  i.e., A = T and G = C.

- *Chargaff's rules,* which account for base equivalence in DNA, are important for the formation of the double helix.

Enumerate the differences between A-, B-, and Z-DNA.

- **A-DNA**
  - Forms at low humidities (< 75%) in microcrystals of DNA in which most of the water has been removed.
  - Is seen in double-stranded RNA and in DNA/RNA hybrids.
  - Is an antiparallel, right-handed double helix.
  - There are 11 base pairs/turn of the helix.
  - The pitch is 2.8 nm or 28Å.
  - The glycoside is in the *anti* conformation.
- **B-DNA**
  - Is the form in which most of the cellular DNA is present.
  - Is an antiparallel, right-handed helix.
  - There are 10.5 base pairs/turn of the helix.
  - The pitch is 3.4 nm or 34Å.
  - The glycoside is in the *anti* conformation.
- **Z-DNA**
  - Is so called because of the zig-zag pattern of the sugar–phosphate backbone.
  - Occurs in polymers with a sequence of alternating purines and pyrimidines, *e.g.,* poly dCG or poly dAC.
  - Is formed from B-DNA at high salt concentration (> 4M NaCl).

- Is the only left-handed helix.
- Has 12 base pairs/turn of the helix.
- Contains only a single groove.
- Plays a regulatory role in gene expression and genetic recombination.

Figure 2.31 shows the A, B and Z forms of DNA.

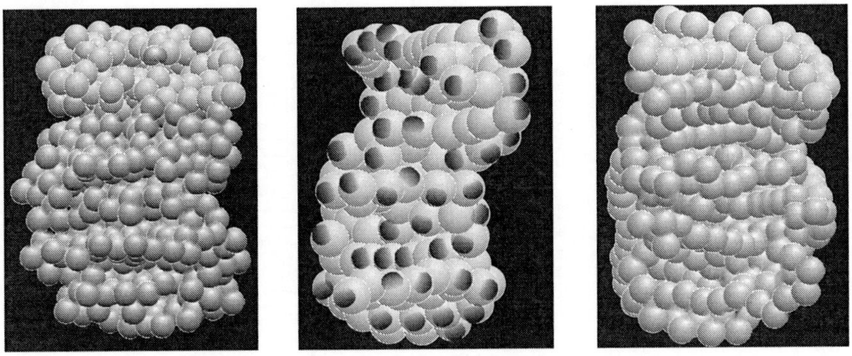

Fig. 2.31 A-, B-, and Z-DNA

❑ What are palindromes?

- *Palindromes* are inverted base sequence repeats with two-fold symmetry (Fig. 2.32).
- They cause intrastrand base pairing due to self-complementarity, forming a *hairpin* structure, if a single DNA strand is involved, or a *cruciform* structure when both strands of the DNA are involved.

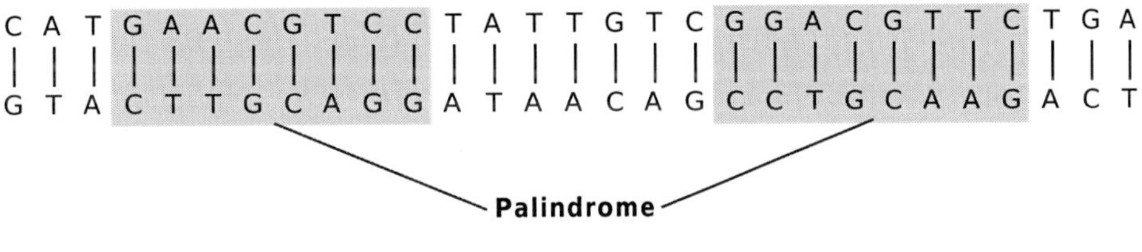

Fig. 2.32 Palindrome

❑ When is a DNA triple helical structure formed?

- A DNA triple helix is formed:
  - when the base pairing ratio of poly A to poly U is 1 : 2.
  - within long sequences containing only purines or pyrimidines.
  - by bases involved in W-C base pairing which form additional H-bonds with functional groups in the major groove.
- Atoms involved in H-bonding in triplex DNA are called *Hoogsteen positions*; these include N^7, O^6, and N^6 of purines.
- DNA triple helix is more stable at low pH.

❑ What is H-DNA?

- *H-DNA* is a triple helical structure that is formed when polypurine/polypyrimidine tracts contain a mirror repeat within the sequence (Fig. 2.33).
- H-DNA may have a role in eukaryotic gene expression.

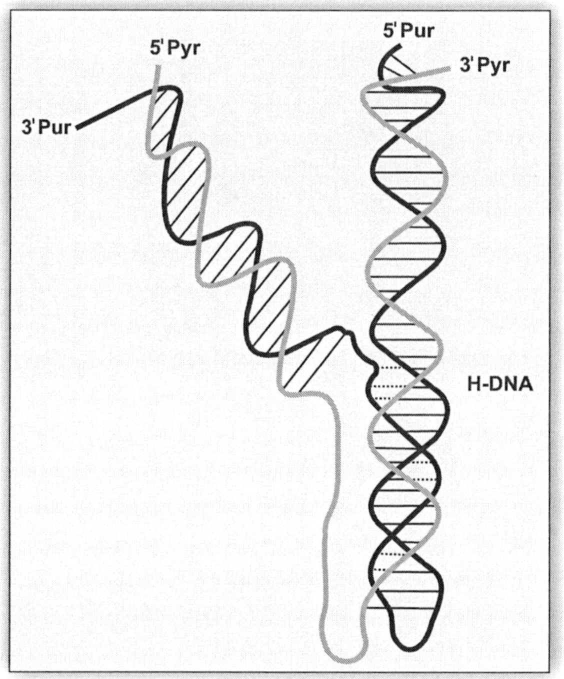

Fig. 2.33 H-DNA structure

❑ When is a quadruple helix formed?

A *quadruple helix* is formed by four chains that contain clusters of G residues such as those found in telomeres of eukaryotic chromosomes.

❑ What is supercoiling?

- Supercoiling refers to the twisting of the DNA double helix in the space around its own axis.
- When DNA is twisted in the same direction as that of the double helix, it is referred to as *positive supercoiling*. This is not seen *in vivo*.
- A DNA molecule that is twisted in the opposite direction to the clockwise turns of the right-handed helix is *underwound*, and the coiling is termed *negative supercoiling*. This occurs when DNA is in the form of a closed circle and if it is bound and stabilized by proteins, thereby restricting free rotation of the strands.
- DNA underwinding decreases the number of helical turns, leading to thermodynamic strain on the molecule.
- Negative supercoiling is important for
  - compaction and packaging
  - transcription
  - replication
  - formation of left-handed Z-DNA.

❑ Write notes on DNA denaturation.

- DNA denaturation refers to the separation of duplex DNA into single strands by breaking H-bonds between the base pairs and destabilizing stacking interactions between successive bases. The covalent backbone is, however, not broken.
- Denaturation can be brought about by
  - heat
  - exposure to high salt concentration
  - very high or very low pH
  - chemicals such as formamide, urea and methanol.
- Denaturation of DNA brought about by an increase in temperature is referred to as *melting*. The temperature at which 50% of the DNA denatured is termed the *melting temperature* or *Tm*. The Tm increases with increasing GC content because more energy is required to break three H-bonds in GC base pairs than two H-bonds in AT base pairs.
- When the temperature is lowered below the Tm, the two complementary strands of DNA spontaneously reassociate into a duplex structure by a process called *renaturation* or *annealing*.
- Denaturation can be detected by an increase in UV absorption at 260 nm, a phenomenon known as *hyperchromic effect*. This is due to the strong UV absorbance of purine and pyrimidine bases.
- Renaturation is accompanied by a decrease in absorbance at 260 nm, a phenomenon termed *hypochromic effect*.

❑ Enumerate the differences between DNA and RNA.

RNA, like DNA, is a polynucleotide in which the mononucleotides are linked by 3' → 5' phosphodiester bonds. However, RNA exhibits the following features that differentiate it from DNA:

1. RNA is mostly a single-stranded molecule unlike DNA, which is mostly double-stranded. Exceptions include
   - Complementary base sequences in single-stranded RNA that base pair to form double-stranded hairpin structures.
   - The *replicative forms* (RFs) of single-stranded RNA viruses such as poliovirus and influenza virus are double-stranded.
2. The sugar residue in RNA is ribose, whereas in DNA it is 2'-deoxyribose.
3. RNA contains the pyrimidine uracil (U) instead of thymine (T).
4. RNA does not show base equivalence unlike DNA. The amount of G is not equal to the amount of C. Similarly, the amount of A is not equal to that of U.
5. RNA undergoes hydrolysis by alkali to form 2', 3' cyclic diesters of the mononucleotides, whereas DNA which lacks a 2'-OH group is not hydrolyzed by alkali.
6. The predominant double-stranded RNA structure is the right-handed A-form, whereas in DNA it is the B-form.
7. The UV absorption at 260 nm (A260) is greater for RNA than for dsDNA. While pure DNA shows A260/A280 of 1.8, the ratio for RNA is 2.0.

❑ What are the major classes of RNA?

There are three major classes of RNA:

- ***Messenger RNA (mRNA)***, which is a product of transcription from DNA that functions as a template for protein synthesis.

- ***Ribosomal RNA (rRNA),*** which is complexed with proteins to form ribosomes, the sites of protein synthesis.
- ***Transfer RNA (tRNA),*** which functions as an adapter in translation.

❑ What are the features of eukaryotic *m*RNA that distinguish it from prokaryotic *m*RNA?

The specific features of eukaryotic *m*RNAs include

1. A 7-methylguanosine triphosphate cap at the 5' end. The 5' cap helps the protein-synthesizing machinery to recognize *m*RNA and also protects against *5'-exonuclease* attack.
2. In prokaryotes, *m*RNA is *polycistronic, i.e.,* a single *m*RNA *encodes* more than one polypeptide whereas in eukaryotes, *m*RNA is *monocistronic, i.e.,* it encodes a single polypeptide. The length of an *m*RNA is determined by the length of the polypeptide it encodes.
3. A polyA tail at the 3' end made up of 20–250 adenine residues. The polyA tail protects against 3'-exonucleases and maintains the intracellular stability of *m*RNA.

❑ Describe the structure and composition of prokaryotic and eukaryotic ribosomes.

The *r*RNA is complexed with proteins to form ribonucleoprotein particles termed *ribosomes* (Fig. 2.34).

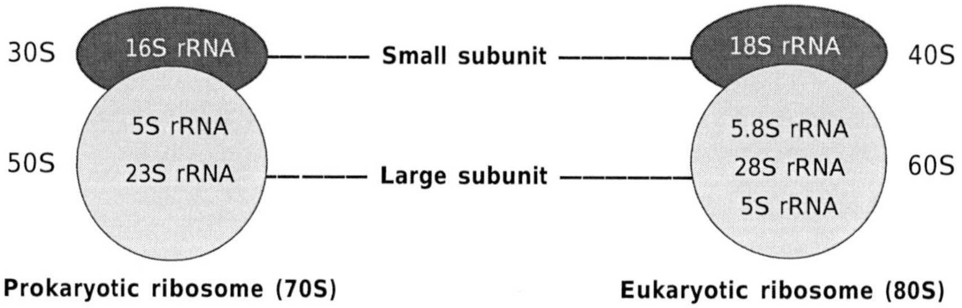

Fig. 2.34 Structures of ribosomes

The ribosomes are found on the surface of the rough endoplasmic reticulum (RER) and are the sites of protein synthesis.

❑ *Prokaryotic ribosomes*

They have a sedimentation velocity of 70S (S is the rate of sedimentation in Svedberg units).
Each 70S ribosome contains

- a larger 50S subunit containing 23S and 5S *r*RNAs and 36 proteins
- a smaller 30S subunit containing a single 16S *r*RNA and 21 proteins.
- a cleft between the two subunits through which *m*RNA passes.

❑ *Eukaryotic ribosomes*

They have a structure similar to the prokaryotic ribosome, but larger in size.
Each 80S ribosome contains:

- A larger 60S subunit containing 5S, 5.8S and 28S *r*RNAs, and 50 polypeptides
- A smaller 40S subunit with 18S *r*RNA and ~30 polypeptides.

- **_Functions_**
  - The *r*RNAs are required for ribosomal assembly and for binding of *m*RNA to ribosomes during translation.
  - *r*RNA molecules such as the *peptidyltransferase* in the large subunit have enzymatic function (*ribozymes*).

Discuss in detail the structure of *t*RNA.

- **_Primary structure_**
  - The length of *t*RNA molecules ranges from 74 to 95 nucleotides.
  - Minor bases, most of which are methylated forms of the normal bases, account for 20% of the total bases in a *t*RNA molecule. Some of the modified bases found in *t*RNAs include 4-thiouridine (S^4U), inosine, 1-methylguanosine (m^1G), N^6-isopentenyladenosine (i^6A), ribothymidine, pseudouridine (φ), and dihydrouridine (DHU). These bases contribute to the characteristic structure of *t*RNA and protect against *nuclease* attack.
- **_Secondary structure_**
  - Resembles a clover-leaf shaped structure due to extensive folding of the primary structure and intrastrand base pairing of complementary bases (Fig. 2.35).

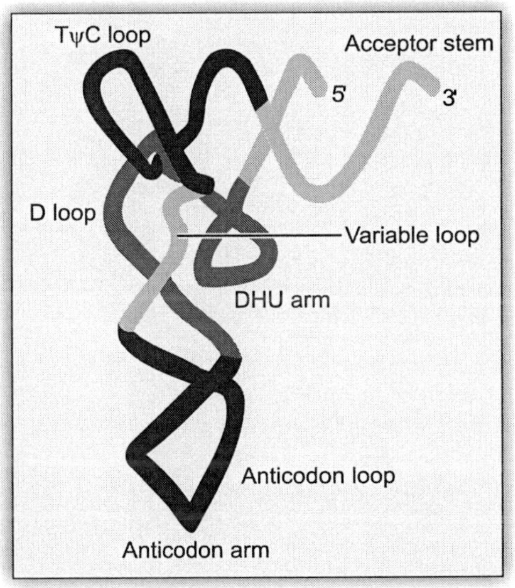

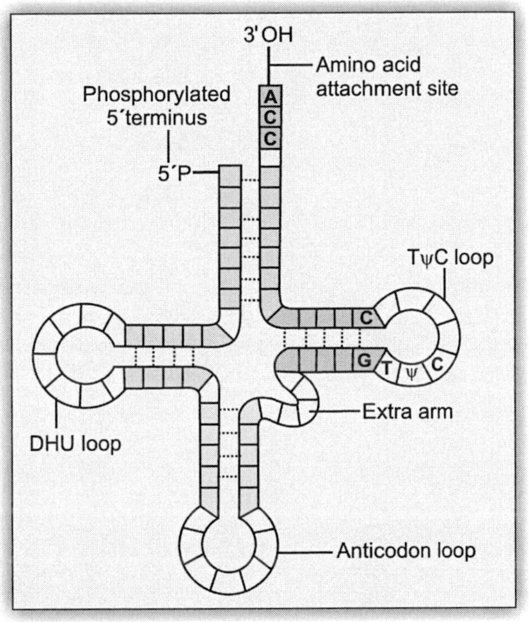

Fig. 2.35 Structure of *t*RNA

  - While the 5' end contains a phosphate group, the 3' end has a CCA terminus.
  - Each *t*RNA molecule has 4 arms:
    a. *Amino acid arm or acceptor arm* with the terminal sequence Pu-CCA at the 3' end. The terminal A residue is esterified to an amino acid in a reaction catalyzed by *aminoacyl tRNA synthetase*.
    b. *TψC arm or T loop* that contains the invariant residues GTψC.
    c. *DHU arm or D loop* containing the modified base dihydrouridine (DHU).
    d. *Anticodon arm* containing a triplet nucleotide sequence termed the anticodon, which is complementary to base pairs with the codon sequence of *m*RNA. One of the bases of the

anticodon that is less specific in its interaction with the corresponding base in the codon is termed the *wobble base* and is usually inosine.

The anticodon is demarcated by a U residue at its 5' end and a purine or a purine derivative at the 3' end.

Some *t*RNAs contain a fifth arm with 3 – 21 residues.

- ***Tertiary structure***
  - Base pairing occurs between residues in the D and T arms that fold the *t*RNA molecule into an L-shaped structure, with the amino acid arm at one end and the anticodon arm at the other.
  - The tertiary structure is stabilized by hydrogen bonding and base stacking interactions.

- ***Functions***

  The *t*RNA molecules function as adaptors for translation of *m*RNA into the amino acid sequence of proteins. Once an amino acid is esterified to *t*RNA, it does not change the specificity of the aminoacyl *t*RNA. There are 32 *t*RNAs for the 20 amino acids and each amino acid may have more than one *t*RNA. For example, serine has 5 *t*RNAs. These are known as *isoaccepting tRNAs*.

Section 2

Enzymes and Vitamins

- Enzymes
- Vitamins and Coenzymes

Chapter 3

Enzymes

❑ What are enzymes?

- Enzymes are biocatalysts that accelerate the rate of biochemical reactions by lowering the activation energy.
- Each enzyme (E) recognizes a specific substrate (S) to which it binds to form an enzyme-substrate (ES) complex that gives rise to the product (P) with regeneration of the enzyme.

$$E + S \rightleftharpoons ES \rightleftharpoons EP \rightleftharpoons E + P$$

- Most enzymes are proteins.
- Some *exceptions* include
  a. RNA molecules that possess catalytic activity. These are known as *ribozymes*.
  b. DNA also exhibits enzyme activity *in vitro*. Such enzymes are called *DNAzymes*.

❑ Define the terms apoenzyme, cofactor, holoenzyme, coenzyme and prosthetic group.

- The protein part of an enzyme is referred to as an *apoenzyme*.
- Most apoenzymes are catalytically active only when bound to a *cofactor*. The cofactor may be an inorganic ion, a coenzyme, or a prosthetic group. Inorganic metal ions such as Fe^{2+} or Mg^{2+} increase catalytic activity.
- A catalytically active enzyme bound to a cofactor is called a *holoenzyme*.
- Low molecular weight complex organic or metallorganic molecules that are bound loosely to the enzyme are called *coenzymes*. Several vitamins such as flavin adenine dinucleotide (FAD), nicotinamide adenine dinucleotide (NAD^+), and pyridoxal phosphate function as coenzymes.
- A *prosthetic group* is a coenzyme that is tightly bound to the enzyme and not released after catalysis (*e.g.*, the heme group of peroxidase).

❑ On what basis are enzymes classified?

- Enzymes are classified based on a systematic scheme developed by the Enzyme Commission (EC) of the International Union of Pure and Applied Chemistry (IUPAC).
- Many enzymes have the suffix *ase* added to the name of the substrate or the reaction they catalyze. For example, arginase acts on the substrate arginine, and urease acts on urea. However, some enzymes such as pepsin do not have the suffix *ase*.

- Enzymes are classified into six major groups based on the reactions they catalyze (Table 3.1).
- Each enzyme is designated by the prefix EC, followed by four numbers indicating the main class, subclass, subsubclass, and the serial number of the enzyme in its subsubclass.

Table 3.1 Major classes of enzymes

EC No.	*Enzyme class*	*Reaction catalyzed*
1.	Oxidoreductase	Oxidation-reduction. One of the substrates is hydrogen or an electron donor.
2.	Transferase	Transfer of chemical group (*e.g.*, amino group)
3.	Hydrolase	Hydrolytic cleavage of C — C, C — N, C — O, and other bonds.
4.	Lyase	Non-hydrolytic cleavage of C — C, C — N, C — O, and other bonds generating double bonds, or addition of groups to a double bond.
5.	Isomerase	Geometrical (spatial) rearrangement of a molecule.
6.	Ligase	Joining of two molecules coupled to hydrolysis of ATP.

❑ Which part of the enzyme is involved in catalytic activity?

Only a small portion of the enzyme consisting of 3 – 12 amino acid residues, organized into a pocket or cleft, comes into direct contact with the substrate. This is known as the *active site* of the enzyme and includes the thiol group of cysteine, the imidazole group of histidine, and the hydroxyl group of serine. The rest of the enzyme molecule is involved in maintaining the 3-dimensional structure of the active site.

❑ Do all enzymes show absolute specificity for their substrates?

- Enzymes vary in the degree of specificity for their substrates.
- While some enzymes show *absolute specificity* for their substrates, others display a lesser degree of substrate specificity. For example, *pyruvate kinase*, which catalyzes the transfer of a phosphate group between phosphoenolpyruvate and adenosine diphosphate (ADP), exhibits a very high degree of specificity for its substrate, whereas *hexokinase* is less substrate specific and catalyzes phosphate group transfer from adenosine triphosphate (ATP) to D-glucose as well as to other sugars such as D-fructose and D-mannose.
- Some enzymes exhibit *group specificity* and act only on particular chemical groups (*e.g.*, *glycosidases* on glycosides). A very high order of group specificity is demonstrated by *chymotrypsin*, which hydrolyses only those peptide bonds in which the carboxyl groups are contributed by aromatic amino acids.
- Enzymes such as *glucose oxidase* show *stereoisomeric specificity*. The enzyme acts only on D-glucose but not on L-glucose and is also capable of differentiating the α and β forms of D-glucose.

❑ Elucidate the mechanisms by which enzymes increase the rate of biochemical reactions.

Enzymes enhance the reaction rate by the following mechanisms:

- ***Proximity:*** An enzyme facilitates close proximity of the reactants at the active site.

- ***Covalent catalysis:*** The nucleophilic groups such as $RCOO^-$, $R–NH_2$, aromatic – OH present in the enzyme attack electrophilic regions of the substrate to form a covalent bond between the enzyme and the substrate. This is seen in the Class 2 or group transfer type of enzymes.
- ***General acid–base catalysis:*** The amino acid side chains of aspartate, glutamate, lysine and histidine function as *acid catalysts* in their protonated *form* and *base catalysts* in the unprotonated forms.
- ***Strain, molecular distortion, and conformational change:*** Both the enzyme and the substrate undergo changes during the reaction. Binding of the substrate is believed to induce a conformational change in the active site to accommodate the substrate. Changes in the enzyme and substrate lead to tighter binding in the transition state than in the ES or EP state. The destabilization of the ES complex results from distortion of bond angles and lengths by electrostatic interactions or desolvation of a charged group in the active site of the enzyme.

❑ What are the factors that influence the rate of an enzyme-catalyzed reaction?

The factors influencing the rate of an enzyme reaction include (i) concentrations of the enzyme and substrates (ii) pH and (iii) temperature.

❑ How does the substrate concentration influence enzyme activity?

- The concentration of the substrate [S] is one of the most important factors that influences the velocity of an enzyme-catalyzed reaction.
- Most enzymes exhibit saturation with the substrate.
- At low substrate concentrations, when most of the enzyme is free and very little is present as the ES complex, the initial reaction velocity v_0 is proportional to the substrate concentration.
- With increasing concentration of the substrate, the enzyme becomes saturated with the substrate, and the velocity of the reaction reaches a limiting value V_{max} (Fig. 3.1).

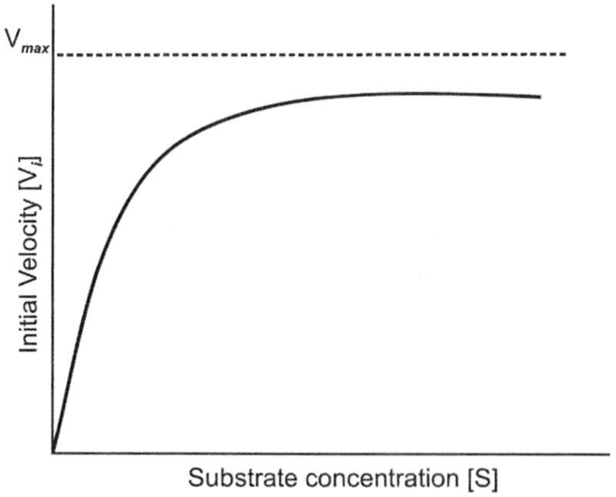

Fig. 3.1 Effect of substrate concentration on enzyme activity

- The relationship between the initial rate of reaction, substrate concentration, and characteristics of the enzyme is expressed by the *Michaelis-Menten (M-M) equation*

$$v_i = \frac{V_{max}[S]}{K_m + [S]},$$

where v_i = initial velocity, V_{max} = maximum velocity, S = substrate concentration, and K_m = Michaelis-Menten constant

- K_m, the *Michaelis–Menten constant*, is the substrate concentration at which the velocity is half maximal. K_m is an index of the affinity of an enzyme for its substrate and is characteristic for an enzyme and a specific substrate. A high K_m value indicates weak interaction between the enzyme and the substrate.

❑ How are K_m and V_{max} evaluated graphically?

The M – M equation, which is a rectangular hyperbola, can be rearranged to yield a straight line. Taking reciprocals on both the sides, we have

$$\frac{1}{v_i} = \frac{1}{V_{max}} + \frac{K_m}{V_{max}} \times \frac{1}{[S]}$$

Now, if $1/v_i$ is plotted against 1/[S], a straight line with a slope of K_m/V_{max} is obtained (Fig. 3.2). This is known as a *Lineweaver-Burk* or *double reciprocal plot*. The values for K_m and V_{max} can be obtained from this plot. The value of V_{max} depends on the amount of enzyme present and is not characteristic for an enzyme.

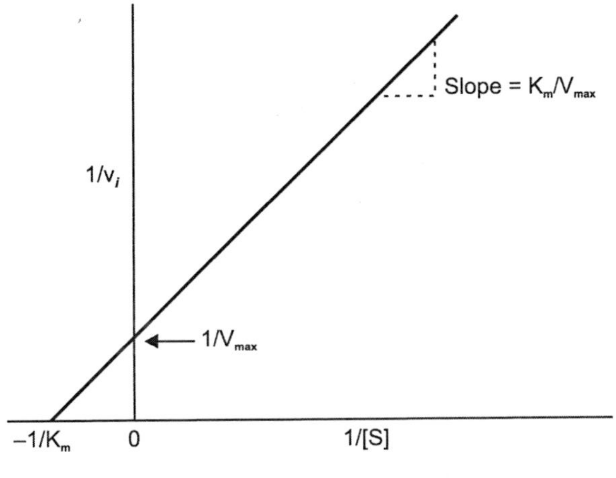

Fig. 3.2 Lineweaver-Burk plot

❑ Explain bisubstrate and multisubstrate enzyme reactions.

- Enzymes that catalyze reactions involving two or more substrates may take the ordered pathway, random-order pathway, or the ping-pong mechanism.
- In the ordered pathway, the first substrate S_1 binds to the enzyme to form a binary complex ES_1, followed by binding of the second substrate S_2 to form the ternary complex ES_1S_2 that generates the products P_1 and P_2.
- In the random-order pathway, the two substrates bind to the enzyme in either orientation.
- The ping-pong mechanism operates in reactions involving bound coenzymes. In this process, the enzyme interconverts between two states E and E'; S_1 bound to the enzyme is converted to P_1 followed by binding of E' to S_2 which is converted to P_2. The enzyme eventually return to its original form.

❑ What is the effect of pH on enzyme activity?

- Enzymes are sensitive to changes in pH.
- The pH at which the enzyme activity is maximal is known as the *optimum pH* and is specific for each enzyme.
- The activity of the enzyme diminishes above or below the optimum pH (Fig. 3.3).

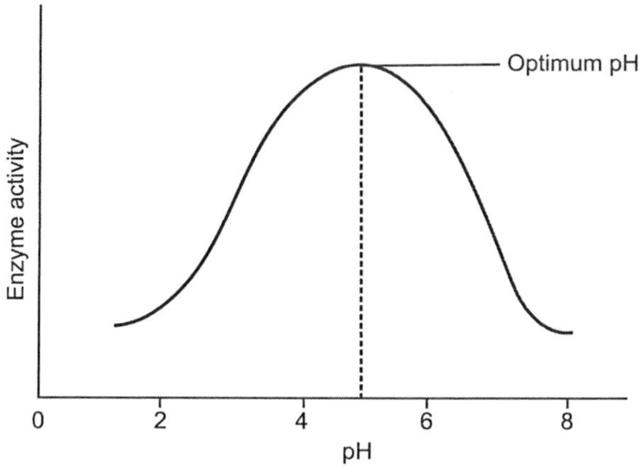

Fig. 3.3 Effect of pH on enzyme activity

❑ What is the effect of temperature on enzyme activity?

- The rate of an enzyme-catalyzed reaction increases with increase in temperature (Fig. 3.4).
- The temperature at which the enzyme activity is maximal is known as the *optimum temperature*. It is specific for each enzyme.

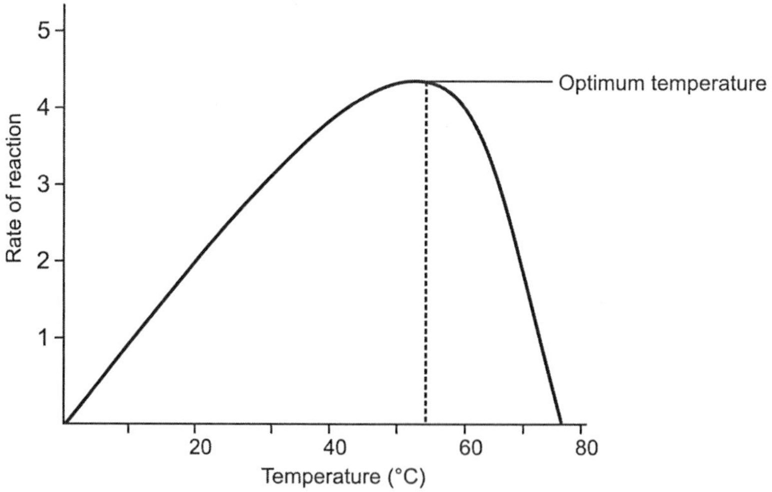

Fig. 3.4 Effect of temperature on enzyme activity

- The enzyme activity doubles with every 10°C rise in temperature (Q10); however, the enzyme protein denatures when the temperature is increased.
- Although the temperature at which different enzymes denature differs, most enzymes show appreciable denaturation above 40°C.

- Enzymes produced by thermophilic organisms such as *Thermus aquaticus* are stable at higher temperatures. These enzymes, known as *extremozymes,* are useful in the polymerase chain reaction (PCR).

❑ What is an enzyme inhibitor?

- A substance that decreases the activity of an enzyme-catalyzed reaction is known as an inhibitor.
- Enzyme inhibitors interfere with the binding of the substrate to the active site or with the formation of products from the ES complex.
- Enzyme inhibition may be *reversible* or *irreversible.*
- Reversible inhibitors react with the enzyme and establish an equilibrium between free and bound forms.
- Irreversible inhibitors bind permanently to the enzyme and modify a functional group such as cysteine or serine, required for catalysis.
- The organophosphorous insecticide diisopropylfluorophosphate (DFP) is a classic example of an irreversible inhibitor. DFP inhibits activity of *acetylcholine esterase* by covalent interaction with a serine residue at the active site leading to perpetual propagation of nerve impulses leading to death.

❑ What are the types of reversible inhibitors?

There are three types of reversible inhibitors: *competitive, non-competitive,* and *uncompetitive.*

◘ *Competitive inhibitor*

A competitive inhibitor

- is structurally similar to the normal substrate.
- competes with the substrate to bind to the active site of the enzyme.
- increases K_m but does not alter V_{max}.
- is prevented from binding to the active site if the substrate is bound to it. Competitive inhibition can, therefore, be reversed by increasing the concentration of the substrate (Fig. 3.2).
- a classic example is malonate, which competes with succinate for *succinate dehydrogenase.*

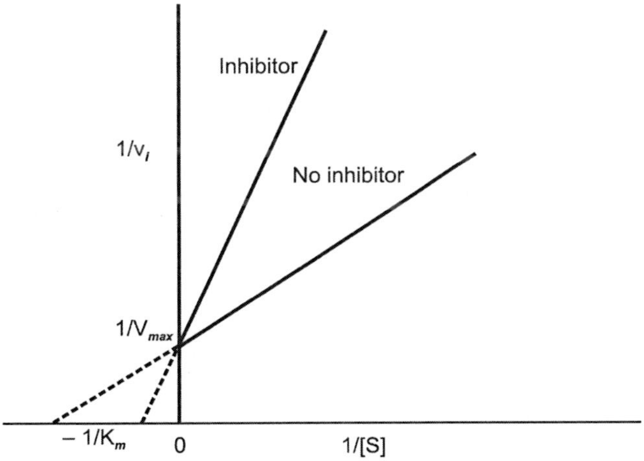

Fig. 3.5 Competitive inhibition

◘ *Non-competitive inhibitor*

A non-competitive inhibitor

- binds to a site other than the active site.

- binds to both the free enzyme as well as the ES complex and lowers V_{max} without affecting the K_m. The inhibition is not reversed by increasing the substrate concentration.
- EDTA, which binds Mg^{2+}, is a non-competitive inhibitor of enzymes requiring Mg^{2+} for their activity.

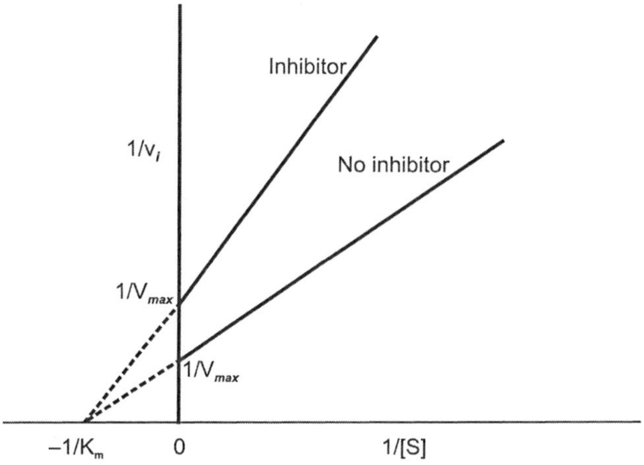

Fig. 3.6 Non-competitive inhibition

- ***Uncompetitive inhibitor***

An uncompetitive inhibitor

- binds only to the ES complex and not to the free enzyme.
- does not affect the velocity of the reaction.
- commonly inhibits reactions in which multiple substrates are involved.

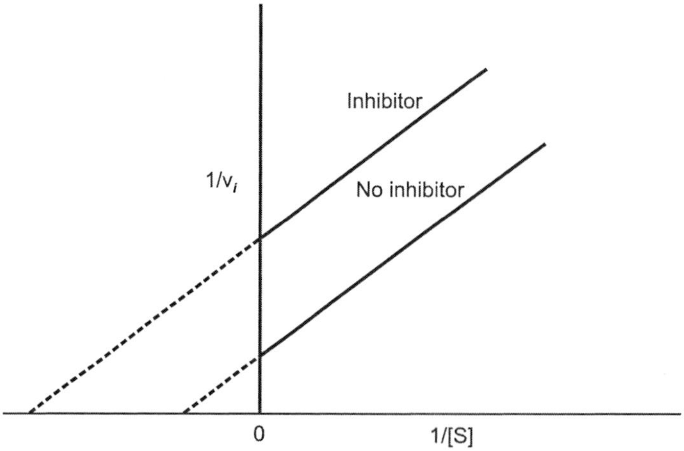

Fig. 3.7 Uncompetitive inhibition

❑ Outline the mechanisms that regulate enzyme activity.

There are several mechanisms that regulate enzyme activity. These include allosteric regulation, reversible covalent modification, proteolytic activation, and regulation of enzyme synthesis and degradation.

Allosteric regulation

- Inhibition of the first enzyme in a metabolic pathway by the end product in the pathway is referred to as *feedback inhibition*.
- In multienzyme systems, the first enzyme in the pathway, that is inhibited by the end product, is called an *allosteric enzyme*.
- A classic example of an allosteric enzyme is *aspartate transcarbamoylase* (*ATCase*) that catalyzes the first step in pyrimidine biosynthesis.
- Allosteric enzymes are large, complex enzymes with one or more polypeptide chains.
- In addition to the active site, allosteric enzymes contain one or more regulatory or allosteric sites for binding a positive or negative modulator.
- Allosteric enzymes do not exhibit M – M kinetics and the substrate saturation curve is *sigmoidal* reflecting co-operative interactions between the multiple subunits.

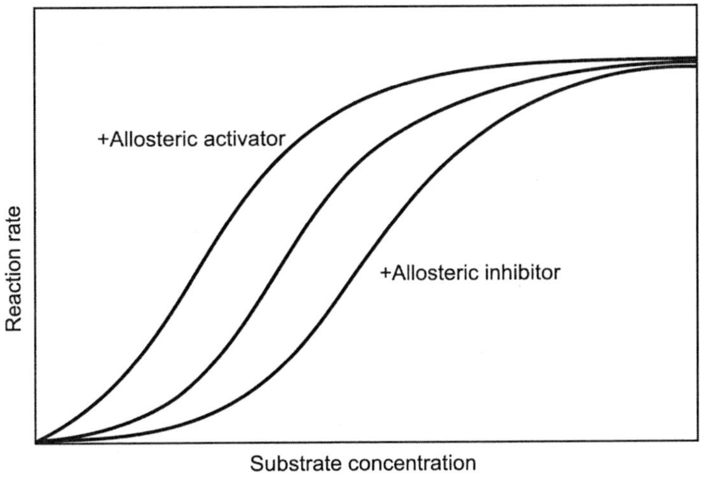

Fig. 3.8 Kinetics of allosteric enzymes

Reversible covalent modification

- Enzymes can be regulated by formation and breaking of a covalent bond between a nonprotein group and the apoprotein.
- The most common covalent modifications include
  a. addition and removal of phosphate groups (phosphorylation and dephosphorylation) brought about by *kinases* and *phosphatases*, respectively
  b. adenylylation
  c. uridylylation
  d. ADP-ribosylation.

Proteolytic activation

- Some enzymes are synthesized as inactive precursors called *zymogens* that are activated by hydrolysis of one or more peptide bonds.
- The proteolytic enzymes trypsin, chymotrypsin, and pepsin are all synthesized from their respective zymogens.

- ***Regulation of enzyme synthesis and degradation***
  - The rate of enzyme synthesis is influenced by induction or repression of the gene encoding the enzyme.
  - The rate of degradation of an enzyme is determined by the time taken for 50% of the enzyme protein to be degraded (half-life).

❑ How is enzyme activity expressed?

Enzyme activity may be expressed in any one of the following ways:

- as SI units of *katals*. A katal refers to the moles of substrate consumed or product formed per second.
- as International Units (IU). An IU of enzyme activity refers to μmoles of substrate consumed or product formed per minute.
- as specific activity that expresses the number of IU per mg protein or katals per kg protein.

❑ What are isoenzymes and isoforms?

- *Isoenzymes* are multiple forms of the same enzyme that catalyze the same reaction but differ in certain physical properties such as heat stability, sensitivity to inhibitors, or electrophoretic mobility. Isoenzymes are encoded at multiple genetic loci or by multiple alleles at a single genetic locus. *Creatine kinase* and *lactate dehydrogenase* exist as isoenzymes.
- *Isoforms* of an enzyme arise as a result of tissue-specific post-translational modifications of the enzyme protein such as partial proteolysis, deamidation, or addition of carbohydrate side chains. For example, the bone and liver alkaline *phosphatase* enzymes are isoforms that differ in their carbohydrate side chains.
- Both isoenzymes and isoforms are useful in clinical diagnosis.

❑ What are the plasma enzymes that are clinically useful?

- Enzymes that are nonfunctional in plasma and present in low concentration in normal plasma, such as *amylase*, *creatine kinase*, *transaminases* and *phosphatases*, are useful in clinical diagnosis as well as in monitoring response to treatment.
- Increased levels of plasma nonfunctional enzymes may be due to
  a. cell damage
  b. cell proliferation
  c. enzyme induction
  d. decreased clearance
  e. obstruction to secretion.

❑ Enumerate the nonspecific causes of altered plasma enzyme levels.

Plasma enzyme assays may be influenced by analytical, artefactual and physiological factors.

- *Analytical factors* include pH, temperature, substrate concentration and enzyme modulators.
- *Artefactual factors* such as *in vitro* hemolysis and delayed separation of plasma from whole blood, increase plasma activities of enzymes such as *aspartate transaminase*.
- *Physiological factors* increase the plasma activities of several enzymes. For example,
  a. age : *aspartate transaminase*
  b. male sex : *γ-glutamyltranspeptidase*
  c. exercise : *creatine kinase*
  d. pregnancy : *alkaline phosphatase*
  e. parturition : *creatine kinase.*

- Ingestion of drugs such as phenobarbitone increases the activity of *γ-glutamyltranspeptidase* by inducing synthesis of the enzyme.

❑ Discuss the clinical significance of plasma amylase.

- α-Amylase found in high concentration in saliva and pancreatic juice, hydrolyses the α-1,4 linkages in starch and glycogen to yield maltose.
- It is the only plasma enzyme found in normal urine.
- Plasma amylase is derived from saliva and pancreatic juice.
- Increased activity of plasma amylase is seen in the following conditions:
  - salivary gland disorders such as mumps, parotitis and salivary calculi
  - acute pancreatitis
  - perforated duodenal ulcer, intestinal obstruction and peritonitis
  - biliary tract diseases such as cholecystitis that result from primary or secondary pancreatic involvement
  - acute oliguric renal failure
  - severe diabetic ketoacidosis
  - morphine administration which induces contraction of the *Sphincter* of *Oddi*.

❑ What is macroamylasaemia?

An increase in *plasma amylase* with decreased urinary excretion despite normal glomerular function is termed *macroamylasemia*. This is believed to result due to a complex formed between amylase and a high molecular weight protein such as IgA or IgG that is not filtered by the glomeruli.

❑ Describe the normal distribution and clinical importance of transaminases.

- *Aspartate transaminase* (AST) and *alanine transaminase* (ALT) catalyze the transfer of amino groups with pyridoxal phosphate functioning as a coenzyme.
- AST (formerly known as *glutamate oxaloacetate transaminase*, GOT) catalyzes the transfer of amino groups of aspartate to 2-oxoglutarate to form glutamate and oxaloacetate.

$$\text{L-aspartate} + \text{2-oxoglutarate} \xrightleftharpoons{\text{AST}} \text{L-glutamate} + \text{oxaloacetate}$$

- ALT (formerly known as *glutamate pyruvate transaminase*, GPT) catalyzes amino transfer from alanine to 2-oxoglutarate to form glutamate and pyruvate.

$$\text{L-alanine} + \text{2-oxoglutarate} \xrightleftharpoons{\text{ALT}} \text{L-glutamate} + \text{pyruvate}$$

Tissue distribution

- AST is present in high concentrations in the heart, skeletal muscle, liver, kidney, and erythrocytes in both mitochondria and cytosol.
- High concentrations of ALT are found in the liver and smaller quantities in the heart, kidney, and skeletal muscle. The enzyme is predominantly cytosolic.

Clinical significance

Plasma transaminases are elevated in the following conditions:

- *Circulatory failure* following cardiac arrest: Both AST and ALT are increased due to hypoxic damage to tissues and decreased rate of clearance.

- *Myocardial infarction*: Plasma AST activity increases 6–8 hours after the onset of chest pain reaching peak values between 24 and 48 hours, with a fall to the normal range by 4–6 days. Plasma ALT activity is however not significantly altered.
- *Liver disease*: Plasma ALT is a more specific indicator of liver damage than AST. Both AST and ALT activities are increased in carcinoma of the liver with the increase in plasma AST greater than ALT since both mitochondrial and cytoplasmic membranes are damaged.
- *Skeletal muscle disease, surgery, trauma, and severe hemolysis*: Only plasma AST is elevated.

❑ Enumerate the causes of increased alkaline phosphatase.

- *Alkaline phosphatase* (ALP) catalyzes the hydrolysis of organic phosphates in the pH range 9.0–10.5.
- The enzyme contains zinc and facilitates metabolite transfer across cell membranes and calcification of bone.
- ALP is present in the osteoblasts of bone, hepatobiliary tract, intestinal epithelium, kidney and placenta.
- Plasma ALP orginates from the bone and liver.
- Plasma ALP activity is elevated in
  a. *Hepatobiliary diseases*: Plasma ALP measurement is useful as a liver function test. High plasma ALP associated with hyperbilirubinaemia is characteristic of cholestasis. A greater increase in enzyme activity is found in extrahepatic cholestasis than in intrahepatic obstruction. In early liver cirrhosis and in hepatocellular carcinoma, plasma ALP is increased without an associated increase in plasma bilirubin.
  b. *Bone disorders*: Plasma ALP activity is increased in conditions associated with increased osteoblastic activity such as Paget's disease of bone, and in osteomalacia and rickets.
  c. Tumors of the bone and liver. A heat stable, placental-like isoenzyme, termed the *Reagan isoenzyme*, is present in the plasma of some patients with carcinoma of the bronchus.

❑ Describe the distribution and clinical significance of *acid phosphatase (ACP)*.

Tissue distribution

- ACP that catalyzes the hydrolysis of organic phosphates at acidic pH is predominantly found in the prostate. Prostatic ACP is termed *tartrate labile ACP* because it is strongly inhibited by tartrate ions.
- It is also present in liver, spleen, semen, erythrocytes, platelets, and bone marrow. The erythrocyte enzyme which is inhibited by formaldehyde is known as the *formol labile ACP*.
- High concentration of ACP in semen is useful in forensic medicine in the investigation of rape.

Factors that influence ACP release into the circulation include

- rectal examination and passage of a catheter, which release the prostatic enzyme due to pressure on prostatic cells
- *in vitro* hemolysis, which releases the erythrocyte enzyme. Since the anticoagulant heparin inhibits ACP activity, only clotted blood should be used for enzyme assay.

Clinical significance

- In metastasizing prostatic carcinoma, both the tartrate labile and the total ACP are elevated.
- Increased activity of non-prostatic ACP is characteristic of Paget's disease of bone, hyperparathyroidism with skeletal involvement, Gaucher's disease, myelocytic leukemia, and malignant invasion of the bones in breast cancer.

❑ Outline the reactions catalyzed, tissue distribution and clinical significance of lactate dehydrogenase.

Reaction catalyzed

- *Lactate dehydrogenase (LD)* catalyzes the oxidation of L-lactate to pyruvate using NAD^+ as the hydrogen acceptor.

$$\underset{\textbf{lactate}}{CH_3-CHOH-COO^-} + NAD^+ \rightleftharpoons \underset{\textbf{pyruvate}}{CH_3-CO-COO^-} + NADH + H^+$$

- The enzyme also catalyzes the oxidation of 2-hydroxybutyrate (2-HBD) to 2-oxobutyrate. This is referred to as *2-HBD activity*.

Tissue distribution

- High concentrations of LD are found in the heart, skeletal muscle, liver, kidney, brain and erythrocytes.
- Plasma LD is derived from erythrocytes and platelets.

Clinical significance

A marked increase in plasma total LD activity is seen in the following conditions:

- Circulatory failure with shock and hypoxia
- Myocardial infarction
- Hematological disorders (megaloblastic anemias, acute leukemias, and lymphomas)
- Toxic hepatitis with jaundice
- Malignancies (Hodgkin's disease, and abdominal and lung cancers)
- Renal infarction.

A moderate increase in total LD activity occurs in

- viral hepatitis
- cirrhosis
- obstructive jaundice
- skeletal muscle diseases.

However, due to lack of tissue specificity, total LD is not useful in clinical diagnosis and LD isoenzymes are preferred.

❑ Give an account of LD isoenzymes and their clinical significance.

LD is a tetramer containing two types of monomer units—the H type and the M type that give rise to 5 isoenzymes (LD-1 to LD-5).

All the 5 LD isoenzymes can be detected in plasma on electrophoresis.

1. LD-1 (HHHH) migrates fastest toward the anode. LD-1 is predominantly found in the heart and erythrocytes and exhibits greater catalytic activity with HBD as a substrate. Plasma HBD, an index of LD-1 activity, rises 12–24 hours after the onset of chest pain in myocardial infarction, reaches a peak value 48–72 hours later, and remains elevated for 7–12 days.
2. LD-2 (HHHM) and LD-3 (HHMM) are increased in megaloblastic anemia and acute leukemia.
3. LD-4 (HMMM)
4. LD-5 (MMMM), which shows the slowest migration on electrophoresis, is present in the liver. It is markedly elevated in liver diseases and in progressive muscular dystrophy.

In addition to the above 5 isoenzymes, two other isoenzymes, LD-X (LDc) and LD-6, have also been identified. LD-X contains four X (or C) subunits and is present in prepubertal human testes, whereas LD-6 is found in the sera of severely ill patients.

❑ What are the causes of increased *creatine kinase* activity?

Reaction catalyzed

Creatine kinase catalyzes the phosphorylation of creatine with ATP as phosphate group donor by the following reaction:

$$\text{Creatine + ATP} \rightleftharpoons \text{Creatine phosphate + ADP}$$

Tissue distribution

The enzyme is predominantly found in cardiac, skeletal and smooth muscle as well as in the brain.

Clinical significance

Very high plasma CK activity is seen in the following conditions:

- Circulatory failure and shock
- Myocardial infarction: enzyme activity rises 4–6 hours after the onset of chest pain, reaches a peak value between 24 and 48 hours and remains elevated for 3–5 days.
- Early stages of muscular dystrophies (*e.g.*, Duchenne muscular dystrophy)
- Necrotizing polymyopathy (acute rhabdomyolysis) due to crush injuries
- Acute postanesthetic stage of malignant hyperpyrexia
- Reye's syndrome, a central nervous system disorder in children.
- Alcohol and drugs such as barbiturates and diethylstilbesterol in pharmacological amounts.

❑ Discuss the isoenzymes of *creatine kinase* (*CK*) and their clinical importance.

- CK is a dimer, and three isoenzymes composed of two monomers, M (for muscle) and B (for brain), have been identified. These include
  - CK-1 (CK-BB), the predominant isoenzyme in the brain that moves most rapidly and migrates in the prealbumin region.
  - CK-2 (CK-MB), present in cardiac muscle that has intermediate mobility and migrates in the α_2-globulin region.
  - CK-3 (CK-MM), present in high concentrations in skeletal and cardiac muscle with the slowest mobility and seen in the γ-globulin fraction.
- All the three isoenzymes are present in the cytosol.
- A mitochondrial isoenzyme CKMt that differs immunologically and by electrophoretic mobility from the other isoenzymes has also been identified.
- Plasma CK is mostly the MM isoenzyme. An increase in plasma CK-MM occurs in the following conditions:
  - Skeletal muscle damage, muscular dystrophy, and myopathy.
  - After a myocardial infarct accompanied by an increase in CK-MM3 : CK MM1 ratio 2–5 hours after the onset of chest pain.
  - Acute cerebrovascular disease or cerebral ischemia
  - Neurosurgical intervention.
- Plasma CK-BB is raised following head injury and in a wide variety of malignancies. An increase in plasma CK-MB in patients with head injury and subarachnoid hemorrhage indicates myocardial damage.

❑ How are the plasma enzymes useful in the diagnosis of myocardial infarction?

- Although plasma enzyme assays do not contribute to diagnosis and management of myocardial infarction in patients with a typical clinical history and ECG changes, they are useful in atypical cases where ECG changes are not characteristic.
- The plasma enzymes of diagnostic value in myocardial infarction include
  - *CK*: Total CK levels rise 4–6 hours after the onset of chest pain reaching peak values in 24–48 hours and remain elevated for 3–5 days after the infarct. The CK-MB isoenzyme is the earliest indicator of myocardial infarction and shows a rise within 3 hours after onset of chest pain. If plasma CK levels fail to fall within the normal time course, either extension of the infarct or a second infarct may be suspected.
  - *LD (or HBD)*: Enzyme activity rises 12–24 hours after the onset of chest pain reaching peak values in 48–72 hours and remains elevated for 7–12 days after the infarct. The LD/HBD ratio is decreased to 0.8–1.2 (normal 1.2–1.6).
  - *AST*: Enzyme activity rises 6–8 hours after the onset of chest pain, reaching peak values in 24–48 hours and remains elevated for 4–6 days after the infarct. Although plasma AST is a sensitive indicator of myocardial infarction, it lacks specificity. Enzyme activity is increased in hepatic congestion following myocardial infarction.
- An increase in plasma levels of cardiac troponin I and T derived from heart muscle is useful as a specific and sensitive marker of myocardial infarction. Furthermore, the rise in plasma troponins remains longer than for plasma enzymes.

❑ What are the plasma enzymes that are valuable in the diagnosis of liver disease?

- Plasma ALT, AST, ALP, and GGT are useful in the diagnosis of liver disease.
- Plasma ALT is a more specific indicator of liver damage than AST.
- Both AST and ALT activities are increased in carcinoma of the liver, with the increase in plasma AST greater than ALT since both mitochondrial and cytoplasmic membranes are damaged.
- An increase in plasma GGT activity seen in all types of liver disease is associated with a rise in plasma ALP.
- High plasma levels of GGT and ALP associated with hyperbilirubinaemia is characteristic of cholestasis.
- In chronic liver disease, plasma GGT levels are permanently raised.
- In early liver cirrhosis and in hepatocellular carcinoma, plasma ALP is increased without an associated increase in plasma bilirubin.

Chapter 4

Vitamins and Coenzymes

❑ How are vitamins classified?

Vitamins are the dietary constituents essential for normal growth and development. These are classified into

- *fat-soluble vitamins* (vitamins A, D, E, and K)
- *water-soluble vitamins* (ascorbate and vitamins of the B-complex).

❑ Outline the sources and requirements of fat-soluble vitamins.

The sources and daily recommended allowances (RDA) of the fat-soluble vitamins are summarized in Table 4.1.

Table 4.1 Sources and requirements of fat-soluble vitamins

Vitamin	*Sources*	*Daily requirements*
A	*Preformed (animal) sources*: Fish liver oil, milk and eggs *Provitamins (plant) sources*: Yellow and green leafy vegetables, carrots, spinach, sweet potatoes, and papaya	1000 Retinol equivalents (RE) for men. 800 RE for women. 1 RE = 1 μg of retinal or 6 μg of β-carotene
D	Fish liver oil, liver, eggs, and butter. Mostly derived by endogenous synthesis by action of UV light on 7-dehydrocholesterol in the skin	200 IU (= 5 μg of cholecalciferol)
E	Wheat germ and corn germ oil, vegetable oils, cereals, legumes, muscle meat, liver, fish, and chicken.	8 – 10 mg
K	Green leafy vegetables (spinach, alfalfa), cauliflower, pulses, cereals, egg yolk, intestinal bacterial synthesis	45 – 80 μg for men, 45 – 65 μg for women

❑ How is vitamin A formed from β-carotene?

β-Carotene is a provitamin that undergoes cleavage to generate two molecules of vitamin A. Figure 4.1 shows the structures of β-carotene, vitamin A aldehyde (retinal), and vitamin A alcohol (retinol).

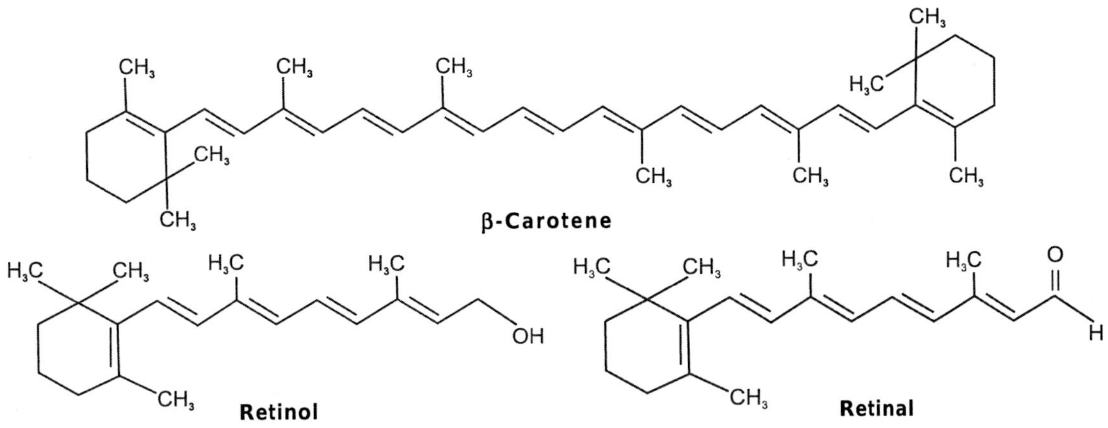

Fig. 4.1 The structures of β-carotene and vitamin A

❑ Enumerate the biological functions of vitamin A.

The biological functions of vitamin A include the following:

- Role in vision
  - Vitamin A plays a crucial role in the visual cycle and is essential for both rod (dim light) and cone (color) vision.
  - When light strikes the retina, rhodopsin (vitamin A containing pigment) is cleaved to form opsin and *all-trans* retinal.
  - Retinol transported from the liver to the retina is converted to *all-trans* retinal by the action of *retinol dehydrogenase.*
  - Isomerization of *all-trans* retinal gives rise to *11-cis* retinal, which combines with the protein opsin to regenerate rhodopsin.
  - The cones contain three retinal containing pigments — porphyrinopsin, iodopsin and cyanopsin sensitive to red, green, and blue respectively.
- Maintenance of normal secretory epithelium
- Normal bone formation and synthesis of mucopolysaccharides in the connective tissue
- Spermatogenesis
- Development and maintenance of placenta.
- Structural integrity and normal permeability of cell membranes and membranes of subcellular organelles such as mitochondria and lysosomes.

❑ What are the manifestations of vitamin A deficiency?

The manifestations of vitamin A deficiency include

1. Night blindness (*nyctalopia*), the earliest sign of vitamin A deficiency.
2. Drying of the conjunctiva and cornea (*xerophthalmia*) and keratinization of epithelium due to decreased mucous secretion. The lachrymal secretion is decreased.
3. Long-standing deficiency causes elevated white patches of keratin debris in the conjunctiva (*Bitot's spots*) characterized by ulceration, infection and scarring of the cornea (*keratomalacia*) resulting in blindness.
4. Hyperkeratinization of the epithelium lining the hair follicle (*follicular hyperkeratosis*).
5. Drying of the skin and atrophy of the sebaceous glands.
6. Defective formation of teeth enamel.
7. Failure in the growth of endochondral bone with imbalance between osteoblasts and osteoclasts, and aberrations in bone shape.

8. Atrophy of germinal epithelium leading to sterility in males and cornification of vaginal epithelium in females.

❑ Is excess consumption of vitamin A toxic?

- Overdosage of vitamin A can cause both acute and chronic hypervitaminosis.
- Toxic symptoms have been reported in the Arctic regions due to ingestion of polar bear liver rich in vitamin A.
- The manifestions of acute toxicity include nausea, vomiting, abdominal pain, drowsiness and headache.
- Chronic hypervitaminosis A is characterized by
  - fatigue, insomnia
  - painful joints, periosteal thickening of long bones, bone pain, bone fragility, and fractures
  - hair loss, desquammation and discoloration of the skin
  - hepatosplenomegaly and liver damage.

❑ How is the metabolically active form of vitamin D formed?

Vitamin D_3 (cholecalciferol) is synthesized endogenously from 7-dehydrocholesterol by ultraviolet (UV) irradiation of the skin. Cholecalciferol is not the metabolically active form of vitamin D. It undergoes hydroxylation in the liver by *25-hydroxylase* to form 25-hydroxycholecalciferol (25-HCC, calcidiol) followed by 1α-hydroxylation in the kidney to generate 1,25-dihydroxycholecalciferol (1,25–DHCC, calcitriol), the metabolically active form of vitamin D (Fig. 4.2).

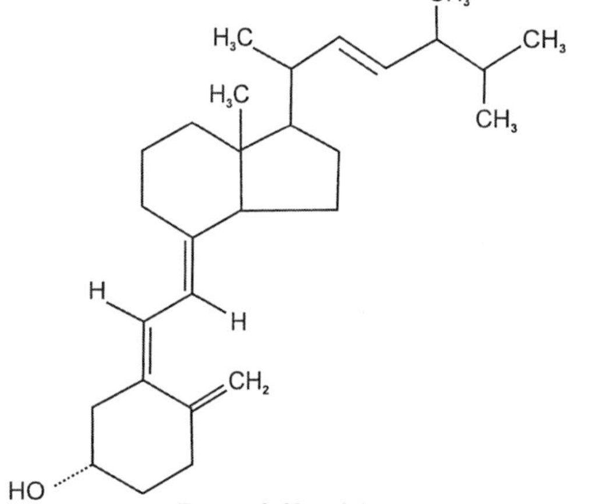

Ergocalciferol (Vitamin D_2)

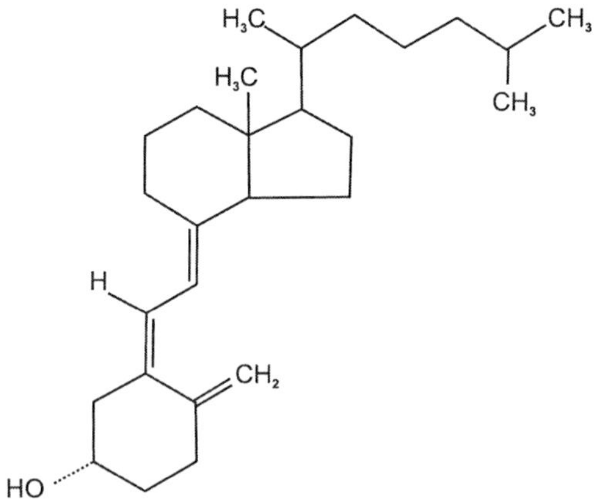

Cholecalciferol (Vitamin D_3)

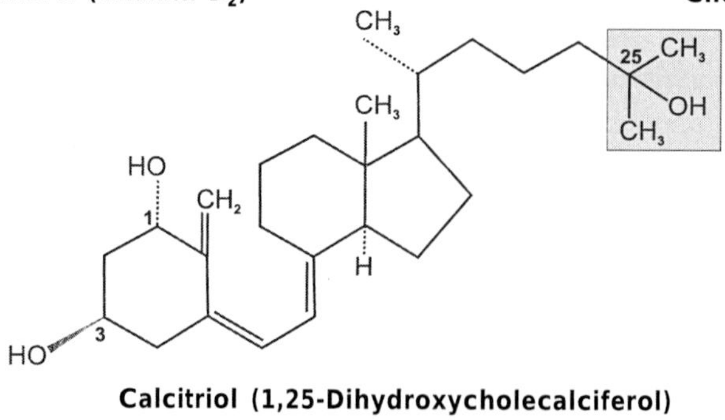

Calcitriol (1,25-Dihydroxycholecalciferol)

Fig. 4.2 The structures of vitamin D

❑ Elucidate the pathogenesis of vitamin D deficiency.

Vitamin D deficiency causes *rickets* in children and *osteomalacia* in adults. This may occur due to the following reasons:

- Inadequate intake
- Decreased endogenous synthesis due to decreased exposure to sunlight
- Malabsorption of vitamin D
- Inherited disorders of vitamin D metabolism due to defects in renal 1α- hydroxylation or mutations in the structure and functions of 1,25 $(OH)_2D$ receptor
- Renal tubular disorders such as *X-linked hypophosphatemia* (vitamin D-resistant rickets) in which there is defective renal reabsorption of phosphate; *de Toni Fanconi syndrome,* characterized by increased excretion of phosphate, glucose, potassium, amino acids and urate; and *renal tubular acidosis,* which causes increased urinary excretion of calcium with dissolution of bone calcium
- Chronic renal failure
- Hypophosphatemia which results from increased phosphate clearance in tumors due to impaired 1α-hydroxylation of 25(OH) vitamin D and phosphate transport.

❑ What are the clinical manifestations of vitamin D deficiency rickets?

The clinical symptoms of rickets include

- Disordered growth of cartilage and bone
- Bone deformities
- Beaded appearance of the ribs (*rachitic rosary*)
- Chest deformities with abnormal prominence of the sternum (*pigeon breast*)
- Bow legs, knock knees, deformities of the spine and pelvis, and susceptibility to fractures in prolonged deficiency
- Tetany in extreme cases

❑ Outline the clinical symptoms of osteomalacia.

- Osteomalacia occurs in pregnant women due to low dietary intake, and in elderly housebound people.
- The clinical manifestations include softness of bone due to progressive decalcification, diffuse skeletal pain, bony tenderness, pain, weakness, and deformities of the bone in the pelvis, ribs, lower lumbar vertebrae and sacrum.

❑ How is vitamin D deficiency treated?

Treatment of vitamin D deficiency is based on the etiology as outlined below:

- Oral administration of vitamin D_2 or D_3, 800–4000 IU (0.02–1.0 mg) for dietary deficiency.
- Administration of calcitriol for intestinal malabsorption.
- Administration of large doses (50,000 IU/d) of vitamin D for X-linked hypophosphatemia.
- Combined administration of oral supplements of phosphate and 1α-hydroxylated derivatives of vitamin D for vitamin D resistant rickets.

❑ Outline the manifestations of hypervitaminosis D.

The clinical manifestations of vitamin D toxicity include nausea, vomiting, anorexia, polydipsia and polyuria, demineralization of bone, and multiple fractures. In severe cases, calcification of soft tissue can result in death.

❑ Enumerate the biological functions of vitamin E.

- Vitamin E or α-tocopherol is a derivative of 6-hydroxychroman with a phytyl side chain (Fig. 4.3).
- The biological functions of vitamin E are mediated by its antioxidant activity and include
  - protection of membranes, cytoskeleton, thiol-rich proteins, LDL, and vitamin A from oxidation
  - prevention of damage to the liver and lungs by toxins and oxidants
  - protection of erythrocytes from hemolysis by oxidizing agents.
- The antioxidant and immunomodulatory effects of vitamin E have been reported to protect cardiovascular disease and cancer.

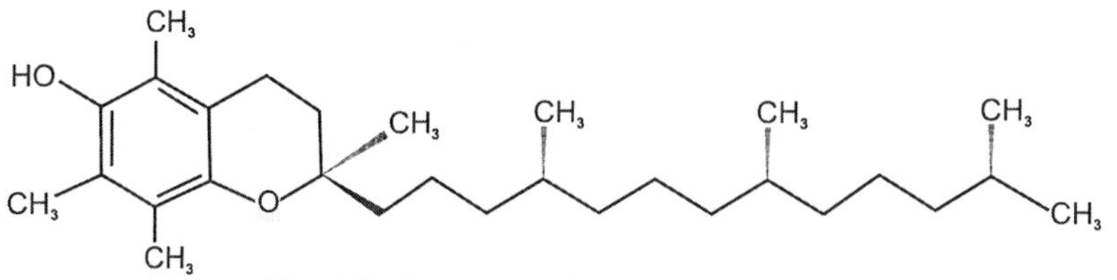

Fig. 4.3 Structure of α-tocopherol (vitamin E)

❑ How is vitamin E deficiency manifested?

Deficiency of vitamin E generally occurs in malabsorption syndromes such as steatorrhea. The clinical symptoms associated with vitamin E deficiency include

- muscular weakness and creatinuria because vitamin E is required for the storage of muscle creatine
- hemolysis and anemia
- neurological manifestations such as areflexia, poor proprioception, and vibratory sense
- liver necrosis.

❑ Describe the structure and biological functions of vitamin K.

Vitamin K contains 2-methyl-1,4-naphthoquinone moiety. While vitamin K_1 has a phytyl radical on position 3, vitamin K_2 has a difarnesyl radical (Fig. 4.4).

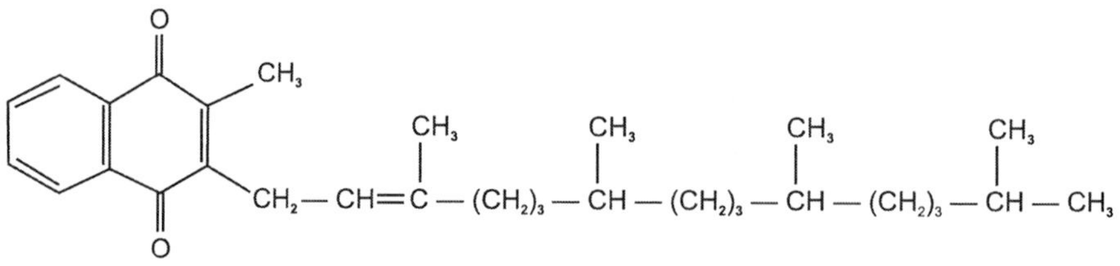

Phylloquinone (Vitamin K_1)

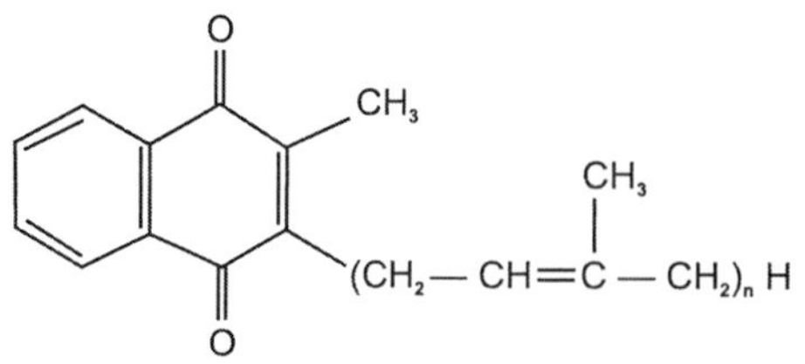

Menaquinone (Vitamin K_2)

Fig. 4.4 The structures of vitamin K

Functions

Vitamin K functions as a coenzyme for the post-translational carboxylation of glutamate residues in several proteins to form *γ-carboxyglutamate* (*Gla*). The Gla residues are important for the following:

1. *Blood coagulation*: Vitamin K is essential for the synthesis of prothrombin and the clotting factors VII, IX, and X in the liver. The Gla residues bind Ca^{2+} and facilitate adsorption to phospholipid membrane surfaces and proteolytic activation.
2. *Bone*: Gla residues in osteocalcin, the most abundant non-collagenous protein in bone, promote tight binding to hydroxyapatite of bone.
3. *Kidney*: Gla-containing proteins may play a role in Ca^{2+} reabsorption.

❑ Enumerate the causes and clinical symptoms of vitamin K deficiency.

Vitamin K deficiency characterized by impaired clotting, increase in prothrombin time, and hemorrhage, is seen in the following conditions:

1. Hemorrhagic disease of the newborn resulting from defective vitamin K transport across the placenta, delay in bacterial colonization of the intestine, or delay in protein synthesis.
2. Malabsorption conditions such as steatorrhea in which absorption of fat soluble vitamins is impaired.
3. Following administration of gut-sterilizing antibiotics that reduce intestinal vitamin K synthesis.

❑ Outline the sources and requirements of water-soluble vitamins.

Table 4.2 summarizes the sources and daily recommended allowances (RDA) of the water-soluble vitamins.

Table 4.2 Sources and requirements of water-soluble vitamins

Vitamin	*Sources*	*Daily requirements*
Ascorbic acid	Citrus fruits, green leafy vegetables, potatoes, tomatoes, and papaya	60 mg
Thiamine	Rice bran, wheat germ, whole cereals, dried yeast, oil seeds, and nuts	1 – 1.5 mg
Riboflavin	Milk, liver, kidney, green vegetables, germinating legumes, yeast, and eggs	1.4 – 1.8 mg in men 1.2 – 1.3 mg in women
Niacin	Yeast, liver, rice polishings, peanuts, whole cereals, lean meat, and fish	15 – 20 mg for men 13 – 15 mg for women
Pyridoxine	Germs of seeds and grains, rice bran, wheat bran, yeast, liver, cereals, and leafy greens	1.4 – 2.0 mg for men 1.4 – 1.6 mg for women
Pantothenic acid	Yeast, liver, kidney, eggs, rice polishings, wheat germ, legumes, oil seeds, milk, and fish	4 – 10 mg for men 4 – 7 mg for women
Biotin	Yeast, liver, rice bran, wheat bran, legumes, cereals, eggs, and milk	100 – 200 μg
Folic acid	Intestinal bacterial synthesis, liver, yeast, and eggs	150 – 200 μg
Vitamin B_{12}	Soil, water, intestinal microorganisms, liver, meat, fish, and eggs	2 μg

❑ Explain the biological functions of ascorbic acid.

Vitamin C (ascorbic acid) is a potent reducing agent that is readily oxidized to dehydroascorbic acid. The structure of ascorbic acid is shown in Figure 4.5.

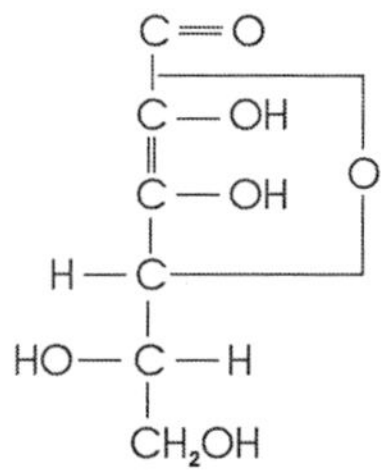

Fig. 4.5 The structure of ascorbic acid

Ascorbic acid is essential for the following functions:

1. Hydroxylation of
   - proline and lysine residues in collagen
   - phenylalanine to tyrosine
   - *p*-hydroxyphenylpyruvic acid to homogentisate
   - tryptamine to 5-hydroxytryptamine.
2. Formation of bone and intercellular substance in the bones, teeth, and cartilage.
3. Iron absorption by reducing Fe^{3+} to Fe^{2+}.
4. Biosynthesis of catecholamines.
5. Antioxidant activity to scavenge reactive oxygen species (ROS) such as hydrogen peroxide, singlet oxygen, superoxide ion and peroxyl radicals. This would help to
   - prevent ROS binding to cellular macromolecules (DNA, RNA, and proteins)
   - maintain sulphydryl compounds such as glutathione in the reduced state
   - protect plasma lipids against lipid peroxidation
   - regenerate vitamin E.

❑ Elucidate the clinical manifestations of ascorbic acid deficiency.

- Ascorbic acid deficiency causes *scurvy* which is characterized by impaired synthesis of collagen and chondroitin sulphate.
- The clinical manifestations of scurvy include
  - sore gums and loosening of teeth
  - swollen joints with joint pain
  - impaired bone formation in children, and osteoporosis and easy fracturability of bones in adults
  - hemorrhages (subcutaneous, periosteal, petechial)
  - poor healing of wounds
  - anaemia.

❑ Discuss the structure and biological functions of thiamin (vitamin B_1).

Figure 4.6 depicts the chemical structures of thiamin and its coenzyme thiamin pyrophosphate (TPP).

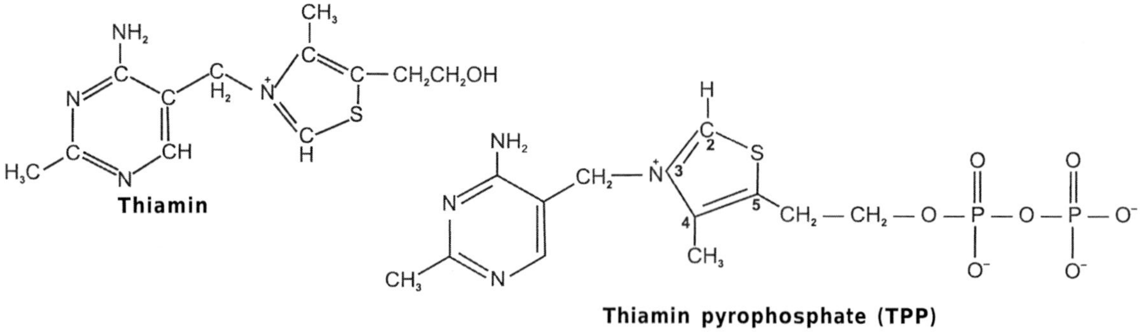

Fig. 4.6 Structures of thiamin and TPP

- TPP is formed by direct transfer of the pyrophosphate group from ATP in a reaction catalyzed by *thiamin diphosphotransferase.*
- Thiamin in the form of its coenzyme TPP is essential for the following reactions:
  1. Oxidative decarboxylation of
     - pyruvate to acetyl CoA
     - 2-oxoglutarate to succinyl CoA
     - branched chain amino acids
  2. Transketolation in the hexose monophosphate (HMP) shunt pathway

❑ What are the types of beriberi?

- Thiamin deficiency causes *beriberi* in areas where polished rice is used as a staple food such as in South Asia, Java and Japan.
- There are two types of beriberi — *dry beriberi* and *wet beriberi.*
- *Dry beriberi* occurs in chronic dietary deficiency of thiamin. The clinical symptoms include the following:
  - anorexia
  - loss of weight
  - muscle wasting, muscular weakness
  - neurological manifestations: anxiety, mental confusion, motor and sensory neuropathy (*Wernicke's encephalopathy*), and peripheral neuritis.
- *Wet beriberi* is seen when thiamin deficiency is severe. It causes
  - neurological symptoms as in dry beriberi
  - cardiovascular manifestations such as enlargement of the heart (cardiomegaly), increase in heart rate (tachycardia), cardiac failure, and edema.

❑ What is Wernicke-Korsakoff syndrome?

- This is seen in alcoholics with chronic thiamin deficiency.
- The syndrome is characterized by a defect in the transketolase enzyme, cardiac and respiratory irregularities, hemorrhagic lesions of the third and fourth ventricles of the brain, weakness, paralysis, retrograde amnesia, and impaired mental function.

❑ What are the coenzyme forms of riboflavin?

- Riboflavin (vitamin B_2) consists of a heterocyclic isoalloxazine ring linked to ribitol, a sugar alcohol.
- The coenzyme forms of riboflavin—*flavin mononucleotide (FMN)* and *flavin adenine dinucleotide (FAD)* are formed by ATP-dependent phosphorylation of riboflavin.
- Riboflavin is converted to FMN in the intestines and to FAD in the liver.
- The chemical structures of riboflavin, FMN, and FAD are shown in Figure 4.7.

❑ Outline the biological functions of riboflavin.

- Riboflavin is necessary for the maintenance of epithelial, mucosal and ocular tissues. Riboflavin in the form of its coenzymes FMN and FAD plays an important role in oxidoreduction reactions.
- FMN is a constituent of L-amino acid dehydrogenase and FAD is a prosthetic group of D-amino acid dehydrogenase, xanthine oxidase, acyl CoA dehydrogenase, etc.

❑ What are the clinical signs and symptoms of riboflavin deficiency?

Ariboflavinosis is characterized by

- Fissures at the corners of the mouth (*cheilosis*)
- Magenta-colored tongue
- Rough scaly skin, seborrheic dermatitis
- Hyperemia of the mucous membranes
- Glossitis
- Angular stomatitis
- Lesions of the genitalia
- Vascularization of the cornea

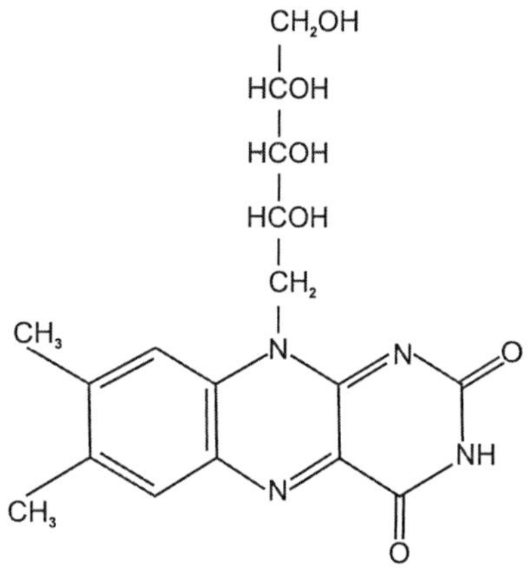

Riboflavin

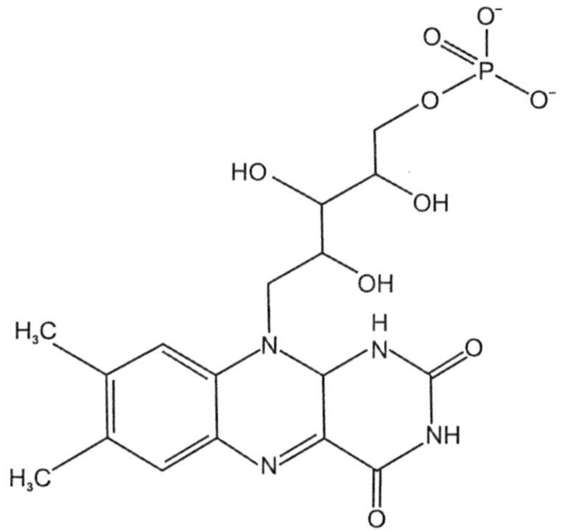

FMN

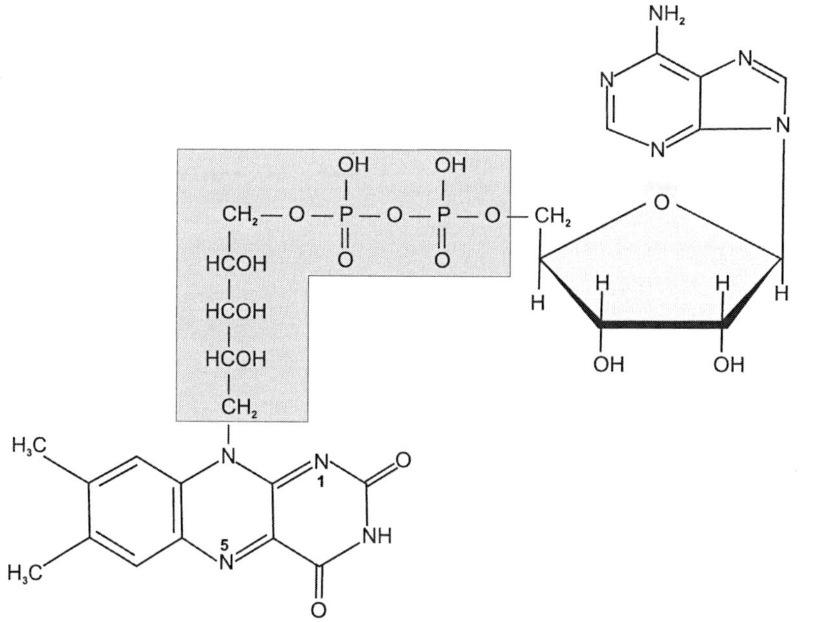

FAD

Fig. 4.7 Chemical structures of riboflavin, FMN and FAD

❑ What are the coenzyme forms of niacin?

The coenzyme forms of nicotinic acid (niacin) are nicotinamide adenine dinucleotide (NAD^+) and $NADP^+$. The chemical structures of niacin and NAD^+ are shown in Figure 4.8.

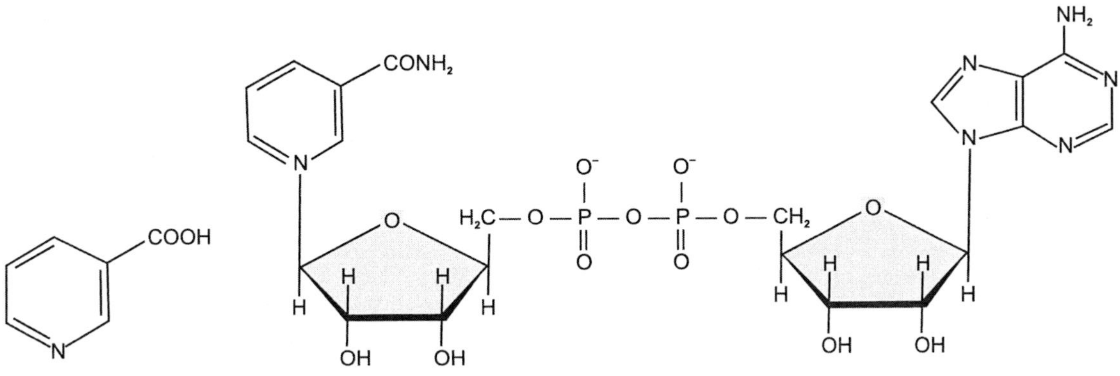

Nicotinic acid **NAD⁺**

Fig. 4.8 Chemical structures of niacin and NAD^+

❑ Elucidate the functions of niacin.

- The coenzymes of nicotinamide, NAD^+ and $NADP^+$, function as electron carriers in redox reactions and are reoxidized by flavoproteins.
- NAD^+ and $NADP^+$ play an important role in glycolysis, oxidative phosphorylation, and in many synthetic and catabolic pathways.
- Nicotinic acid reduces the hepatic secretion of VLDL and hence the plasma levels of VLDL and LDL.

❑ What are the causes and clinical manifestations of niacin deficiency?

Niacin is formed from tryptophan (60 mg of tryptophan produces 1 mg of niacin). Niacin deficiency causes *pellagra*, which is

- Common in areas where the major cereal diet is maize or jowar, which is low in tryptophan.
- Induced by the anti-tuberculosis drug, isonicotinic acid hydrazide (isoniazid, INH).
- Seen in two inborn errors of metabolism:
  i. Hartnup disease in which intestinal absorption of tryptophan is low.
  ii. Carcinoid syndrome, characterized by increased channeling of tryptophan to serotonin.

The clinical symptoms of pellagra include the *three Ds*:

- *Dermatitis,* which can lead to pigmentation, thickening, and roughness of the skin.
- *Diarrhea,* caused due to inflammation of the gastrointestinal mucosal membranes.
- *Dementia,* caused as a result of irritability and depression.

Other clinical features include nausea, vomiting, glossitis, stomatitis, and vaginitis.

❑ Describe the structure and biological functions of pyridoxine (vitamin B_6)

Structure

- Pyridoxine includes three groups of substances — pyridoxal, pyridoxamine, and pyridoxol.

- The active forms of the vitamin are pyridoxal phosphate and pyridoxamine phosphate.
- The chemical structures of different forms of vitamin B_6 are depicted in Figure 4.9.

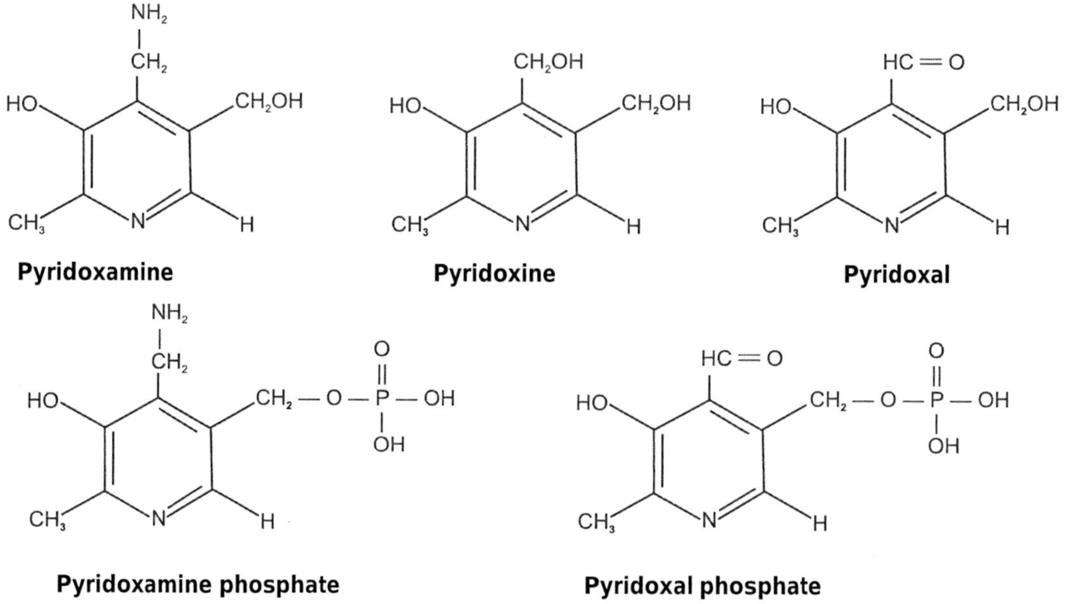

Fig. 4.9 Structures of vitamin B_6

Biological Functions

Pyridoxal phosphate functions as a coenzyme for the following reactions:

1. *Transamination,* an important step in the metabolism of amino acid nitrogen. During amino group transfer, pyridoxal phosphate undergoes reversible conversion to pyridoxamine phosphate.
2. *Decarboxylation* of tyrosine, arginine, and glutamate to form the corresponding amines. Decarboxylation of glutamate generates gamma-aminobutyric acid (GABA), a neurotransmitter.
3. *Amino acid transport.*
4. *Tryptophan metabolism,* as a coenzyme for *kynureninase,* an enzyme involved in the biosynthesis of niacin from tryptophan. B_6 deficiency can cause niacin deficiency.
5. *Cysteine metabolism,* for transulphuration and desulphuration.
6. *Deamination* of serine to pyruvate.
7. *Racemization* of D-amino acids to L-amino acids.
8. *Glycogen phosphorylase*
9. Synthesis of porphyrins and heme.
10. Formation of serotonin, epinephrine, GABA, and histamine. Vitamin B_6 is essential for the maintenance and integrity of neuronal tissue.
11. *Synthesis of coenzyme A* from pantothenic acid.
12. *Antibody synthesis.*

❑ What are the symptoms of pyridoxine deficiency?

The clinical symptoms of pyridoxine deficiency include

- rough skin
- sore tongue and mucosal lesions

- hypochromic, microcytic, sideroblastic anemia
- neurological changes, such as hyperirritability, peripheral neuropathy, epileptiform seizures
- gastrointestinal distress

❑ Discuss the structure and biological functions of pantothenic acid.

Structure

1. Pantothenic acid is a component of coenzyme A.
2. The structures of pantothenic acid and CoA are shown in Figure 4.10.

Biological Functions

Coenzyme A plays an important role in the metabolism of carbohydrates, lipids, amino acids, and steroid hormones. The reactions include

1. *Formation of acetyl CoA,* which can be channelized into the TCA cycle, acetylcholine synthesis, cholesterol and steroid hormone synthesis, or ketone body formation.
2. *Heme biosynthesis* via formation of succinyl CoA.
3. *Fatty acid metabolism.* The first step in fatty acid oxidation catalyzed by *thiokinase* requires CoA. Phosphopantetheine is the prosthetic group in acyl carrier protein (ACP), a coenzyme in fatty acid biosynthesis.
4. *Catabolism of branched chain amino acids,* which involves formation of CoA intermediates.

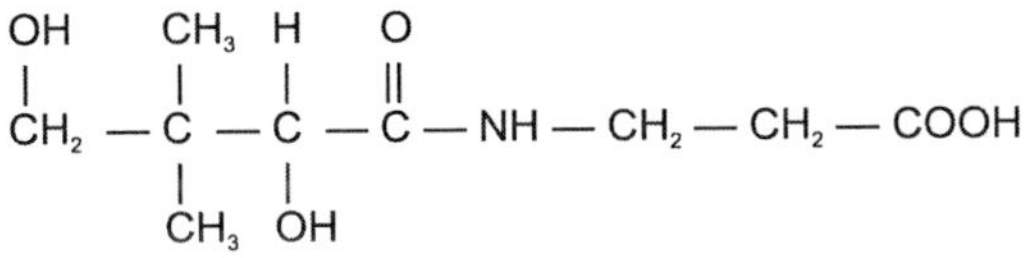

Pantothenic acid

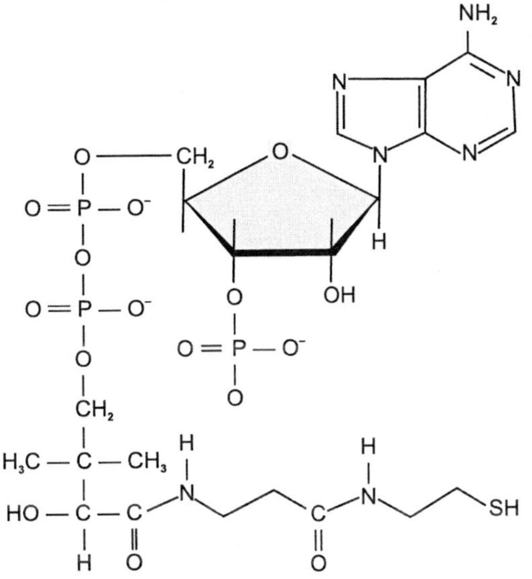

CoA

Fig. 4.10 Structures of pantothenic acid and CoA

❑ Elucidate the biological functions of biotin.

- The structure of biotin is depicted in Figure 4.11.
- Biotin functions as a prosthetic group for carboxylases and is attached to the enzyme by an amide bond between the carboxyl group of biotin and the terminal ε-amino group of a lysine residue of the enzyme, forming a biotinyl enzyme.

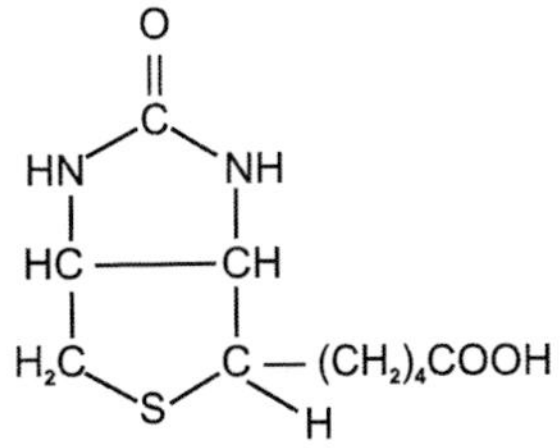

Fig. 4.11 Structure of biotin

- The enzymes for which biotin functions as a cocarboxylase include
  - *acetyl CoA carboxylase*, which catalyzes the conversion of acetyl CoA to malonyl CoA during fatty acid biosynthesis.
  - *pyruvate carboxylase*, which converts pyruvate to oxaloacetate during gluconeogenesis.
  - *3-methylcrotonyl CoA carboxylase*, which is involved in leucine catabolism.
  - *propionyl CoA carboxylase* that catalyzes the formation of methylmalonyl CoA from propionyl CoA.

In addition, biotin is required for fixation of CO_2 in C-6 of the purine nucleus during *de novo* purine biosynthesis.

❑ What are the causes and manifestations of biotin deficiency?

Causes of dietary deficiency of biotin include
- Administration of gut-sterilizing antibiotics.
- Ingestion of large amounts of raw egg white which contains *avidin*, an antivitamin of biotin that has high affinity for biotin and binds to biotin, thereby preventing intestinal absorption.

Clinical symptoms of biotin deficiency include anorexia, nausea, muscle pain and seborrheic dermatitis.

❑ Explain the structure of folate and its role in 1-carbon metabolism.

Folic acid (folate) consists of a pteridine nucleus, *p*-aminobenzoic acid (PABA), and glutamate (Fig. 4.12).

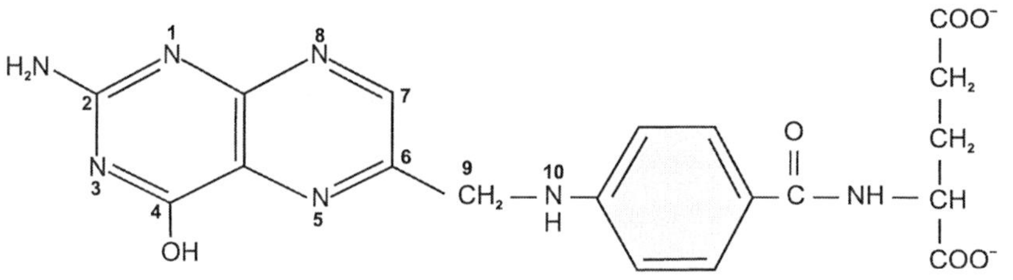

Fig. 4.12 The structure of folate

Folate in the form of its active coenzymes N^{10}-formyl H_4 folate, N^5, N^{10}-methenyl H_4 folate and N^5-formimino H_4-folate plays an important role in the transfer of the 1-carbon moieties formyl (— CHO), formate (— $HCOO^-$), methyl (— CH_3), hydroxymethyl (— CH_2OH), and formimino (— CH = NH —) groups.

The role of folate coenzymes in 1-carbon metabolism includes the following:
- Provision of carbon atoms 2 and 8 of the purine ring during *de novo* purine biosynthesis.
- Methylation of dUMP to dTMP.
- Formation of serine from glycine.
- Methyl group transfer from homocysteine to form methionine.
- Choline biosynthesis.

❑ What are the clinical symptoms of folate deficiency?

Folate deficiency, generally, occurs as part of a combined deficiency of folate, vitamin B_{12}, ascorbic acid and iron.

The clinical manifestations include

- Glossitis and cheilosis
- Gastrointestinal disturbances
- Defective erythropoiesis
- Failure of growth
- Megaloblastic anemia
- Leukocytopenia.

❑ Describe the structure, transport and biological functions of vitamin B_{12}.

- Vitamin B_{12} (cobalamin), also known as the *extrinsic factor of Castle*, is available in the form of cyanocobalamin.
- The coenzyme form (cobamide coenzyme) has a porphyrin-like *corrin* ring system surrounding a central divalent cobalt (Fig. 4.13).
- Vitamin B_{12} is absorbed in the ileum complexed with intrinsic factor secreted by the gastric mucosa, and is absorbed into the portal blood.
- The vitamin is transported by transcobalamin II in the portal blood and by transcobalamin I in the plasma.

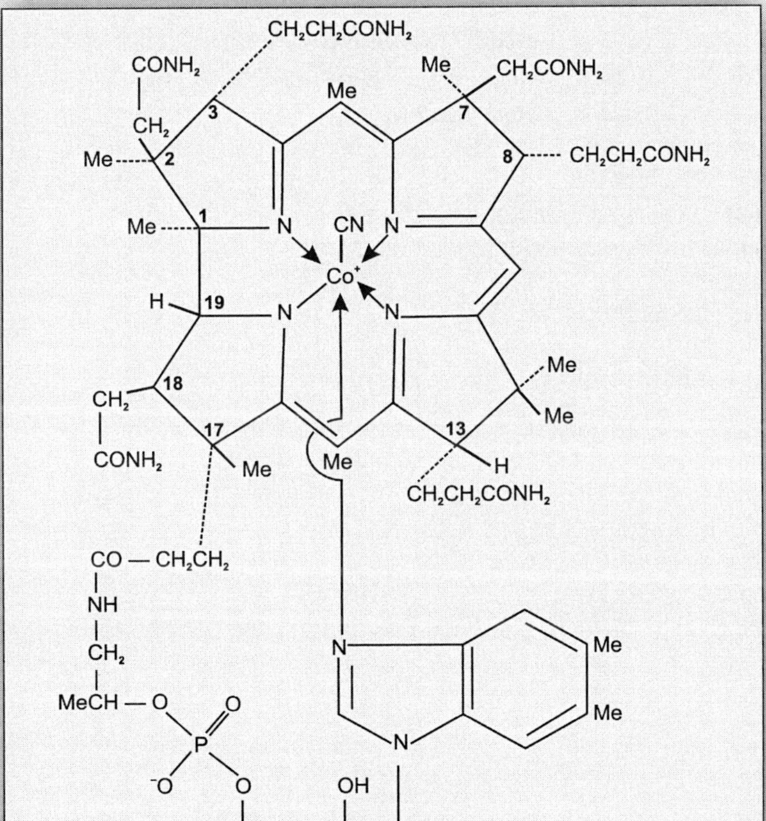

Fig. 4.13 Structure of cyanocobalamin

- Vitamin B_{12} is converted to the metabolically active forms — 5-deoxyadenosylcobalamin and methylcobalamin in the liver, bone marrow and reticulocytes.
- The biological functions of vitamin B_{12} include
  - DNA synthesis together with folic acid
  - Maintenance of myelin sheath and epithelial cells
  - Cycling of tetrahydrofolate back into the folate pool
  - Amino acid metabolism for the
    1. methylation of homocysteine to methionine catalyzed by homocysteine H_4 folate *methyltransferase* with methylcobalamin as a cofactor.
    2. conversion of methylmalonyl CoA to succinyl CoA catalyzed by methylmalonyl CoA *isomerase* with 5-deoxyadenosyl cobalamin as a cofactor.

Enumerate the clinical symptoms of vitamin B_{12} deficiency.

- Vitamin B_{12} deficiency causes pernicious anemia associated with atrophy of the gastric fundus and achlorhydria.
- Pernicious anemia is caused by the absence of intrinsic factor and consequent defective vitamin B_{12} absorption. Strict vegetarians (vegans) are susceptible to vitamin B_{12} deficiency.
- Pernicious anemia is characterized by
  - macrocytic, megaloblastic anemia with decrease in erythrocytes and hemoglobin
  - subacute combined degeneration of the spinal cord due to demyelination and paraesthesia (numbness and tingling of the extremities)
  - lesions of the skin and mucous membranes
  - glossitis
  - methylmalonic aciduria.

What are antivitamins? Give examples.

Antivitamins are antagonists of vitamins that block their functions. Examples include the following

- Dicumarol, salicylates and tocopheryl quinone, which are vitamin K antagonists.
- Isoniazid (isonicotinic acid hydrazide, INH), an anti-tuberculosis drug, is an antivitamin of B_6. It forms a hydrazone complex with PLP and promotes its excretion. INH causes deficiency manifestations of vitamin B_6. High doses of INH can cause neurological changes and affect tryptophan metabolism.
- Avidin, an antivitamin of biotin, binds to biotin preventing its absorption from the intestine.

Section 3

Bioenergetics and Metabolism

- Bioenergetics and Biologic Oxidation
- Metabolism of Carbohydrates
- Metabolism of Lipids
- Metabolism of Amino Acids and Nonprotein Nitrogen
- Metabolism of Purines and Pyrimidines
- Metabolism of Porphyrins and Bile Pigments
- Metabolism of Calcium, Phosphate, and Magnesium
- Trace Elements

Chapter 5

Bioenergetics and Biologic Oxidation

❑ State the laws of thermodynamics.

- The *first law of thermodynamics* states that the total energy of a system, plus that of its surroundings, remains constant. This law forms the basis of conservation of energy and for the principle that energy cannot be created or destroyed.
- The *second law of thermodynamics* states that the total entropy of a system must increase if a process is to occur spontaneously.

❑ Define free energy and entropy.

- *Free energy* is a type of energy that is capable of doing work at constant temperature and pressure.
- *Entropy* refers to energy in a state of randomness or disorder. Entropy is maximum in a system as it approaches equilibrium.

❑ Give the mathematical relationship between free energy and entropy.

Under conditions of constant temperature and pressure, the relationship between free energy change (ΔG) of a reacting system and the change in entropy (ΔS) is given by the equation

$$\Delta G = \Delta H - T\Delta S$$

where, ΔH is change in *enthalpy* (heat) and T is the absolute temperature.

As the reaction proceeds toward equilibrium, entropy increases and free energy decreases. ΔH is negative when it loses heat to the surroundings (*exergonic*) and positive when it absorbs heat from the surroundings (*endergonic*).

❑ What is the standard free energy change for biochemical reactions?

The standard free energy change ($\Delta G^{\circ\prime}$) is constant under standard conditions of temperature, pressure, and concentrations of reactants and products, at a pH of 7.0. It is calculated using the following relationship:

$$\Delta G^{\circ\prime} = -2.303\ RT \log K'_{eq}$$

where, R is the gas constant and T is the absolute temperature.

When $\Delta G^{\circ\prime}$ is negative, products contain less free energy than that of reactants and the reaction proceeds forward to form products.

When $\Delta G^{\circ\prime}$ is positive, the reaction proceeds in the reverse direction.

❑ Outline the role of ATP in the energy transfer process.

- ATP plays a central role in transferring energy from reactions that generate energy to the reactions that require energy.
- ATP is referred to as a *high-energy phosphate compound*. The standard free energy change during hydrolysis of ATP to ADP and inorganic phosphate (Pi) is –7.3 kcal/mol.
- Compounds such as phosphoenolpyruvate whose standard free energy of hydrolysis is greater than that of ATP, are referred to as *super high-energy phosphate compounds*.
- Compounds such as glucose 6-phosphate with free energy of hydrolysis lower than that of ATP, are referred to as *low-energy phosphate compounds*.
- ATP serves as an intermediate carrier of phosphate groups from super high-energy phosphate compounds to low-energy phosphate compounds. This involves the following steps:
  - transfer of a phosphate group from super high-energy phosphate compounds generated during catabolism to ADP to form ATP in a reaction catalyzed by a *kinase* enzyme
  - transfer of the terminal phosphate group from ATP to an acceptor molecule by another specific *kinase* to form a low-energy phosphate compound

  Figure 5.1 illustrates the ATP/ADP cycle.

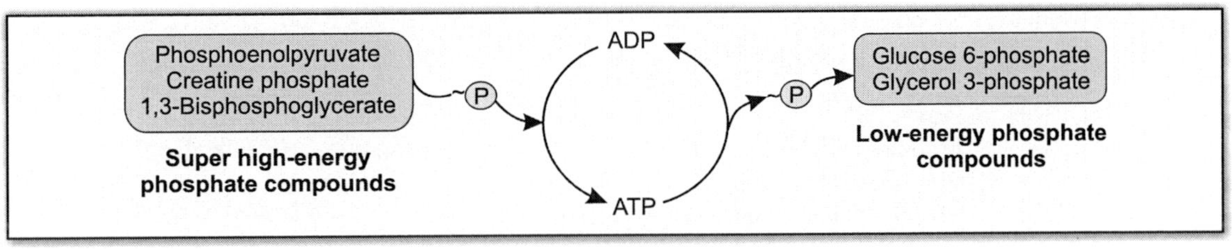

Fig. 5.1 The ATP/ADP cycle

❑ Give a detailed account of the formation, utilization and breakdown of ATP.

- ***Formation of ATP***

  The sources of ATP include
  - *Glycolysis* in which there is a net generation of two high-energy phosphates in the form of ATP molecules.
  - *Oxidative phosphorylation*, which accounts for the highest quantitative source of energy in aerobic organisms.
  - *Substrate-level phosphorylation* in the citric acid cycle in which one ATP is generated directly in the reaction catalyzed by *succinate thiokinase*.
- ***Utilization of ATP***

  ATP is utilized for the following:
  - *Muscle contraction* and *relaxation*.
  - *Cell division*, for the separation of chromosomes by the spindle apparatus.
  - *Active transport* against the concentration gradient. A classic example is the transport of Na^+ and K^+ across the plasma membrane in a process catalyzed by Na^+K^+*-ATPase*.

- *Bioluminescence*, a phenomenon in which the energy derived from ATP hydrolysis is used to activate firefly luciferin by *luciferase* to form oxyluciferin with emission of light.
- *Biosynthesis* of carbohydrates, lipids, proteins and nucleic acids.

- ***Breakdown of ATP***
  ATP can undergo loss of either an orthophosphate to form ADP and Pi, or a pyrophosphate to generate AMP and PPi. The pyrophosphate is subsequently hydrolyzed by *pyrophosphatase* to form two molecules of Pi.

❑ Define oxidation and reduction.

Oxidation refers to the removal of electrons whereas reduction is the gain of electrons.

❑ What are redox reactions?

- Chemical reactions in which electrons are transferred from one molecule to another are referred to as *oxidation – reduction reactions* or *redox reactions.*
- The electron donating molecule in such reactions is termed a *reducing agent* or *reductant* and the electron accepting molecule as the *oxidizing agent* or *oxidant.*
- Reducing and oxidizing agents function as *conjugate reductant-oxidant pairs* (*redox pairs*). For example,

$$Fe^{2+} \text{ (electron donor)} \rightleftharpoons \text{Electron} + Fe^{3+} \text{ (electron acceptor)}$$

- Electrons flow from a relatively electronegative conjugate redox pair to more electropositive electron acceptors, due to loss of free energy.
- The standard free energy change of an electron transfer reaction is given by

$$\Delta G^{\circ\prime} = - n7\Delta E_0^{\prime}$$

  where, $\Delta G^{\circ\prime}$ is the standard free energy change in calories, n is the number of electrons transferred, 7 is a constant (Faraday), and $\Delta E_0^{\prime}$ is the difference between standard potential of the electron donor system and that of the electron acceptor system.
- Redox reactions occur only in the presence of a catalyst or an enzyme.

❑ What are oxidoreductases? Describe the different classes of oxidoreductases.

Enzymes that catalyze oxidation-reduction reactions are termed *oxidoreductases.* These are classified into four groups:

1. *Oxidases*: These enzymes catalyze the removal of hydrogen from a substrate using only O_2 as hydrogen acceptor. They contain copper and form H_2O as a product except *uricase* and *monoamine oxidase,* which form H_2O_2. *Cytochrome oxidase* (*cytochrome aa*$_3$), the terminal component of the respiratory chain, is a classic example of an *oxidase.*

$$AH_2 + \tfrac{1}{2} O_2 \longrightarrow A + H_2O$$

2. *Dehydrogenases*: They may be aerobic or anerobic.
   a. *Aerobic dehydrogenases*
      - These are flavoprotein enzymes containing FMN or FAD as prosthetic groups.
      - These catalyze the removal of hydrogen from a substrate using either oxygen or artificial substances, such as methylene blue, as hydrogen acceptor to form H_2O_2.

$$AH_2 + O_2 \longrightarrow A + H_2O_2$$

- These enzymes include L-*amino acid dehydrogenase, xanthine dehydrogenase* and *aldehyde dehydrogenase.*

b. *Anaerobic dehydrogenases,* which catalyze the removal of hydrogen from a substrate but are unable to use oxygen as hydrogen acceptor. These are of three types:
  - *Nicotinamide-dependent coenzymes* (NAD^+ or $NADP^+$). While NAD-linked dehydrogenases catalyze redox reactions in catabolic pathways such as glycolysis and the citric acid cycle, NADP-linked dehydrogenases play a role in synthetic pathways such as lipogenesis.
  - *Riboflavin-linked dehydrogenases* (FMN or FAD). These enzymes are involved in electron transport in the respiratory chain. Examples include *NADH dehydrogenase* and *succinate dehydrogenase.*
  - *Cytochromes* (except cytochrome oxidase). These are iron-containing hemoproteins that function as electron carriers from flavoproteins to cytochrome oxidase in the respiratory chain by undergoing transition between Fe^{3+} and Fe^{2+} states.

3. *Hydroperoxidases*: These are the enzymes that utilize H_2O_2 or an organic peroxide as substrate. These include the following:

a. *Peroxidase* present in milk, leukocytes, and other tissues, catalyzes the reaction

$$H_2O_2 + AH_2 \longrightarrow 2H_2O + A$$

b. *Glutathione peroxidase* present in the erythrocytes that contains selenium as a prosthetic group. It is involved in the detoxification of H_2O_2 and lipid hydroperoxides, and protection of membrane lipids and hemoglobin.

c. *Catalase,* which is a hemoprotein present in blood, bone marrow, mucous membrane, kidney and liver. It catalyzes the following reaction:

$$2H_2O_2 \longrightarrow 2H_2O + O_2$$

4. *Oxygenases*: These are the enzymes that catalyze the direct transfer and incorporation of O_2 into the substrate. These are classified into two groups:

a. *Dioxygenases* (*e.g., homogentisate dioxygenase*) that catalyze incorporation of both the atoms of O_2 into the substrate.

$$A + O_2 \longrightarrow AO_2$$

b. *Monooxygenases* or *mixed function oxidases* (*e.g., cytochrome P450*) that catalyze incorporation of only one atom of O_2 into the substrate, while the other atom is reduced to H_2O.

$$A\text{-}H + O_2 + BH_2 \longrightarrow A\text{-}OH + H_2O + B$$

❑ Give an account of cytochrome P450 monooxygenases.

- The *cytochrome P450 monooxygenases* (CYPs) belong to a superfamily of membrane-bound hemoproteins.
- Drugs such as phenobarbital induce CYP synthesis.
- These enzymes are found together with cytochrome b_5 in the liver microsomes.
- The CYP enzymes play an important role in the detoxification of xenobiotics such as benzopyrene, aniline and morphine. Detoxification involves hydroxylation that increases solubility and, therefore, excretion.
- CYP enzymes are involved in the metabolism of steroid hormones in the mitochondrial fraction of steroidogenic tissues such as adrenal cortex, testes, ovaries and placenta.

- CYP enzymes are also involved in the metabolism of eicosanoids and lipid-soluble biological signaling molecules.
- In the kidney, 1-hydroxylation of 25-hydroxy vitamin D_3 is CYP-dependent.
- Hydroxylation reactions in bile acid synthesis in the liver are CYP-dependent.

❑ Outline the role of the respiratory chain.

- The *respiratory chain* or the *electron transport chain* (ETC) is a redox gradient of electron carrier molecules located in the inner membrane of the mitochondria in eukaryotes, and the plasma membrane in prokaryotes.
- All the energy released from the oxidation of carbohydrates, fatty acids and proteins is channeled into the ETC as reducing equivalents from NADH to the final acceptor, oxygen, to form H_2O.
- The components of the ETC are capable of accepting and donating electrons, and releasing free energy in the form of ATP by *oxidative phosphorylation* at three energy-conserving sites in the chain.
- In eukaryotes, electron transport coupled to oxidative phosphorylation, is responsible for reoxidizing NADH and $FADH_2$ from glycolysis, citric acid cycle, and fatty acid oxidation. During this process, the energy released is trapped as ATP.

❑ Enumerate the various components of the electron transport chain.

- Electrons flow from the electronegative component NADH, through a series of electron carriers, to the more electropositive oxygen.
- The components of the electron transport chain are organized sequentially in the order of increasing redox potential (Fig. 5.2).

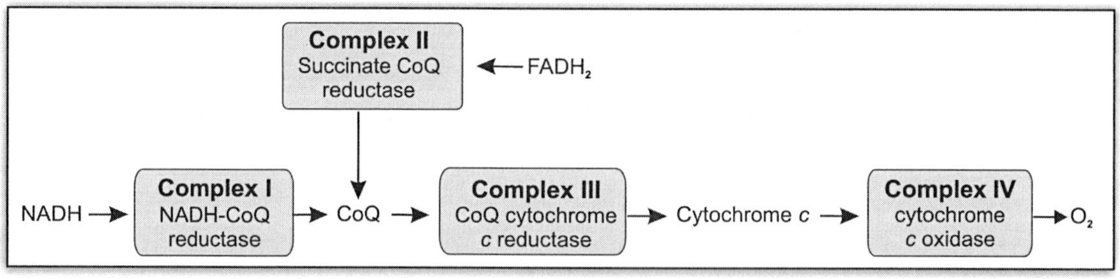

Fig. 5.2 Components of the electron transport chain

- The protein components of the ETC exist as four large protein complexes embedded in the inner mitochondrial membrane and include
  1. *Complex I (NADH-ubiquinone reductase)*
  2. *Complex II (Succinate-ubiquinone reductase)*
  3. *Complex III (Ubiquinone-cytochrome c reductase)*
  4. *Complex IV (Cytochrome c oxidase).*

❑ Describe the flow of electrons from NADH to oxygen through the components of the electron transport chain.

- Electrons are transported from NADH to oxygen, as indicated in Figure 5.2.
- Electron pairs that arise from the action of *NAD(P)-linked dehydrogenases* on different substrates are collected by $NAD(P)^+$ to form NADH.

- NADH passes the electrons to *Complex I* (*NADH-ubiquinone reductase*), which contains FMN as prosthetic group and several iron-sulfur proteins. In the process,
  - FMN is converted to $FMNH_2$ on accepting the electrons.
  - Electrons from $FMNH_2$ are transferred to the iron atoms in the FeS centers, which accept and donate electrons by undergoing Fe^{3+} to Fe^{2+} transitions.
- Electrons pass from *Complex I to ubiquinone* (*coenzyme Q, CoQ*), which is converted to ubiquinol or $CoQH_2$.
  - CoQ is a fat-soluble quinone with a long isoprenoid side chain.
  - CoQ is the entry point for reducing equivalents from substrates that are linked to the ETC via *flavoprotein-linked dehydrogenases*.
  - The $FADH_2$ formed is reoxidized to FAD by donating two electrons to *Complex II* (*succinate-ubiquinone reductase*), which passes the electrons to CoQ to form ubiquinol.
- Ubiquinol passes the electrons to *Complex III* (*ubiquinone-cytochrome c reductase*), which contains cytochrome *b*, cytochrome c_1, and an FeS protein.
  - Cytochromes are heme-containing proteins with a central iron atom that alternates between Fe^{3+} and Fe^{2+} states during electron transfer.
- Electrons pass from *Complex III* to *cytochrome c*.
  - Cytochrome *c* is a peripheral membrane protein loosely bound to the outer surface of the inner mitochondrial membrane.
  - On accepting an electron from Complex III, cytochrome *c* undergoes Fe^{3+} to Fe^{2+} transition.
- Cytochrome *c* binds to and donates an electron to *Complex IV* (*Cytochrome c oxidase*).
  - Complex IV contains two cytochromes viz., cytochrome *a* and a_3, each of which is bound to a copper atom which undergoes Cu^+– Cu^{2+} transitions, in addition to iron atoms which cycle between Fe^{3+} and Fe^{2+} states during electron transfer.
- Four electrons pass from *Complex IV* to molecular oxygen to form two molecules of H_2O.

❑ What is meant by oxidative phosphorylation?

Oxidative phosphorylation refers to the synthesis of ATP coupled to oxidation of NADH and $FADH_2$ by electron transport through the respiratory chain.

❑ What are the sites in the electron transport chain that are coupled to ATP synthesis?

- Oxidative phosphorylation occurs in three energy-conserving segments of the ETC. These include
  - *site I*, the span between NADH and CoQ
  - *site II*, the span between cytochrome *b* and cytochrome *c*
  - *site III*, the span between cytochrome aa_3 and O_2.
- Electron pairs derived from *NAD-linked dehydrogenases* pass through all the three sites and generate a maximum of 3 ATPs, whereas electron pairs derived from *flavin-linked dehydrogenases* yield only 2 ATPs because they enter at ubiquinone, which is beyond site I.
- Research shows that these numbers are not integers and values of 2.5 and 1.5 have been arrived at for ATP generation from NAD- and FAD-linked dehydrogenases respectively.

❑ How does the chemiosmotic theory explain the mechanism of oxidative phosphorylation?

According to the chemiosmotic theory proposed by Mitchell

1. A high-energy state couples electron transport with ADP phosphorylation.
2. The coupling mechanism requires an intact mitochondrial membrane.

3. The components of the ETC function as an active transport system to *pump* H^+ ions from the mitochondrial matrix across the inner membrane into the intermembrane space, generating a H^+ gradient that renders the matrix more alkaline and the intermembrane space more acidic.
4. The protons flow back into the mitochondrion by the action of *F_0F_1ATPase* that drives the synthesis of ATP.
5. Approximately 3 ATP molecules are formed for each NADH oxidized, and 2 ATPs for every $FADH_2$ oxidized.

❑ Write a short note on F_0F_1 *ATPase*.

- The enzyme *F_0F_1 ATPase* or *ATP synthase*, located in the inner mitochondrial membrane, catalyzes ATP synthesis during electron transport.
- *ATP synthase* has two components — F_0 and F_1.
- F_0 has a knob-like structure containing 10 – 14 subunits attached by a stalk to F_1 embedded in the inner membrane.
- F_1 contains 9 subunits and has *ATPase* activity. However, *F_1ATPase* cannot synthesize ATP without the presence of F_0.
- F_0, which spans the membrane, forms the proton channel. Rotation of F_0 caused by the flow of protons is suggested to drive ATP synthesis by the F_1 complex.

❑ Explain respiratory control of electron transport.

- Normally, electron transport is tightly coupled to ATP synthesis and electrons do not flow through the ETC without simultaneous phosphorylation of ADP to ATP.
- When ADP levels are low, the rate of respiration declines and there is no phosphorylation.
- This phenomenon in which the rate of electron transport is controlled by the concentration of ADP is termed *respiratory control.*
- This ensures that electron flow occurs only when ATP synthesis is required.

❑ Write a note on inhibitors of the electron transport chain.

Inhibitors of the electron transport chain act at three loci:
- Barbiturates such as *amobarbital* and the fish poison *rotenone* inhibit electron transport from NADH to ubiquinone.
- *Antimycin A* and *dimercaprol* inhibit the ETC between cytochrome *b* and cytochrome *c*.
- Hydrogen sulfide, carbon monoxide, and cyanide inhibit cytochrome oxidase.

❑ Explain the mechanism of action of uncouplers and inhibitors of oxidative phosphorylation, and ionophores.

◘ *Uncouplers*

- Uncoupling agents such as 2,4-dinitrophenol allow electron transport to occur but prevent phosphorylation of ADP.
- These agents carry H^+ ions across the inner mitochondrial membrane and dissipate the proton gradient.
- The energy obtained by uncoupling is released as heat.

- ***Inhibitors***
  - Inhibitors such as *oligomycin* prevent both stimulation of oxygen consumption by ADP and phosphorylation of ADP.
  - These agents do not directly inhibit the components of the ETC.
  - They function by preventing the ATP synthesizing machinery from utilizing the high-energy intermediate state generated by electron transport.
  - Uncouplers promote breakdown of the high-energy intermediate state and relieve inhibition.
- ***Ionophores***
  - These agents promote the breakdown of the high-energy intermediate state by forming lipid-soluble complexes in the presence of monovalent cations that enables passage through the mitochondrial membrane. For example, the antibiotic *valinomycin* requires K^+ for its inhibitory effect.
  - These agents inhibit oxidative phosphorylation by diverting the energy derived from electron transport for pumping cations into the mitochondrial matrix rather than for ATP synthesis.

Chapter 6

Metabolism of Carbohydrates

❑ How are dietary carbohydrates digested?

- ***Dietary carbohydrate*** is in the form of
  - starch (predominant)
  - glycogen, sucrose, lactose, galactose and fructose (in lesser amounts).
- ***Digestion of starch and glycogen***
  - by salivary *α-amylase* in the mouth is not significant due to short contact time
  - does not occur in the stomach because amylolytic enzymes are absent
  - occurs mainly in the small intestine by pancreatic amylase.
- ***Pancreatic amylase***
  - cleaves internal $\alpha 1 \rightarrow 4$ glycosidic linkages in starch and glycogen
  - cannot hydrolyse $\alpha 1 \rightarrow 6$ linkages or $\alpha 1 \rightarrow 4$ bonds around the branch points
  - hydrolyses amylose to maltose and maltotriose
  - hydrolyses amylopectin to limit dextrins.
- ***Digestion of oligosaccharides, limit dextrins, and disaccharides***
  - Oligosaccharides resulting from *α-amylase* action, are digested in the small intestine by *oligosaccharidases.*
  - Limit dextrins are digested by the sequential action of *maltase* on $\alpha 1 \rightarrow 4$ linkages and isomaltase on $\alpha 1 \rightarrow 6$ linkages.
  - Dietary disaccharides are enzymatically hydrolyzed to generate the corresponding monosaccharide units.

Enzymatic hydrolysis of dietary disaccharides

Sucrose + H_2O	⟶	D-glucose + D-fructose
Lactose + H_2O	⟶	D-glucose + D-galactose
Maltose + H_2O	⟶	2 D-glucose

- ***End products of carbohydrate digestion:*** Glucose, fructose, and galactose.

❑ Explain the mechanism of absorption of carbohydrates.

- The end products of carbohydrate digestion are absorbed into circulation and transported to the liver.
- Fructose is absorbed by *facilitated diffusion* by a sodium-dependent transporter, GLUT5.
- Glucose and galactose are absorbed by *carrier-mediated active transport* by means of a *sodium symport system.*
- Transport of glucose and galactose depends on their lumenal concentrations. Thus,
  - when the concentrations are high, after digestion, transport occurs down the gradient into the cell.
  - when the concentrations fall, the sugars are actively transported against the concentration gradient into the cell.
- *Active transport* is accomplished by hydrolysis of ATP by *Na^+K^+-ATPase*, which is inhibited by *ouabain*, a cardiac glycoside, and by *phlorizin*, an inhibitor of glucose reabsorption in the renal tubules.
- Glucose absorption
  - 60% is absorbed into the portal circulation by a serosal carrier
  - 25% is absorbed by passive diffusion
  - 15% of glucose is transported back into the lumen.
- In the liver, fructose and galactose are converted to glucose.

❑ What is dietary fiber?

- Cellulose which contains $\beta 1 \rightarrow 4$ linkages and plant pentosans cannot be digested by the mammalian gastrointestinal tract due to lack of enzymes that hydrolyze these carbohydrates.
- Bacteria in the intestines partially degrade pentosans and polysaccharides releasing carbon dioxide, alcohol and organic acids.
- The organic acids stimulate peristalsis, while cellulose functions as bulk or roughage.
- These undigestible carbohydrates that aid digestion are referred to as *dietary fiber.*

❑ Give a brief overview of the metabolism of glucose.

- Glucose formation occurs by two pathways:
  - *glycogenolysis*
  - *gluconeogenesis.*
- Glucose is utilized in the following manner:
  - conversion to glycogen and storage in the liver (*glycogenesis*)
  - conversion to triglycerides, which are transported by circulation to the adipose tissue and stored.
  - transport to the muscles where it is oxidized or stored as muscle glycogen.
  - oxidation by:
    1. the *hexose monophosphate* (*HMP*) *shunt pathway* or *pentose phosphate pathway.*
    2. the *Embden–Meyerhof glycolytic pathway* to pyruvate. Pyruvate can be completely oxidized by the citric acid cycle or converted to lactate, fatty acids, and amino acids.

Figure 6.1 gives an overview of glucose metabolism.

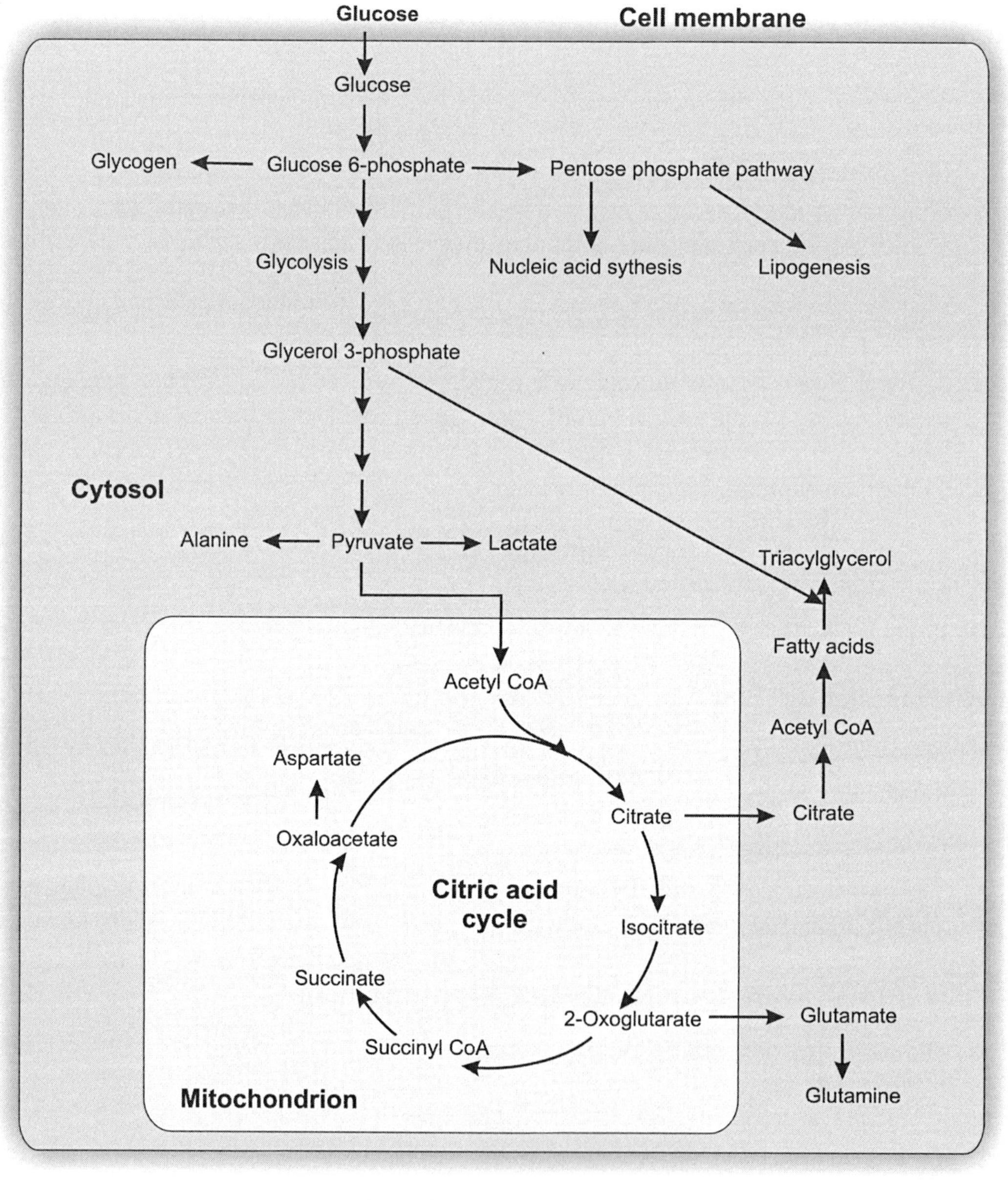

Fig. 6.1 Overview of glucose metabolism

❑ Describe the pathway of glycolysis.

Glycolysis is the pathway of glucose breakdown to pyruvate.

- ***Subcellular site*:** Cytoplasm of all living cells.
- ***Purpose***
  - energy production
  - provision of intermediates for biosynthetic pathways.

- ***Steps:***
  1. Phosphorylation of glucose to glucose 6-phosphate (G6P). This is catalyzed by *glucokinase* in the liver and *hexokinase* in other tissues.
  2. Isomerization of G6P to fructose 6-phosphate by *phosphoglucoisomerase.*
  3. Phosphorylation of fructose 6-phosphate to fructose 1,6-bisphosphate by *phosphofructokinase*, an allosteric enzyme.

  4, 5. Cleavage of fructose1,6-bisphosphate by *aldolase* to glyceraldehyde 3-phosphate (G3P) and dihydroxyacetone phosphate (DHAP), which are interconverted by *triose phosphate isomerase.*

  6 – 8. Conversion of glyceraldehyde 3-phosphate to phosphoenolpyruvate via formation of the following intermediates: 1,3-bisphosphoglycerate, 3-phosphoglycerate, and 2-phosphoglycerate in reactions catalyzed by *G3-P dehydrogenase, phosphoglycerate kinase* and *phosphoglycerate mutase* respectively.

  9. Conversion of 2-phosphoglycerate to phosphoenolpyruvate catalyzed by *enolase.* This enzyme requires Mg^{2+} or Mn^{2+} and is inhibited by fluoride. This forms the basis for the inhibition of glycolysis by the addition of fluoride during glucose estimation.
  10. Conversion of phosphoenolpyruvate to pyruvate. Two molecules of pyruvate are formed from one molecule of glucose.

- ***Energetics***

  Two ATPs are utilized and four ATPs are synthesized. The net yield of ATPs for every molecule of glucose oxidized to two molecules of pyruvate is two (Fig. 6.2).

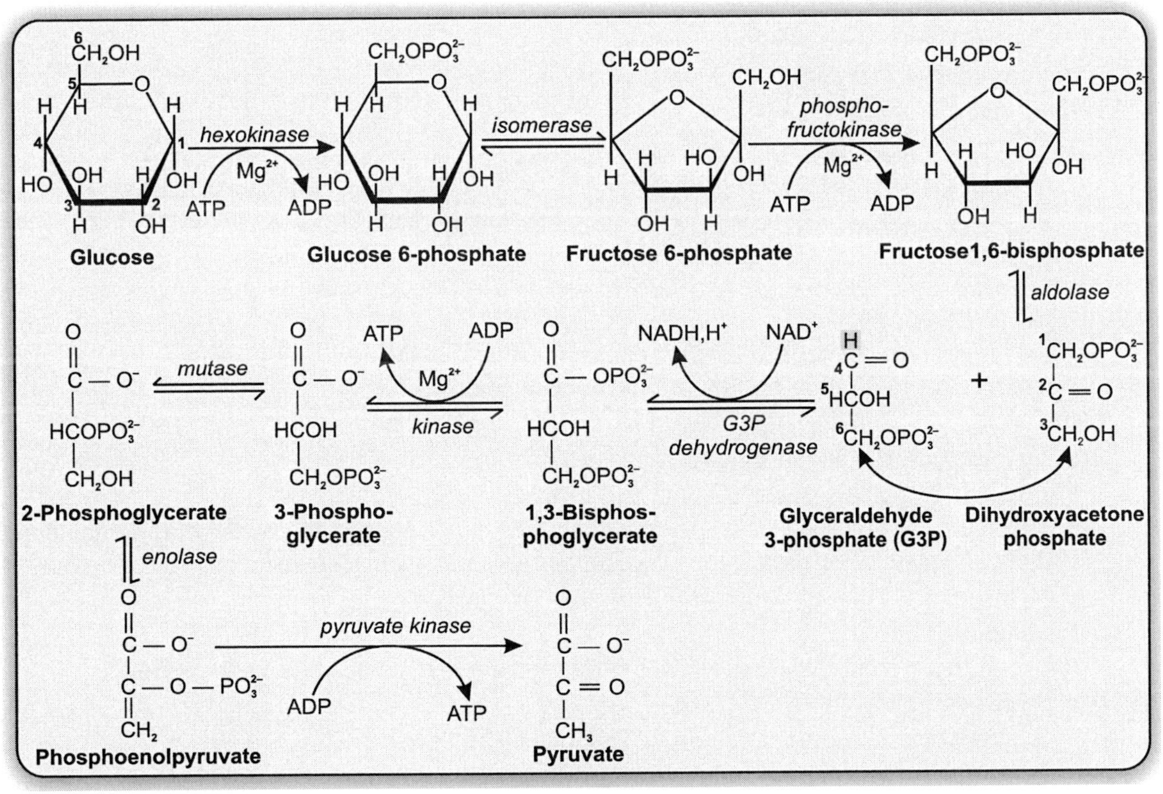

Fig. 6.2 Glycolysis

❑ What is the fate of pyruvate formed by the glycolytic pathway?

The fate of pyruvate depends on the availability of oxygen (Fig. 6.3). Thus,

- Under anaerobic conditions:
  - Pyruvate is converted to lactate by *lactate dehydrogenase (LD)*. Oxidized nicotinamide adenine dinucleotide (NAD^+) regenerated in the reaction allows glycolysis to continue.
  - In yeast, pyruvate is converted to ethanol by *alcohol dehydrogenase.*
  - The net yield of ATP is 2.
- Under aerobic conditions:
  - Pyruvate is oxidized to carbon dioxide and water in the TCA cycle.
  - The net yield of ATP for the complete oxidation of glucose to carbon dioxide and water is 38.

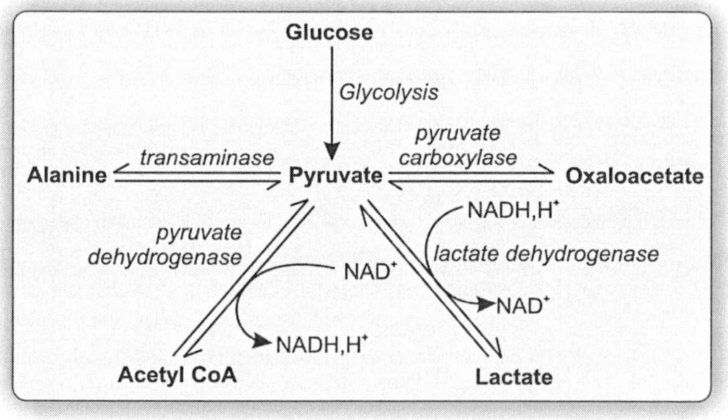

Fig. 6.3 Fate of pyruvate

❑ Give a brief overview of the citric acid cycle.

The *citric acid cycle* is also known as the *tricarboxylic acid (TCA) cycle* or the *Krebs cycle.*

***Subcellular site*:** Mitochondria.

***Starting compound*:** Acetyl CoA

Reactions

- The cycle consists of eight reactions.
- Acetyl CoA condenses with oxaloacetate to form citrate, which undergoes a series of dehydrogenations and decarboxylations to regenerate oxaloacetate releasing carbon dioxide (Fig. 6.4).

Energetics

- *Substrate level phosphorylation*: One molecule of GTP is formed in the reaction catalyzed by *succinyl CoA synthetase.*
- *Oxidative phosphorylation*: The reducing equivalents NADH and $FADH_2$ released are oxidized to water by the respiratory chain and ATP is generated.

Functions

- Energy production in the form of ATP
- Provision of precursors for biosynthetic pathways
- Participation in gluconeogenesis, deamination, transamination and fatty acid synthesis (amphibolic). When intermediates in the cycle are used for biosynthesis, ancillary metabolic pathways (*anaplerotic pathways*) replenish the cycle with these intermediates.

Regulation

- [ATP]/[ADP] and [NADH]/[NAD^+] ratios regulate several enzymes in the cycle.
- *Allosteric regulation*
  - citrate synthase inhibition by ATP and long chain fatty acyl CoA
  - isocitrate dehydrogenase activation by ADP and inhibition by ATP and NADH.
- Pyruvate dehydrogenase inhibition by acetyl CoA and inactivation by phosphorylation.

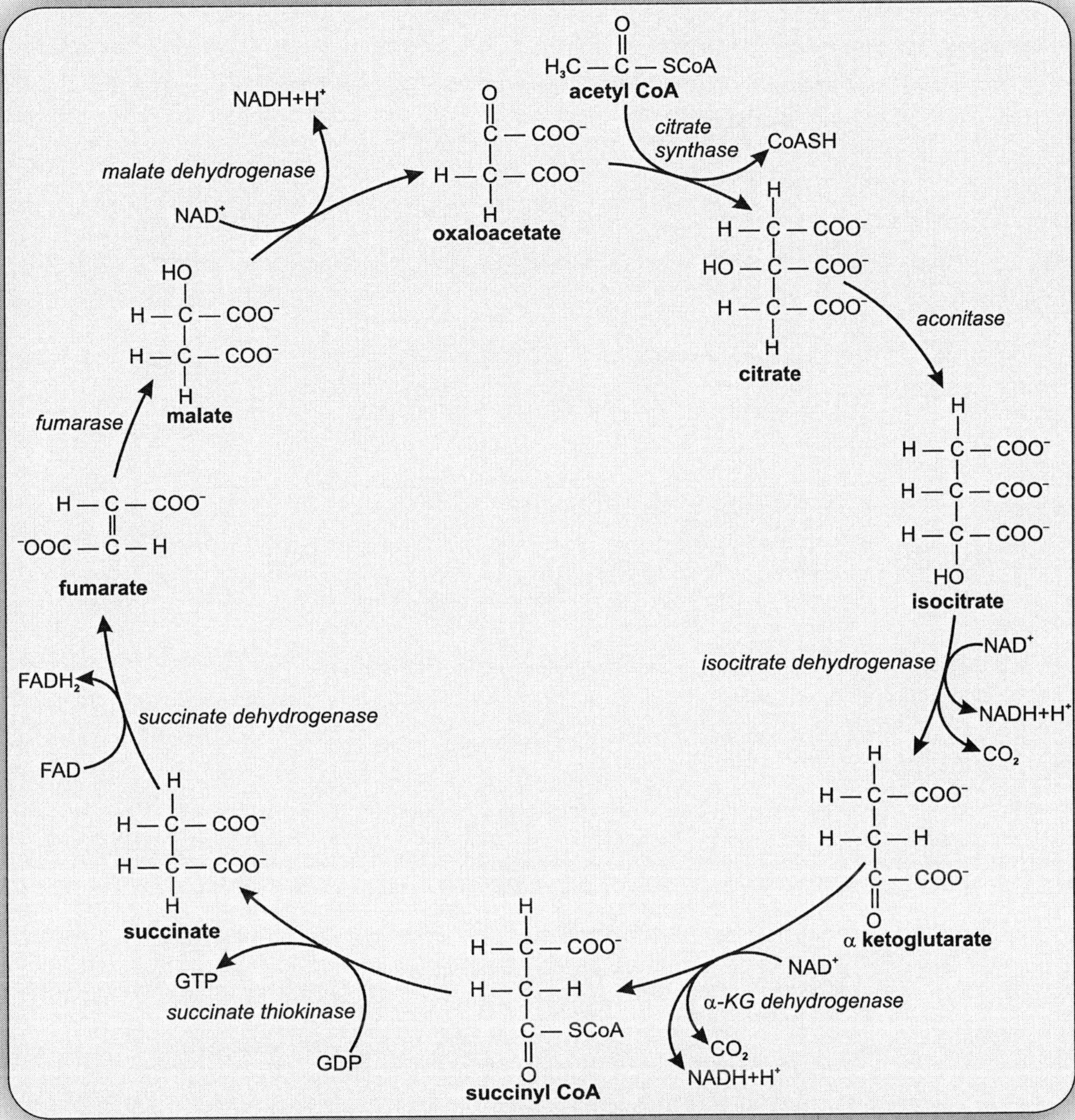

Fig. 6.4 The citric acid cycle

❑ What is gluconeogenesis?

- Gluconeogenesis is the synthesis of glucose from non-carbohydrate sources such as pyruvate, lactate, glycerol, alanine, amino acids, and citric acid cycle intermediates.
- ***Function***: Maintenance of blood glucose during starvation and muscular exercise.
- ***Localization***: The pathway takes place mostly in the liver and to a lesser extent in the kidney. The gluconeogenic enzymes are predominantly cytosolic except *pyruvate carboxylase* (mitochondrial matrix) and glucose 6-phosphatase (smooth endoplasmic reticulum).

❑ What are the irreversible steps in glycolysis? How are they bypassed in gluconeogenesis?

Irreversible steps

Gluconeogenesis involves reversal of glycolysis (Fig. 6.5) except for three irreversible steps, which include the reactions catalyzed by

- *pyruvate kinase*
- *phosphofructokinase*
- *hexokinase.*

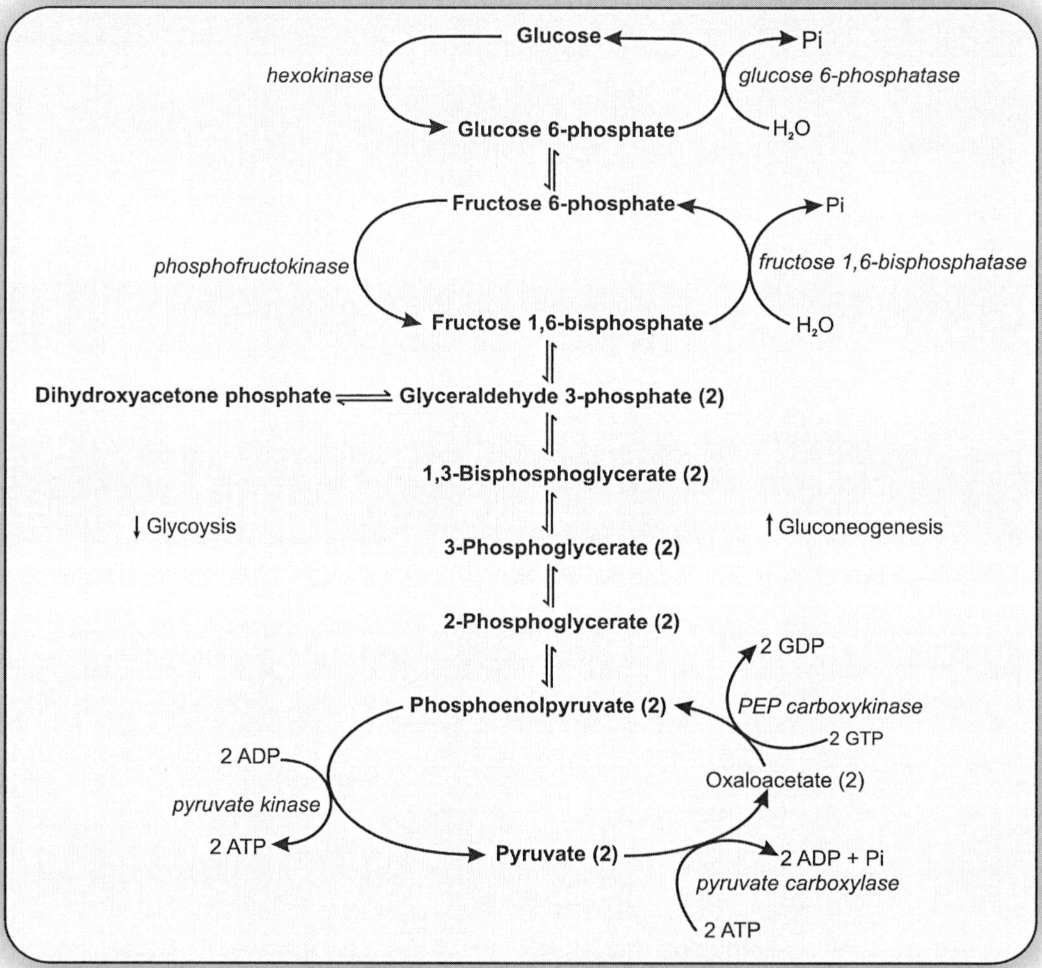

Fig. 6.5 The gluconeogenic pathway

Bypass reactions

The irreversible steps in glycolysis are reversed as detailed below:

1. Bypassing the pyruvate kinase reaction involves the following sequence of steps:
    - conversion of pyruvate to oxaloacetate by the mitochondrial enzyme *pyruvate carboxylase.*

- since oxaloacetate cannot cross the mitochondrion, it is converted to malate by *malate dehydrogenase.*
- malate traverses the mitochondrion and is converted to oxaloacetate in the cytosol.
- decarboxylation and phosphorylation of oxaloacetate by *PEP carboxykinase* in the cytosol to form phosphoenolpyruvate (PEP).
- conversion of PEP to fructose 1,6-bisphosphate by reversal of the glycolytic reactions.

2. Bypassing the *phosphofructokinase* reaction involves
   - dephosphorylation of fructose 1,6-bisphosphate by *fructose 1,6-bisphosphatase* to fructose 6-phosphate.
   - conversion of fructose 6-phosphate to glucose 6-phosphate.
3. Bypassing *hexokinase* reaction involves
   - dephosphorylation of glucose 6-phosphate by *glucose 6-phosphatase* to glucose.

Energetics: 6 molecules of ATP are utilized to synthesize one molecule of pyruvate.

❑ What are the cycles that fuel gluconeogenesis?

Gluconeogenesis is fuelled by

The glucose-alanine cycle

- The cycle operates during starvation.
- Alanine is transported from the muscle to the liver.
- In the liver, alanine is converted to pyruvate and eventually to glucose.
- Glucose is transported from the liver to the muscle where it is converted to pyruvate, which undergoes transamination to alanine perpetuating the cycle.

The glucose-alanine cycle is summarized in Figure 6.6.

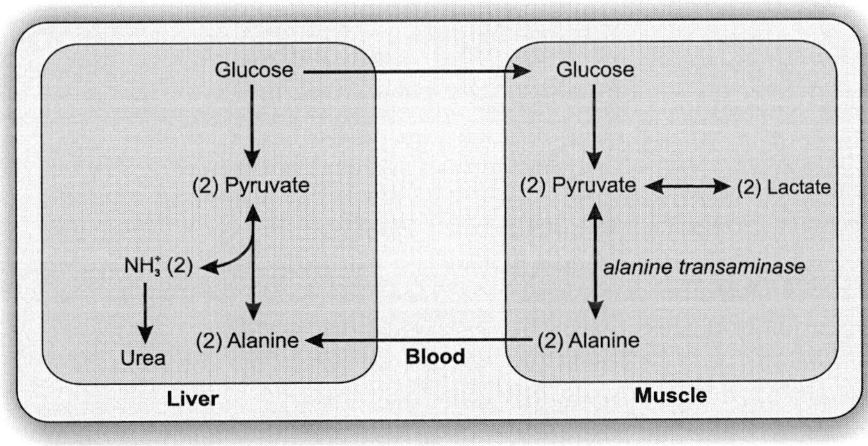

Fig. 6.6 The glucose-alanine cycle

The Cori cycle

- This cycle operates during strenuous muscular exercise.
- Pyruvate formed by glycolysis is converted to lactate.
- Lactate is transported from the muscle by the circulation to the liver where it is converted to glucose.
- Glucose released from the liver into circulation is taken up by the muscle.

The Cori cycle is summarized in Figure 6.7.

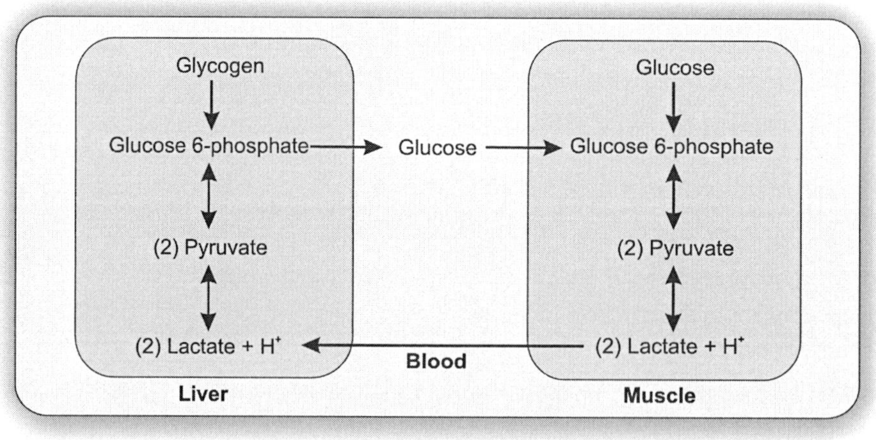

Fig. 6.7 The Cori cycle

❑ Outline the functions of the hexose monophosphate (HMP) shunt pathway.

The HMP shunt pathway, also known as the *pentose phosphate pathway,* produces two biologically important products:

- Ribose 5-phosphate, a constituent of nucleic acids and nucleotide coenzymes.
- Reduced nicotinamide adenine dinucleotide (NADPH) involved in:
  - biosynthesis of fatty acids and steroids
  - formation of reduced glutathione (GSH) from oxidized glutathione (GSSG) by *glutathione reductase.*

❑ Briefly elucidate the steps in the HMP shunt pathway.

The HMP shunt pathway is briefly described in Figure 6.8.

Site: Adipose tissue, adrenal cortex, and mammary gland.

Subcellular site: Cytosol

Steps: There are *three* main stages in the HMP pathway:

1. Conversion of glucose 6-phosphate to ribulose 5-phosphate by oxidative reactions, with generation of two ATP molecules.
2. Isomerization of ribulose 5-phosphate to ribose 5-phosphate.
3. Conversion of ribose 5-phosphate to fructose 6-phosphate and glyceraldehyde 3-phosphate by *transketolase* and *transaldolase.* This links the pathway to glycolysis.

Clinical significance: Deficiency of *glucose-6-phosphate dehydrogenase* (*G-6-PD*) leads to impairment in NADPH generation with consequent decrease in *glutathione peroxidase* activity that leads to lipid peroxidation, loss of erythrocyte membrane integrity, and hemolytic anemia.

❑ How is galactose metabolized?

- ***Source of galactose*:** Intestinal hydrolysis of dietary lactose, the disaccharide present in milk.
- ***Importance of galactose*:** Essential for the formation of certain glycoproteins, cerebrosides and of milk during lactation.

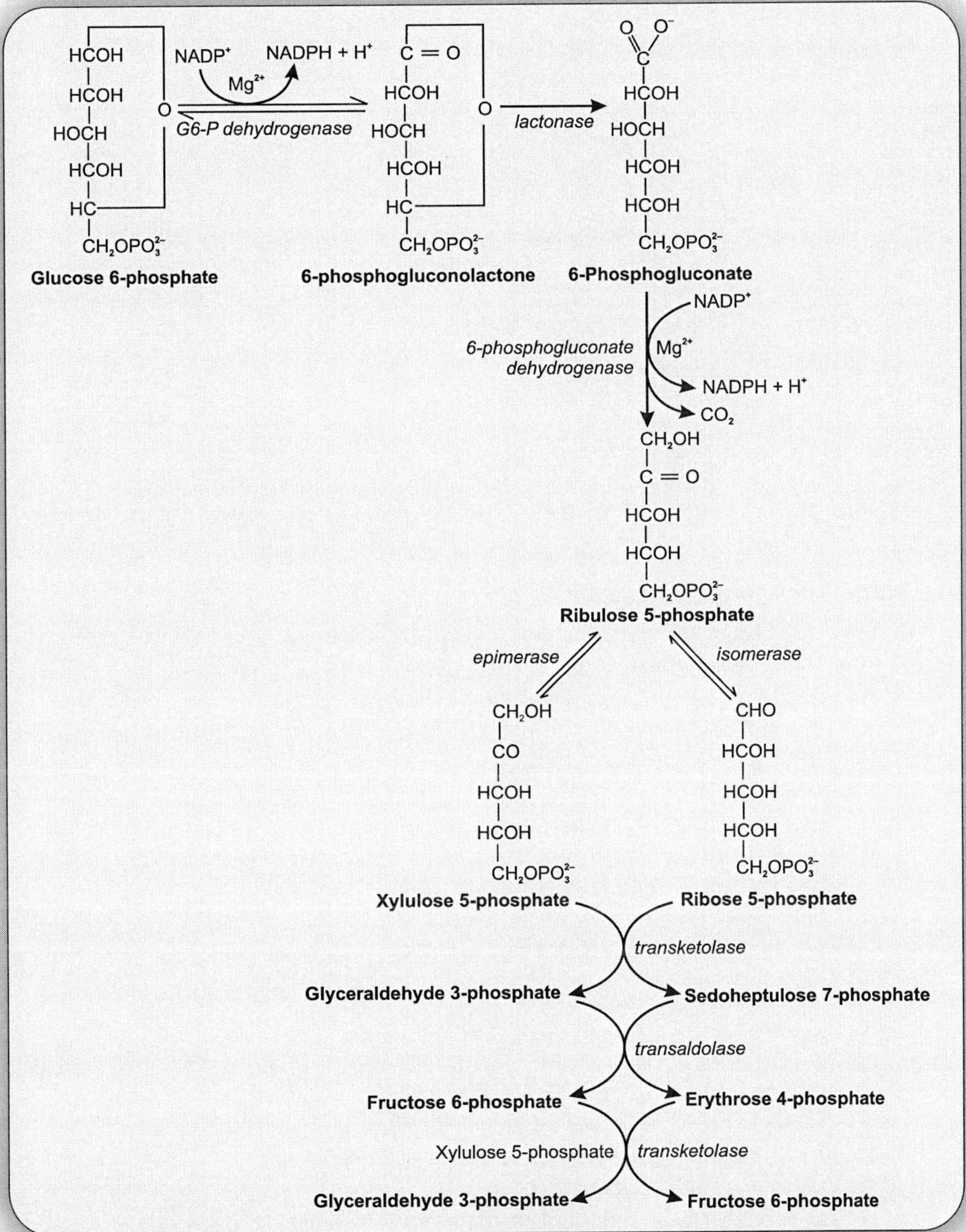

Fig. 6.8 The HMP shunt pathway

- ***Galactose metabolism*:** Galactose enters glycolysis via the galactose–glucose interconversion pathway (Fig. 6.9).

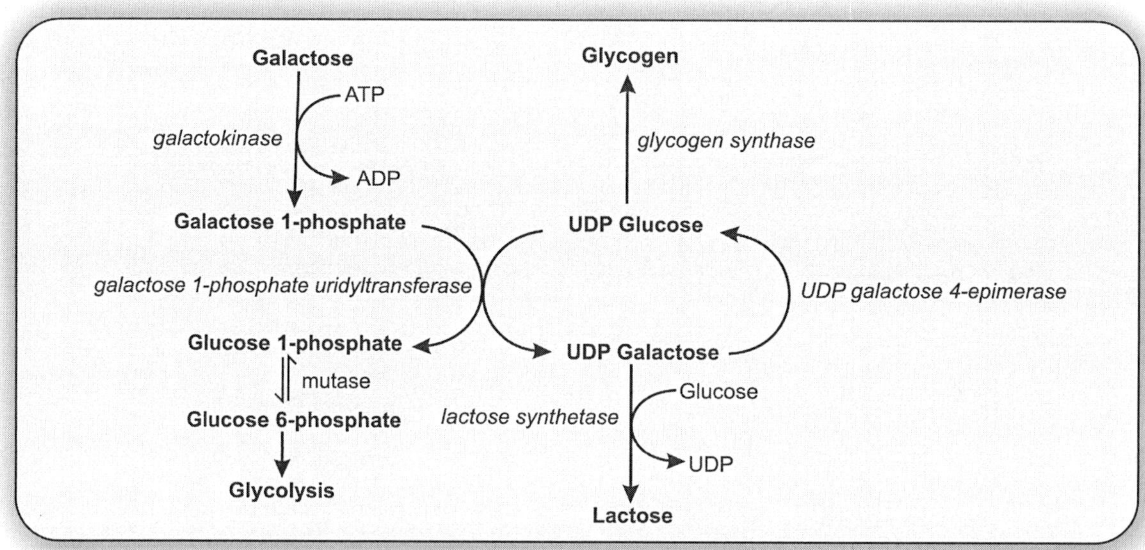

Fig. 6.9 Metabolism of galactose

Write notes on galactosemia.

Galactosemia encompasses three autosomal recessive inborn errors of galactose metabolism:

1. ***Galactokinase deficiency***, which is characterized by
   - Galactose accumulation in the blood and tissues
   - Cataract formation due to reduction of galactose by *aldose reductase* to *galactitol* (*dulcitol*) to which the lens is impermeable.
2. ***Classic galactosemia***
   - *Incidence*: 1 in 50,000 births
   - *Enzyme deficiency*: Deficiency of *galactose-1-phosphate uridyltransferase* (*GALT*) leads to tissue accumulation of galactose and galactose-1-phosphate.
   - *Clinical and biochemical features*
     - Vomiting and diarrhea with failure to thrive
     - Hypoglycemia.
     - Jaundice, hepatomegaly, and cirrhosis.
     - Irreversible mental retardation after 6–12 months
     - Renal tubular damage and generalized aminoaciduria (*Fanconi syndrome*) due to galactose-1-phosphate deposition in the kidney.
     - Bacterial sepsis with *E.coli*, leading to death.
3. ***Uridine diphosphate (UDP)-galactose-4-epimerase deficiency***
   - Very rare and asymptomatic
   - The enzyme may be deficient in erythrocytes but is found in liver.
4. ***Management*:** All three types of galactosemia can be treated by elimination of galactose from the diet (*i.e.*, milk and milk products).

❑ How is fructose metabolized?

- In the adipose tissue and muscle, fructose is converted by *hexokinase* to fructose 6-phosphate which enters the glycolytic pathway (Fig. 6.10).
- In the liver, fructose is metabolized by the fructose 1-phosphate pathway involving the conversion of fructose to fructose 1-phosphate and then to glyceraldehyde 3-phosphate.

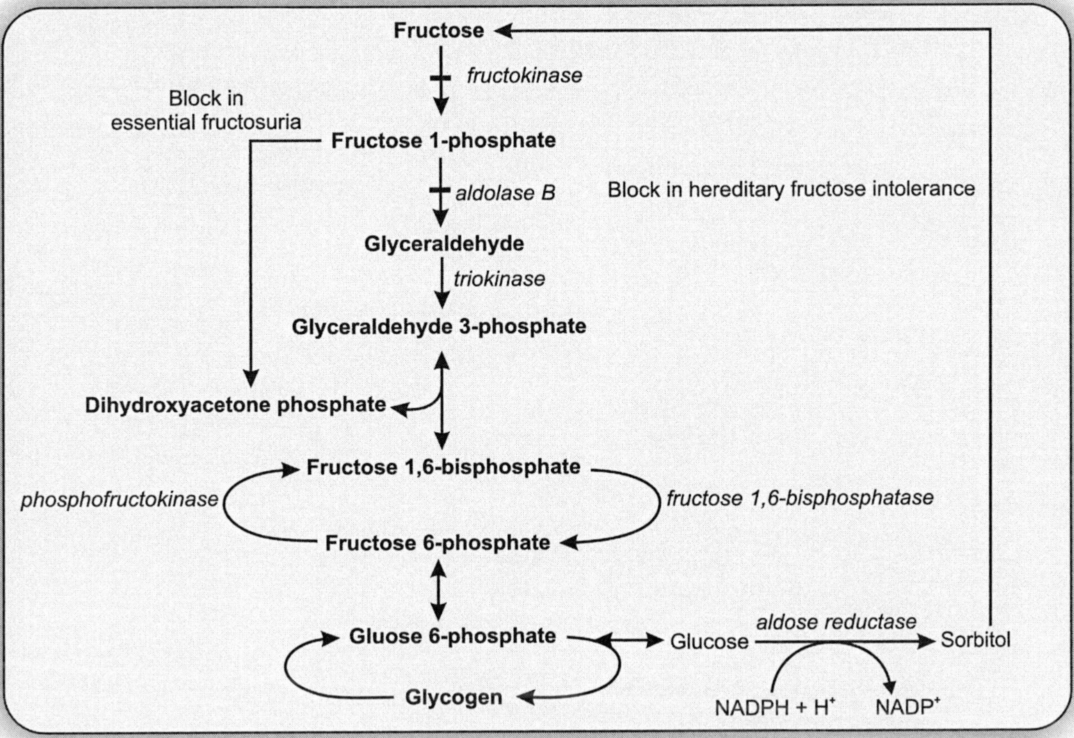

Fig. 6.10 Metabolism of fructose

❑ What are the disorders of fructose metabolism?

There are three major disorders of fructose metabolism:

1. *Essential fructosuria* due to lack of *fructokinase*. A low fructose diet is recommended.
2. *Fructose 1,6-bisphosphatase deficiency* that prevents gluconeogenesis. Blood sugar maintenance is dependent on exogenous glucose. The manifestations include lactic acidosis, hyperventilation, hypoglycemia, ketosis, and coma.
3. *Hereditary fructose intolerance*, an autosomal recessive disorder due to *fructose 1-phosphate aldolase* (*aldolase B*) deficiency. The enzyme block causes
   - accumulation of fructose-1-phosphate in tissues
   - liver damage and jaundice due to fructose 1-phosphate deposition that can progress to cirrhosis and ascites
   - renal tubular damage
   - hypoglycemia due to inhibition of glycogenolysis

A low fructose diet is recommended.

❑ How is glycogen metabolized?

Glycogen is formed from glucose in the muscle and liver by *glycogenesis*. While muscle glycogen is used only for local needs, liver glycogen is broken down (*glycogenolysis*) to glucose, which is released into circulation and used as an energy source by all tissues. Glycogen metabolism is summarized in Figure 6.11.

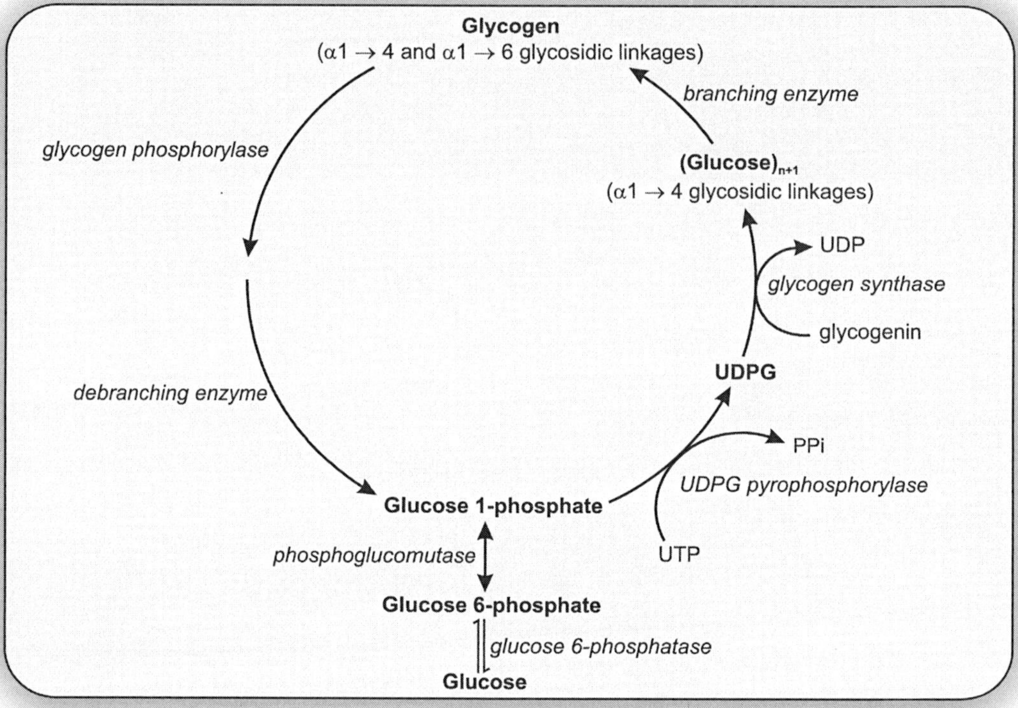

Fig. 6.11 Glycogen metabolism

❑ Explain the pathway of glycogen synthesis.

Glycogenesis is the pathway of glycogen synthesis from glucose.
Site: Muscle and liver

Sequence of reactions:

1. Phosphorylation of glucose by *glucokinase* in the liver and *hexokinase* in the muscle to form glucose-6-phosphate (G-6-P).
2. Conversion of G-6-P to glucose-1-phosphate by *phosphoglucomutase.*
3. Formation of uridine diphosphate glucose (UDPGlc) by reaction of glucose-1-phosphate with uridine triphosphate (UTP) catalyzed by *UDPGlc pyrophosphorylase.*
4. Successive transfer of glucosyl residues from UDPGlc to a pre-existing glycogen molecule or a *'primer'* attached to a protein backbone. This is the rate-limiting reaction catalyzed by *glycogen synthase.*
5. Branching of the glycogen molecule that involves transfer of part of the chain containing six glucose residues to a neighboring chain catalyzed by a *branching enzyme* [*amylo(1 → 4) → (1 → 6)-transglucosidase*]. This forms a branch point with α1 → 6 linkage.

❑ Explain the process of glycogenolysis.

Glycogenolysis, the process of glycogen breakdown to glucose, involves the following sequence of reactions (Fig. 6.11):

1. Sequential removal of the terminal glycosyl residues in the glycogen molecule by *glycogen phosphorylase* until a glycogen chain containing approximately four glucose residues on either side of a branch point is left.
2. Removal of branches in the glycogen molecule by the debranching enzyme system, a polypeptide with two enzyme activities:
   - transfer of a trisaccharide unit from one branch to the chain terminus of another branch by *(α1 $\rightarrow$4)$\rightarrow$(α1 $\rightarrow$4) glucan transferase (glycosyl transferase)*, exposing the 1$\rightarrow$6 branch point.
   - hydrolysis of the 1 $\rightarrow$ 6 linkage by *amylo (1,6) glucosidase*.
3. Conversion of glucose-1-phosphate formed by the action of glycogen phosphorylase to glucose-6-phosphate by *phosphoglucomutase*.
4. Dephosphorylation of glucose-6-phosphate by *glucose-6-phosphatase* to form glucose, which diffuses into the blood. The enzyme is absent in the muscle.

❑ How is glycogen metabolism regulated?

- Glycogen metabolism is regulated by
  - allosteric mechanisms
  - covalent modifications
  - hormones.
- *Glycogen synthase* is present in two forms:
  - active dephosphorylated form (a)
  - inactive phosphorylated form (b).
- Glycogen phosphorylase exists in two forms:
  - active phosphorylated form (*phosphorylase a*)
  - inactive dephosphorylated form (*phosphorylase b*).
- Conversion of phosphorylase *b* to phosphorylase *a* is accomplished by *phosphorylase b kinase* which also exists in an active phosphorylated form and an inactive dephosphorylated form.
- Glucagon and epinephrine increase blood sugar by simultaneous activation of phosphorylase and inactivation of glycogen synthase. Insulin exerts the opposite effect.

A schematic representation of the regulation of glycogen metabolism is shown in Figure 6.12.

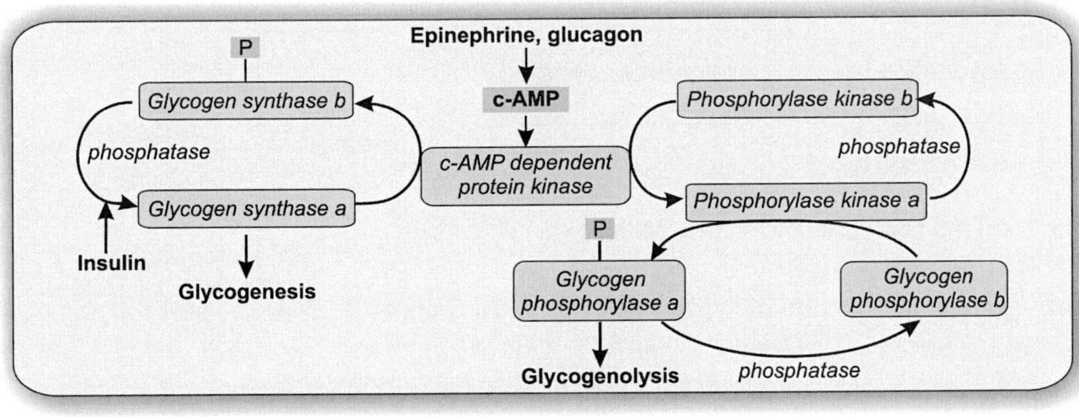

Fig. 6.12 Regulation of glycogen metabolism

❑ Give a brief account of glycogen storage diseases.

Glycogen storage diseases (*GSD, glycogenoses*) are a group of genetic disorders of glycogen metabolism (glycogenesis and glycogenolysis). The different types of glycogenoses are summarized in Table 6.1.

Table 6.1 Glycogen storage diseases

Type/disease	*Enzyme defect*	*Clinical symptoms*	*Biochemical findings*	*Management*
Ia von Gierke's disease	Glucose-6-phosphatase	Short stature, delayed adolescence, hypoglycemia, hepatomegaly, bleeding, anemia	Increased glycogen and lactate, lactic acidosis. Hyperlipidemia with increase in cholesterol and triglycerides Hyperuricemia	Frequent feeding with complex polysaccharides. Lactose and sucrose restriction. Allopurinol for treatment of hyperuricemia
Ib	Glc-6-phosphate-translocase	Similar to those in type Ia in addition, neutropenia and recurrent infection	Similar to Ia	Similar to Ia
II Pompe's disease	Lysosomal-α-glucosidase	Skeletal muscle hypotonia and weakness, cardiac enlargement, and hepatomegaly	Excessive glycogen accumulation. Increased creatine kinase	Treatment ineffective
III Cori's disease limit dextrinosis	Debranching enzyme	Hypoglycemia. hepatomegaly, mild myopathy	Accumulation of branched glycogen with abnormal structure. High cholesterol and triglyceride levels. Increased aspartate transaminase activity	Frequent feeding
IV Andersen's disease, amylopectinosis	Branching enzyme	Hepatomegaly, failure to thrive, cirrhosis, myotonia, and weakness	Accumulation of abnormal glycogen with few branch points	No effective treatment
V Mc Ardle's disease	Myophosphorylase	Pain, cramps, and myoglobinuria after strenuous exercise	Increase in muscle glycogen, serum creatine kinase. Blood lactate fails to increase after ischemic excercise test	Avoidance of strenuous exercise
VI Hers' disease	Liver phosphorylase	Hepatomegaly, hypoglycemia	Increase in liver glycogen content	Similar to type III
VII Tarui's disease	Phosphofructokinase	Similar to type V. In addition, hemolytic anemia	Similar to type V	Similar to type V
VIII	Liver phosphorylase kinase	Hepatomegaly, hypoglycemia	Increase in liver glycogen with normal structure	Similar to type III

❑ Explain the formation and utilization of lactate.

▫ *Lactate formation*

- Lactate is formed from pyruvate as the end product of the anaerobic metabolism of glucose by the action of *lactate dehydrogenase* (*LD*).

$$\text{Pyruvate} + \text{NADH} + \text{H}^+ \longrightarrow \text{Lactate} + \text{NAD}^+$$

- *Sites of synthesis*: Erythrocytes, skin, intestine, brain, renal medulla, retina, and muscle.

- ***Lactate utilization***
  - Liver removes 53% of the total lactate by gluconeogenesis.
  - Kidney removes 30% of the lactate load by metabolizing lactate. The kidney reabsorbs the filtered lactate which, therefore, does not appear in the urine.

Enumerate the causes of lactic acidosis.

- Normal blood levels: 0.4 – 1.8 mmol/L (4 – 16 mg/dL)
- Lactic acidosis results from an imbalance between the rate of production and the rate of utilization of lactate.
- It can occur with or without tissue hypoxia.
- Lactic acidosis associated with tissue hypoxia occurs in the following conditions:
  - During strenuous muscle contraction when the rate of glycolysis exceeds the availability of oxygen, leading to increased lactate generation and transient lactic acidosis
  - Shock due to sepsis, myocardial infarction, and hemorrhage
  - Respiratory arrest and pulmonary diseases
  - Severe anemia and carbon monoxide poisoning because of impaired oxygen delivery to tissues.
- Lactic acidosis without tissue hypoxia occurs in the following conditions:
  - Malignancy due to overproduction of lactate and poor tissue perfusion
  - Diabetes mellitus in which pyruvate oxidation is inhibited
  - Liver diseases in which lactate metabolism is impaired, leading to lactate accumulation
  - Drugs and toxins such as the following:
    i. Oral hypoglycemic agents such as biguanides, which enhance glycolysis and inhibit TCA cycle, and gluconeogenesis from lactate.
    ii. Salicylates, by impairment of electron transport and oxidative phosphorylation.
    iii. Ethanol, by increasing the $NADH/NAD^+$ ratio which inhibits gluconeogenesis from lactate and enhances anaerobic glycolysis.
  - Enzyme defects such as
    i. Glucose-6-phosphatase deficiency in *von Gierke's disease*, which causes hypoglycemia with breakdown of muscle glycogen to lactate.
    ii. Deficiency of fructose 1,6-bisphosphatase and pyruvate carboxylase that inhibit gluconeogenesis.

Chapter 7

Metabolism of Lipids

❑ How are dietary lipids digested and absorbed?

- ***Dietary lipid*** is in the form of
  - Triacylglycerols (predominant)
  - Phospholipids
  - Cholesterol and cholesteryl esters.
- ***Digestion and absorption in the stomach***
  - Triacylglycerol is hydrolysed by *gastric lipase* to form free fatty acids (FFA) and 1,2-diacylglycerol.
  - Short and medium chain fatty acids are absorbed into the portal vein through the stomach wall.
  - Larger chain fatty acids enter into the duodenum.
- ***Digestion and absorption in the small intestinal lumen***
  - Triacylglycerols are hydrolysed at positions 1 and 3 by *pancreatic lipase* to form FFA and 2-monoacylglycerol (major products), and glycerol (small amounts).
  - Phospholipids are acted upon by *phospholipase A_2* to form lysophospholipids and fatty acids.
  - Cholesteryl esters are hydrolysed by *cholesterol ester hydrolase*.
  - Bile salts play an important role in lipid digestion and absorption by
    a. Emulsification and solubilization of lipids by detergent action
    b. Formation of *mixed micelles* containing fatty acids, monoacylglycerol, cholesterol, phosphatidylcholine, and fat-soluble vitamins.
- ***At the intestinal membrane***
  - FFA and 2-monoacylglycerol passively diffuse from the micelles into the intestinal mucosal cell.
  - Free cholesterol also crosses the intestinal membrane and enters the intestinal mucosal cell.
- ***In the small intestinal mucosal cell***
  - Long chain fatty acids are re-esterified to triacylglycerols which together with phospholipids, cholesterol, cholesterol esters, and protein are incorporated into chylomicrons.
  - The chylomicrons are secreted into the lymph from where they enter the blood *via* the thoracic duct.
  - Medium and short-chain fatty acids, and glycerol being water-soluble are absorbed directly into the portal circulation.

The digestion and absorption of lipids is summarized in Figure 7.1.

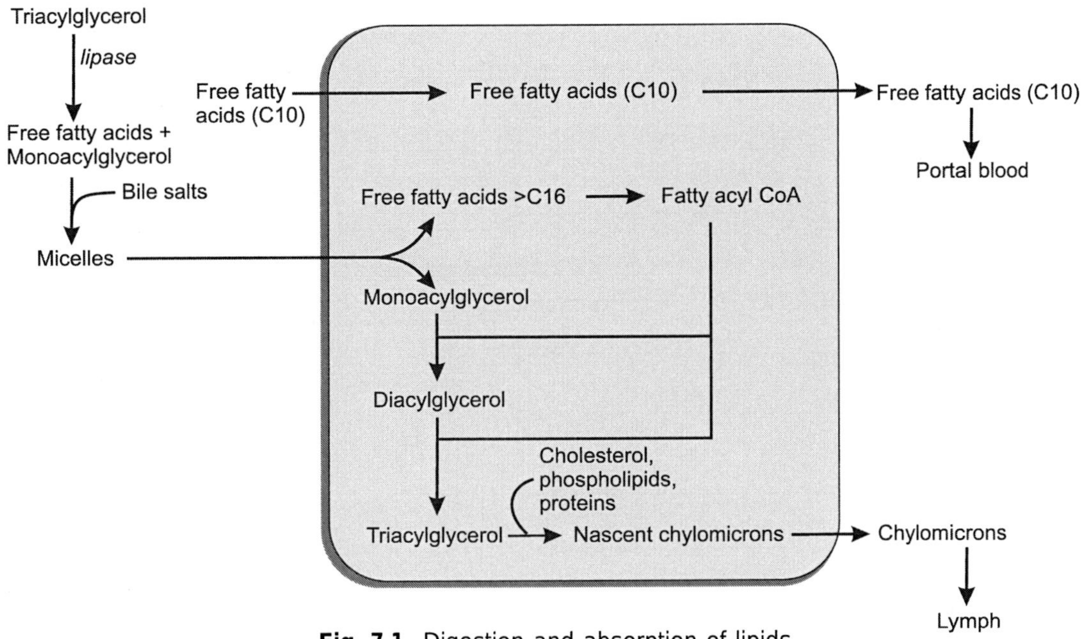

Fig. 7.1 Digestion and absorption of lipids

❑ How does the gastrointestinal hormone cholecystokinin facilitate lipid digestion?

Cholecystokinin stimulates

- gallbladder contraction and releases bile into the small intestine.
- secretion of *pancreatic lipase.*

❑ What is the role of bile in lipid digestion and absorption?

Bile promotes emulsification of lipids by the detergent action of bile salts, which form mixed micelles with fatty acids, monoacylglycerol, cholesterol, phosphatidylcholine, and fat-soluble vitamins.

❑ Explain the steps in the β-oxidation pathway.

β-Oxidation refers to the successive removal of 2-carbon units from fatty acids to acetyl CoA, which is completely oxidized to carbon dioxide and water *via* the TCA cycle.

***Subcellular site*:** Mitochondrial matrix

***Steps*:** Before the oxidation reactions, the fatty acids are activated and transported to the mitochondria. The following sequence of events are involved:

1. *Activation of fatty acids* to fatty acyl CoA by *acyl CoA synthase* (*fatty acid thiokinase*).

$$\text{RCOOH + ATP + CoASH} \rightleftharpoons \text{R–COS CoA + AMP + PPi}$$

2. *Transfer of fatty acyl CoA* by *carnitine acyltransferase I* to carnitine. Carnitine is required as an acyl group carrier because the mitochondrion is impermeable to fatty acyl CoA. The acyl group from acylcarnitine is transferred to CoA in the mitochondrial matrix by *carnitine acyltransferase II.*

3. *β-Oxidation* involves repeated sequence of the following four reactions (Fig. 7.2):
   a. *Oxidation* of the acyl CoA to *trans* Δ^2-enoyl CoA by FAD-dependent *dehydrogenase*.
   b. *Hydration* of the trans-Δ^2-enoyl CoA to 3-hydroxyacyl CoA by *enoyl CoA hydratase*.
   c. *Oxidation* of 3-hydroxyacyl CoA to 3-ketoacyl CoA by NADH-dependent *dehydrogenase*.
   d. *Thiolysis* (cleavage) of 3-ketoacyl CoA to acetyl CoA by *β-ketothiolase*.

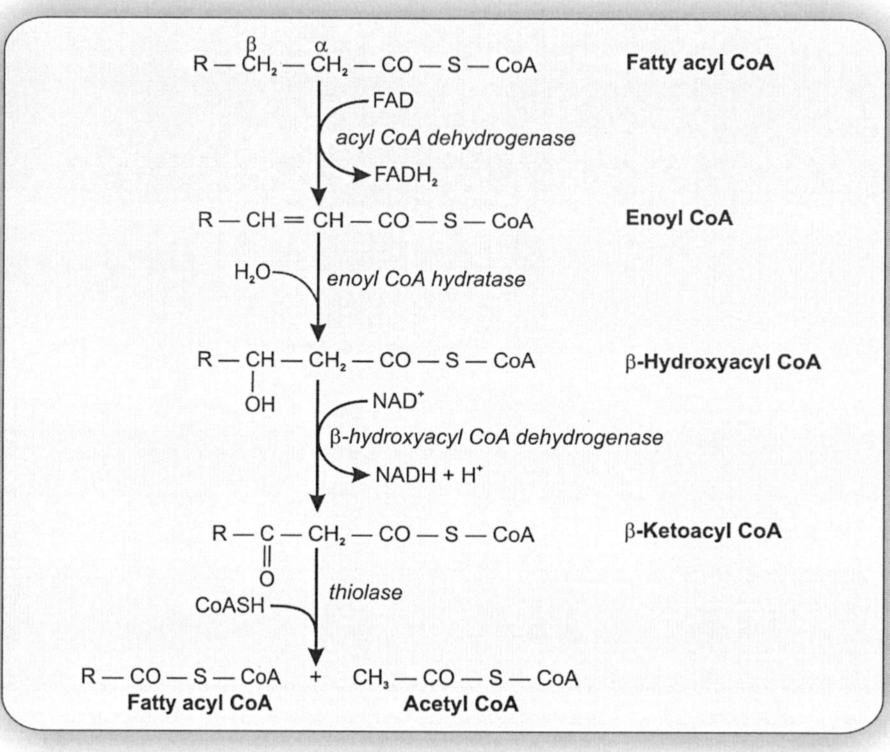

Fig. 7.2 β-Oxidation of fatty acids

Energetics: Degradation of palmitoyl CoA, a C16 saturated acyl CoA, involves seven rounds of β-oxidation with a net energy yield of 129 ATPs.

However, assuming that oxidative phosphorylation produces 1.5 ATP per $FADH_2$ and 2.5 ATP per NADH oxidized, the net yield of ATP is 108.

❑ What are ketone bodies?

Ketone bodies are acetoacetate, β-hydroxybutyrate, and acetone that are formed from acetyl CoA. These are used as energy sources by extrahepatic tissues such as the heart and kidney.

❑ Outline the process of ketogenesis.

Ketogenesis refers to the formation of ketone bodies. It occurs in the liver when the amount of acetyl CoA formed exceeds the rate of entry into the TCA cycle.

Steps:
1. Condensation of two molecules of acetyl CoA to form acetoacetyl CoA
2. Formation of 3-hydroxy-3-methylglutaryl CoA (HMG CoA) by condensation of acetoacetyl CoA with another molecule of acetyl CoA
3. Cleavage of HMG CoA to form Δ-3-hydroxybutyrate or decarboxylation to acetone.

Figure 7.3 illustrates the process of ketogenesis.

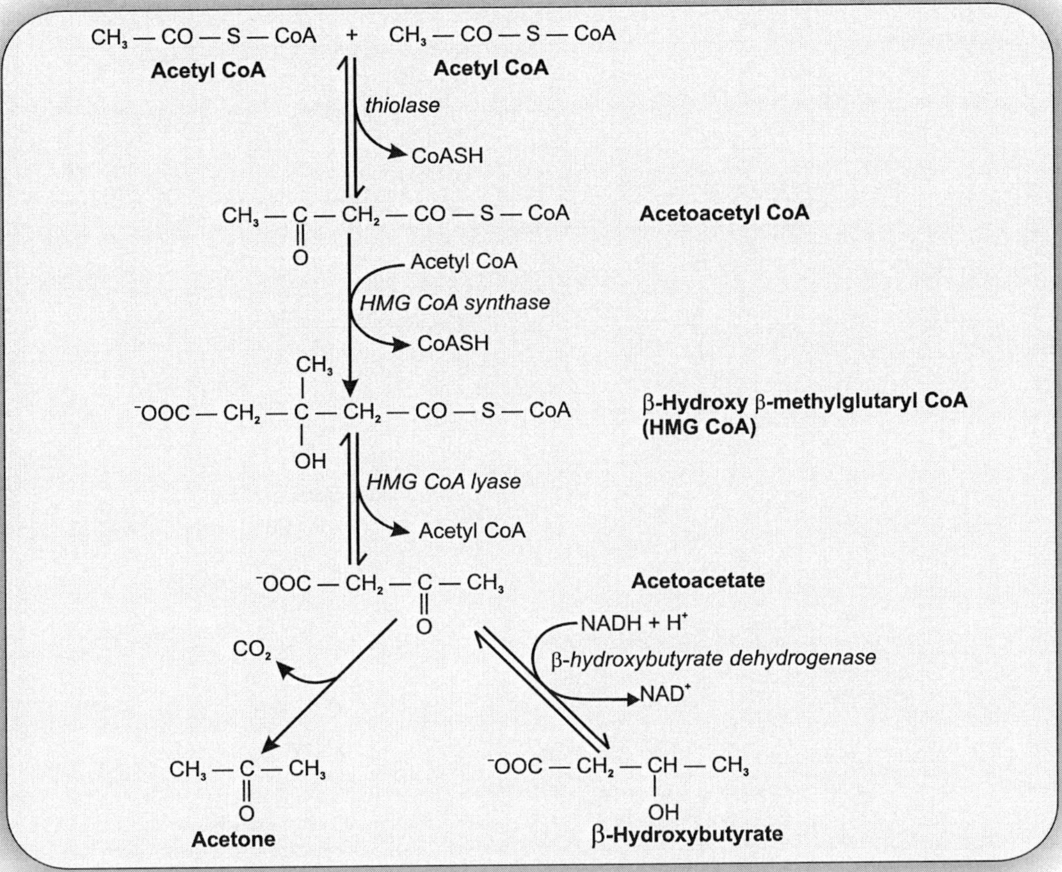

Fig. 7.3 Ketogenesis

In uncontrolled diabetes, ketogenesis exceeds the rate of utilization by extrahepatic tissues leading to increased levels of ketone bodies in the blood (*ketonemia*) and excretion in the urine (*ketonuria*) with acetone odor in the breath.

❑ Discuss the biosynthesis of fatty acids.

The synthesis of fatty acids involves the condensation of 2-carbon units in the form of acetyl CoA. During this synthesis, the fatty acids are covalently linked to an acyl carrier protein (ACP).

***Sites of synthesis*:** Liver, kidney and adipose tissue

***Subcellular site*:** Cytosol

***Precursor*:** Acetyl CoA

***Requirements*:** Fatty acid synthase complex, NADPH as a reductant

Transfer of acetyl CoA from the mitochondrion to the cytosol (Fig. 7.4): The inner mitochondrial membrane is not permeable to acetyl CoA. The following reactions facilitate acetyl CoA transport:

1. Condensation of acetyl CoA with oxaloacetate to form citrate
2. Transport of citrate to the cytosol
3. Cleavage of citrate by *ATP-citrate lyase* to regenerate acetyl CoA and oxaloacetate
4. Re-entry of oxaloacetate to the mitochondria after conversion to malate, generating NADPH.

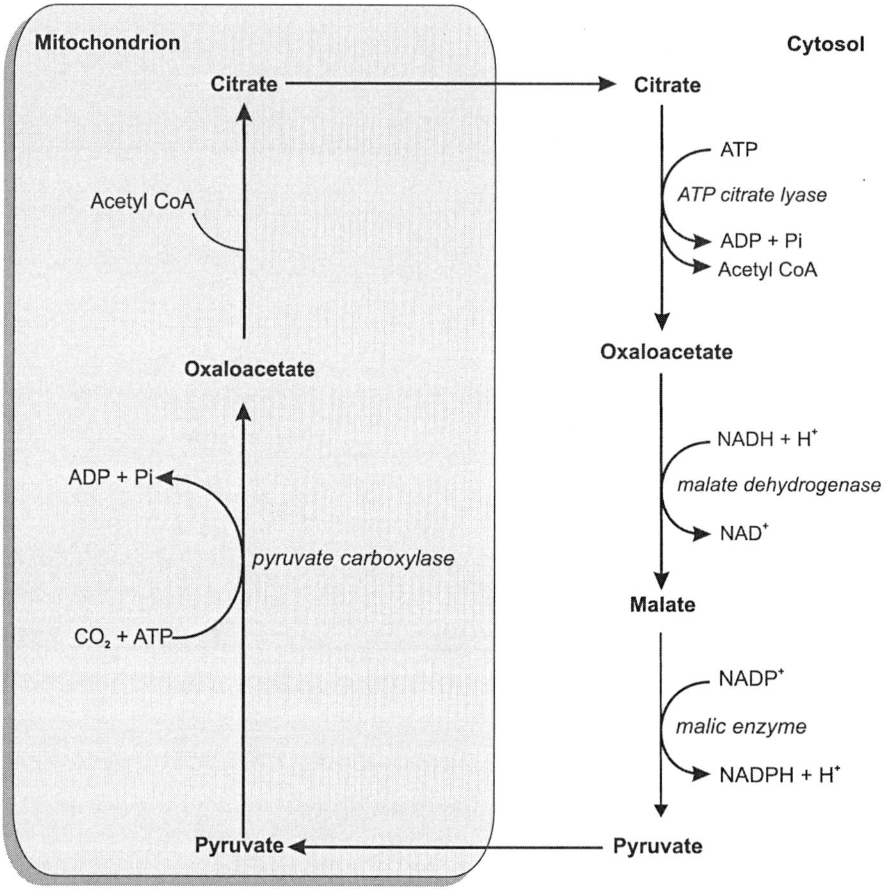

Fig. 7.4 Transfer of acetyl CoA from the mitochondrion to the cytosol

Steps in biosynthesis

1. Carboxylation of acetyl CoA to malonyl CoA by *acetyl CoA carboxylase* that contains biotin as a prosthetic group.
2. Conversion of acetyl CoA and malonyl CoA to their corresponding ACP-derivatives catalyzed by *acetyl transacylase* and *malonyl transacylase*, respectively.
3. Elongation cycle involves a repeated sequence of four reactions:
   a. *Condensation* of acetyl-ACP and malonyl-ACP to yield acetoacetyl ACP.
   b. *Reduction* of acetoacetyl-ACP to 3-hydroxybutyryl ACP by NADPH.
   c. *Dehydration* to crotonyl ACP.
   d. *Reduction* of crotonyl ACP to butyryl-ACP by NADPH.

Thus, successive addition of two-carbon units forms the 16C palmitoyl-ACP. Hydrolysis of palmitoyl-ACP by a *thioesterase* generates palmitate and ACP. The overall reaction is as follows:

$$8\text{ Acetyl CoA} + 7\text{ATP} + 14\text{ NADPH} + 14\text{ H}^+ \longrightarrow \text{Palmitate} + 14\text{ NADP}^+ + 8\text{ CoA} + 6\text{ H}_2\text{O} + 7\text{ ADP} + 7\text{ Pi}$$

4. *Desaturation*: Palmitate undergoes chain elongation to form C18 stearate. Palmitate and stearate are desaturated in the smooth endoplasmic reticulum to form palmitoleate and oleate, respectively.

Figure 7.5 illustrates the various steps of fatty acid biosynthesis.

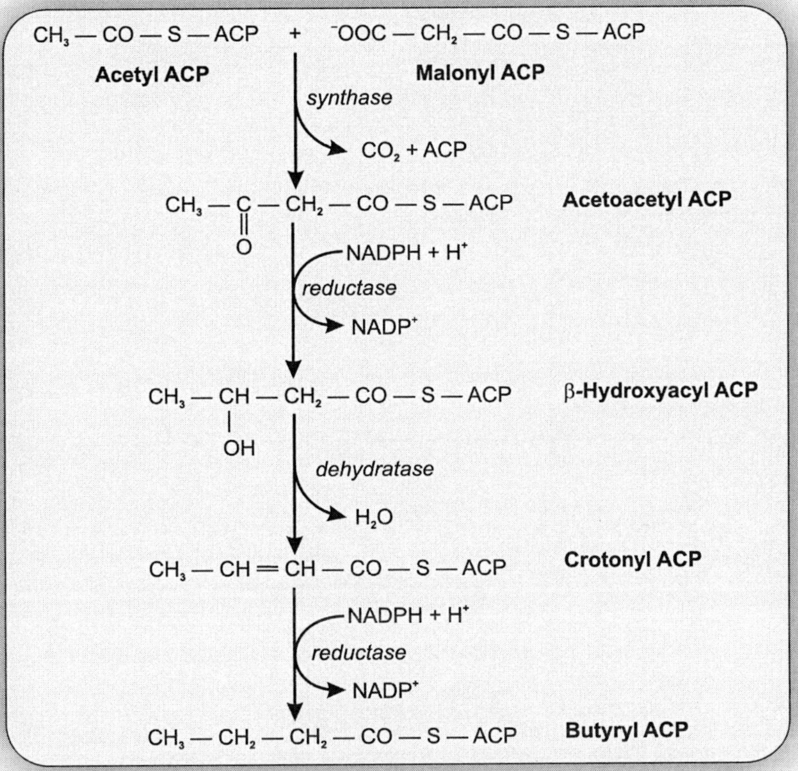

Fig. 7.5 Steps in fatty acid biosynthesis

Regulation: Lipogenesis is regulated by the following factors:

- *Hormones*: Insulin stimulates fatty acid synthesis, whereas glucagon and epinephrine inhibit fatty acid synthesis.
- *Acetyl CoA carboxylase*: This is an allosteric enzyme, which is stimulated by citrate and inhibited by long-chain acyl CoA molecules.
- *Nutritional state*: Fatty acid synthesis is decreased by caloric restriction and increased in the well-fed state.

❑ Briefly explain the metabolism of triacylglycerols.

Biosynthesis

1. Addition of fatty acyl CoA to glycerol 3-phosphate gives rise to lysophosphatidate and subsequently phosphatidate.
2. Dephosphorylation of phosphatidate to diacylglycerol (DAG).
3. Acylation of DAG to triacylglycerol.

Figure 7.6 illustrates the biosynthesis of triacylglycerol.

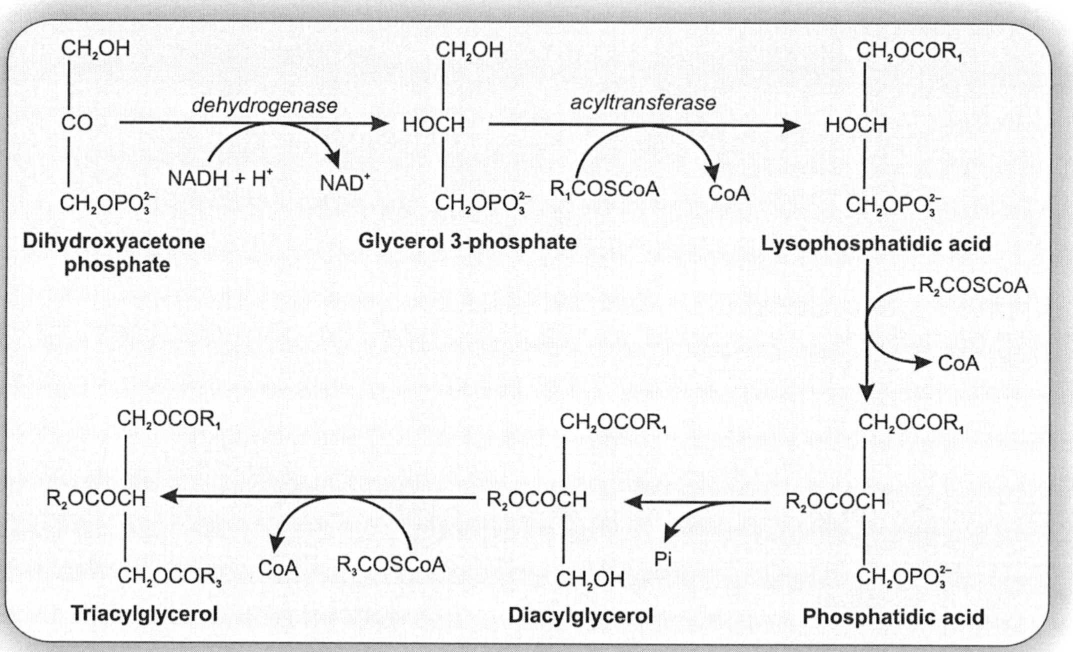

Fig. 7.6 Biosynthesis of triacylglycerol

Catabolism

1. Hydrolysis of triacylglycerols by *lipases* to form free fatty acids and glycerol.
2. β-oxidation of free fatty acid.
3. Conversion of glycerol to dihydroxyacetone phosphate, which enters glycolysis.

❑ Elucidate the metabolism of phospholipids.

Biosynthesis

- Phosphatidylcholine, phosphatidylethanolamine, and phosphatidylserine are formed from DAG.
- Phosphatidylinositol and cardiolipin are formed from phosphatidate.

The biosynthesis of phospholipids is summarized in Figure 7.7.

Catabolism

Phospholipids are degraded by *phospholipases* as follows (Fig. 7.8):

- *Phospholipase* A_1 cleaves the bond in position 1 of phospholipids.
- *Phospholipase* A_2 hydrolyses the bond in position 2 of phospholipids.

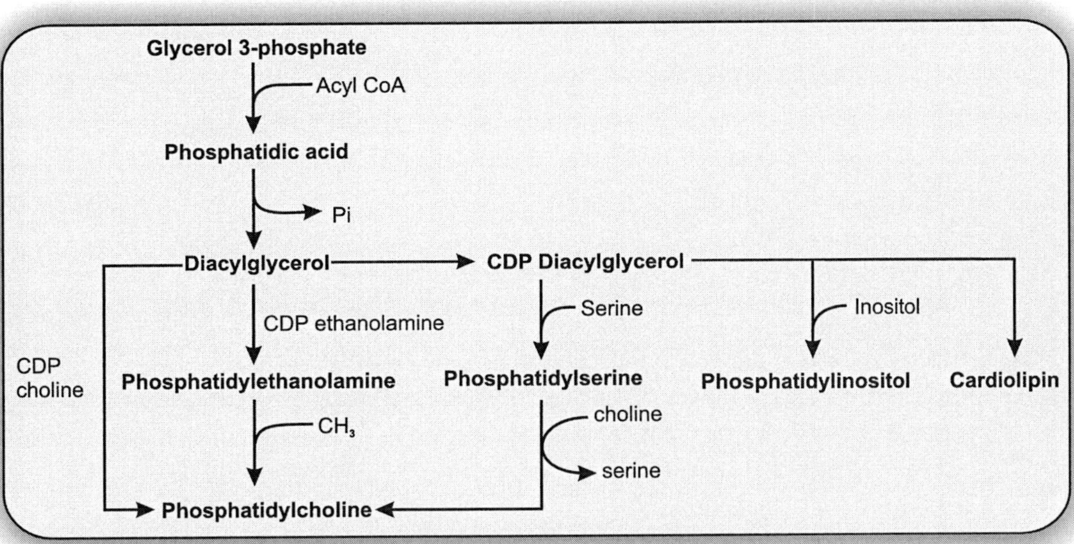

Fig. 7.7 Biosynthesis of phospholipids

3. *Phospholipase B* attacks both the acyl groups.
4. *Phospholipase C* hydrolyses the bond in position 3 to liberate 1,2-diacylglycerol and a phosphoryl base.
5. *Phospholipase D* hydrolyses the nitrogenous base from phospholipids. It is found in plants and plays a role in mammalian signal transduction.

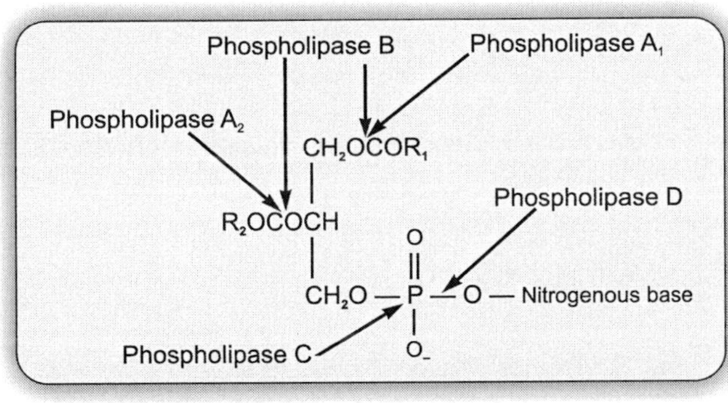

Fig. 7.8 Degradation of phospholipids by phospholipases

❑ Give a brief account of the biosynthesis of sphingolipids.

- Ceramide is the precursor for both glycosphingolipids (cerebrosides) and gangliosides.
- Glucosylceramide (GlcCer), found in extraneural tissue, is the precursor of other complex glycosphingolipids or gangliosides.
- Galactosylceramide (GalCer) present in myelin forms sulphogalactosylceramide (sulphatide) on reaction with phosphoadenosine-5'-phosphosulphate (PAPS).
- Gangliosides are formed by the step-wise addition of uridine diphosphate glucose (UDPGlc), uridine diphosphate galactose (UDPGal), and sialic acid (usually N-acetylneuraminic acid) to ceramide.

The biosynthesis of sphingolipids is summarized in Figure 7.9.

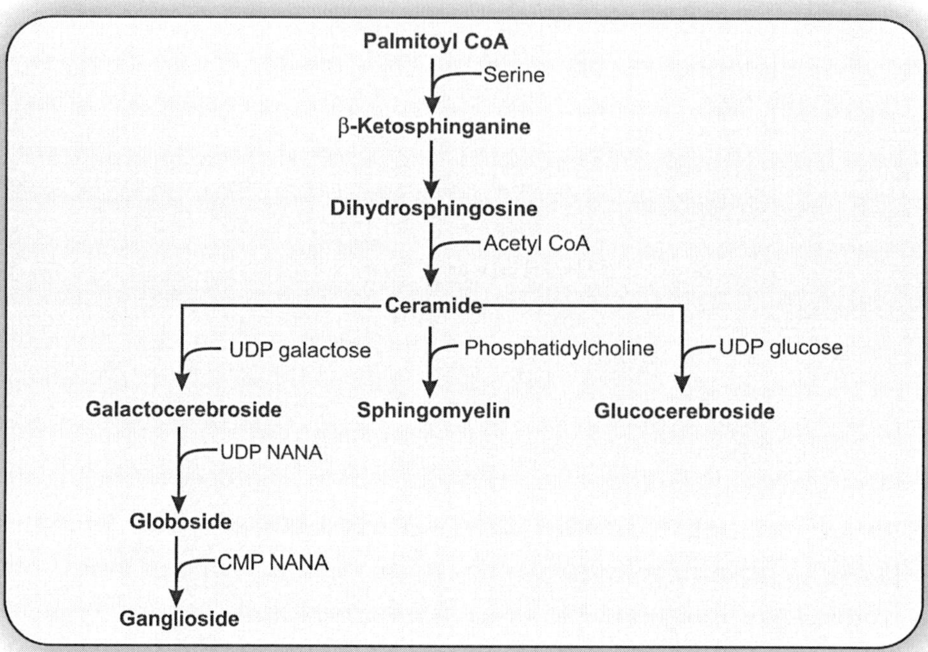

Fig. 7.9 Biosynthesis of sphingolipids

❑ Discuss the metabolism of cholesterol.

Biosynthesis

Dietary cholesterol constitutes less than 50% of the body cholesterol. Body cholesterol is mostly obtained by biosynthesis from acetyl CoA.

Site of synthesis: Microsomal and cytosol fractions of all nucleated cells. The liver and intestines synthesize 10% of the total cholesterol.

Steps involved:

1. Condensation of two molecules of acetyl CoA to form acetoacetyl CoA.
2. Formation of HMG CoA by combination of acetyl CoA with acetoacetyl CoA.
3. Reduction of HMG CoA to mevalonate by *HMG CoA reductase*. This is the committed step in cholesterol biosynthesis. The enzyme is inhibited by cholesterol and by the fungal compounds lovastatin and compactin.
4. Conversion of mevalonate to the 5-carbon isoprene compounds 3-isopentenyl pyrophosphate and dimethylallyl pyrophosphate.
5. Formation of C10 geranyl pyrophosphate by condensation of the 5-carbon isoprene compounds.
6. Addition of another molecule of isopentenyl pyrophosphate to geranyl pyrophosphate to form C15 farnesyl pyrophosphate.
7. Condensation of two molecules of farnesyl pyrophosphate to form C30 squalene.
8. Cyclization of squalene to lanosterol.

9. Loss of three methyl groups from lanosterol to form cholesterol (Fig. 7.10).

Excretion

- Excess cholesterol is excreted as bile salts.

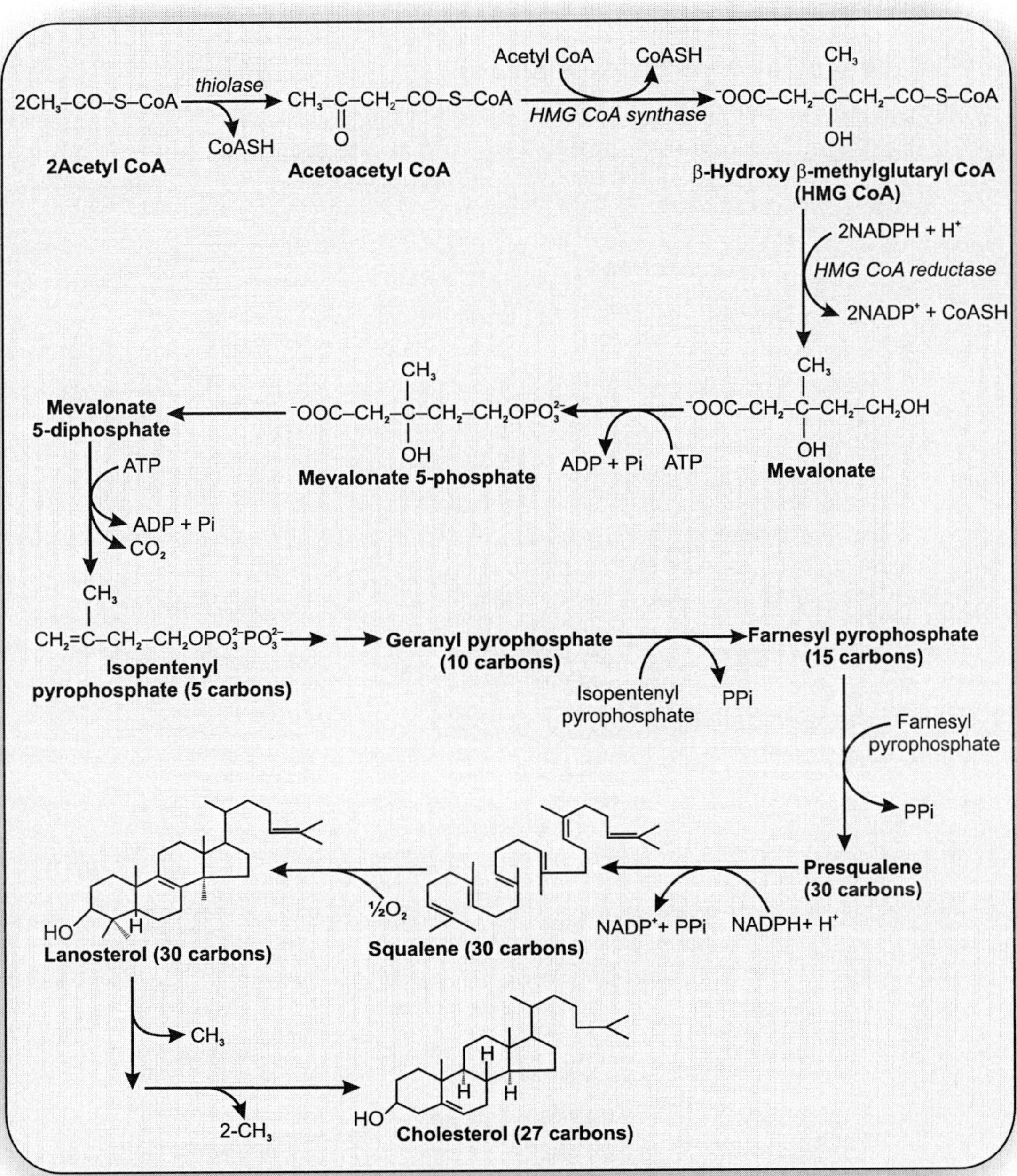

Fig. 7.10 Biosynthesis of cholesterol

- Bile salt synthesis, which takes place in the liver, involves the following steps:
  - conversion of cholesterol to cholyl CoA
  - conjugation with glycine or taurine to form glycocholate or taurocholate, respectively.
- Bile salts are absorbed into the portal circulation and returned to the liver by the *enterohepatic circulation.*

❑ What are the functions of lipoproteins?

The major functions of lipoproteins include the following:

1. Transport of dietary fat from the intestine to various tissues.
2. Transport of triglycerides from the liver to extrahepatic tissues for storage or oxidation.
3. Removal of excess cholesterol from extrahepatic tissues to the liver for excretion.

❑ How are chylomicrons formed?

Formation of chylomicrons involves the following steps:

1. Absorption of fatty acids and cholesterol (products of lipid digestion), and cholesterol from the bile into the intestinal epithelial cells.
2. Re-esterification of fatty acids and cholesterol to triglycerides and cholesterol ester, respectively.
3. Addition of phospholipids.
4. Incorporation of apoproteins A and B: Apoproteins synthesized in the rough endoplasmic reticulum are incorporated into chylomicrons in the smooth endoplasmic reticulum.
5. Addition of carbohydrate residues as the chylomicrons pass through the Golgi apparatus.
6. Release of chylomicrons from the intestinal cell by *reverse pinocytosis.*
7. Secretion into the lymphatic system draining the intestine and entry into the systemic circulation *via* the thoracic duct.
8. Transfer of apo C and apo E from HDL to the "*nascent*" chylomicron particles.

❑ Explain the metabolism of chylomicrons.

- ***Sites of metabolism***: Adipose tissue and skeletal muscle.
- ***Steps*** (Fig. 7.11):

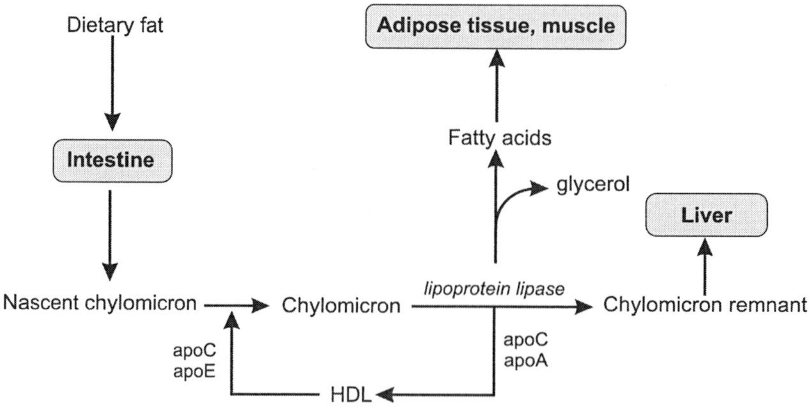

Fig. 7.11 Metabolism of chylomicrons

1. Hydrolysis of triglycerides in chylomicrons by *lipoprotein lipase* to form *chylomicron remnants*. While glycerol removed from triglycerides enters glycolysis, most of the FFA are taken up by adipose tissue and muscle cells.
2. Uptake of chylomicron remnants (enriched in cholesterol, cholesterol esters, and apoproteins B48 and E) by the liver by receptor-mediated endocytosis.
3. Hydrolysis of triglycerides and phospholipids in chylomicron remnants by *hepatic lipase*.
4. Conversion of cholesterol in chylomicron remnants to bile acids and excretion in the bile. A small amount of cholesterol is excreted without conversion to bile acids. Cholesterol is also distributed to extrahepatic tissues.

❑ Describe the metabolism of VLDL.

1. VLDL is a triglyceride-rich lipoprotein synthesized in the liver. It also contains phospholipids, cholesterol, cholesterol ester, and apoprotein B-100. The apoproteins C and E found in VLDL are transferred from HDL.
2. VLDL is secreted from the hepatic parenchymal cells into the circulation.
3. VLDL is metabolized in extrahepatic tissues (chiefly, adipose tissue and muscle) by *lipoprotein lipase* which catalyzes hydrolysis of triglycerides to form *VLDL remnants* known as *intermediate-density lipoproteins (IDL)*. The fatty acids generated are used for triglyceride synthesis within the adipose tissue.
4. IDL are taken up by the liver *via* the *LDL receptors* or transformed to LDL in the plasma.
5. During the conversion of IDL to LDL, the remaining triglycerides and all the apoproteins except apoB100 are removed.

The metabolism of VLDL is presented in Figure 7.12.

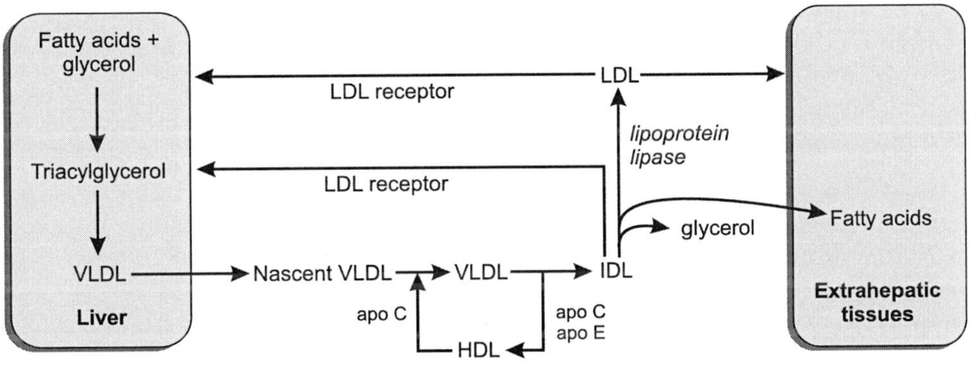

Fig. 7.12 Metabolism of VLDL and LDL

❑ Discuss the metabolism of LDL.

- LDL is a cholesterol-rich lipoprotein formed from VLDL *via* IDL.
- It accounts for 70% of plasma cholesterol.
- Plasma LDL and cholesterol levels are regulated by LDL receptors.
- An increase in intracellular cholesterol levels
  - reduces the number of LDL receptors
  - decreases LDL uptake by cells
  - increases plasma LDL levels.

- About 70 – 80% of LDL is taken up by receptors in
  - the liver (abundant), where it is used for the synthesis of bile acids and free cholesterol.
  - extrahepatic tissues (adrenal cortex, lymphocytes, and kidneys), where it is hydrolysed by *lysosomal cholesteryl esterase* (*acid lipase*) to liberate cholesterol, which is used for membrane formation and steroid hormone synthesis.
- The remaining LDL is degraded by the reticuloendothelial system.

❑ Outline the role of the liver in cholesterol metabolism.

The liver plays an important role in cholesterol metabolism. It
- takes up cholesterol from the diet and from lipoproteins *via* LDL receptors,
- synthesizes cholesterol endogenously, and
- excretes cholesterol in the bile.

❑ Discuss the metabolism of HDL.

- HDL is synthesized and secreted from the liver and intestine as *nascent HDL* into the circulation. Nascent HDL is a discoid particle that contains phospholipids, free cholesterol, and apoproteins.
- Apoproteins C and E are transferred to nascent HDL after it enters circulation.
- Nascent HDL is converted to plasma HDL_3 by uptake of
  - lipids and apoproteins from chylomicrons and VLDL
  - cholesterol from extrahepatic tissues.
- HDL_3 is the major site of action of *lecithin cholesterol acyltransferase* (*LCAT*) that converts phospholipids and free cholesterol to cholesterol ester and lysolecithin. As a consequence,
  - lysolecithin is transferred to plasma albumin
  - cholesterol ester moves to the core of HDL_3 to form a spherical pseudomicellar structure.
- HDL_3 is converted to HDL_2 by transfer of
  - cholesterol ester from HDL_3 to chylomicrons and VLDL
  - triglycerides from chylomicrons and VLDL to HDL_3.

  Due to these transfers, HDL_2 becomes larger and enriched in triglycerides.
- HDL_2 becomes more dense and re-forms HDL_3 by
  - hydrolysis of triacylglycerols and phospholipids by *hepatic lipase*
  - cholesterol release for uptake by the liver.
- HDL degradation occurs in the liver and intestine. The kidneys remove apoA-I dissociated from HDL.

Figure 7.13 displays schematic representation of HDL metabolism.

❑ What are the functions of HDL?

The functions of HDL include
1. Provision of apo C and E to chylomicrons, and VLDL.
2. Transfer of cholesterol esters to chylomicrons, and VLDL.
3. *Reverse cholesterol transport*, which refers to transport of cholesterol from peripheral tissues to the liver for excretion *via* bile into the feces. In addition to HDL, VLDL, LDL, LCAT and lipid transfer proteins play a role in reverse cholesterol transport.

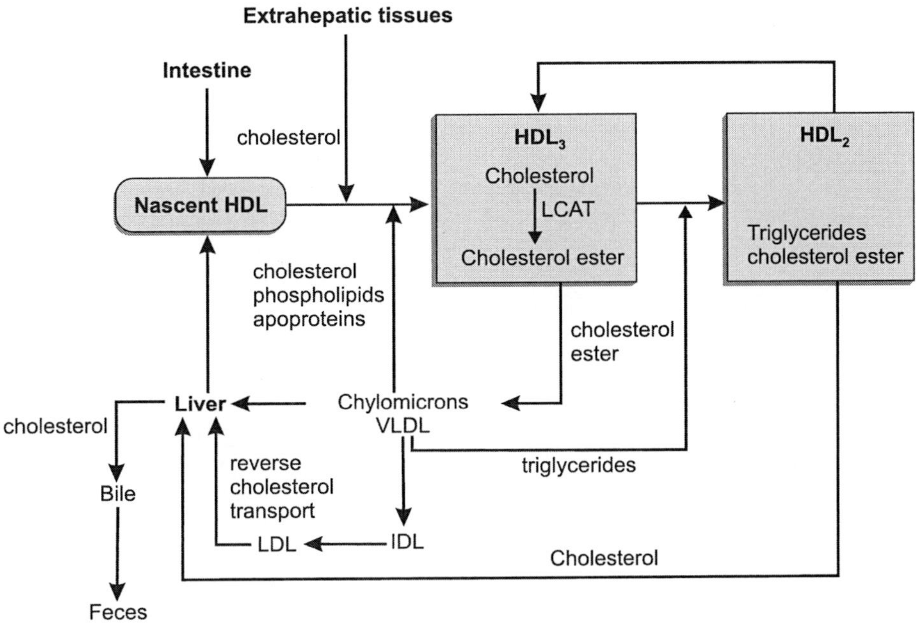

Fig. 7.13 Metabolism of HDL

❑ What are hyperlipoproteinemias?

The hyperlipoproteinemias may be

- *Primary hyperlipoproteinemias*, which are of five types:
  - type I characterized by deficiency of lipoprotein lipase
  - type II due to a defect in the LDL receptor gene
  - type III with a defect in the clearance of chylomicron and VLDL remnants
  - type IV characterized by elevation of plasma VLDL
  - type V where three different lipoprotein patterns (IIa, IIb, and IV) are seen.
- *Secondary hyperlipoproteinemias*, which arise due to
  - diabetes mellitus
  - Cushing's syndrome
  - alcohol consumption
  - nephrotic syndrome.

❑ Give a brief account of type I hyperlipoproteinemia.

- *Type I hyperlipoproteinemia or familial lipoprotein lipase deficiency* arises due to a deficiency of *lipoprotein lipase* or *apoprotein CII* that leads to
  - a block in the metabolism of chylomicrons
  - massive accumulation of chylomicrons in plasma, especially after a high fat diet.
- ***Clinical symptoms*** as a consequence of triglyceride deposition include
  - *eruptive xanthomas* (crops of small, itchy, yellow nodules on the skin)
  - hepatomegaly, splenomegaly
  - foam cell infiltration of the bone marrow
  - *lipemia retinalis* (paleness of the retina)
  - abdominal pain due to pancreatitis.

- ***Diagnosis***
  - lipemic fasting plasma with a creamy layer
  - electrophoretic pattern: increase in chylomicrons (Fig. 7.14).
- ***Treatment***
  Fat-free diet.

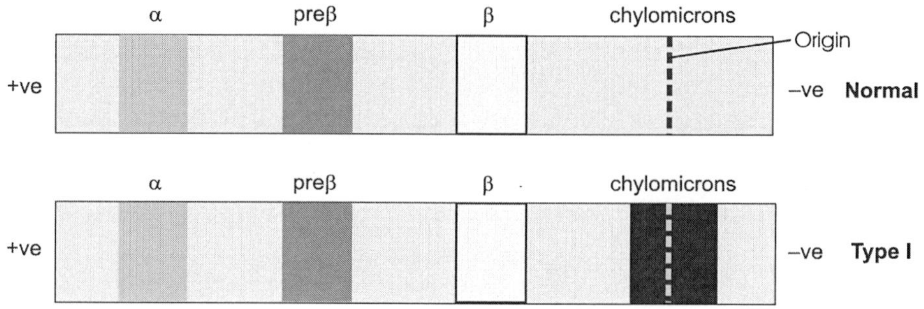

Fig. 7.14 Plasma electrophoretic pattern in type I hyperlipoproteinemia

Explain the biochemical defect, clinical symptoms, and management of type II hyperlipoproteinemia.

- ***Biochemical lesion*:** Type II hyperlipoproteinemia, also known as familial hypercholesterolemia, is characterized by defective LDL receptors or mutation in the ligand region of apo B-100. In type IIa, plasma LDL is increased, whereas in type IIb, both LDL and VLDL are elevated (Fig. 7.15).

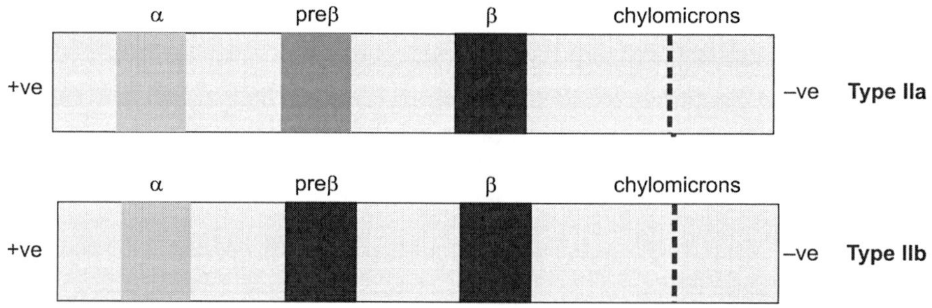

Fig. 7.15 Plasma electrophoretic pattern in type II hyperlipoproteinemia

- ***Clinical symptoms*** include
  - Tendon xanthomas due to cholesterol ester deposition in tissue macrophages.
  - Xanthelasma due to cholesterol deposition in the soft tissue of the eyelid.
  - Premature atherosclerosis and myocardial infarction due to LDL infiltration into arteries.
- ***Management***
  - Restriction of dietary cholesterol
  - Replacement of saturated fats with polyunsaturated fats
  - Administration of bile acid binding resins (*e.g.*, cholestyramine) and inhibitors of HMG CoA (*e.g.*, lovastatin).

❑ Discuss the biochemical lesion, clinical features, diagnosis, and management of type III hyperlipoproteinemia.

- ***Synonyms***: *Familial dysbetalipoproteinemia, broad β disease*
- ***Transmission***: Single-gene mutation as well as additional environmental and/or genetic factors. Obesity, alcoholism, diabetes, and hypothyroidism are implicated in disease expression.
- ***Biochemical lesion***: Defect in the clearance of chylomicrons and VLDL remnants due to an abnormality in apo E. This leads to increase in cholesterol and triglycerides in plasma and tissues.
- ***Clinical features***: Xanthomas and atherosclerosis of peripheral and coronary arteries.
- ***Diagnosis***
  - broad β band on lipoprotein electrophoresis (Fig. 7.16).
  - increase in total cholesterol and triglycerides.

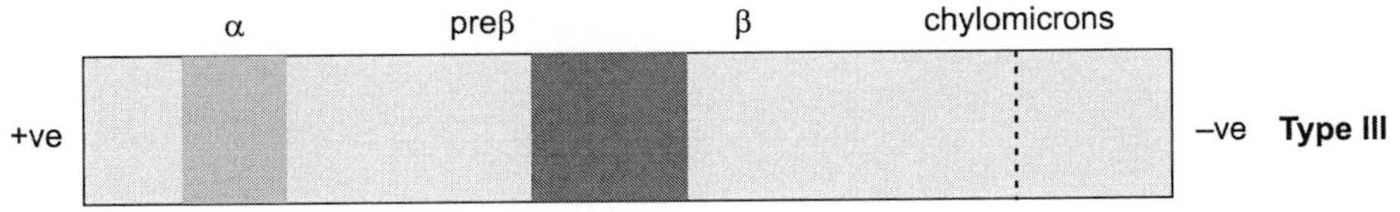

Fig. 7.16 Plasma electrophoretic pattern in type III hyperlipoproteinemia

- ***Treatment***
  - administration of hypolipidemic drugs (*e.g.*, clofibrate)
  - diet control and insulin treatment in patients with obesity and diabetes mellitus.
  - administration of levothyroxine if hypothyroidism is present.

❑ Give an account of type IV hyperlipoproteinemia.

Type IV hyperlipoproteinemia (*familial hypertriglyceridemia*) is characterized by

1. Increase in plasma VLDL and triglycerides
2. Increase in the pre-β fraction on electrophoresis (Fig. 7.17)
3. The triad of obesity, hyperglycemia and hyperinsulinemia
4. Hypertension and hyperuricemia that can predispose to atherosclerosis.

Exacerbating conditions include diabetes mellitus, hypothyroidism, alcohol, and oral contraceptives.

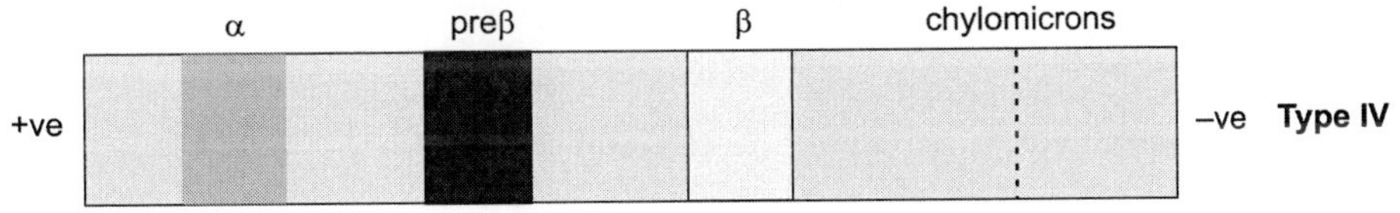

Fig. 7.17 Plasma electrophoretic pattern in type IV hyperlipoproteinemia

Treatment involves control of exacerbating conditions, restriction of saturated fat and calories, and administration of clofibrate or gemfibrozil if dietary measures fail.

❑ Briefly explain the biochemical and clinical findings in type V hyperlipoproteinemia.

Type V hyperlipoproteinemia (*familial combined hyperlipidemia*) is characterized by

- Three different lipoprotein patterns in affected subjects
  - hypercholesterolemia (IIa)

- hypertriglyceridemia (IV)
- both hypercholesterolemia and hypertriglyceridemia (IIb)

- Below normal plasma HDL levels
- Premature atherosclerosis.

Figure 7.18 shows the plasma electrophoretic pattern in type V hyperlipoproteinemia.

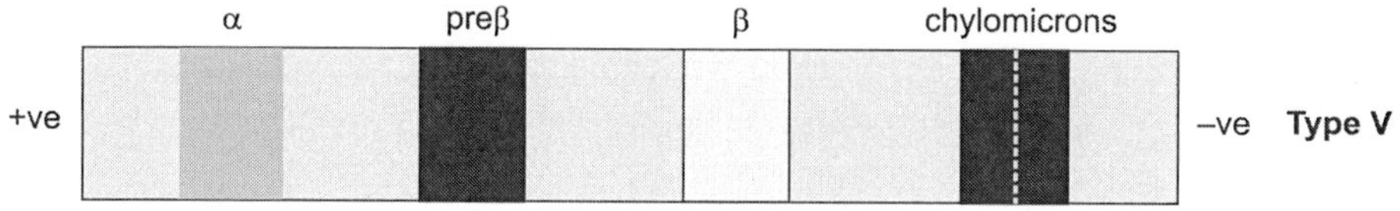

Fig. 7.18 Plasma electrophoretic pattern in type V hyperlipoproteinemia

Treatment

- Restriction of dietary fat and cholesterol
- Avoidance of alcohol and oral contraceptives
- Drugs to control increase in triglycerides and cholesterol.

❑ Give a brief account of familial hyperalphalipoproteinemia.

Familial hyperalphalipoproteinemia is a rare condition characterized by

- Increased plasma HDL that is predominantly genetic. Pesticides, alcohol, and oestrogens have also been implicated.
- Normal LDL, VLDL, and triglyceride levels.
- Protection against myocardial infarction.
- Slightly increased longevity.

❑ What are the consequences of familial lecithin: cholesterol acyltransferase (LCAT) deficiency?

The consequences of familial LCAT deficiency include the following:

- Decrease in plasma cholesterol esters and lysolecithin
- Accumulation of excess unesterified cholesterol in plasma and tissues
- Increase in plasma VLDL, free cholesterol
- Inhibition of reverse cholesterol transport
- Presence of HDL as nascent discoid HDL in stacks or rouleaux
- Presence of lipoprotein X, an abnormal LDL fraction
- Corneal opacity
- Renal damage
- Hemolytic anemia
- Premature atherosclerosis

❑ Name the conditions where hypolipoproteinemia is seen.

Hypolipoproteinemia is seen in two rare, autosomal recessive disorders presenting in childhood:

1. ***Abetalipoproteinemia***, characterized by the following sequence of events:
   - Defect in apo B formation
   - Impaired synthesis of chylomicrons, LDL, and VLDL
   - Impaired lipid transport in intestine and liver

- Accumulation of acylglycerols
- Very low plasma cholesterol and triglycerides
- Clinical symptoms such as steatorrhoea, neuropathy, and thorny red cells (*acanthocytosis*).

2. ***Tangier Disease***, characterized by the following sequence of events:
   - Abnormal apo A with increased HDL catabolism
   - Low plasma cholesterol ester with accumulation in the reticuloendothelial system.
   - Clinical features such as large orange tonsils, corneal opacity, and enlargement of liver and lymph nodes.

❑ What is fatty liver?

Fatty liver refers to excessive accumulation of lipid in the liver that causes fibrotic changes that can eventually lead to cirrhosis and liver failure.

❑ Enumerate the causes of fatty liver.

Fatty liver may be caused by one or more of the following:

1. *Increased plasma fatty acids* due to increased mobilization from adipose tissue or increased catabolism of chylomicrons and VLDL. Triglycerides accumulate in the liver because the rate of fatty acid influx exceeds the rate of lipoprotein synthesis.
2. *Increased hepatic fatty acid synthesis.* This is seen in the well-fed state when lipogenesis is increased and plasma fatty acids are low.
3. *Metabolic block in lipoprotein synthesis* can occur at the following levels:
   - Apoprotein synthesis
   - Provision of phospholipids
   - Secretory mechanism
   - Glycosylation of lipoproteins
   - Conjugation of lipid with apoprotein

❑ What are the major factors associated with fatty liver?

1. Chronic alcoholism: This is the most important factor in fatty liver pathogenesis. Metabolism of alcohol increases $NADH/NAD^+$ ratio that leads to
   a. decreased fatty acid oxidation
   b. decreased TCA cycle activity
   c. increased lipogenesis and cholesterol synthesis
   d. increased esterification of fatty acids to triglycerides.
2. Nutritional factors such as
   a. starvation and high fat diet which increase plasma fatty acids
   b. deficiency of methionine and choline, which are lipotropic agents essential for provision of phospholipids
   c. *Jamaican vomiting sickness*, caused by ingestion of unripened *ackee* fruit, which contains *hypoglycin A*, an inhibitor of β-oxidation.
3. Antibiotics such as puromycin that inhibit apoprotein synthesis.
4. Orotic acid that interferes with glycosylation in the Golgi apparatus.
5. Carbon tetrachloride that interferes with lipoprotein metabolism by
   a. blocking apoprotein synthesis
   b. conjugation of lipid with apoprotein
   c. lipoprotein secretion.
6. *Others*: Diabetes mellitus, Cushing's syndrome, and obesity.

❑ What are lipotropic factors?

A lipotropic factor is one that prevents or reverses the development of fatty liver. Examples include choline and methionine. Choline, which is synthesized using the labile methyl groups of methionine, is required for the formation of phospholipids that are components of lipoproteins.

❑ What is atherosclerosis?

Atherosclerosis is the term used for thickening and hardening of the intimal layers of the walls of the larger arteries (abdominal aorta, coronary and cerebral arteries). Atherosclerotic lesions are classified into *fatty streaks, fibrous plaques*, and *complicated lesions.*

❑ Give a brief account of the metabolism of the artery wall.

The artery wall is metabolically active. The cells of the arterial wall

- Synthesize collagen, elastin, and glycosaminoglycans
- Metabolize glucose
- Synthesize fatty acids, phospholipids, triglycerides, and cholesterol
- Facilitate the entry of plasma lipoproteins by receptor-mediated endocytosis and lipoprotein cholesterol by receptor-independent pathways, with utilization of lipids and accumulation of cholesterol ester.

❑ Enumerate the risk factors implicated in the pathogenesis of atherosclerosis.

The risk factors associated with atherosclerosis include the following:

1. ***Ageing*** that causes changes in the arterial wall such as thickening and progressive accumulation of cholesterol ester (chiefly cholesterol linoleate) and phospholipids (mainly sphingomyelin).
2. ***Male gender***, men are at higher risk than premenopausal women who are protected by higher levels of HDL.
3. ***Family history*** of IHD due to inheritance of risk factors such as hypertension, diabetes mellitus, hyperlipidemia, etc.
4. ***Diet*** rich in saturated fat, cholesterol, and total calories, which are associated with hyperlipidemia, hypertension, obesity, and diabetes mellitus.
5. ***Hypertension*** that causes injury *via* mechanical stress on the endothelial cells, altering endothelial permeability and enhancing lysosomal enzyme activity.
6. ***Hyperlipidemia*** that causes injury to the endothelium and atheromatous lesions. The plasma lipid levels considered to be hyperlipidemic include
   a. triglyceride concentration greater than 2.8 mmol/L (250 mg/dL)
   b. total plasma cholesterol higher than 6.2 mmol/L (240 mg/dL)
   c. plasma LDL cholesterol levels >160 mg/dL. High plasma LDL facilitates LDL entry into the artery wall. Oxidized LDL is more accessible to arterial wall macrophages and causes formation of foam cells.
   d. plasma HDL concentration <35 mg/dL predisposes to atherosclerosis. HDL stimulates cholesterol efflux from arterial wall cells. Estrogens, exercise, and moderate alcohol intake increase HDL, while androgens and cigarette smoking reduce HDL cholesterol.
7. ***Hyperglycemia in diabetes mellitus*** that can cause atherogenesis by the following mechanisms:
   a. accumulation of sorbitol (formed by the polyol pathway), which retains water and decreases oxygenation in the arterial wall.

   b. glycosylation of proteins and apolipoproteins with disruption of normal arterial wall cell function. For example, glycosylated LDL undergoes oxidation and delivers cholesterol more efficiently than native LDL.
8. ***Abdominal obesity*** that causes insulin resistance, hyperinsulinemia, increased hepatic VLDL synthesis and plasma triglycerides and cholesterol.
9. ***Hyperinsulinemia***, which causes
   a. increased endogenous lipid synthesis
   b. proliferation of arterial smooth muscle cells
   c. enhanced binding of LDL and VLDL with decreased binding of HDL to cells
10. ***Smoking,*** which causes injury to the endothelial cells due to
    a. nicotine which can be converted to cholesterol
    b. formation of carboxyhemoglobin that decreases oxygen availability to tissues, impairs LDL catabolism causing LDL accumulation in arterial wall cells.
11. ***Other factors*** such as sedentary life style, stress, and anxiety.

❑ Elucidate the management of atherosclerosis.

Prevention of atherosclerosis can be achieved by risk factor reversal, which involves management by diet and drugs.

- ***Dietary management*** by
  - Reduced intake of saturated fat, cholesterol, total calories, and animal fats. Substitution of vegetable oils containing *polyunsaturated fatty acids* (*PUFA*) is beneficial.
  - Avoidance of alcohol if hypertriglyceridemia is present; in the absence of hypertriglyceridemia, alcohol in moderate amounts is beneficial.

    Red wine, which contains resveratrol and other antioxidants, has been shown to reduce the risk of IHD.
  - Regular physical exercise as an adjunct to dietary management.
- ***Drug treatment*** when LDL cholesterol exceeds 4.1 mmol/L (160 mg/dL). The drugs include
  - ***Fibric acid derivatives,*** such as clofibrate, decrease VLDL synthesis and increase lipoprotein lipase activity.
  - ***Nicotinic acid***, lowers VLDL and LDL, and elevates HDL cholesterol.
  - ***Bile acid-binding resins*** such as cholestyramine increase sterol excretion, LDL catabolism, and decrease LDL cholesterol.
  - ***HMG CoA reductase inhibitors*** such as lovastatin block HMG CoA reductase, and reduce LDL cholesterol levels by upregulation of LDL receptors.
  - ***Estrogens*** such as estradiol are useful in postmenopausal women with hypercholesterolemia.

❑ Give an account of sphingolipidoses.

- *Sphingolipidoses* (lipid storage diseases, lipidoses) are a group of inherited lysosomal storage diseases that are progressive and fatal.
- These are characterized by
  - accumulation of complex ceramide-containing lipids due to a deficiency of the lysosomal enzyme responsible for degrading the lipid.
  - clinical symptoms that result from damage caused by the stored lipid, which include neurological dysfunction, skeletal abnormalities, and visceromegaly.
  - infantile, juvenile and adult forms.

- ***Diagnosis***: Enzyme assay on fibroblasts, serum, and leukocytes.

Types: The different types of sphingolipidoses and the corresponding enzyme defect are summarized in Table 7.1.

Table 7.1 Enzyme defect in different sphingolipidoses

Sphingolipidosis	*Enzyme defect*
GM1 gangliosidosis	β-Galactosidase
GM2 gangliosidosis	
Tay-Sachs disease	Hexosaminidase A
Sandhoff disease	Hexosaminidase A and B
AB variant of GM2 gangliosidosis	Protein factor required for hexosaminidase activity
Krabbe galactosylceramide lipidosis	Galactosylceramide β-galactosidase
Metachromatic leukodystrophy	Arylsulphatase A or cerebroside sulphatase
Niemann Pick disease	Sphingomyelinase
Gaucher disease	β-Glucocerebrosidase
Fabry disease	α-Galactosidase A
Farber's disease	Ceramidase
Fucosidosis	α-Fucosidase

❑ What is GM1 gangliosidosis?

- ***Inheritance***: Autosomal recessive trait
- ***Enzyme defect***: *β-Galactosidase*
- ***Clinical symptoms***:
  - *Infantile form*: Mental retardation, seizures, coarse facies, early blinding, retinal cherry red spots, corneal clouding and skeletal dysplasia.
  - *Juvenile form*: Neurological impairment, seizures, and milder skeletal and ocular abnormalities.
  - *Adult form*: Less severe symptoms, including spasticity, ataxia, and minimal bony abnormalities.

❑ Give an account of GM2 gangliosidosis

- ***Enzyme defect***: *Hexosaminidase*
- ***Variants***: *Tay-Sachs disease, Sandhoff disease, AB variant of GM2 gangliosidosis*
- ***Tay-Sachs disease***
  - *Enzyme defect*: *Hexosaminidase A.* Ashkenazi Jews are at high risk.
  - *Clinical symptoms*
    i. *Infantile form*: Delayed development, progressive neurological deterioration, seizures, cherry red spots, and augmented startle response to sound.
    ii. *Juvenile-onset type*: Dementia, seizures, and ocular abnormalities.
    iii. *Adult form*: Slowly progressive with ataxia, lower motor neuron disease, and recurrent psychoses.
- ***Sandhoff disease***
  - *Enzyme defect*: *Hexosaminidase A and B*
  - *Clinical symptoms*: Mental retardation, seizures, and blindness.

- ***AB variant of GM2 gangliosidosis***, due to a deficiency of a protein factor required for the activity of the enzyme.

❑ Elucidate the enzyme defect and clinical findings in Krabbe galactosylceramide lipidosis and metachromatic leukodystrophy.

- ***Krabbe galactosylceramide lipidosis***
  - ***Enzyme defect***: *Galactosylceramide β-galactosidase*. Only the infantile form has been recognized.
  - ***Clinical symptoms***: Neurological deterioration due to storage of globoid cells in the brain. Death occurs within 1 – 2 years of onset
- ***Metachromatic leukodystrophy***
  - ***Enzyme defect***: *Arylsulphatase A* or *cerebroside sulphatase*
  - ***Clinical symptoms***: Abnormalities in gait, hypotonia, ataxia, psychosis, and dementia.

❑ Discuss the different types of Niemann Pick disease.

There are three types of Niemann Pick diseases:

- *Type A disease*, which is due to a deficiency of *sphingomyelinase*, and is characterized by failure to thrive, hepatosplenomegaly, neurological impairment, and retinal cherry red spots.
- *Type B disease*, also due to deficiency of *sphingomyelinase*, is more benign and is characterized by hepatosplenomegaly without neurological symptoms.
- *Type C disease*, is not associated with deficiency of sphingomyelinase. It is characterized by massive accumulation of cholesterol in lysosomes and progressive neurologic deterioration in childhood.

❑ Outline the enzyme defect, clinical features, and management of Gaucher disease.

- ***Enzyme defect***: *β-Glucocerebrosidase* (*β-glucosidase*) that results in glucosylceramide lipidosis.
- ***Clinical symptoms***:
  - *Infantile form*: Hepatosplenomegaly and neurologic deterioration leading to early death.
  - *Juvenile form*: Milder symptoms.
  - *Adult Gaucher* (also known as type I): Non-neuronopathic form prevalent among Ashkenazi Jews (1 in 1000 births) that presents with splenomegaly, anemia, thrombocytopenia, leukopenia, bone pain, pathological fractures, hepatic dysfunction, and pulmonary hypertension.
- ***Management***:
  - Partial or total splenectomy
  - Enzyme replacement therapy with *aglucerase*, a modified form of *β-glucosidase* prepared from human placenta.

❑ Write a short note on Fabry disease.

- ***Inheritance***: X-linked disorder.
- ***Enzyme defect***: *α-Galactosidase A* leading to accumulation of the trihexoside, galactosylgalactosylglucosylceramide.

- ***Clinical symptoms*** such as:
  - Painful neuropathy
  - Angiokeratoma
  - Decreased sweating
  - Corneal opacity
  - *Cardiovascular manifestations*: Cardiac arrhythmias, lipid deposition in the myocardium resulting in myocardial infarction
  - *Renal manifestations*: Lipid deposition in the kidney with progressive renal impairment and renal failure.

The symptoms are severe in hemizygous males and milder in heterozygous females.

- ***Management***
  - Administration of phenytoin for painful neuropathy
  - Dialysis or transplantation for renal complications.

❑ Briefly explain the abnormality in Farber's disease and fucosidosis.

Farber's disease is characterized by
- deficiency of *ceramidase*
- clinical manifestations such as mental retardation, dermatitis, and skeletal abnormalities.

Fucosidosis is characterized by
- deficiency of *α-fucosidase*
- accumulation of glycolipids, glycopeptides, and oligosaccharides
- clinical symptoms such as mental retardation, thick skin, and muscle spasticity.

Chapter 8

Metabolism of Amino Acids and Nonprotein Nitrogen

❑ Briefly explain protein digestion in the stomach and intestine.

- Normal protein intake: Approximately 1 g/kg body weight, which forms 11% of the energy requirements.
- Proteins derived from the diet, digestive secretions, and desquammated cells undergo proteolytic digestion to form amino acids.

Protein digestion in the stomach

- Protein digestion begins in the stomach by the action of *pepsin* secreted by the chief cells of the gastric mucosa as the inactive zymogen *pepsinogen.*
- Conversion of pepsinogen to pepsin
  a. involves removal of 42 amino acids from the N-terminal end
  b. requires pepsin (*autocatalysis*) and gastric HCl.
- Cleavage sites of pepsin are the peptide bonds in which the carboxyl group is contributed by the aromatic amino acids (phe, tyr, trp) and leucine.

Protein digestion in the intestine

- Partially digested material from the stomach is acted upon by pancreatic secretion containing:
  a. bicarbonate, which neutralizes acidic chyme from the stomach and provides a slightly alkaline pH optimum for pancreatic enzyme activity.
  b. proteolytic enzymes (present as zymogens) such as
     - *endopeptidases* (trypsin, chymotrypsin, and elastase)
     - *exopeptidases* (carboxypeptidases A and B).
- Trypsinogen is converted to trypsin by:
  a. *enteropeptidase* (*enterokinase*) present in the intestinal brush border membrane
  b. autocatalysis.
- Trypsin converts all other zymogens to their active forms.
- *Proteases* produced by the intestinal epithelial cells complete the digestion of dietary proteins to amino acids.
- *Aminopeptidases* are exopeptidases that remove amino acids from the N-terminal end of peptides, residue by residue.
- *Dipeptidases* and *tripeptidases* act on dipeptides, and tripeptides respectively, to form amino acids.

❑ How are amino acids absorbed?

Amino acids, the end products of protein digestion, are absorbed by

- The *Na^+ symport system*, a major transport system for amino acids from the intestinal lumen into the blood. Transport is catalyzed by *Na^+K^+ATPase* and involves ATP hydrolysis.
- The *γ-glutamyl cycle* that involves
  - reaction of glutathione with an amino acid in the intestinal lumen to form γ-glutamyl amino acid that is transported across the cell membrane to the cytoplasm with the release of the amino acid.
  - the γ-glutamyl moiety, is used to resynthesize glutathione.
- *Na^+-independent carriers*: These transport neutral and lipophilic amino acids (phenylalanine and leucine) and cationic amino acids (lysine).

❑ Give a brief overview of the metabolism of amino acids.

Metabolism of amino acids involves the following:

- ***Metabolism of amino acid nitrogen***
  - during biosynthesis, the amino acid nitrogen is added to the carbon skeletons and removed during catabolism.
  - the amino acid nitrogen is converted to urea in mammals.
- ***Metabolism of the carbon skeletons***
  - biosynthesis of the non-essential amino acids involves obtaining the primary amino groups and the carbon skeletons.
  - the primary amino groups of all the amino acids are obtained by transamination of glutamate.
  - the carbon skeletons are derived from glycolysis (gly, ala, ser, and cys) or the citric acid cycle (asp, asn, glu, gln, pro, and arg).
  - the amino acid tyrosine is derived by hydroxylation of phenylalanine.

❑ What are the stages in the conversion of amino acid nitrogen to urea?

There are four stages in the conversion of amino acid nitrogen to urea. These include

- transamination
- oxidative deamination
- ammmonia transport
- urea cycle.

❑ What is transamination?

- Transamination involves interconversion of a pair of amino acids and a pair of keto acids. It is catalyzed by *transaminases* (*aminotransferases*) with pyridoxal phosphate (PLP) as the coenzyme (Fig. 8.1).

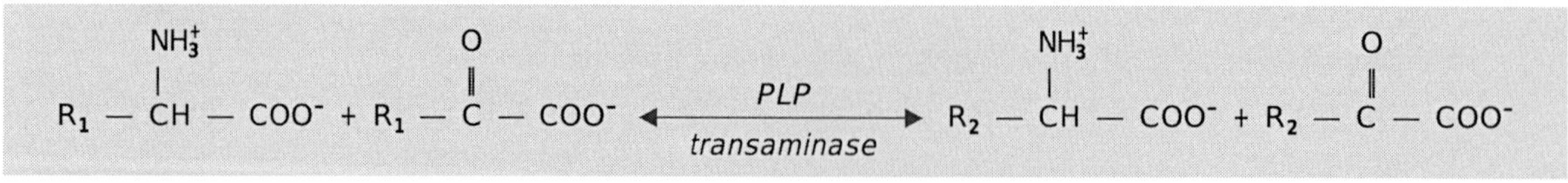

Fig. 8.1 Transamination reaction

- PLP serves as a carrier of amino groups and is involved in the formation of an enzyme-bound Schiff base intermediate.

- Amino acids undergo transamination to form glutamate because L-glutamate is the only amino acid that can undergo oxidative deamination efficiently.
- The amino acids that do not undergo transamination include lysine, threonine, proline, and hydroxyproline.

❑ What is the fate of glutamate formed by transamination?

- Glutamate formed by transamination undergoes *oxidative deamination* catalyzed by *glutamate dehydrogenase* (*GDH*) (Fig. 8.2).

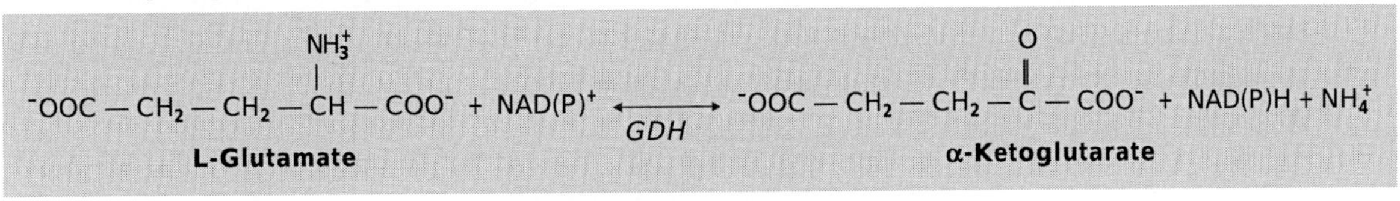

Fig. 8.2 Oxidative deamination

- GDH
  - is ubiquitous in all mammalian tissues
  - uses either NAD^+ or $NADP^+$ as oxidant
  - in the liver is an allosteric enzyme that is activated by ADP and inhibited by ATP, GTP, and NADH. It participates in both amino acid catabolism and biosynthesis.

❑ Explain the role of L-amino acid oxidases.

- L-amino acid oxidases are involved in the removal of NH_3 from α-amino acids.
- These enzymes are auto-oxidizable flavoproteins present in the mammalian liver and kidney that convert amino acids to keto acids *via* an imino acid intermediate with liberation of ammonium ion.
- The reduced flavin is reoxidized by molecular oxygen to H_2O_2, which is hydrolysed to O_2 and H_2O by the enzyme *catalase* (Fig. 8.3).

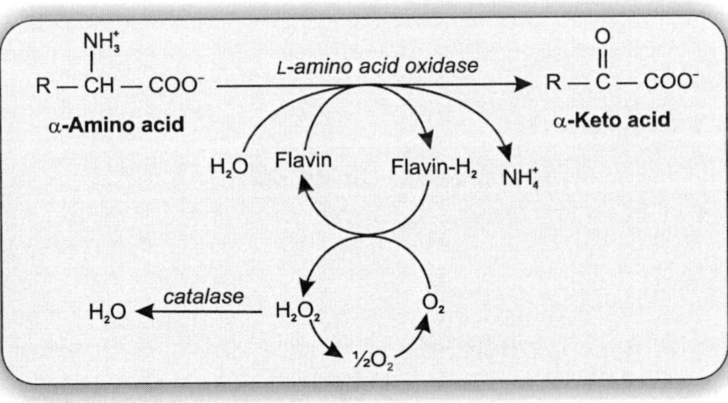

Fig. 8.3 Reaction catalyzed by L-amino acid oxidases

❑ How is ammonia transported to the liver for the biosynthesis of urea?

- Ammonia formed by transamination and oxidative deamination is highly toxic.
- It is removed from the blood by the liver by conversion to glutamine and glutamate.

- The enzymes *glutamine synthetase* and *glutaminase* catalyze the interconversion of free ammonium ion and glutamine (Fig. 8.4).

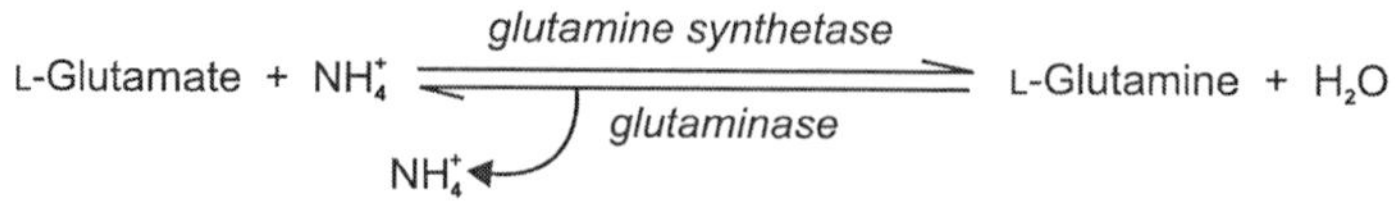

Fig. 8.4 Interconversion of glutamate and glutamine

- Glutamate is converted to glutamine by *glutamine synthetase*, a mitochondrial enzyme present in high concentration in the kidney.
- Ammonia is detoxified in the brain by conversion to glutamine.
- The brain forms glutamate from α-ketoglutarate.
- Removal of the amide nitrogen of glutamine is catalyzed by the enzyme *glutaminase*.

Explain the reactions of the urea cycle.

- Urea is the end product of protein metabolism.
- *Site of synthesis*: Predominantly in the liver
- *Subcellular site*: Mitochondria and cytosol
- ***Steps***
  1. Formation of carbamoylphosphate from ammonia formed by the deamination of amino acids and in a reaction catalyzed by *carbamoyl phosphate synthase* (CPS-I). This is the rate-limiting step. The reaction requires two molecules of ATP and N-acetylglutamate as an allosteric activator.
  2. Transfer of the carbamoyl moiety from carbamoyl phosphate to ornithine to form citrulline in a reaction catalyzed by *ornithine transcarbamoylase*.
  3. Condensation of citrulline with aspartate to form argininosuccinate coupled to ATP hydrolysis and catalyzed by *argininosuccinate synthase*.
  4. Cleavage of argininosuccinate by *argininosuccinase* to form arginine and fumarate.
  5. Cleavage of arginine by *arginase* to regenerate ornithine and release urea.

Figure 8.5 shows the reactions of the urea cycle.

What is the fate of urea?

1. Urea synthesized by the liver enters the circulation.
2. It is completely filtered by the renal glomeruli and excreted in the urine.
3. About 30 g of urea is excreted per day in the urine. The excretion of urea is decreased in acute renal impairment.

Briefly describe the disorders of the urea cycle.

Inherited disorders in the 5 enzymes involved in the urea cycle lead to ammonia intoxication. The clinical symptoms include

- decreased appetite
- lethargy
- mental retardation
- coma, and death.
- vomiting
- hypotonia
- convulsions

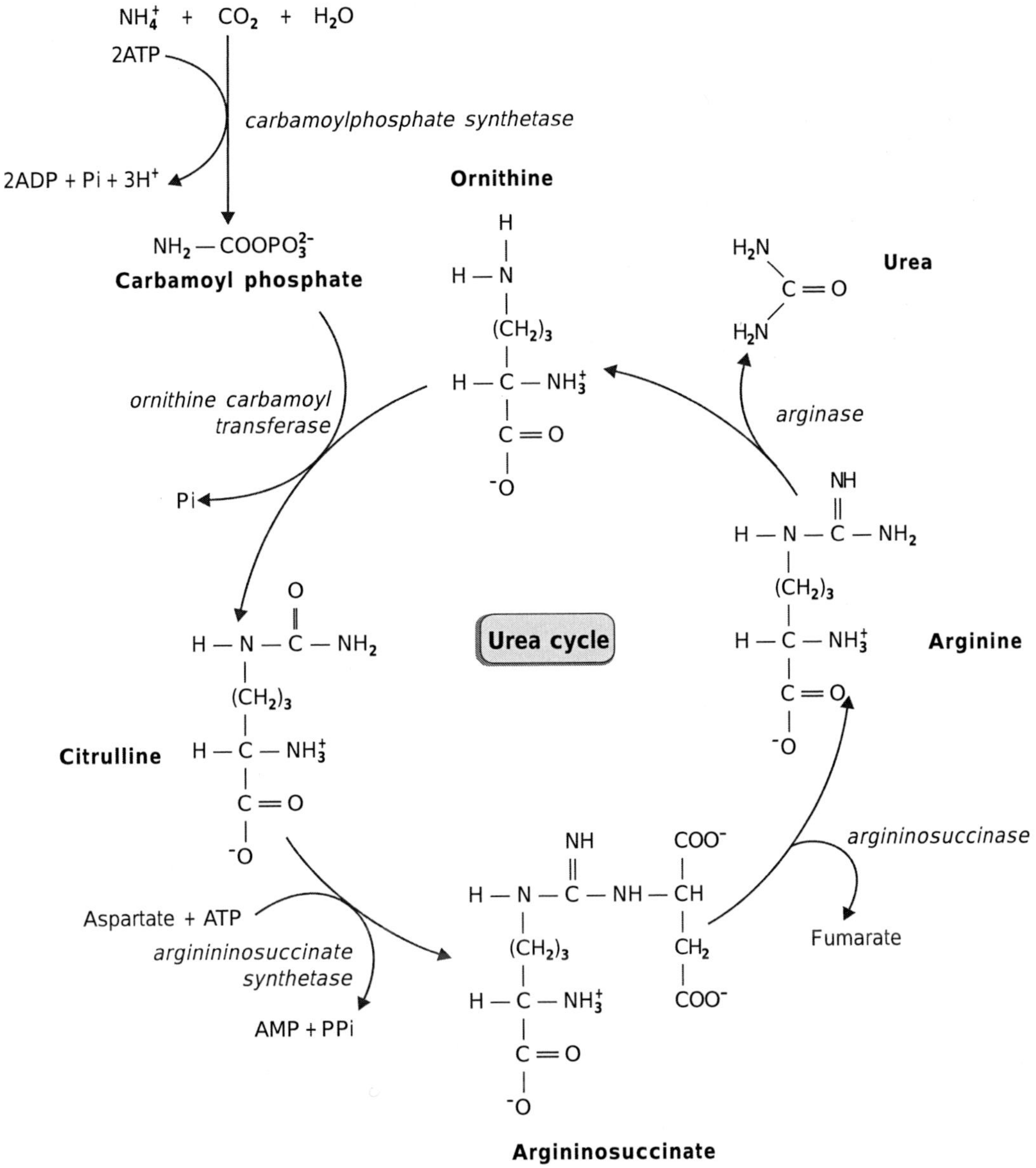

Fig. 8.5 The urea cycle

❑ Enumerate the causes of uremia.

- Normal plasma urea nitrogen level: 8–25 mg/dL (2.9–8.9 mmol/L)
- Plasma urea levels are increased due to a high protein diet.
- *Uremia* refers to high plasma urea concentration.
- Based on the underlying cause, it is classified into prerenal, renal, and postrenal uremia.
- ***Prerenal uremia*** results from a decrease in plasma volume due to depletion of salt and water. This leads to a decrease in blood pressure with reduced renal blood flow and glomerular filtration rate (GFR), and retention of urea.

Prerenal uremia is seen in the following conditions:

- Addisonian crisis
- Congestive cardiac failure
- Dehydration as in vomiting and diarrhea
- Diabetic ketoacidosis
- Hematemesis
- Hemorrhage into the gastrointestinal tract with digestion of protein, absorption of the amino acids, and deamination in the liver
- Hypercatabolic states such as fever, wasting diseases, and trauma
- Shock due to burns, hemorrhage, or toxemia of pregnancy.

- ***Renal uremia*** is associated with increase in plasma urea levels. In renal failure, renal plasma flow is decreased with decrease in GFR and increase in plasma urea together with plasma creatinine (*azotemia*).
- ***Postrenal uremia*** occurs due to obstruction to urine flow, which is seen in the following conditions:
  - benign prostatic hypertrophy
  - urethral stricture
  - urinary calculi
  - malignant tumors of the bladder.

❑ How are amino acids biosynthesized?

- Protein biosynthesis requires all the 20 amino acids.
- While microorganisms and plants synthesize all the 20 amino acids found in proteins, mammals are capable of synthesizing only the 11 *nutritionally non-essential amino acids*.
- Biosynthesis of the non-essential amino acids involves obtaining
  - the primary amino groups by transamination of glutamate
  - the carbon skeletons derived from
    i. glycolysis (gly, ala, ser, and cys)
    ii. the citric acid cycle (asp, asn, glu, gln, pro, and arg).

Tyrosine is formed by hydroxylation of phenylalanine.

Amino acids obtained from glycolysis

The amino acids obtained from glycolysis include serine, glycine, cysteine, and alanine (Fig. 8.6).

1. Oxidation of 3-phosphoglycerate, an intermediate in the glycolytic pathway, followed by transamination and dephosphorylation gives serine.
2. Transfer of a methylene group from serine to tetrahydrofolate yields glycine.
3. Methylation of serine by S-adenosylmethionine (SAM) as methyl group donor forms cysteine.
4. Transamination of pyruvate generates alanine.

Amino acids formed by TCA cycle intermediates

These include aspartate, asparagine, glutamate, glutamine, proline, and arginine (Fig. 8.7).

1. Transamination of oxaloacetate gives aspartate.
2. Amidation of aspartate forms asparagine.
3. Transamination or oxidative deamination of α-ketoglutarate produces glutamate.
4. Amidation of glutamate generates glutamine.

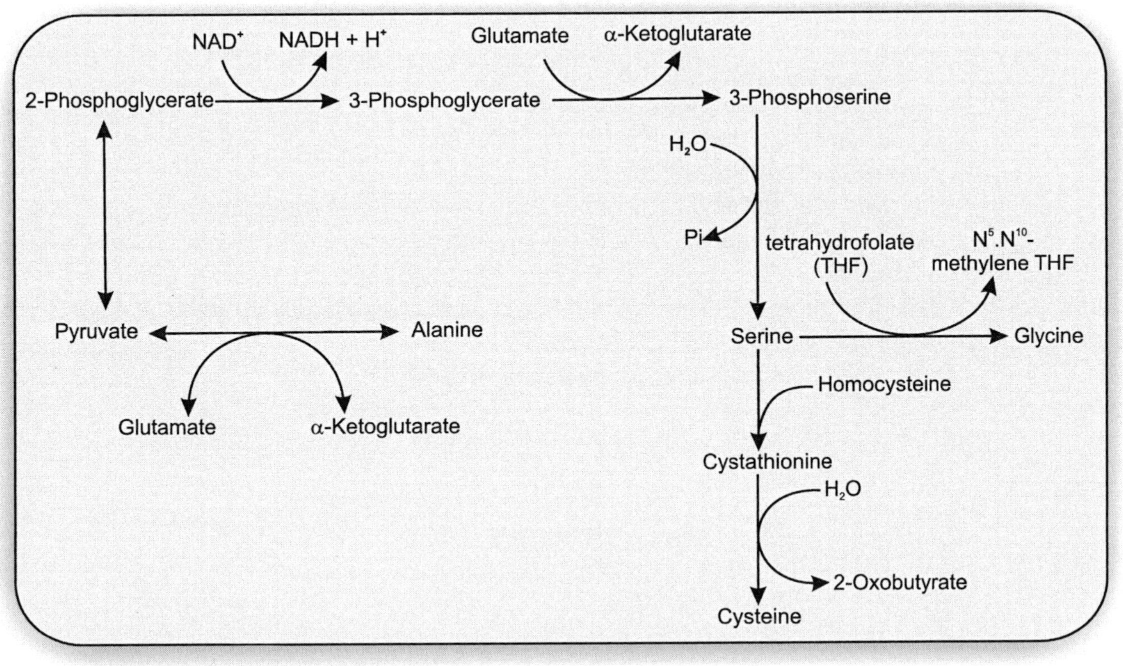

Fig. 8.6 Amino acids obtained from glycolysis

5. Reduction of glutamate to glutamate semialdehyde (GSA) followed by cyclization forms proline.
6. Ornithine formed from GSA by transamination is hydrolysed by arginase to form arginine.

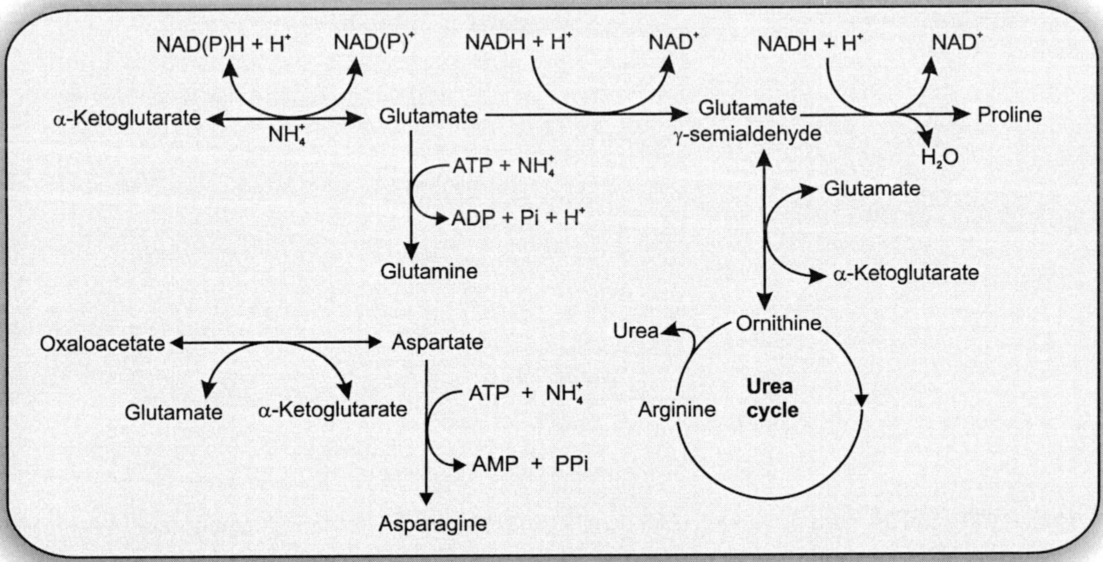

Fig. 8.7 Amino acids obtained from TCA cycle intermediates

❑ ***Discuss the catabolism of the carbon skeletons of amino acids.***

The carbon skeletons of the 20 amino acids are catabolized to pyruvate or TCA cycle intermediates.

- ***Amino acids converted to pyruvate*** include serine, glycine, cysteine, and alanine (Fig. 8.8).
  1. Serine is converted to pyruvate and NH_4^+ in a reaction catalyzed by *serine dehydratase.*
  2. Glycine has three metabolic fates:
     a. Formation of serine on reaction with methylene tetrahydrofolate.
     b. Formation of glyoxylate, which is converted to oxalate, or degraded to carbon dioxide and water.
     c. Degradation to CO_2 and NH_4^+ in a reaction requiring tetrahydrofolate and NAD^+.
  3. Cysteine on desulphuration gives pyruvate.
  4. Alanine undergoes transamination to form pyruvate.

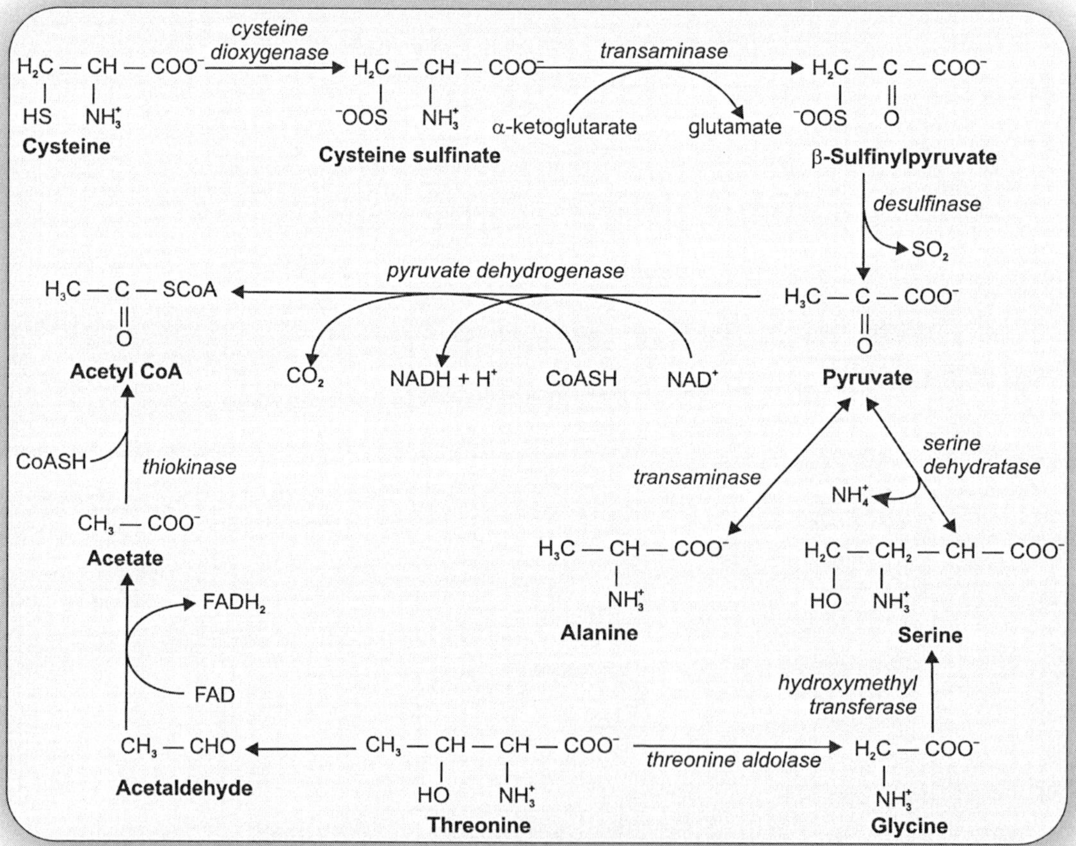

Fig. 8.8 Catabolism of amino acids to pyruvate

- ***Amino acids converted to TCA cycle intermediates.*** These can be categorized into four groups (Fig. 8.9):
  1. *Amino acids converted to α-ketoglutarate* include glutamine, glutamate, proline, arginine, and histidine.

a. Glutamate formed by *glutaminase* from glutamine undergoes transamination or deamination to α-ketoglutarate.
b. Proline forms α-ketoglutarate after oxidation to glutamate.
c. Arginine forms ornithine, which undergoes transamination to GSA that is oxidized to glutamate.
d. Histidine is converted to formiminoglutamate (figlu), which forms glutamate by transfer of the formimino group to tetrahydrofolate.

2. *Amino acids converted to succinyl CoA* include threonine, methionine, valine, and isoleucine. These amino acids form succinyl CoA via the intermediates propionyl CoA and methylmalonyl CoA. Threonine forms α-ketobutyrate that undergoes oxidative decarboxylation to propionyl CoA.
3. *Amino acids converted to oxaloacetate* include aspartate and asparagine. Asparagine forms aspartate, which undergoes transamination to oxaloacetate.
4. *Amino acids converted to fumarate* include phenylalanine and tyrosine. Phenylalanine undergoes hydroxylation to form tyrosine, which is catabolized eventually to fumarate and acetoacetate.

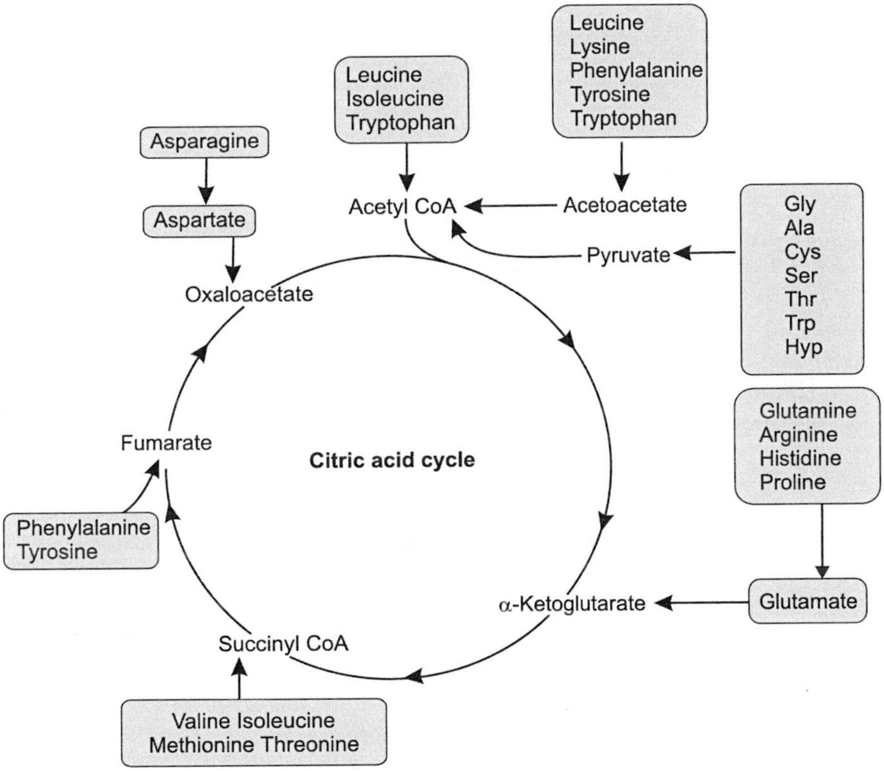

Fig. 8.9 Catabolism of amino acids to TCA cycle intermediates

- ***Amino acids converted to acetyl CoA or acetoacetate*** include
  1. Threonine, isoleucine, lysine, and tryptophan (acetyl CoA)
  2. Phenylalanine and tyrosine (acetoacetate)
  3. Leucine (acetyl CoA and acetoacetate).

❑ What are glucogenic amino acids?

- Amino acids that are converted to compounds that can synthesize glucose by gluconeogenesis (pyruvate, α-ketoglutarate, succinyl CoA, oxaloacetate, and fumarate) are termed *glucogenic* amino acids.
- Fourteen amino acids other than leucine, lysine, phenylalanine, tyrosine, tryptophan, and isoleucine are purely glucogenic.

❑ What are ketogenic amino acids?

- Amino acids that are catabolized to acetoacetate and acetyl CoA that form ketone bodies, are termed *ketogenic* amino acids.
- Only leucine and lysine are solely ketogenic.
- Phenylalanine, tyrosine, tryptophan and isoleucine are both glucogenic and ketogenic.

❑ Briefly explain the metabolism of phenylalanine and tyrosine.

Metabolism of phenylalanine involves the following steps:

1. Conversion of phenylalanine to tyrosine by the enzyme *phenylalanine hydroxylase*, a mixed function oxygenase present in the mammalian liver. The reaction involves incorporation of one atom of molecular oxygen into the para position of phenylalanine, and reduction of the other atom to water. Tetrahydrobiopterin, which functions as a cofactor, is maintained in the active tetrahydro form by NADPH-dependent *dihydropteridine reductase*.
2. Transamination of tyrosine to *p*-hydroxyphenylpyruvate.
3. Oxidation of *p*-hydroxyphenylpyruvate to homogentisate with ascorbate as a coenzyme.
4. Cleavage of homogentisate by *homogentisate oxidase* to generate fumarate and acetoacetate.

Tyrosine is also metabolized by *tyrosine hydroxylase* and *tyrosinase* to form DOPA, melanin, and catecholamines.

❑ Discuss the biochemical lesion, clinical manifestations, diagnosis, and management of classic phenylketonuria.

Classic phenylketonuria is also known as type I phenylketonuria (PKU).
Inheritance: Autosomal recessive with a frequency of 1 in 10,000 births.

- ***Biochemical lesion***: Absence of *phenylalanine hydroxylase*
  Consequence: Due to the enzyme block, phenylalanine is not converted to tyrosine. It accumulates in circulation and is channelled to alternative catabolic pathways to form phenylpyruvate, phenyllactate, and phenylacetate (Fig. 8.10). Phenylalanine and its metabolites are excreted in the urine. Phenylacetate is conjugated with glutamine in the liver and is excreted as phenylacetylglutamine in the urine.
- ***Clinical manifestations***
  1. Impaired growth
  2. Intellectual impairment, mental retardation develops between 4 and 6 months
  3. *Neurological manifestations*: Psychomotor irritability, tremors, and convulsions during the first few weeks of life
  4. Feeding problems and vomiting
  5. *Dermatological manifestations*:
     a. Reduced pigmentation due to inhibition of melanin synthesis. The patients have pale skin, fair hair, and blue eyes.

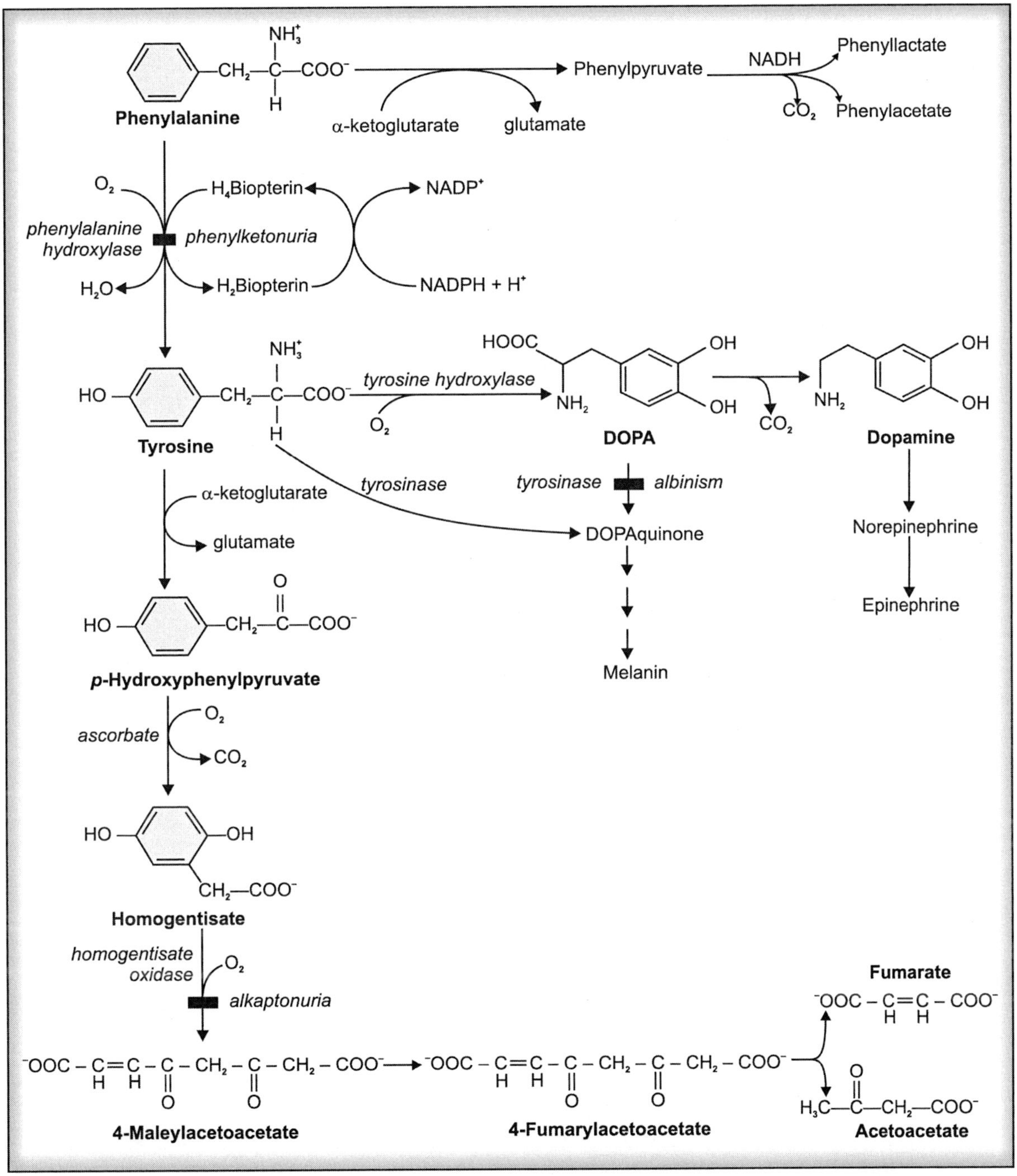

Fig. 8.10 Metabolism of phenylalanine and tyrosine

b. Rashes
c. Inflammation
d. Generalized eczema.

6. Mousy odor of skin, hair, and urine due to phenylacetate accumulation.

- ***Diagnostic tests***
  1. Determination of phenylalanine concentration in blood obtained from a heel prick 6–10 days after birth.
  2. Estimation of urine phenylpyruvate.
  3. Plasma tyrosine estimation.
  4. Mass screening of infants 4–6 days after birth by microbiological *Guthrie test* using a strain of *Bacillus subtilis* that grows only under conditions of excess phenylalanine.
  5. Analysis of the gene for phenylalanine hydroxylase for prenatal diagnosis.
- ***Management***
  1. Dietary restriction of phenylalanine
  2. Dietary supplementation with tyrosine.

❑ Briefly explain the different types of tyrosinemia.

Tyrosinemia Type I

- ***Enzyme defect***: Deficiency of *fumarylacetoacetate hydrolase* and/or *maleylacetoacetate hydrolase*
- ***Biochemical findings***
  - High plasma and urine tyrosine levels
  - High plasma methionine levels
  - Increased urinary excretion of dihydroxyphenylalanine (DOPA) and other tyrosine metabolites
  - Increase in platelet count
  - Prolonged prothrombin time
  - High α-fetoprotein levels.
- ***Clinical manifestations***

  In the acute form:
  - Failure to thrive
  - *Gastrointestinal manifestations*: diarrhea, vomiting, hepatosplenomegaly
  - *"Cabbage-like"* odor
  - *Renal manifestations*: nephropathy with Fanconi syndrome
  - In untreated cases, death from hepatic failure in 6 – 8 months.

  In the chronic form. The symptoms are milder and death may occur by age ten.
- ***Management***: By diet low in tyrosine, phenylalanine, and methionine.

Tyrosinemia Type II (Richner–Hanhart Syndrome)

- ***Enzyme defect***: Deficiency of hepatic *tyrosine transaminase* that converts tyrosine to *p*-hydroxyphenylpyruvate.
- ***Biochemical findings***:
  - Increased levels of tyrosine in plasma and urine
  - Increased urinary excretion of N-acetyltyrosine, tyramine, and *p*-hydroxyphenylacetate.
- ***Clinical manifestations*** include
  - *Ocular lesions*: Keratitis and conjunctivitis
  - *Skin lesions*: Hyperkeratosis
  - *Neurological symptoms*: Self-mutilation, abnormalities in fine co-ordination, and moderate mental retardation.
- ***Management***: Restriction of tyrosine and phenylalanine in the diet.

Neonatal Tyrosinemia

- ***Enzyme defect***: Deficiency of *p-hydroxyphenylpyruvate hydroxylase* that converts *p*-hydroxyphenylpyruvate to homogentisate.
- ***Biochemical findings***:
  - Increased plasma phenylalanine and tyrosine
  - Increased excretion of tyrosine, *p*-hydroxyphenylpyruvate, *p*-hydroxyphenylacetate, *p*-hydroxyphenyllactate, tyramine, and N-acetyltyrosine in the urine.
- ***Management***: Intake of ascorbic acid, a cofactor for the reaction.

❑ Give an account of the enzyme defect, clinical features, diagnosis, and management of alkaptonuria.

- ***Inheritance***: Autosomal recessive with a frequency of 1 in 2 – 5 million live births.
- ***Enzyme defect***: Deficiency of *homogentisate 1,2-dioxygenase* that catalyzes the conversion of homogentisic acid to maleylacetoacetic acid.
- ***Consequences***:
  - Homogentisate accumulates in cells and body fluids, with excessive excretion in the urine.
  - Formation of *alkapton*, a gray to bluish-black pigment, by oxidation of homogentisate by *polyphenol oxidase* to benzoquinone acetate which undergoes polymerization and gets deposited in connective tissues (*ochronosis*).
  - This can lead to degenerative changes in the cartilage, resulting in arthritis.
  - Conversion of homogentisate to alkapton is accelerated under alkaline conditions.
- ***Clinical manifestations*** include
  1. *Characteristic triad*
     a. Ochronotic pigmentation: there is pigmentation of heart valves, larynx, tympanic membranes, and skin
     b. Degenerative arthritis with pain and stiffness
     c. Darkening of urine on standing.
  2. Slight discoloration of the sclerae and ears (earliest symptom)
  3. Thickening of the cartilages of the ears
  4. Pigmented renal or prostatic calculi
  5. Degenerative cardiovascular disease (in later stages).
- ***Diagnostic tests***
  1. *Clinical diagnosis*: Characteristic triad of ochronotic pigmentation, degenerative arthritis, and darkening of urine on standing.
  2. *Biochemical test*: Detection of homogentisate in urine by the formation of a purple-black color on addition of ferric chloride and confirmation by chromatography, enzymatic and spectrophotometric measurements.
  3. *Radiological diagnosis*: Degeneration and calcification of intervertebral disks with narrowing of intervertebral spaces in X-rays of the lumbar spine.
- ***Management***
  1. Dietary restriction of tyrosine and phenylalanine
  2. Symptomatic treatment of arthritis
  3. Supplementation of vitamin C (required for homogentisate oxidase activity) to prevent the oxidation and polymerization of homogentisic acid.

❑ Write a short note on albinism.

- ***Inheritance***: Autosomal recessive trait
- ***Enzyme defect***: Deficiency of *tyrosinase*
- ***Types***
  1. Type I: Deficiency of *tyrosinase* in the melanosomes within the melanocytes of the epidermis
  2. Type II: *Tyrosinase* activity is normal. Absence of *permease* blocks the transport of tyrosine into the melanosomes
  3. Type III: Block in the pathway of melanin biosynthesis distal to the formation of dihydroxyphenylalanine (DOPA) quinone.
- ***Clinical manifestations***
  1. Reduced pigmentation of skin with increased susceptibility to skin cancers
  2. Decreased pigmentation of the iris resulting in photosensitivity
  3. Decreased pigmentation of the hair.

❑ How is tryptophan catabolized?

Catabolism of tryptophan involves the following reactions (Fig. 8.11):

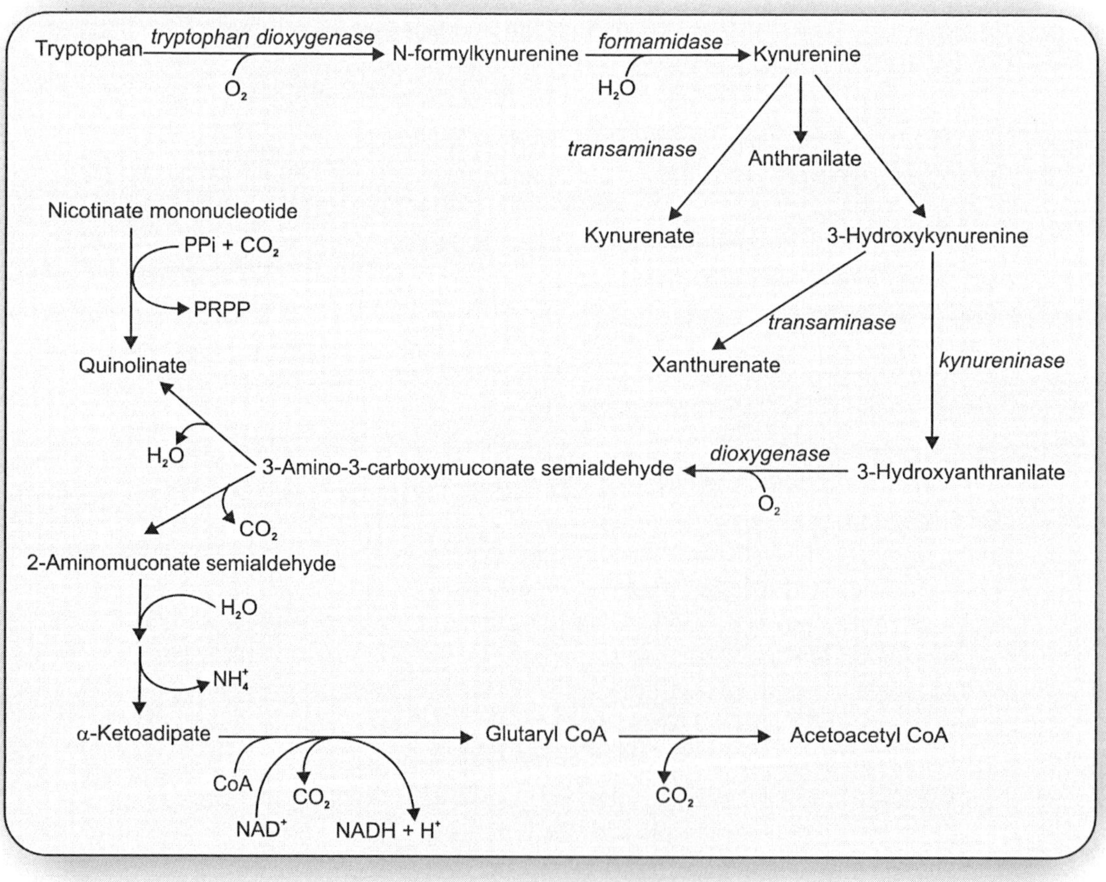

Fig. 8.11 Catabolism of tryptophan

1. Formation of kynurenine from tryptophan catalyzed by *tryptophan oxygenase*
2. Conversion of kynurenine to 3-hydroxykynurenine
3. Hydrolytic cleavage of the alanine side chain and the aromatic ring in 3-hydroxykynurenine to form 2-amino-3-carboxymuconate-6-semialdehyde (ACS)
4. Decarboxylation of ACS followed by oxidation to form 2-aminomuconate
5. Reduction of 2-aminomuconate to α-ketoadipate
6. Catabolism of α-ketoadipate to acetyl CoA
7. Spontaneous cyclization of some amount of ACS forms quinolinate, an intermediate in the biosynthesis of nicotinamide.

❑ Write a short note on Hartnup disease.

Hartnup disease is a rare, autosomal recessive disorder of tryptophan metabolism named after the first patient.

- ***Biochemical lesion***
  Defect in tryptophan transport across the intestine and the renal tubules due to a mutation of shared neutral amino acid transport protein.
- ***Consequences***
  - Conversion of unabsorbed tryptophan and indole compounds, which are absorbed by the intestine and excreted in the urine.
  - Excessive urinary loss of tryptophan leads to low plasma tryptophan levels.
  - Niacin formation from tryptophan is significantly reduced.
- ***Clinical manifestations***
  Pellagra-like symptoms develop due to niacin deficiency. The symptoms include the following
  - *Dermatological manifestations*: Rough skin and red scaly rash on exposed areas.
  - *Neurological manifestations*: Cerebellar ataxia, mild emotional lability, frank delirium, and mental confusion.
- ***Diagnosis***
  Aminoaciduria with increased excretion of leucine, isoleucine, valine, glutamine, asparagine, phenylalanine, tyrosine, tryptophan, threonine, serine, alanine, and histidine.
- ***Management***
  - High protein diet
  - Oral administration of niacin to alleviate pellagra-like symptoms.

❑ How are branched chain amino acids metabolized?

Metabolism of branched chain amino acids involves the following steps shown in Figure 8.12:

1. Transamination of all the three branched chain amino acids to form the corresponding α-keto acids.
2. Oxidative decarboxylation of α-keto acids in a reaction catalyzed by branched chain *α-keto acid dehydrogenase*, a mitochondrial multienzyme complex, to form acyl CoA thioesters.
3. Dehydrogenation of the acyl CoA thioesters to the corresponding α,β-unsaturated acyl CoA thioesters.
4. The unsaturated acyl CoA thioesters follow three different pathways to form different end products.
   a. β-Methylcrotonyl CoA from leucine forms acetoacetate and acetyl CoA.
   b. Methacrylyl CoA from valine generates succinyl CoA.
   c. Tiglyl CoA from isoleucine gives rise to propionyl CoA and acetyl CoA.

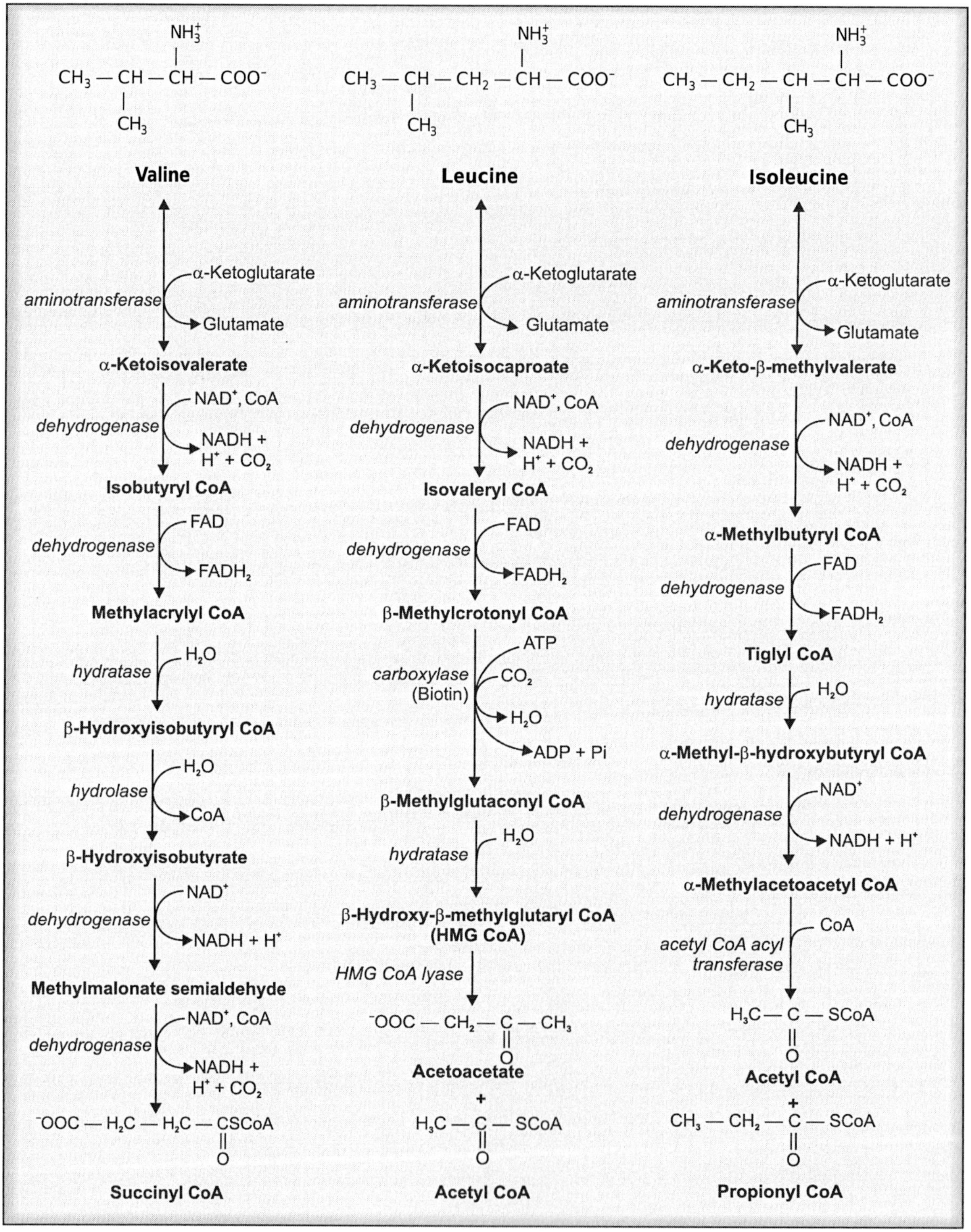

Fig. 8.12 Catabolism of branched chain amino acids

❑ Discuss the biochemical lesion, clinical features, diagnosis, and management of maple syrup urine disease (MSUD).

Maple syrup urine disease, also known as *branched-chain ketonuria*, is inherited as an autosomal recessive trait that occurs with a frequency of 5 – 10 per million live births. The disease is manifested in the first week of life.

Enzyme defect: Deficiency or absence of *α-keto acid decarboxylase* that converts keto acids resulting from deamination of the branched chain amino acids to their corresponding acyl CoA thioesters and CoA. Due to the enzyme block, branched chain amino acids and their corresponding oxo acids accumulate in plasma followed by their excretion in the urine.

Clinical manifestations include

1. Vomiting
2. Hypoglycemia and brain damage due to increased levels of leucine
3. *Neurological manifestations*: Ataxia, convulsions, impaired synthesis of proteins, lipoproteins and myelin, and progressive degeneration of the nervous system. If untreated, death occurs by the end of the first year of life.

Diagnosis

1. Increased plasma and urine levels of branched chain amino acids and their corresponding keto acids
2. Detection of enzyme defect in leukocytes and cultured skin fibroblasts.

Management: Elimination of branched chain amino acids from the diet.

❑ How is methionine metabolized?

Metabolism of methionine involves the following steps:

1. Condensation of methionine with ATP to form S-adenosylmethionine (SAM).
2. SAM donates a methyl group to an acceptor to form S-adenosylhomocysteine.
3. Hydrolysis of S-adenosylhomocysteine to form homocysteine.
4. Condensation of homocysteine with serine to form cystathionine in a reaction catalyzed by *cystathionine β-synthase*.
5. Hydrolytic cleavage of cystathionine to form cysteine and α-ketobutyrate and eventually propionyl CoA.

The pathway converts homocysteine to cystathionine and serine to cysteine, and reconverts homocysteine to methionine.

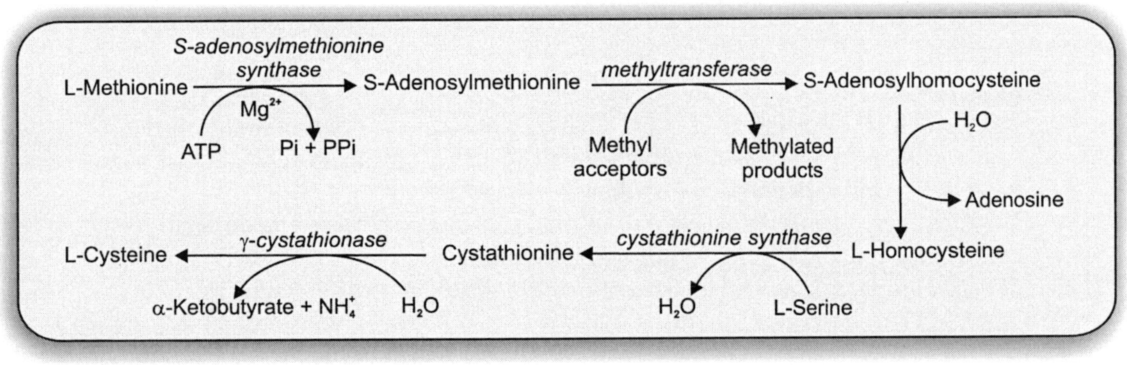

Fig. 8.13 Catabolism of methionine

❑ Give an account of homocystinurias.

The homocystinurias are disorders in the metabolism of sulphur-containing amino acids characterized by increased plasma levels of homocysteine and excessive urinary excretion.

Types

Different types of homocystinuria with distinct enzyme defects have been recognized. These include

Type I: *Cystathionine β-synthase* deficiency

Type II: *N^5, N^{10}-methylenetetrahydrofolate reductase* deficiency

Type III: Decreased activity of *N^5-methyltetrahydrofolate-homocysteine transmethylase* due to defective synthesis of methyl cobalamin

Type IV: Decreased activity of *N^5-methyltetrahydrofolate* due to reduced intestinal absorption of cobalamin.

Homocystinuria Type I

- ***Enzyme defect***: Deficiency of *cystathionine β-synthase*, a pyridoxal phosphate-dependent enzyme, that catalyzes condensation of homocysteine with serine to form cystathionine.
- ***Consequences***
  - Accumulation of homocysteine and methionine in cells and body fluids due to the enzyme block.
  - Impaired synthesis of cysteine and cystine and low plasma levels of these amino acids.
- ***Clinical manifestations***
  Homocysteine blocks cross-linking of collagen leading to the following effects:
  - Skeletal abnormalities such as osteoporosis
  - Ocular complications such as dislocated optic lenses, resulting in glaucoma and impaired visual acuity
  - Vascular complications such as thromboses, occlusion of coronary, renal, and cerebral arteries, and tissue infarction
  - Mental retardation and convulsions.
- ***Diagnosis***
  - Detection of homocysteine by
    a. the cyanide nitroprusside test
    b. 1-dimensional paper chromatography with iodoplatinate staining or by high-voltage electrophoresis.
  - Assay of enzyme activity in tissue extracts.
- ***Treatment***
  Pyridoxine treatment or methionine restricted cysteine-supplemented diet based on the patient response.

Homocystinuria Type II

- ***Enzyme defect***: Deficiency of *N^5,N^{10}-methylenetetrahydrofolate reductase* that catalyzes the formation of methylenetetrahydrofolate involved in the conversion of homocysteine to methionine. The enzyme is also required for DNA and RNA synthesis.
- ***Consequences***: Decrease in enzyme activity leads to low levels of methionine, impaired nucleic acid synthesis, and central nervous system dysfunction.
- ***Clinical manifestations*** include severe neurological abnormalities such as:
  - Cerebral atrophy
  - Retarded development
  - Mild psychiatric problems in adults.
  - Behavioral disturbances (catatonia)
  - Death in the first year of life

- ***Diagnosis***
  - Increased homocysteine levels with low levels of methionine in body fluids
  - Enzyme assay in tissues.
- ***Treatment***

  Folate administration.

Discuss the biochemical lesion, clinical manifestations, diagnosis, and management of cystinuria.

Cystinuria is an autosomal recessive disorder characterized by excessive urinary excretion of the dibasic amino acids cystine, ornithine, arginine, and lysine (COAL is a useful mnemonic aid).

- ***Biochemical lesion***
  - Defective transport of the dibasic amino acids across the renal tubules and intestinal mucosa. These amino acids may share a common transport system.
  - Increased urinary concentrations of cystine, which is relatively insoluble and causes precipitation of cystine, and formation of renal calculi.
- ***Types***
  - *Type I*: It is characterized by defective intestinal transport of cystine, ornithine, arginine, and lysine in homozygotes, and normal excretory patterns of these amino acids in heterozygotes.
  - *Type II*: The intestinal transport of lysine is absent, while cystine transport is normal. The cystine-basic aminoaciduria is lower in heterozygotes compared to homozygotes.
  - *Type III*: The intestinal transport of the four amino acids is functional. The cystine-basic aminoaciduria in heterozygotes is of a lower magnitude than that seen in Type II.
- ***Clinical manifestations***

  Recurrent urolithiasis causes
  - hematuria
  - obstructive uropathy
  - progressive renal insufficiency.
  - renal colic
  - infection
- ***Diagnosis***
  - Detection of cystine in urinary tract stones by the nitroprusside test
  - Examination of the amino acid excretion pattern by chromatography or electrophoresis.
- ***Management*** involves decreasing the urinary concentration of cystine by
  - Increasing fluid intake
  - Alkalinization of urine to enhance the solubility of cystine
  - Administration of D-penicillamine to form penicillamine-cystine mixed disulphide, which has greater solubility.

Discuss the pathogenesis, clinical features, diagnosis, and management of cystinosis.

Cystinosis is a rare, autosomal recessive disorder of cystine transport, characterized by cystine accumulation in the lysosomes, especially in the reticuloendothelial system, cornea, conjunctiva, lymph nodes, and leukocytes.

Infantile, juvenile, and adult forms have been recognized.

Clinical manifestations

- *In the infantile form*, the following are seen:
  - Growth retardation

- Fever
- Polyuria, dehydration, and metabolic acidosis.
- *Renal manifestations*: Cystine accumulation in the kidney leads to
  - i. generalized proximal tubular dysfunction (*Fanconi syndrome*) with hypophosphatemia, hyperphosphaturia, vitamin D-resistant rickets, renal glycosuria, and generalized aminoaciduria.
  - ii. narrowing and shortening of the proximal tubule with typical swan neck deformity.
  - iii. pyelonephritis and interstitial fibrosis leading to glomerular insufficiency.
- *Ocular manifestations*: Deposition of cystine crystals in the cornea and conjuctiva leading to photophobia and retinal degeneration.
- Death before age 10 due to uremia or infections.

- *In the adult form*, only ocular abnormalities are seen.
- *In the juvenile form*, clinical presentation is intermediate of the infantile and adult forms.

Diagnosis

1. Detection of cystine crystals in the cornea, leukocytes, or in biopsies of rectal mucosa.
2. Quantitative estimation of cystine in the peripheral blood lymphocytes or cultured fibroblasts.
3. Prenatal diagnosis in cultured amniotic fluid.

Management

1. Prevention of fluid, electrolyte, and pH imbalance
2. Maintenance of adequate fluid intake to prevent dehydration
3. Administration of large doses of calcium, phosphate, and vitamin D to treat rickets.

Enumerate the specialized products formed from amino acids.

The specialized products formed from amino acids are detailed below:

1. Glycine is required for the formation of δ-aminolevulinic acid, a key intermediate in heme biosynthesis.
2. Serine is a constituent of phosphatidylserine.
3. Cysteine contributes to the – SH group of glutathione, acyl carrier protein, and coenzyme A.
4. Methionine forms S-adenosylmethionine, a methyl group donor, for various biochemical reactions.
5. Aspartate is required for the biosynthesis of urea, purine, and pyrimidine nucleotides.
6. Glutamate forms several biologically important compounds such as
   - glutathione, an important cellular reductant and antioxidant
   - tetrahydrofolate, involved in 1-carbon metabolism
   - γ-aminobutyric acid, a neurotransmitter.
7. Glutamine is necessary for the biosynthesis of purine and pyrimidine nucleotides.
8. Arginine is essential for the biosynthesis of creatine.
9. Histidine is essential for the synthesis of
   - formimino-FH_4 which has a role in 1-carbon metabolism
   - histamine, required for gastric acid secretion and contraction of smooth muscle.
10. Tyrosine gives rise to several biologically important compounds such as
    - melanin
    - thyroid hormones
    - catecholamines.
11. Tryptophan forms the neurotransmitter serotonin (5-hydroxytryptamine) and nicotinamide.

❑ What is nonprotein nitrogen?

Nonprotein nitrogen (NPN) refers to a group of compounds that include

- amino acids
- urea, which constitutes about 50% of the total NPN in blood
- uric acid
- creatine
- creatinine.

❑ What is the difference between creatine and creatinine?

Creatine is methylguanidoacetate, which is essential for muscle contraction, whereas creatinine, an internal anhydride of creatine, is a waste product that is excreted in the urine.

❑ Outline the steps in the biosynthesis of creatine.

Creatine is formed from the amino acids — glycine, arginine, and methionine by the following sequence of reactions:

1. Transamidination of arginine to glycine in the kidney to form guanidoacetate (glycocyamine).
2. N-methylation of guanidoacetate in the liver to form creatine. The methyl group donor is S-adenosylmethionine (SAM).
3. Transport of creatine by circulation to the skeletal muscle.
4. Phosphorylation of creatine to creatine phosphate in a reaction catalyzed by *creatine kinase* in the skeletal muscle accompanied by ATP hydrolysis.

Figure 8.14 shows the steps involved in the biosynthesis of creatine.

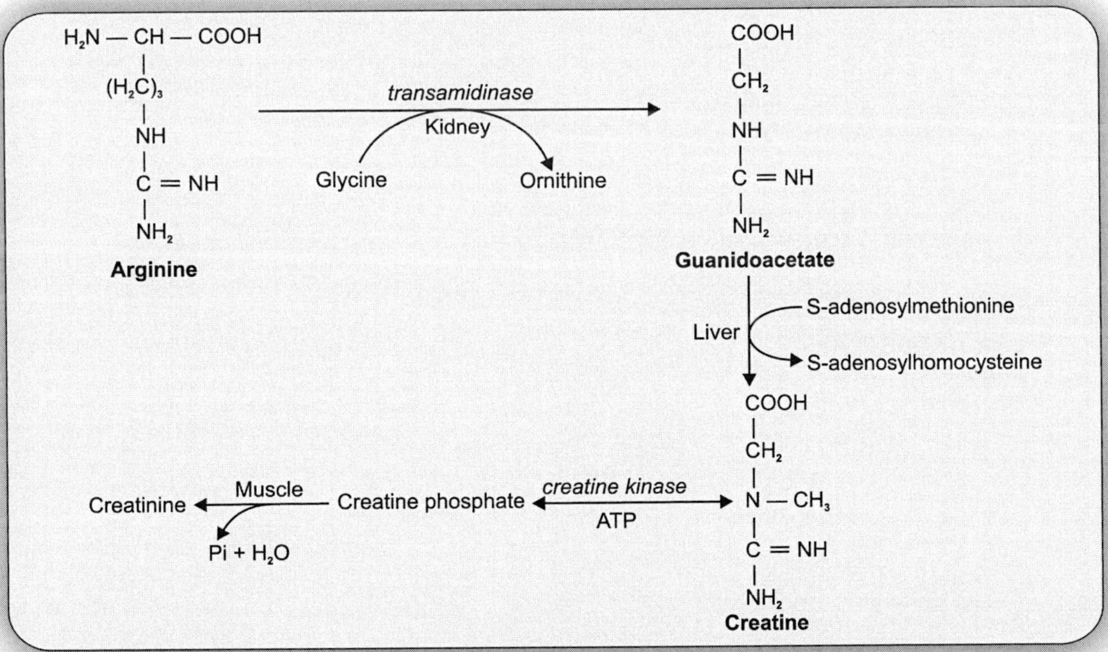

Fig. 8.14 Biosynthesis of creatine

❑ How is creatine metabolized?

- Creatine phosphate is a high-energy phosphate storage compound that provides energy for muscle contraction.
- The enzyme *creatine kinase* acts on creatine phosphate to form creatine with simultaneous phosphorylation of ADP to ATP.
- Rephosphorylation of creatine occurs during aerobic glycolysis.
- Creatine and creatine phosphate are spontaneously converted to the internal anhydride creatinine.
- The amount of creatinine synthesized is related to the total muscle mass. About 2% of the total creatine is transformed to creatinine every day.
- Although both creatine and creatinine are filtered by the glomeruli, creatine is reabsorbed by the tubules, whereas creatinine is secreted by the tubules and is excreted in the urine.

❑ Write notes on plasma and urinary creatinine.

- It is a normal constituent of urine
- Normal daily urinary excretion is 1–2 g
- Normal plasma creatinine is 0.6–1.3 mg/dL (55–120 mmol/L)
- Plasma creatinine level is a reliable index of kidney function
- Creatinine clearance is used as a measure of GFR
- Plasma creatinine is increased in renal failure.

❑ Enumerate the physiological and pathological causes of creatine excretion in urine.

- Creatine excretion occurs normally
  - in minute amounts in adult males
  - during fasting
  - on a low carbohydrate diet
  - in childhood
  - in pregnancy.
- Increased urinary creatine excretion is seen in
  - muscular disorders such as myasthenia gravis, and muscular dystrophies
  - fever
  - diabetes mellitus
  - hyperthyroidism.

Chapter 9

Metabolism of Purines and Pyrimidines

❑ How are nucleic acids and nucleoproteins digested and absorbed?

▫ *Digestion*

- Digestion of nucleoproteins occurs in the stomach and intestines, releasing histones and protamines.
- Nucleic acid digestion begins in the duodenum.
- DNA and RNA are hydrolysed by the action of pancreatic *ribonuclease (RNase)* and *deoxyribonuclease* (*DNase*), respectively to form oligonucleotides.
- The oligonucleotides are hydrolysed by *diesterases* to mononucleotides.
- Mucosal *nucleotidases* act on mononucleotides to form nucleosides and orthophosphate.

▫ *Absorption*

- The nucleosides are absorbed by passive diffusion.
- A mucosal *pyrimidine nucleosidase* releases pyrimidines, which are absorbed by diffusion or by specific transport pathways.
- The enzyme *xanthine oxidase* oxidizes the purines hypoxanthine and xanthine to uric acid, which is absorbed either by diffusion or by carrier-mediated transport.

❑ Give an account of the *de novo* pathway of purine biosynthesis.

The *de novo* pathway for the biosynthesis of purine nucleotides is a multistep process. The various steps involved (Fig. 9.1) are discussed below.

1. *Formation of phosphoribosyl pyrophosphate* (*PRPP*) from ribose-5-phosphate and ATP by *PRPP synthase*. PRPP, a key metabolite in nucleotide metabolism, is also required for the biosynthesis of nucleotide coenzymes and in the purine salvage pathway.
2. *Formation of phosphoribosylamine* from the reaction of PRPP with glutamine is the rate-limiting step in purine biosynthesis catalyzed by *PRPP glutamyl amidotransferase*. The enzyme is regulated by PRPP, and the end products, inosine monophosphate and other nucleotides.
3. *Formation of inosine 5'-monophosphate* (*IMP*) is a ten-step process involving sequential addition of glycine, N^5, N^{10}-methenyl H_4folate, glutamine, carbon dioxide, aspartate, and N^{10}-formyl H_4folate.
4. *Formation of adenosine and guanosine 5'-monophosphates* (*AMP and GMP*). Addition of aspartate to IMP followed by removal of fumarate forms AMP. Oxidation of IMP to xanthosine 5'-monophosphate (XMP) followed by addition of an amino group from glutamine leads to the formation of GMP.

5. *Formation of nucleoside diphosphates* involves phosphate group transfer from ATP catalyzed by the enzyme *nucleoside monophosphate kinase.*
6. *Phosphorylation of GDP to GTP* is accomplished by *nucleoside diphosphate kinase.*
7. *Phosphorylation of ADP to ATP* involves oxidative phosphorylation and reactions of glycolysis and citric acid cycle.

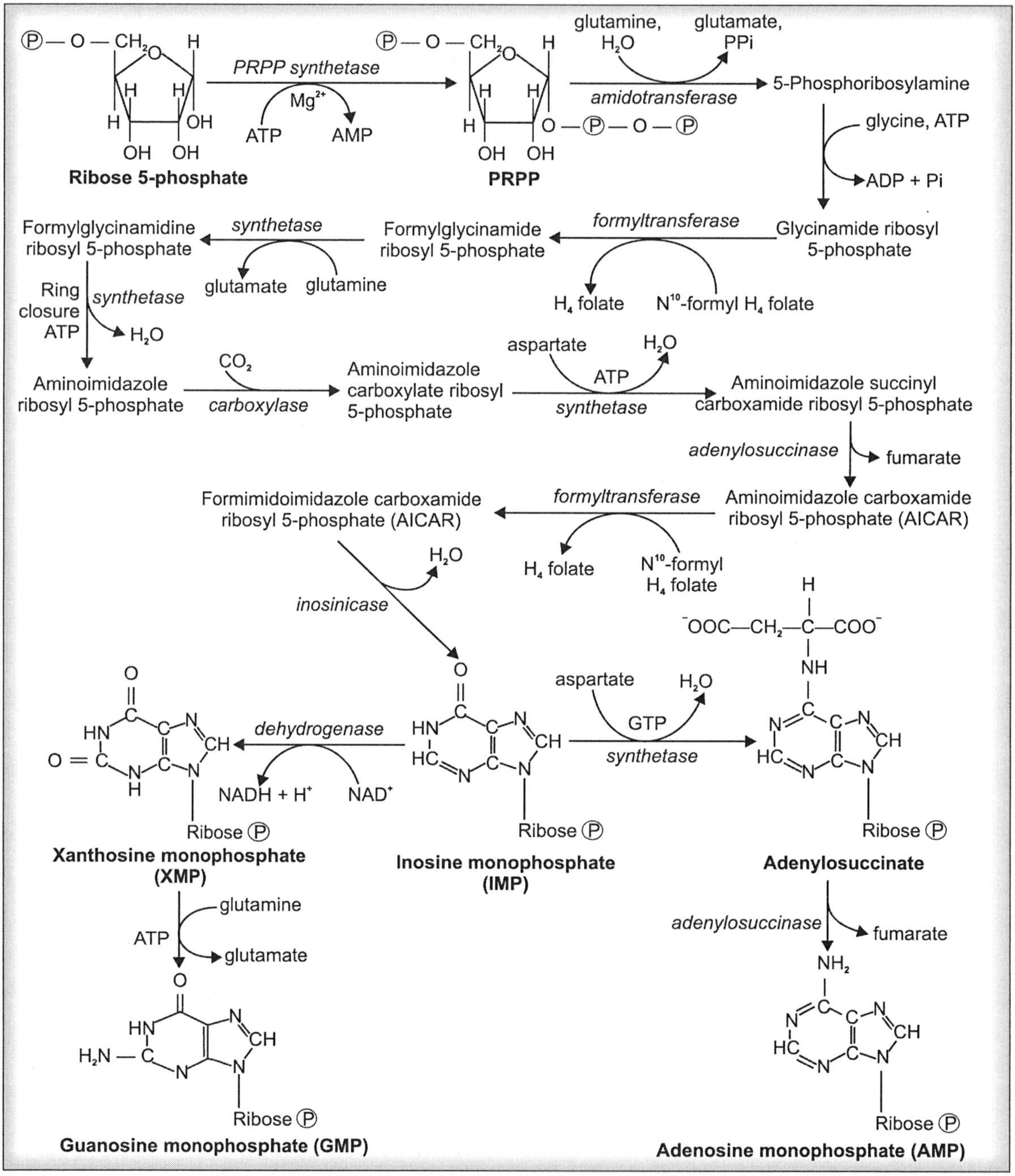

Fig. 9.1 *De novo* pathway of purine biosynthesis

Regulation

Feedback inhibition of the first two enzymes in the pathway — *PRPP synthetase* and *PRPP amidotransferase* by IMP, GMP, and AMP.

❑ Explain the salvage pathway for purine nucleotide synthesis.

The salvage pathway involves conversion of purines and their ribo- and deoxyribonucleosides to mononucleotides (Fig. 9.2). The major reactions include the following:

- Formation of IMP and GMP by combination of hypoxanthine and guanine, respectively, with PRPP in a reaction catalyzed by *hypoxanthine guanine phosphoribosyl transferase* (*HGPRT*).
- Formation of AMP from adenine in a reaction catalyzed by the enzyme *adenine phosphoribosyl transferase* (*APRT*).

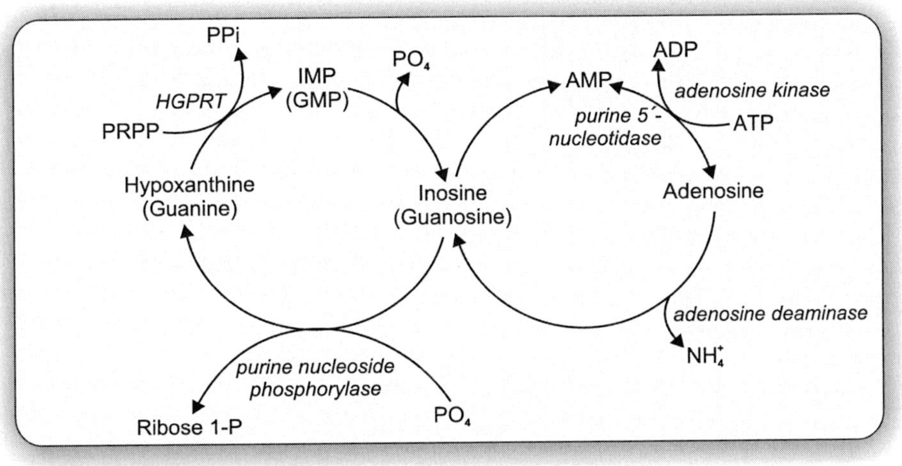

Fig. 9.2 Purine salvage pathway

Regulation is by feedback inhibition by end products. A decrease in activity of the salvage pathway increases PRPP levels leading to stimulation of *de novo* purine synthesis.

❑ How are deoxyribonucleotides formed?

Deoxyribonucleoside diphosphates (dNDPs) are formed by reduction of purine and pyrimidine ribonucleotides with the help of the *ribonucleotide reductase complex*, which requires thioredoxin, thioredoxin reductase, and NADPH (Fig. 9.3).

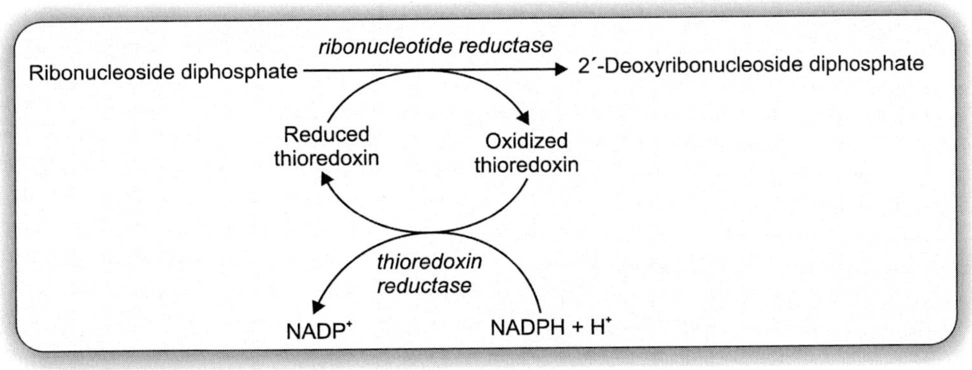

Fig. 9.3 Formation of deoxyribonucleotides

❑ Explain the catabolism of purines.

In humans, the end product of purine metabolism is uric acid. The process involves

1. Removal of 5'-phosphates from nucleotides to form the corresponding nucleosides.
2. Deamination of adenosine by *adenosine deaminase* to inosine.
3. Conversion of inosine and guanosine to hypoxanthine and guanine, respectively by removal of the ribose moiety as ribose 1-phosphate in a reaction catalyzed by *purine nucleoside phosphorylase*.
4. Deamination of guanine to xanthine by *guanase*.
5. Oxidation of hypoxanthine to xanthine by *xanthine oxidase*.
6. Oxidation of xanthine to uric acid by *xanthine oxidase*. Inhibitors of xanthine oxidase such as *allopurinol* are used in the management of patients with hyperuricemia and gout.

The catabolism of purines is shown in Figure 9.4.

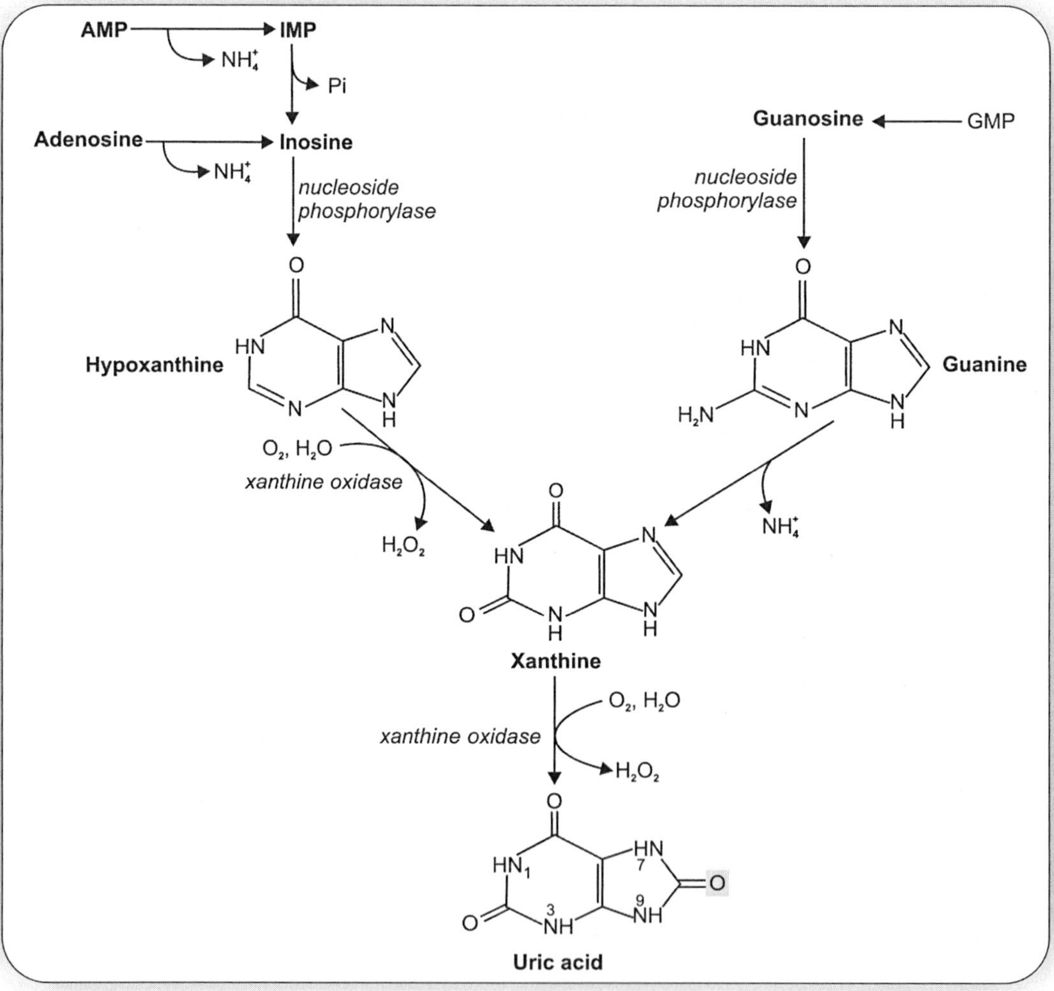

Fig. 9.4 Catabolism of purines

❑ What is the fate of uric acid formed by purine catabolism?

The concentration of urate in the body (miscible urate pool), which is about 1200 mg, depends on the balance between urate produced and urate excreted.

- *Urate production* depends on factors such as
  - purine intake
  - *de novo* biosynthesis
  - salvage pathway activity
  - catabolism.
- *Urate excretion* occurs *via*
  - the kidneys (75%): About 400 – 600 mg/d of uric acid is excreted in the urine. Urate is filtered by the glomerulus, and completely reabsorbed by the tubules. Urine urate is mostly due to tubular secretion.
  - degradation by colonic bacteria by *uricolysis* (25%).
- *Plasma urate*
  - Is predominantly present as monosodium urate
  - Normal levels:
    - 0.12 – 0.42 mmol/L (2.0 – 7.1 mg/dL) in men
    - 0.09 – 0.36 mmol/L (1.5 – 6.1 mg/dL) in women.

❑ Discuss the metabolism of pyrimidines.

Biosynthesis

1. Formation of carbamoyl phosphate from CO_2, glutamine, and ATP in a reaction catalyzed by cytosolic *carbamoyl phosphate synthase II.*
2. Condensation of carbamoyl phosphate with aspartate to form carbamoyl aspartate by *aspartate transcarbamoylase.*
3. Formation of dihydroortic acid from carbamoyl aspartate.
4. Conversion of carbamoyl aspartate to orotic acid.
5. Transfer of a ribose phosphate moiety to orotic acid to generate orotidine-5'-monophosphate (OMP) in a reaction catalyzed by *orotate phosphoribosyltransferase.*
6. Decarboxylation of OMP to produce uridine 5'-monophosphate (UMP).
7. Formation of cytidine triphosphate (CTP) by amination of UTP.
8. Formation of thymidine monophosphate (TMP) by methylation of dUMP (Fig. 9.5).

Catabolism

Pyrimidines are eventually broken down to water soluble end products. These include

1. CO_2, NH_3, and β-alanine from cytosine and uracil
2. β-aminoisobutyrate from thymine.

❑ What are antimetabolites?

Antimetabolites are compounds that inhibit specific steps in the biosynthesis of purines and pyrimidines, thereby inhibiting DNA synthesis.

Antimetabolites, useful in the chemotherapy of cancer, include

1. *5-Fluorouracil* (5-FU), an analog of uracil, which is converted to 5-fluorodeoxyuridine 5'-monophosphate (FdUMP) that blocks the synthesis of dTMP by interacting with thymidylate synthase and N^5, N^{10}-methylene tetrahydrofolate.
2. *Methotrexate,* a folate analog, that blocks dTMP synthesis by inhibiting dihydrofolate reductase.
3. *6-Mercaptopurine* and *thioguanine,* which exert feedback inhibition on PRPP *amidotransferase.*

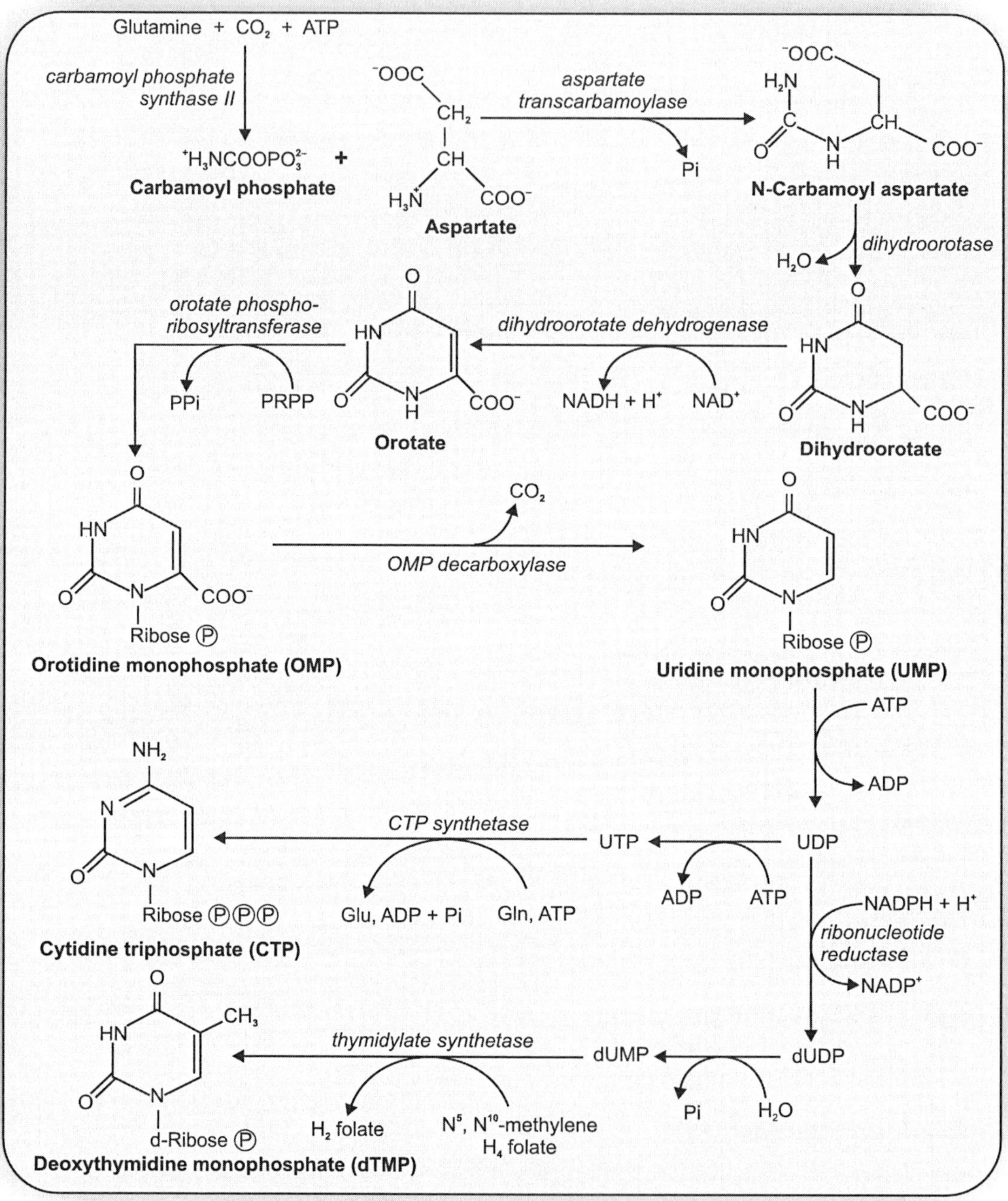

Fig. 9.5 Biosynthetic pathway for pyrimidines

❑ Enumerate the causes of hyperuricemia.

1. Hyperuricemia refers to a plasma urate concentration that is greater than 7.0 mg/dL (420 mmol/L).
2. Hyperuricemia may be asymptomatic or associated with the clinical symptoms of *gout*.

- ***Causes of hyperuricemia***
  Hyperuricemia may result from urate overproduction, decreased excretion, or a combination of both the mechanisms.
- ***Increased urate production*** occurs due to
  1. Purine-rich diets such as *'sweet breads'* (thymus and pancreas), organ meats (liver, kidney, heart), poultry, and fish.
  2. Abnormalities of the following enzymes involved in purine biosynthesis:
     a. Increased activity of *PRPP synthase* due to enzyme overactivity, insensitivity to feedback inhibition by purine nucleotides, or low K_m for substrate can increase PRPP levels and stimulate *PRPP amidotransferase*.
     b. *HGPRT* deficiency that may be complete as in *Lesch-Nyhan syndrome*, or partial as in *Kelley Seegmiller syndrome*. This leads to PRPP accumulation with increased channeling into the *de novo* purine biosynthetic pathway.
  3. Increased nucleic acid turnover due to
     a. enhanced tissue growth as in leukemias
     b. trauma with tissue damage
     c. following radiotherapy and chemotherapy for cancer.
- ***Decreased urate excretion*** as in
  1. Chronic renal failure with reduced GFR.
  2. Competitive inhibition of renal tubular secretion of urate by high plasma levels of
     a. ketone bodies as in starvation and diabetic ketoacidosis
     b. lactate in lactic acidosis
     c. salicylates
- ***Combined mechanisms***: Hyperuricemia due to a combination of urate overproduction and decreased excretion occurs in
  1. *von Gierke's disease* (*glycogen storage disease, type I*) in which glucose-6-phosphate that accumulates due to glucose-6-phosphatase deficiency is channelized into
     a. the pentose phosphate pathway, increasing PRPP synthesis, and therefore, urate overproduction.
     b. anaerobic glycolysis, with increase in lactate. Hyperlactacidemia blocks tubular secretion of urate and inhibits urate excretion.
  2. *Hereditary fructose intolerance* characterized by *fructose-1-phosphate aldolase* deficiency that causes fructose-1-phosphate accumulation, ATP depletion, increased purine nucleotide catabolism, and hyperuricemia. There is also urate retention consequent to lactic acidosis and renal tubular acidosis.
  3. Chronic alcoholism, which causes
     a. increased ATP breakdown and urate production
     b. increased production of lactate, which inhibits tubular secretion of urate and renal urate excretion.

❑ Elucidate the consequences of hyperuricemia.

The consequences of hyperuricemia include

- ***Acute gouty arthritis***
  - Characterized by severe joint pain with swelling and redness that results from deposition of sodium urate crystals in and around joints of the feet, causing an acute inflammatory reaction.

- Urate crystals
  - interact with polymorphonuclear leukocytes to release lysosomal enzymes, reactive oxygen species, leukotrienes, and prostaglandin metabolites
  - activate inflammatory processes and the complement pathway.
- ***Chronic gout***
  Acute gouty arthritis progresses to *chronic gout*, when aggregated deposits of sodium urate crystals referred to as *'tophi'* cause deformation and destruction of hard and soft tissues.
- ***Renal damage and nephrolithiasis***
  - Precipitation of urate in the kidneys causes renal damage.
  - Sodium urate crystals form proximal to the site of acidification and uric acid crystals distal to the site of acidification.

❑ Discuss the diagnostic tests and management of hyperuricemia.

Diagnostic tests

- Estimation of plasma and urine uric acid.
- Assay of PRPP and HGPRT activities in erythrocytes.
- Analysis of renal calculi.
- X-ray and analysis of synovial fluid.
- Examination of sodium urate crystals in polymorphonuclear leukocytes and synovial fluid. The crystals are negatively birefringent when visualized under a polarizing microscope, yellow and needle shaped when viewed parallel to the plane of polarized light, and blue when viewed perpendicular to it.

Management

Treatment of gout involves the following measures:

- Dietary restriction of purine.
- Restriction of alcohol.
- *Uricosuric drugs* (salicylates, probenecid, and sulfinpyrazole)
  - reduce plasma urate levels by enhancing renal urate excretion.
  - cause nephrolithiasis, which can be countered by administration of drugs in low doses, increased fluid intake, and alkalinization of urine.
- *Allopurinol* (4-hydroxy-pyrazolopyrimidine) the most widely used antihyperuricemic agent
  - is a structural analog of hypoxanthine
  - is a competitive *suicide inhibitor* of *xanthine oxidase*
  - forms oxypurinol by oxidation which binds tightly to the enzyme preventing it from oxidizing xanthine or hypoxanthine.
  - reduces PRPP levels and decreases *de novo* purine biosynthesis.
- *Colchicine*, a non-steroidal anti-inflammatory drug, reduces inflammation by inhibiting crystal-induced chemotactic factor (CCF) release from leukocytes.

❑ What is pseudogout?

Pseudogout is an inflammatory joint disease that is similar to gout but is characterized by deposition of calcium pyrophosphate crystals in joint cavities and normal plasma urate levels. The calcium pyrophosphate crystals are positively birefringent and appear as short blunt rods, rhomboids, and cuboids when observed under a polarizing microscope.

❑ Discuss the enzyme defect and clinical manifestations in Lesch-Nyhan syndrome.

Lesch-Nyhan syndrome is a rare, inborn error of purine metabolism inherited as a recessive X-linked trait.

Enzyme defect: Complete deficiency of HGPRT, a key enzyme in the purine salvage pathway.

Consequences
- Hypoxanthine and guanine are not salvaged
- IMP and GMP levels are decreased
- PRPP levels are increased
- *De novo* purine biosynthesis is increased, leading to urate overproduction.

Clinical and biochemical features
- Hyperuricemia with gout and urinary lithiasis
- *Neurological manifestations*: self-mutilation, aggressive behavior, mental retardation, athetosis, and spastic paraplegia

Management: Treatment with allopurinol controls hyperuricemia but not the neurological symptoms.

❑ Write a short note on hypouricemia.

- Hypouricemia refers to a plasma urate concentration of less than 2.0 mg/dL.
- It is a relatively rare condition that does not cause any clinical symptoms.
- Hypouricemia may result from decreased urate production, increased excretion, or a combination of both mechanisms.
- Hypouricemia in Dalmatian dogs is due to an inability to reabsorb uric acid.
- Conditions associated with hypouricemia include
  - xanthinuria
  - uricosuric drugs (*e.g.*, aspirin)
  - hepatic cirrhosis
  - inappropriate vasopressin secretion
  - total parenteral nutrition
  - neoplastic disease
  - diabetes mellitus
  - defects in renal tubular transport.

❑ What is xanthinuria?

- Xanthinuria is a rare, autosomal recessive, inborn error of purine metabolism characterized by a complete deficiency of hepatic *xanthine oxidase.*
- Due to the enzyme block, there is accumulation of hypoxanthine and xanthine with excretion in the urine.
- Xanthine lithiasis may be present.
- Plasma and urine urate levels are low.

❑ Explain the consequences of adenosine deaminase (ADA) deficiency.

- ADA catalyzes the formation of inosine from adenosine.
- Deficiency of ADA leads to the following consequences:
  - Accumulation of adenosine and deoxyadenosine, which are converted to their corresponding nucleotides in leukocytes.
  - Increase in dATP which inhibits *ribonucleotide reductase* and blocks DNA synthesis and proliferation of leukocytes.
  - *Severe combined immunodeficiency (SCID)* with impairment of both cellular and humoral immunity, recurrent infection, and failure to thrive; the condition is usually fatal.

- *Management* involves
  - bone marrow transplantation
  - enzyme replacement therapy
  - gene therapy.

Explain the biochemical lesion and clinical findings in hereditary orotic aciduria.

There are two types of orotic aciduria.

Type I orotic aciduria

It is more common and is inherited as an autosomal recessive trait.

- *Enzyme defect*: Deficiency of both *orotate phosphoribosyl transferase* and *OMP decarboxylase* with accumulation of orotate and OMP and excretion in urine. Deficiency of UMP and other pyrimidine nucleotides which inhibit nucleic acid synthesis.
- *Clinical features* include
  - megaloblastic anemia
  - immune deficiency
  - failure to thrive
  - orotic acid crystalluria.
- *Management*: Oral administration of uridine supplies pyrimidine nucleotides and depresses orotate synthesis by feedback inhibition.

Type II orotic aciduria

- *Inheritance*: Rare, autosomal recessive trait.
- *Enzyme defect*: Deficiency of *OMP decarboxylase*. Orotidine is the major product excreted in the urine.
- *Clinical symptoms* include
  - orotidinuria
  - orotic aciduria
  - megaloblastic anemia.
- *Management*: Administration of oral uridine.

Chapter 10

Metabolism of Porphyrins and Bile Pigments

❑ What are porphyrins?

- Porphyrins are derivatives of *porphin* containing four pyrrole rings linked by methenyl (= CH –) bridges (Fig. 10.1).
- There are eight hydrogen atoms in porphyrins which are substituted by side chains such as
  - acetate (– CH_2COOH; A)
  - propionate (– CH_2 – CH_2 – COOH; P)
  - methyl (– CH_3; M)
  - vinyl (– CH = CH_2; V).
- Porphyrins are classified based on the arrangement of the substituents into type I and type III.
- In type I porphyrins, the substituents are arranged in a symmetric manner.
- In type III porphyrins, the arrangement of the substituents is asymmetric. Type III porphyrins are widely distributed and abundant in nature.

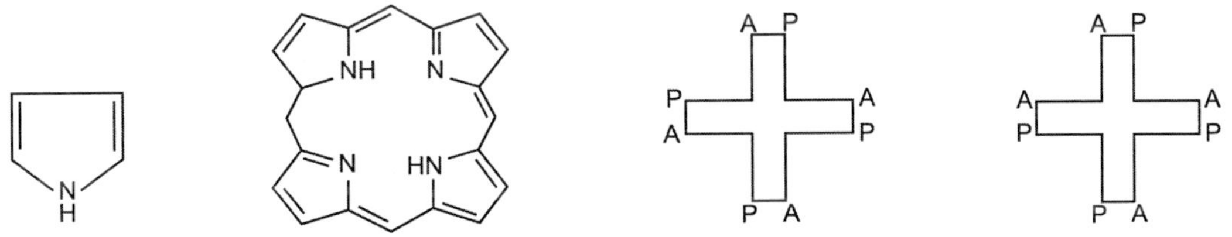

Pyrrole **Porphin** **Uroporphyrin I** **Uroporphyrin III**

Fig. 10.1 Structure of porphyrins (A = acetate, P = propionate)

❑ What is heme?

Heme is a metalloporphyrin that contains iron in the ferrous form.

❑ List the various heme pigments.

Heme pigments that serve various biological functions include

1. Hemoglobin, the pigment responsible for the transport of blood gases
2. Myoglobin that stores oxygen in muscle
3. Cytochromes, the constituents of the electron transport chain involved in energy capture.

❑ Explain the structure and functions of hemoglobin.

- Hemoglobin, the red pigment in erythrocytes, is made up of four molecules of heme and the protein, globin.
- The heme moiety is a tetrapyrrole called *protoporphyrin IXa* that is linked to ferrous ion (Fe^{2+}) (Fig. 10.2).
- The globin portion in heme consists of two pairs of polypeptide chains —2α and 2β chains, each of which binds one molecule of heme.
- Hemoglobin serves to transport oxygen, carbon dioxide and protons in blood.

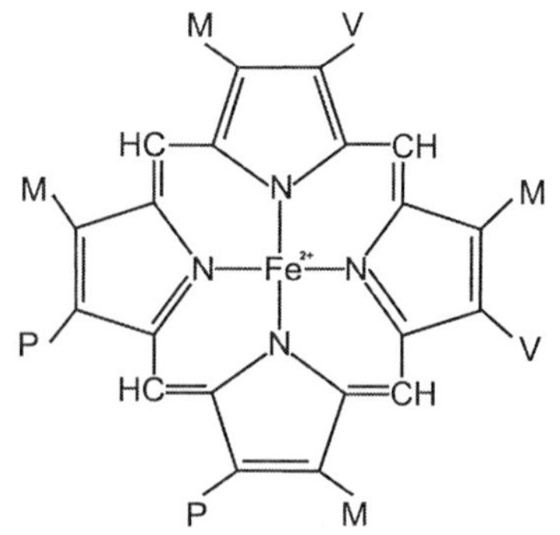

Fig. 10.2 The structure of heme (M = methyl, P = propionate, V = vinyl)

❑ Explain the transport of oxygen by hemoglobin.

- In the lung, oxygen diffuses into the red blood cells (RBCs) and combines with hemoglobin to form oxygenated hemoglobin (HbO_2).
- Four molecules of oxygen combine with hemoglobin to form Hb_4O_8. Binding of one molecule of O_2 favors uptake of additional O_2 molecules by a phenomenon called *positive cooperativity*. This gives rise to the characteristic sigmoid shape of the oxygen dissociation curve (Fig. 10.3).
- A shift of the oxygen dissociation curve to the left, as in fetal hemoglobin, (HbF) increases the affinity of hemoglobin for O_2, whereas a rightward shift decreases the affinity of hemoglobin for O_2 resulting in O_2 release to the tissues.
- Factors that cause a shift of the O_2 dissociation curve to the right include
  - decrease in blood pH
  - increase in temperature
  - increase in CO_2 tension
  - increased concentration of 2,3-bisphosphoglycerate (BPG).

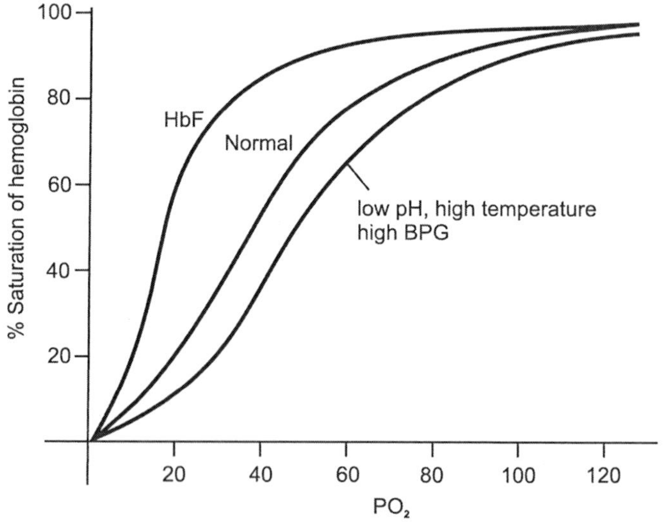

Fig. 10.3 The oxygen dissociation curve

❑ What is Bohr effect?

The binding of carbon dioxide to hemoglobin with simultaneous deoxygenation of hemoglobin is known as *Bohr effect*. This is responsible for oxygenation of hemoglobin in the lungs and deoxygenation in the tissues.

❑ Elucidate the role of 2,3-bisphosphoglycerate (BPG) in oxygen transport.

- BPG formed in the RBCs during glycolysis binds to deoxyhemoglobin and shifts the O_2 dissociation curve to the right.

- HbF, which has high affinity for O_2, binds poorly to BPG.
- Acidosis inhibits red cell glycolysis and lowers BPG concentration.
- Factors that increase BPG concentration include
  - hormones (thyroid hormones, growth hormones)
  - exercise
  - tissue hypoxia by increasing O_2 delivery to tissues.

How is CO_2 transported?

More than 90% of CO_2 in the blood is transported by hemoglobin.

The sequence of events in the *tissues* is as follows:

- CO_2 produced by oxidation diffuses into the RBCs where it is hydrated by *carbonic anhydrase* (*carbonate dehydratase*) to carbonic acid (H_2CO_3).
- H_2CO_3 dissociates to form bicarbonate (HCO_3^-) and hydrogen (H^+) ions.
- HCO_3^- ions diffuse into plasma.
- H^+ is taken up by hemoglobin (*Bohr effect*).
- O_2 is released into the tissues (*isohydric shift*).
- Cl^- diffuses from the plasma into the RBCs in exchange for HCO_3^- to maintain electrochemical neutrality (*chloride shift*).

In the lungs

- CO_2 from RBCs diffuses through plasma into the alveoli due to lower PCO_2.
- O_2 from the alveoli diffuses into the RBCs and combines with Hb to form HbO_2.
- Plasma HCO_3^- enters into the erythrocytes and combines with H^+ to form H_2CO_3.
- H_2CO_3 is dehydrated to CO_2 by *carbonic anhydrase.*
- CO_2 diffuses from RBCs through plasma into alveolar space and is exhaled out.

Net result

O_2 is transported from the lungs to the tissues and CO_2 from the tissues to the lungs (Fig. 10.4).

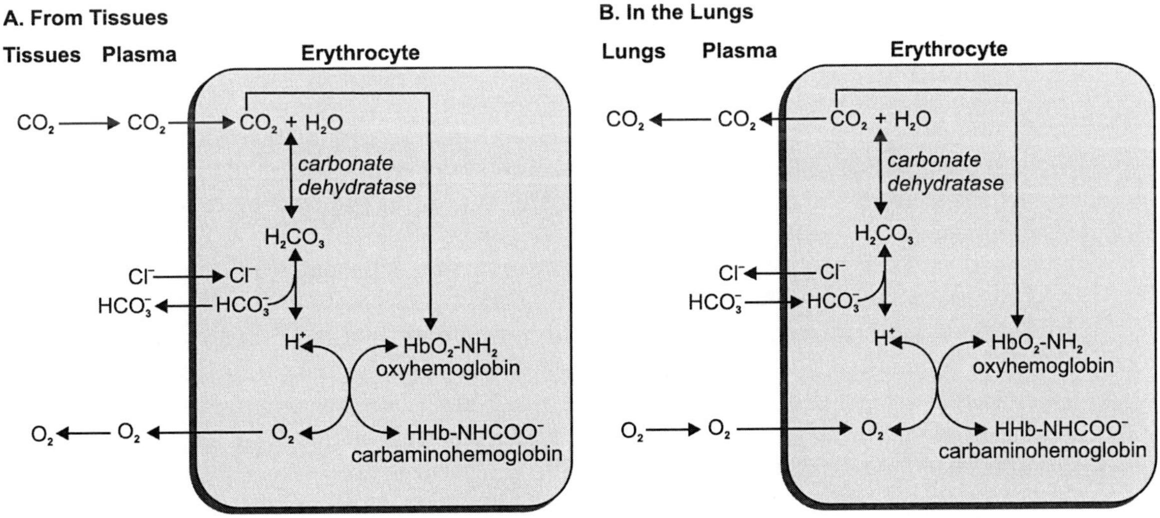

Fig. 10.4 The mechanism of CO_2 transport

❑ Outline the steps in heme biosynthesis.

- Heme biosynthesis occurs in all mammalian cells except erythrocytes, which lack mitochondria.
- The intermediates in the heme biosynthetic pathway are unstable, colorless compounds called porphyrinogens, which are readily oxidized by light to their corresponding porphyrins that are excreted in the urine and feces.
- Heme biosynthesis takes place both in the mitochondria and the cytosol by the following series of enzyme-catalyzed reactions (Fig. 10.5):

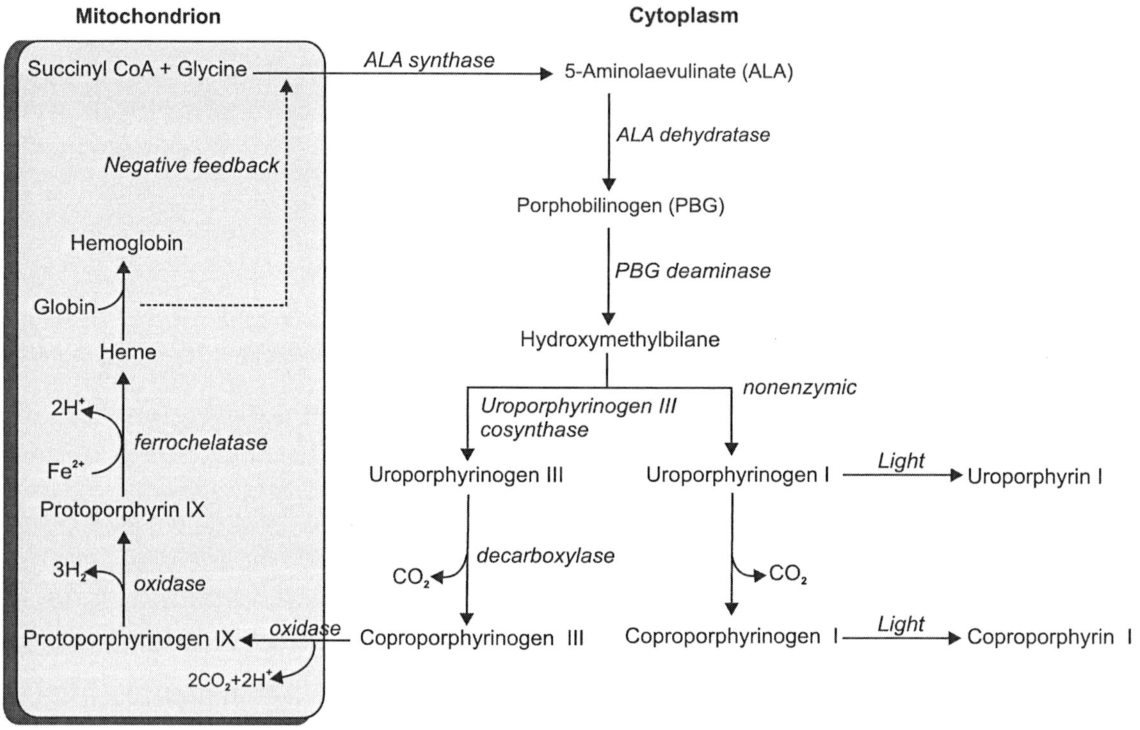

Fig. 10.5 The biosynthesis of heme

1. Condensation of succinyl CoA (active succinate) with glycine followed by decarboxylation to form 5-aminolevulinate (ALA) catalyzed by *ALA synthase*. This is the rate-limiting step which is subject to feedback inhibition by heme. The enzyme is induced by drugs, pesticides, and carcinogens.
2. Condensation of two molecules of ALA to form porphobilinogen (PBG) by the action of *ALA dehydratase* (*PBG synthase*).
3. Condensation of four molecules of PBG to form uroporphyrinogen in a reaction catalyzed by *PBG deaminase*. Type III uroporphyrinogen, the major isomer, is formed by the action of *uroporphyrinogen III cosynthase* while the type I isomer is formed non-enzymatically.
4. Decarboxylation of uroporphyrinogen to coproporphyrinogen in a reaction catalyzed by *uroporphyrinogen decarboxylase*.
5. Conversion of coproporphyrinogen III to protoporphyrinogen IX by *coproporphyrinogen oxidase* in the mitochondria.
6. Conversion of protoporphyrinogen IX to protoporphyrin IX by *protoporphyrinogen oxidase*.

7. Incorporation of ferrous ion (Fe^{2+}) into protoporphyrin by *ferrochelatase* to form heme.
8. Formation of hemoglobin by combination of four molecules of heme with two α and two β globin chains.

Porphyrinogens, intermediates formed during heme biosynthesis, are oxidized by light to porphyrins and excreted in the urine and feces.

❑ How is hemoglobin catabolized?

The catabolism of hemoglobin involves the following steps:

1. Breakdown of erythrocytes after a life span of 120 days by the reticuloendothelial system with the release of hemoglobin
2. Formation of heme and globin from hemoglobin
3. Breakdown of the globin portion to amino acids
4. Dissociation of iron from heme
5. Reutilization of amino acids and iron
6. Degradation of the porphyrin moiety by the reticuloendothelial system (specialized cells in the liver, spleen, and bone marrow) by *heme oxygenase and NADPH-cytochrome c reductase* in the microsomes to form *biliverdin*, which is reduced to *bilirubin* by *bilirubin reductase*
7. Excretion of bile pigments via bile in the urine and feces.

❑ Explain the metabolism of bilirubin.

▫ *Formation*

- Bilirubin is formed by hemoglobin breakdown to heme and globin in the reticuloendothelial cells.
- Globin is degraded to amino acids that enter the amino acid pool and are reutilized. The heme moiety is broken down to iron and porphyrin.
- Iron is reutilized.
- The porphyrin portion is converted to biliverdin, which is reduced to bilirubin.

▫ *Transport*

Bilirubin being water-insoluble is transported in the plasma bound to albumin.

▫ *Conjugation in the liver*

- Bilirubin leaves albumin and is taken up by *ligandin* (*Y protein*) at the surface of the liver.
- In the smooth endoplasmic reticulum, bilirubin is made water-soluble by conjugation with glucuronate in a 2-step process catalyzed by the enzyme *uridyldiphosphate (UDP)-glucuronyl transferase*.
- The first step forms bilirubin monoglucuronide, which forms bilirubin diglucuronide on addition of a second molecule of glucuronide.
- Bilirubin diglucuronide is secreted into bile canaliculi and thence into the bile by an active process.

▫ *In the intestine*

- Bilirubin diglucuronide is deconjugated by bacterial *β-glucuronidases* and reduced by fecal flora to colorless compounds called *urobilinogens*.
- Urobilinogen is oxidized to *urobilin* and excreted in the feces (100 – 200 mg or 170 – 339 μmol per day). This pigment accounts for the characteristic color of the feces.
- Some amount of urobilinogen is reabsorbed by the portal circulation, transported to the liver, and reexcreted into the bile (*enterohepatic circulation*).

- *Urinary excretion*
  - A small amount of urobilinogen enters the systemic circulation and is excreted in the urine (1 – 3 mg/day or < 5 μmol/day).
  - Conversion of urobilinogen to urobilin is responsible for the characteristic yellow color of the urine.
- *Plasma bilirubin*

  It is unconjugated, protein-bound, not filtered by the glomeruli, and does not appear in the urine. Figure 10.6 summarizes the metabolism of bilirubin.

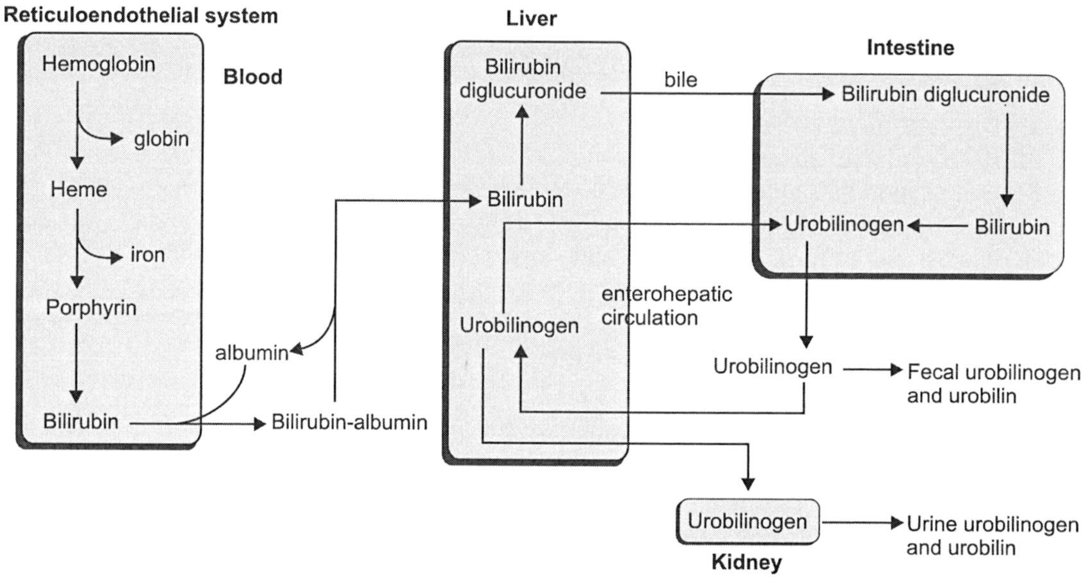

Fig. 10.6 Metabolism of bilirubin

❑ What are hemoglobinopathies?

Hemoglobinopathies are genetic abnormalities in hemoglobin synthesis. These are of two types:

1. Substitutions of one or more amino acids in the α- or β-chain of hemoglobin. For example, the hemoglobin variant *HbS* in *sickle cell anemia* in which glutamate in position 6 of the β-chain is substituted by valine associated with altered solubility properties, deformation of red cells, and anemia.
2. Inherited defects in the rate of synthesis of the α- or β-chains of hemoglobin (*thalassemias*). While deficient formation of the α-chain results in production of abnormal β_4 (HbH) or γ_4 (Hb Bart's) chains, total absence of α-chain synthesis is incompatible with life. Deficiency of β-chain synthesis may lead to a compensatory increase in other chains such as in HbA_2 and HbF.

❑ What are porphyrias?

- Porphyrias are inherited or acquired disorders caused by specific enzyme defects in heme biosynthesis.
- The enzyme defect decreases heme synthesis with increase in *ALA synthase* activity leading to overproduction, accumulation, and excretion of porphyrins and/or their precursors.

❑ How are porphyrias classified?

- Porphyrias are classified into *hepatic* and *erythropoietic*, based on the site of the enzyme defect.
- Based on the clinical manifestations, porphyrias are classified into *acute*, characterized by neurologic symptoms; and *chronic*, characterized by cutaneous photosensitivity.
- There are four types of hepatic porphyria — intermittent acute porphyria, variegate porphyria, hereditary coproporphyria, and porphyria cutanea tarda.
- Erythropoietic porphyrias are of two types — congenital erythropoietic porphyria and protoporphyria.

❑ Discuss the biochemical defect, clinical symptoms, and management of intermittent acute porphyria (IAP).

Biochemical defect: Deficiency of *PBG deaminase* with additional acquired deficiency of *steroid 5α-reductase.*

This leads to

- decreased synthesis of heme
- derepression of *ALA synthase*
- overproduction and urinary excretion of ALA and PBG
- polymerization of PBG to porphobilin and porphyrin, with darkening of urine on standing.

Clinical manifestations

1. *Gastrointestinal*: Colicky abdominal pain, vomiting, constipation.
2. *Neurological*: Peripheral neuropathy, convulsions, depression, hysteria, and psychosis. Peripheral neuropathy may be precipitated by agents that decrease heme, such as barbiturates, oral contraceptives and alcohol.
3. *Cardiovascular*: Sinus tachycardia and hypertension.

Management

1. Avoidance of precipitating factors
2. Maintenance of fluid and electrolyte balance
3. Adequate intake of carbohydrates
4. Physiotherapy
5. Hematin infusion to lower *ALA synthase* activity.

❑ Outline the biochemical lesion, clinical manifestations, and management of variegate porphyria.

***Biochemical defect*:** Deficiency of *porphyrinogen oxidase.* This leads to

- a partial block in the conversion of protoporphyrinogen to heme
- increase in *ALA synthase* activity
- overproduction of all intermediates proximal to the block
- excretion of ALA, PBG, uroporphyrin, and coproporphyrin in urine, and uroporphyrin, coproporphyrins, and protoporphyrin in the feces.

***Clinical manifestations*:** Gastrointestinal and neurological symptoms as in IAP. In addition, there is cutaneous photosensitivity.

***Management*:** Hematin infusion.

❑ Discuss the biochemical findings in hereditary coproporphyria.

Biochemical defect: Partial deficiency of *coproporphyrinogen oxidase* that converts coproporphyrinogen III to protoporphyrinogen IX. This leads to

- derepression of *ALA synthase*
- excretion of coproporphyrinogen III in large quantities in the urine and feces.

The clinical symptoms and treatment are similar to those described for IAP.

❑ Elucidate the biochemical lesion, clinical manifestations, and management of porphyria cutanea tarda.

Biochemical lesion: 50% decrease in *uroporphyrinogen decarboxylase* that converts uroporphyrinogen to coproporphyrinogen. In *type I*, the enzyme deficiency is in the liver; whereas in *type II*, the enzyme is deficient in all tissues. As a result of the enzyme block, there is accumulation of uroporphyrinogen and intermediates proximal to the block with deposition in the skin and excretion in the urine.

Clinical manifestations

- Liver disease and hepatic siderosis due to alcohol and iron overload
- *Cutaneous symptoms*: Photosensitivity, facial hyperpigmentation, ulcerative lesions of the face and forearms, and erythema, progressing to vesicles and bullae.

Management:

- Removal of precipitating factors
- Avoidance of direct sunlight
- Use of barrier creams to protect the skin
- Removal of excess iron by venesection.

❑ Give an account of erythropoietic porphyrias.

There are two types of erythropoietic porphyria:

1. *Congenital erythropoietic porphyria* characterized by
   - deficiency of uroporphyrinogen III *cosynthetase*
   - increase in *ALA synthase* activity in erythropoietic tissues
   - overproduction of ALA, PBG
   - excretion of the unphysiologic type I porphyrinogens
   - red color of urine.

   Clinical manifestations are similar to those seen in IAP. In addition, cutaneous photosensitivity is also seen.

   Management involves avoidance of exposure to sunlight.
2. *Protoporphyria* characterized by
   - partial deficiency of *ferrochelatase*
   - increased amounts of protoporphyrin IX in the erythrocytes, plasma, and feces
   - red fluorescence of reticulocytes and skin
   - mild photosensitivity and acute urticaria on exposure to sunlight
   - liver failure due to hepatocellular damage.

Chapter 11

Metabolism of Calcium, Phosphate, and Magnesium

❑ Explain the distribution of calcium in the body.

Calcium is a major constituent of bones and teeth. Approximately 25,000 mmol (1 kg) calcium is present in the body of which 99% is present in the skeleton. The extracellular fluid (ECF) contains about 22.5 mmol calcium of which 9 mmol is present in the plasma.

❑ Enumerate the functions of calcium.

Calcium performs the following functions in the body:

- Component of bones and teeth
- Nerve impulse conduction
- Second messenger of hormone action
- Membrane formation and permeability
- Blood coagulation
- Muscle contraction
- Release and activation of enzymes

❑ What are the sources and daily requirements of calcium?

Milk, cheese, cabbage and mustard greens are rich sources of calcium, while egg yolk, cauliflower, beans, oyster, and shrimp are good sources.

The recommended daily allowances (mg/day) for calcium are as follows:

Infants	0 – 12 months	360 – 540
Children	1 – 10 years	800
Adults	11 – 18 years	1200
	19 – 50 years	800
	> 51 years	800
Pregnancy and lactation		1200

❑ Discuss the mechanism of maintenance of calcium homeostasis.

Calcium homeostasis is maintained by regulating the balance between the amount absorbed from the diet and the amount excreted from the body.

Absorption

- About 6 – 12 mmol (0.25 – 0.50 g) of calcium is absorbed in the upper small intestine from the diet every day.
- Absorption is higher in children, and during pregnancy and lactation.

- Calcium absorption is increased by
  - *Calcitriol* (*1,25-dihydroxycholecalciferol* (*1,25(OH)$_2$D$_3$*), the active metabolite of vitamin D which
    i. stimulates synthesis of calcium-binding proteins that facilitate active transport of calcium across the intestinal mucosa,
    ii. regulates synthesis of calcium-dependent *ATPases* involved in active calcium transport, and
    iii. alters viscosity of intestinal membrane cells.
  - Ascorbic acid
  - Lactate
  - Protein
  - Low pH.
- Calcium absorption is inhibited by
  - dietary factors, such as oxalates, phytates, and phosphates which form calcium salts in the intestine
  - steatorrhea in which excess fatty acids in the intestinal lumen form insoluble calcium soaps
  - alkalinity.

Excretion

- About 20 mmol (0.8 g) of calcium is excreted in the feces every day.
- Approximately 2.5 – 7.5 mmol/day (0.1 – 0.3 g) of calcium is excreted in the urine.
- Trace amounts are excreted in the sweat and through the skin.

❑ What are the forms in which calcium is present in plasma?

Normal range of plasma calcium: 2.1 – 2.6 mmol/L (8.5 – 10.5 mg/dL). Plasma calcium is maintained by mobilization from or deposition in bone.

There are three forms of calcium in the plasma:

1. *Protein-bound calcium* is unionized and accounts for less than half of the total calcium concentration. About 80% of this fraction is bound to albumin and 20% to γ-globulins.
2. *Ionized calcium,* (*free calcium, Ca^{2+}*), which constitutes more than half of the total calcium, is the physiologically active form of calcium.
3. Complexed with anions such as citrate, phosphate, and bicarbonate (small amounts).

❑ How is phosphate distributed in the body?

Approximately 1 kg of phosphorous is present in a normal healthy adult of which

- 80 – 85% is present in the bone
- 10% is in combination with proteins, carbohydrates, lipids, and other molecules in blood and muscle
- 10% is distributed among various biomolecules.

❑ Enumerate the functions of phosphate.

Phosphate performs the following functions:

1. *Structural role*: Constituent of bones, teeth, cell membranes, and nerve tissue.
2. Component of phosphoproteins, phospholipids, nucleic acids, and nucleotides.
3. Component of coenzymes (*e.g.*, NADP).

4. Buffering action.
5. Energy transfer (as ATP and creatine phosphate).

❑ What are the sources and daily requirements of phosphate?

Milk, cheese, egg yolk, meat, fish, fowl, legumes, whole nuts, and whole grain cereals are rich sources of phosphate.

The recommended daily allowances (mg/day) for phosphate are as follows:

Infants	0 – 12 months	240 – 360	**Adults**	11 – 18 years	1200
Children	1 – 10 years	800		19 – 50 years	800
				> 51 years	800
				Pregnancy and lactation	1200

❑ Explain how phosphate is absorbed in the intestine.

- Dietary phosphate in the form of inorganic phosphate (Pi) or organic phosphate is converted to Pi in the gastrointestinal tract and absorbed in the small intestine by active transport.
- The ratio of calcium and phosphate in the diet (optimal 1 : 1) influences the absorption of phosphate as well as calcium. An excess of dietary calcium increases excretion of phosphate, and vice versa.

❑ How is phosphate excreted?

- Approximately two-third of dietary phosphate is excreted in the urine and only one-third through the feces.
- About 85% of plasma phosphate filtered by the glomerulus is reabsorbed in the tubules by active transport.
- Phosphate excretion is low in glomerular damage and high in renal tubular disorders.
- Parathyroid hormone (PTH) inhibits tubular reabsorption of phosphate.

❑ What are the forms in which phosphate is present in plasma?

- Normal plasma [Pi]: 0.8 – 1.4 mmol/L (2.5 – 4.5 mg/dL).
- Plasma phosphate is maintained by mobilization from bone when plasma concentrations are low, and deposition in bone when plasma levels are increased.
- Plasma phosphate is present mostly as orthophosphate, with HPO_4^{2-} and $H_2PO_4^-$ in a ratio of 4 : 1.
- Small amounts of plasma phosphate are also present as pyrophosphate, hexose phosphate, triose phosphates.
- Plasma also contains organic phosphate in the form of phospholipids.
- Plasma calcium and phosphate show a reciprocal relationship. A decrease in plasma [Pi] is accompanied by an increase in plasma [Ca^{2+}].
- The product of the concentrations of total calcium and phosphate is 2.8, and that of ionized calcium and phosphate is approximately 1.2.

❑ Discuss the distribution, functions and metabolism of magnesium.

□ ***Distribution***

- Approximately 1000 mmol of magnesium is present in a normal healthy adult, of which 50% is in the bone, 25% in the muscle and 25% in soft tissues.
- The ECF contains 15 – 29 mmol of magnesium, of which 0.7 – 1.0 mmol/L (1.6 – 2.4 mg/dL) is present in the plasma.

□ ***Functions***

- Cofactor for over 300 enzymes in the form of Mg^{2+}-ATP (*e.g.*, enzymes involved in glycolysis, protein synthesis, and ion transport)
- Maintenance of structural integrity of ribosomes, nucleic acids, and some proteins
- Role in permeability and electrical properties of membranes.

□ ***Metabolism***

- Normal daily intake of 10 – 12 mmol is greater than that required for maintaining magnesium balance (~8 mmol/day).
- Excess is excreted in the urine.
- Urinary excretion is increased by:
  i. expansion of ECF volume
  ii. hypercalcemia
  iii. hypermagnesemia.
- Parathyroid hormone stimulates renal tubular reabsorption of magnesium. Aldosterone increases renal excretion of magnesium.

❑ Describe the composition and formation of bone.

□ Bone contains calcium and phosphate in the form of hydroxyapatite [3 $Ca_3(PO_4)_2$ $Ca(OH)_2$]. Sodium, magnesium, potassium, carbonate, fluoride, and citrate are found deposited on the hydroxyapatite crystals.

□ Bone formation (*mineralization*) is governed by *osteoblasts*, which have the following functions:

- synthesis of a collagenous organic matrix termed *osteoid*
- secretion of alkaline phosphatase (ALP) involved in hydroxyapatite formation
- secretion of *osteocalcin*, a γ-carboxyglutamate (Gla) containing protein that plays a role in bone mineralization.

□ Bone resorption (*demineralization*) is governed by *osteoclasts*, a type of macrophage that destroys the matrix.

□ Bone is in a state of dynamic equilibrium with bone mineralization and demineralization occurring continuously.

❑ Enumerate the hormones that regulate plasma calcium and phosphate.

Plasma calcium and phosphate are regulated mainly by the combined action of three hormones:

□ *Parathyroid hormone* (*PTH*), secreted by the chief cells of the parathyroid glands in response to low plasma ionized calcium levels,

- stimulates formation of calcitriol
- increases osteoclastic bone resorption
- increases intestinal absorption of calcium and phosphate

- stimulates renal tubular reabsorption of calcium
- promotes phosphate excretion.

- *Calcitriol* (*1,25-dihydroxycholecalciferol*), the active form of vitamin D,
  - increases intestinal absorption of calcium and phosphate
  - stimulates calcium mobilization from bone
  - increases renal tubular reabsorption of calcium and phosphate.
- *Calcitonin*, synthesized by parafollicular or C cells of the thyroid gland,
  - inhibits bone resorption
  - decreases renal tubular reabsorption of calcium and phosphate.

A decrease in plasma calcium stimulates PTH secretion and inhibits calcitonin secretion. PTH stimulates 1α-hydroxylase activity, thereby mediating calcitriol synthesis.

Other hormones

In addition to PTH, calcitriol, and calcitonin, other hormones also influence calcium and phosphate metabolism. These include

- Thyroxine, which stimulates osteoclast activity
- Glucocorticoids that influence the production of bone matrix due to their protein-catabolizing action
- Growth hormone, which stimulates cartilage growth
- Estrogens and androgens that stimulate matrix production.

❑ Discuss the synthesis and biological actions of parathyroid hormone.

Synthesis

- Parathyroid hormone is synthesized as a large precursor, *preproPTH* containing 115 amino acids.
- PreproPTH is converted to the 90-residue proPTH by *peptidases.*
- ProPTH moves from the endoplasmic reticulum to the Golgi complex where six amino acids are removed at the N-terminal end to form PTH.
- PTH secreted by the chief cells of the parathyroid glands is a single polypeptide chain containing 84 amino acid residues. The first 34 amino acids in the N-terminal region of the hormone are responsible for the biological activity of PTH.
- PTH secretion is stimulated by a fall in plasma ionized calcium, whereas an increase in calcium levels inhibits secretion.

Biological actions

The target organs for PTH are bone, kidney, and the gastrointestinal tract. PTH exerts the following biological effects in a process mediated by cAMP:

- ***In the bone***, PTH
  - Increases resorption with loss of proteoglycan from bone matrix
  - Promotes dissolution of matrix with release of calcium, phosphate, hydroxyproline, and hydroxyl ions into the ECF
  - Inhibits collagen synthesis in osteoblasts
  - Increases osteoclastic osteolysis
  - Promotes maturation of precursor cells into osteoblasts and osteoclasts
  - Causes mobilization of Ca^{2+} from bone indirectly by stimulating 1α-hydroxylation of 25-hydroxycholecalciferol in the kidney.

- ◻ ***In the kidney,*** PTH
  - Promotes 1α-hydroxylation of 25-hydroxycholecalciferol to form calcitriol
  - Decreases renal tubular reabsorption of phosphate, and promotes phosphate excretion
  - Increases renal tubular reabsorption of calcium.
- ◻ ***In the gastrointestinal tract,*** PTH
  - Increases calcium and phosphate absorption via calcitriol.

❑ Give an account of the synthesis and biological actions of calcitriol.

Synthesis

Synthesis of calcitriol (1,25-dihydroxycholecalciferol), the biologically active form of vitamin D, involves the following steps:

1. Formation of cholecalciferol (vitamin D_3) from 7-dehydrocholesterol by the action of sunlight on skin.
2. Hydroxylation of cholecalciferol in the liver, catalyzed by the enzyme *25-hydroxylase* to form *25-hydroxy* D_3 *(calcidiol*).
3. Hydroxylation of calcidiol in the kidney to form calcitriol, catalyzed by *1α-hydroxylase.* Low plasma calcium and phosphate and high PTH levels stimulate enzyme activity. Calcitriol exerts feedback inhibition on enzyme activity. Hormones such as oestrogen, prolactin, growth hormone, and cortisol increase enzyme activity.

   The synthesis of calcitriol is shown in Figure 11.1.

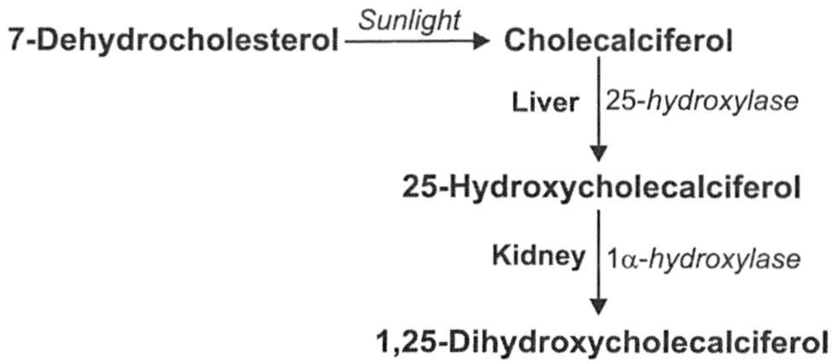

Biological actions

Calcitriol exerts the following biological effects:

- ◻ ***In the gastrointestinal tract***, it
  - Increases absorption of calcium and phosphate by the intestinal mucosal cells *via* synthesis of *calbindin D,* a calcium-binding protein in the enterocytes.
- ◻ ***In the bone***, it
  - Stimulates calcium mobilization together with PTH.
  - Increases synthesis of ALP and osteocalcin, a calcium-binding protein.
- ◻ ***In the kidney***, it
  - Increases renal tubular reabsorption of calcium and phosphate and decreases their excretion in urine.

Mechanism of action

Calcitriol mediates its biological actions by binding to cytosolic receptors followed by binding of the hormone–receptor complex to the nucleus and stimulating transcription and synthesis of a calcium-binding protein and calcium *ATPase.*

❑ Elucidate the biological actions of calcitonin.

Calcitonin exerts the following biological effects on the target organs — the bone and kidney, by stimulation of *adenylate cyclase* and formation of cAMP.

- ***In the bone***, it
  - Inhibits bone resorption by decreasing the mobilization of calcium and phosphate
  - Decreases osteoclastic activity
  - Decreases ALP activity
  - Inhibits the action of calcitriol permitting increased uptake of calcium from the gut.
- ***In the kidney***, it
  - Decreases tubular reabsorption of calcium and phosphate, and increases their urinary excretion.

❑ Enumerate the causes of hypercalcemia.

Hypercalcemia is seen in the following conditions:

1. *Malignancy* due to
   - bone destruction by metastasis
   - increased synthesis of calcitriol
   - release of circulating hypercalcemic factors.

   In tumors of the lung, kidney, and squamous cell carcinoma, the gene for *parathyroid hormone related peptide* (*PTHRP*) is induced. PTHRP, a peptide hormone activates the same receptors as PTH.
2. *Primary hyperparathyroidism* commonly due to
   - adenoma of the parathyroid
   - *multiple endocrine neoplasia* (*MEN*) such as tumors of the pituitary, thyroid, adrenal medulla, and pancreatic islets.
3. *Hypervitaminosis D* due to
   - excess intake as a result of overmedication
   - increased synthesis of 1,25 $(OH)_2D_3$ as in sarcoidosis.
4. *Tertiary hyperparathyroidism* as in patients with end-stage renal failure, characterized by autonomous PTH secretion.
5. *Paget's disease* due to increased bone turnover.
6. *Familial hypocalciuric hypercalcemia* inherited as an autosomal dominant trait characterized by high levels of plasma calcium, low phosphate, and normal or slightly elevated PTH levels.

❑ Outline the clinical manifestations, laboratory findings, and management of hypercalcemia.

Clinical Manifestations

- *Neuromuscular*: Muscle weakness, tiredness, depressed neuromuscular excitability, decreased concentration, and coma (in severe cases).
- *Gastrointestinal*: Anorexia, nausea, vomiting, constipation, abdominal pain, and peptic ulceration.
- *Renal*: Hypercalciuria, polyuria, dehydration, renal calculi, and renal failure (in severe cases).
- *Cardiovascular*: Cardiac arrhythmias, hypertension, and cardiac arrest (in severe cases).
- *Other symptoms*: Bone pain, metastatic calcification, corneal calcification, and vascular calcification.

Laboratory findings

- Plasma phosphate is
  - low in primary hyperparathyroidism and malignancy
  - high in cases of vitamin D intoxication, high bone turnover, and renal failure.
- Plasma ALP is raised, especially in malignancy.
- Plasma PTH is inappropriately high in parathyroid related hypercalcemia.

Management

Drugs available for the treatment of hypercalcemia include

- Bisphosphonates that bind to the hydroxyapatite of bone and inhibit calcium mobilization from bone.
- Oral sodium phosphate that forms calcium phosphate and prevents calcium absorption from the gut. This may cause metastatic calcification if there is renal impairment.
- Calcitonin in severe cases.
- Mithramycin, which inhibits calcium mobilization from bone, is used in malignancy.
- Steroids for hypercalcemia of malignancy, vitamin D intoxication, and sarcoidosis.

❑ What are the causes of hypocalcemia?

Hypocalcemia is seen in

- *Vitamin D deficiency* due to
  - reduced dietary intake as in elderly debilitated persons
  - inadequate synthesis
  - malabsorption as in steatorrhea
  - defective metabolism as in
    - *vitamin D-dependent rickets type I* with impaired 1α-hydroxylation of 25-hydroxy-cholecalciferol
    - *renal rickets (vitamin D-resistant rickets)*, characterized by hypophosphatemia, hypocalcemia, and decreased intestinal absorption of calcium and phosphorous.
- Primary hypoparathyroidism which may be
  - *hereditary* as in *Di George syndrome,* characterized by congenital absence of the parathyroid glands
  - *acquired* (as in case of surgery, radiation, iron overload, and autoimmune disease).
- *Pseudohypoparathyroidism* in which plasma calcium levels are low despite high PTH levels.
- *Secondary hyperparathyroidism* as in patients with chronic renal disease and vitamin D deficiency. Although low plasma calcium causes appropriate PTH secretion, inadequate calcitriol synthesis causes resistance to calcium mobilization from bone.
- *Magnesium deficiency* because magnesium is required for PTH secretion.
- *Neonatal hypocalcemia* due to delayed PTH secretion secondary to maternal hypercalcemia.

❑ Elucidate the clinical symptoms, laboratory findings and management of hypocalcemia.

Clinical manifestations

- *Neuromuscular manifestations*
  - Increased neuromuscular excitability
  - Muscle cramps and spasms
  - Irritability
  - Carpopedal spasm
  - Tetany
  - Stupor
  - Numbness
  - Paraesthesia

- Convulsions
- Chvostek's sign positive (contraction of facial muscle on tapping facial nerve).

- *Cardiovascular manifestations*
  - Prolonged QT interval
  - Prolonged coagulation time
  - Trousseau's sign positive (carpal spasm when sphygmomanometer cuff applied to the upper arm is inflated midway between systolic and diastolic pressure for 3 minutes).
- *Other manifestations*
  - malabsorption
  - cataracts
  - basal ganglia calcification.

Laboratory findings

- In primary hypoparathyroidism and pseudohypoparathyroidism
  - high plasma phosphate
  - low plasma PTH
  - normal plasma calcitriol.
- In secondary hyperparathyroidism
  - low plasma phosphate
  - high PTH
  - low calcitriol.

Management

- Oral vitamin D supplements in asymptomatic and mild hypocalcemia.
- Calcium supplements following thyroidectomy or parathyroidectomy if there is tetany. Administration of intravenous calcium gluconate for severe tetany.
- Calciferol treatment if the parathyroid glands are permanently damaged.
- Oral aluminium hydroxide for hypocalcemia with hyperphosphatemia in renal disease with osteomalacia. Aluminium hydroxide binds phosphate in the intestinal lumen and lowers plasma phosphate. This can be followed by calciferol treatment.

❑ Discuss the causes, consequences, and management of hyperphosphatemia.

Causes

Hyperphosphatemia may be due to

- *Excessive intake* as a result of
  - feeding undiluted cow's milk to infants
  - excessive intravenous administration of phosphate
  - dietary laxatives and enemas.
- *Increased intestinal absorption* as in hypervitaminosis D
- *Redistribution from cells to ECF* as in
  - acromegaly
  - bone destruction
  - chemotherapy
- *Increased tissue catabolism* as in
  - hematological malignancies
  - diabetic ketoacidosis.

- *PTH abnormalities* such as
  - hypoparathyroidism
  - pseudohypoparathyroidism.
- *Decreased renal excretion* as in
  - glomerular failure
  - hypoparathyroidism

Consequences

- Inhibition of *1α-hydroxylase* in the kidney, with decreased synthesis of calcitriol.
- Combination of phosphate with calcium leading to metastatic calcium deposition in tissues, and hypocalcemia.

Management

- Treatment of the underlying cause.
- Oral administration of aluminium, calcium, or magnesium to bind phosphate and prevent intestinal absorption.

❑ Outline the causes, consequences, and management of hypophosphatemia.

Causes

Hypophosphatemia may be due to

- *Decreased intake* as a result of
  - malnutrition
  - inadequate intravenous fluid replacement
  - imperfect dialysis.
- *Inadequate intestinal absorption*
  - phosphate binding agents (magnesium and aluminium hydroxide)
- *Redistribution from ECF to cells*
  - into bone during healing of rickets and after parathyroidectomy
  - into somatic cells during acute treatment of diabetes mellitus
- *Increased loss* as in
  - severe diarrhea
  - hyperparathyroidism
- *Others*
  - vitamin D deficiency
  - recovery phase of diabetic ketoacidosis
  - alcohol withdrawal
  - respiratory alkalosis.

Consequences

Severe hypophosphatemia can cause

- muscular weakness
- central and peripheral neurological disorders due to ATP depletion
- rickets and osteomalacia

Management: Enteral or parenteral administration of phosphate.

❑ Enumerate the causes and consequences of hypermagnesemia.

Hypermagnesemia is seen in acute and chronic renal failure due to a defect in glomerular filtration.

Consequences of hypermagnesemia include

- weakness
- mental confusion
- nausea
- impairment of neuromuscular transmission
- respiratory paralysis, cardiac arrest, and renal failure in severe cases.

Management: Involves intravenous administration of calcium, and dialysis if there is renal failure.

❑ Give an account of the causes, clinical features, and management of hypomagnesemia.

Causes

Hypomagnesemia may result from

- *Inadequate intestinal absorption* as in
  - severe malnutrition
  - malabsorption
  - chronic diarrhea
  - post-gastrointestinal surgery.
- *Increased renal loss* due to
  - diuretics
  - cisplatin-induced nephropathy
  - hyperaldosteronism
  - mineralocorticoid excess.
- *Redistribution from plasma to cells* as in
  - acute pancreatitis
  - post parathyroidectomy
  - during recovery phase of ketoacidosis
- *Others*
  - alcoholism
  - cirrhosis
  - fistulae.

Clinical Features

Hypomagnesemia can decrease PTH secretion leading to

- Hypocalcemia
- Delirium
- Tremor
- Tetany
- Cardiac arrhythmias.
- Muscular weakness
- Ataxia
- Choreiform movements
- Convulsions

Management

- Oral supplementation of magnesium
- Intravenous administration in severe deficiency with malabsorption.

❑ Describe the etiopathogenesis, clinical features, diagnosis, and management of osteoporosis.

Pathogenesis

- Osteoporosis results from a decrease in the amount of mineralized bone due to an imbalance between bone formation and resorption.
- Osteoporosis may be
  - *Primary*, due to ageing. The disease is more common in postmenopausal women because estrogen deficiency after menopause leads to increased bone resorption.
  - *Secondary,* due to endocrine abnormalities (hypogonadism, Cushing's syndrome, diabetes mellitus), malignancies, and calcium malabsorption.
- The disease is characterized by pain, deformities, and increased susceptibility to fractures.

Diagnosis

- Increased urinary excretion of calcium and hydroxyproline in rapidly progressing disease
- Increased urinary excretion of peptides involved in cross-linking mature collagen (*pyridinium cross links*)
- Dual energy X-ray absorptiometry (DEXA), a useful imaging technique.

Management

- Hormone replacement therapy (HRT)
- Calcium intake
- Avoidance of alcohol
- Vitamin D, bisphosphonates and fluoride.

Chapter 12

Trace Elements

❑ What are trace elements?

- Trace elements are minerals that are essential in the diet in amounts less than 100 mg per day.
- These include iron, iodine, zinc, copper, manganese, cobalt, chromium, molybdenum, fluoride, and selenium.

❑ Explain the distribution of iron in the body.

Iron is present in the body in two forms:

- Heme (iron-porphyrin) in hemoglobin, myoglobin, cytochromes, catalase, and peroxidase.
- Non-heme iron, which is protein-bound, and includes storage (ferritin, hemosiderin) and transport forms of iron and flavoproteins.

In addition, iron is also present in interstitial fluid (labile pool), and in plasma.

❑ Outline the sources and daily requirements of iron.

Sources: Organ meats (liver, heart, kidney, and spleen), egg yolk, enriched bread, cereals, raisins, apricots, dates, nuts, fish, molasses, legumes, beans, and spinach.

Daily requirements

Children	10 – 15 mg
Adult men and postmenopausal women	10 mg
Premenopausal women	18 mg

Since only 1 mg/d of iron is lost in a normal, healthy adult male, the daily requirement for iron is minimal. In females, iron requirement is high due to menstrual loss of ~16 – 32 mg per month. Iron requirement is also high in pregnancy, lactation, in the first two years of life, and in puberty.

❑ Explain the mechanism of iron absorption.

- Dietary iron, which is in the ferric form (Fe^{3+}), is reduced to the ferrous state (Fe^{2+}) in the stomach by gastric HCl and dietary reductants such as ascorbic acid and cysteine.
- Fe^{2+} in the stomach is prevented from precipitating in the intestine by combining with *gastroferrin*, amino acids, or fructose.
- Iron is absorbed in the proximal small intestine.
- Heme iron is absorbed intact, the heme degraded, and the iron released.

- Non-heme iron absorbed in the Fe^{2+} state is oxidized to Fe^{3+} form, which is bound by an intracellular carrier protein.
- Only 10% of the dietary iron is absorbed and a very little is excreted.
- The iron requirements of the body are met by heme degradation.
- Iron absorption is inhibited by
  - phytate, phosphate, and oxalate which form insoluble complexes
  - malabsorption syndromes such as steatorrhea
  - gastrectomy or surgical removal of the intestine.
- Iron absorption is increased in
  - women and children
  - iron deficiency anemia
  - pernicious and hypoplastic anemias.

❑ How is iron transported in blood?

Iron absorbed from the intestinal mucosal cells in ferrous state (Fe^{2+}) into the portal blood is converted by ceruloplasmin into the ferric form (Fe^{3+}) and incorporated into *transferrin.*

Transferrin (siderophyllin)

- binds two atoms of Fe^{3+} per molecule.
- has a total iron-binding capacity (TIBC) of about 300 µg/dL (54 µmol/L), of which, only one-third (~100 µg/dL) is saturated and the remaining two-third of the protein is iron-free [unsaturated iron-binding capacity (UIBC)]
- transports iron to cells that synthesize heme proteins (hemoglobin, myoglobin, cytochromes).
- binds to a receptor, the iron-transferrin-receptor complex, which is taken up by endocytosis, the iron internalized, and transferrin returned to circulation.

Approximately 80% of the iron delivered to cells is utilized for hemoglobin synthesis and the rest stored.

❑ Enumerate the conditions in which plasma iron levels are altered.

- Plasma iron is *decreased* in
  - iron-deficiency anemia
  - chronic inflammatory diseases
  - infection.
- Plasma iron is *increased* in
  - megaloblastic anemia, with deficiency of vitamin B_{12} and folate and inability of the marrow to utilize iron
  - hemoglobinopathies, due to hemolysis
  - hepatitis, with release of iron from the liver
  - cirrhosis
  - oral overdose of iron.

❑ What are the storage forms of iron?

The important intracellular iron storage proteins are

- *Ferritin,* present in the intestine, liver, spleen, and bone marrow. It is an acute phase protein that is increased in inflammatory conditions, iron overload, liver disease, and malignancy due to release of the protein from tissues. Plasma ferritin is low in iron deficiency.

- *Hemosiderin*, a water-insoluble protein found mainly in the reticuloendothelial system. It has lower mobilization than that of ferritin.

❑ Enumerate the causes and consequences of iron deficiency.

▫ *Causes*
- Inadequate intake: High cereal, low meat diet.
- Malabsorption due to achlorhydria, diarrhea, steatorrhea, and surgical resection of the stomach or intestine.
- Excessive loss of blood.

▫ *Consequences*
- Hypochromic, microcytic anemia.
- Decreased erythrocyte formation and hemoglobin synthesis.
- Low ferritin stores.
- High plasma transferrin levels.

❑ Enumerate the causes, clinical consequences, biochemical findings and management of iron overload.

▫ *Causes*

Iron overload may result from

1. Excessive parenteral administration due to multiple blood transfusions. The excess iron is deposited as hemosiderin in the liver and spleen, leading to *hemosiderosis* or *acquired hemochromatosis*.
2. Increased intestinal absorption due to
   - Dietary overload. This condition is seen among *Bantu* tribes of Africa due to consumption of corn, which is low in phosphate content, cooked in iron pots (*Bantu siderosis*).
   - Anemia associated with ineffective erythropoiesis
   - Idiopathic hemochromatosis, a common inherited disease in Caucasians.

▫ *Clinical consequences*

Hemosiderosis occurs due to an increase in iron stores in the form of hemosiderin that may cause parenchymal damage, leading to the clinical condition known as *hemochromatosis* that is characterized by
- hepatomegaly, hepatic fibrosis, and cirrhosis
- pancreatic islet cell destruction and diabetes mellitus associated with bronzed pigmentation of the skin (*bronze diabetes*)
- cardiac arrhythmias and cardiac failure
- hypogonadism.

▫ *Biochemical findings*
- High plasma iron concentration (>100 mg/dL) with very high percentage saturation of transferrin
- Elevated plasma ferritin
- Increased amounts of stainable iron in liver biopsy specimens.

▫ *Management*: Removal of iron by repeated venesection.

❑ What is Bantu siderosis?

Dietary iron overload seen in *Bantu* tribes of Africa due to consumption of food high in corn, low in phosphate content, and cooked in iron pots, causes increased iron absorption and increase in hemosiderin (*Bantu siderosis*).

❑ What is hemochromatosis?

- Hemochromatosis is a clinical condition in which there is an increase in iron stores as hemosiderin (hemosiderosis).
- Hemochromatosis causes parenchymal damage in several tissues such as the liver and pancreas.

❑ What is bronze diabetes?

Bronze diabetes is a condition characterized by pancreatic islet cell destruction and diabetes mellitus associated with bronzed pigmentation of the skin resulting from iron overload and increase in hemosiderin (hemochromatosis).

❑ Describe the distribution of zinc.

- Appreciable amount of zinc is present in bones, skin, teeth, prostate, spermatozoa, epididymis, liver, and kidney.
- The enzyme *carbonic anhydrase* accounts for most of the zinc present in the erythrocytes.
- Zinc induces the synthesis of the protein *metallothioneine*, which binds zinc and functions as a storage depot for zinc.

❑ Enumerate the biological functions of zinc.

Zinc performs a wide range of functions that include

- growth
- reproduction
- wound healing
- cofactor for enzymes such as *alcohol dehydrogenase, carbonic anhydrase, pyruvate carboxylase,* cytosolic *superoxide dismutase,* and *DNA polymerase.*
- maintenance of plasma vitamin A concentrations and mobilization of vitamin A from the liver; thus, zinc has a role in the visual cycle.
- storage and release of insulin by forming zinc – insulin complex.

❑ Outline the causes and clinical manifestations of zinc deficiency and zinc excess.

Zinc deficiency is caused by

1. poor absorption due to consumption of diets high in phytic acid.

   Acrodermatitis enteropathica, a rare, autosomal recessive disease, characterized by severe zinc deficiency resulting from poor intestinal absorption, that is seen in infants after weaning from breast milk. Picolinic acid in breast milk is believed to aid zinc absorption.
2. excessive excretion as in alcoholism, trauma, surgery, chronic renal disease, and proteinuria
3. parenteral nutrition
4. chronic liver disease
5. malignancies.

The ***clinical manifestations*** of zinc deficiency include

1. growth retardation
2. delayed wound healing
3. hypogonadism.

Zinc excess

The clinical manifestations of zinc excess are nausea, vomiting, gastric ulcer, pancreatitis, respiratory distress, and pulmonary fibrosis.

❑ Briefly explain the transport and metabolic fate of copper.

- About 90–95% of the total plasma copper is incorporated into ceruloplasmin and released into circulation.
- Plasma copper, being protein bound, is not excreted in the urine. It is excreted into the bile and thence into the feces.

❑ Enumerate the biological functions of copper.

Copper plays a vital role in metabolic reactions as a component of the following enzymes:

1. *Cytochrome c oxidase*, an important component of the electron transport chain.
2. Cytosolic *superoxide dismutase* that scavenges superoxide, a toxic reactive oxygen species.
3. *Ceruloplasmin*, a Cu^{2+} binding plasma protein.
4. *Lysyl oxidase* that acts on lysine to form an aldehyde.
5. *Tyrosinase*, which catalyzes the synthesis of dopa and melanin.
6. *Dopamine hydroxylase*, a mixed function oxidase involved in norepinephrine synthesis.
7. *Monoamine oxidase* that catalyzes oxidative deamination of catecholamines.
8. *Diamine oxidase* that converts aliphatic diamines to amino aldehydes.

❑ What are the consequences of copper deficiency?

Deficiency of copper can lead to the following consequences:

- Impaired release of iron from ferritin into plasma due to decreased ceruloplasmin synthesis. This results in decreased heme synthesis and hypochromic microcytic anemia.
- Decrease in *cytochrome oxidase* with decrease in ATP synthesis and impaired synthesis of phospholipids required for myelin formation.
- Decreased amine oxidase activity, which impairs cross-linking of bone collagen and aortic elastin, leading to bone and cardiovascular disorders.

❑ Write notes on Wilson's disease.

- ***Inheritance***: Autosomal recessive trait.
- ***Biochemical defect***
  - Accumulation of excess copper due to reduced excretion into the bile.
  - Plasma ceruloplasmin is low, with reduced incorporation of copper into ceruloplasmin. There is increased excretion of copper in the urine.
- ***Clinical consequences***
  - Deposition of copper in the liver and the lenticular nucleus of the brain, leading to hepatolenticular degeneration.
  - Cirrhosis with elevated plasma levels of liver marker enzymes (*aminotransferases*).
  - Deposition of copper in the Descement's membrane of the cornea (*Kayser–Fleischer rings*).
  - Accumulation of copper in the kidney, leading to renal tubular damage and increased urinary excretion of amino acids, peptides, and glucose.
  - Hemolytic anemia.

- ***Diagnosis***
  - Plasma ceruloplasmin level <200 mg/L (20 mg/dL)
  - Plasma copper >250 µg/g dry weight
  - Presence of Kayser–Fleischer rings.
- ***Treatment***: Oral administration of penicillamine.

❑ Elucidate the physiological functions of manganese.

Manganese is required for the following functions:

1. Bone structure
2. Reproduction
3. Central nervous system function
4. Component of enzymes such as:
   - *superoxide dismutase*
   - *pyruvate carboxylase*
   - *diamine oxidase.*
5. Activation of the enzymes
   - *glycosyl transferases*
   - *arginase*
   - *phosphatases*
   - *cholinesterase.*
6. Oxidative phosphorylation
7. Blocking lipid peroxidation.

❑ Outline the biological importance of iodine.

- Iodine is a constituent of thyroid hormones.
- Deficiency of iodine leads to cretinism in children and myxoedema in adults, whereas iodine excess causes thyrotoxicosis.

❑ What are the manifestations of manganese deficiency and excess?

- Manganese deficiency causes bleeding and increase in prothrombin time.
- Excess manganese induces neuropsychiatric signs and symptoms.

❑ Write a note on the physiological role and clinical significance of cobalt.

1. Cobalt as a constituent of vitamin B_{12} is required for the normal formation of erythrocytes and for the activity of the enzymes *methylmalonyl CoA mutase, methyltetrahydrofolate oxidoreductase, homocysteine methyltransferase*, and *ribonucleotide reductase.*
2. Cobalt deficiency is associated with vitamin B_{12} deficiency.
3. Excess cobalt causes polycythemia, inhibition of iodine uptake, and development of goitre.

❑ What is the biological role of molybdenum?

Molybdenum is a component of the enzymes *xanthine oxidase, aldehyde oxidase,* and *sulphite oxidase.*

❑ What are the functions of selenium?

Selenium is required for the following functions:

1. Normal growth and fertility.
2. Protection against toxic free radicals: Selenium together with vitamin E forms an antioxidant system that prevents free radical induced membrane damage. As a component of *glutathione peroxidase*, selenium protects membrane lipids against peroxidative damage.
3. Immune response.
4. Synthesis of ubiquinone and mitochondrial ATP.
5. Protection against carcinogens.
6. Detoxification of heavy metals such as mercury.

❑ What are the clinical symptoms of selenium deficiency and excess?

◘ *Deficiency*

- Selenium deficiency has been reported in Keshan province in China due to low concentrations of selenium in the soil.
- *Keshan disease* is characterized by myopathy, cardiac enlargement, and arrhythmias.

◘ *Excess*

- Selenium excess causes alopecia, dermatitis, lassitude, and irritability.

❑ Explain the physiological role and clinical significance of chromium.

- Chromium, a cofactor for insulin, potentiates insulin in stimulating glucose utilization and plays a role in glucose tolerance in humans.
- Chromium deficiency causes impaired growth and glucose tolerance.
- Chromium toxicity, which is an occupational hazard, can cause dermatitis, kidney failure, and lung cancer.

❑ Outline the physiological and clinical significance of fluoride.

- ◘ Fluoride in trace amounts is essential for the development of teeth and bones, and protection against dental caries and osteoporosis.
- ◘ Excess fluoride in drinking water causes *dental fluorosis* (*mottled enamel*), and in severe cases, loss of enamel with *'pitting'* of the teeth.
- ◘ Chronic fluoride toxicity
  - increases bone density,
  - induces hypercalcification of the bones of the limbs, spine, and pelvis. Collagen in the bone and the ligaments of the spine may also undergo calcification.

Section 4

Endocrinology and Nutrition

- Hormone Classification and Mechanism of Action
- Hormones of the Hypothalamus and the Pituitary
- Thyroid Hormones
- Adrenal Hormones
- Gonadal Hormones
- Gastrointestinal and Pancreatic Hormones
- Elements of Nutrition

Chapter 13

Hormone Classification and Mechanism of Action

❑ What is a hormone?

A hormone is a substance secreted by an endocrine gland and transported by circulation to another tissue or gland whose function it regulates.

❑ What are trophic hormones?

1. Hormones secreted by the anterior pituitary that stimulate synthesis and secretion of hormones from target endocrine glands are termed *trophic hormones*.
2. Adrenocorticotrophic hormone (ACTH), the anterior pituitary hormone, is a trophic hormone that regulates the secretion of hormones from the adrenal cortex.

❑ How do hormones mediate their biological effects?

1. Hormones mediate their biological effects by binding to specific *receptors* on target cells.
2. The *receptors* are proteins that have high affinity for the hormone. They contain two functional domains — a recognition domain that binds the hormone, and a coupling domain that couples the bound hormone to an intracellular function (signal transduction).

❑ How are hormones classified?

1. Based on their chemical structure, hormones are classified into
   - proteins or peptides (*e.g.*, insulin)
   - amino acid derivatives (*e.g.*, catecholamines)
   - fatty acid derivatives (*e.g.*, eicosanoids)
   - cholesterol derivatives (*e.g.*, steroids)
   - gases (*e.g.*, nitric oxide).
2. Based on their solubility, hormones are categorized into
   - *hydrophilic hormones* (*e.g.*, catecholamines, polypeptide, and protein hormones).
   - *lipophilic hormones* (*e.g.*, steroid hormones).
3. Based on the site of synthesis, hormones can be classified into
   - *endocrine hormones* that are synthesized by endocrine glands and transported to their target cells by circulation.
   - *paracrine hormones* that are synthesized close to their target tissues.
   - *autocrine hormones* that exert their effects on the cells synthesizing them.
4. Based on the location of the receptors and signal transduction, hormones are classified into
   - *group I hormones* that bind to intracellular receptors and transduce the signal via the hormone–receptor complex.

- *group II hormones* that bind to cell surface receptors and mediate their effects *via* intermediary molecules called second messengers.

❑ What are the salient features of group I hormones?

The group I hormones

- are lipophilic.
- are transported in circulation bound to proteins.
- include the steroid hormones, calcitriol, retinoic acid, and the thyroid hormones.
- diffuse through the plasma membrane of target cells and bind to specific high affinity receptors in the cytosol or the nucleus. The hormone – receptor complex binds to a specific region on DNA termed the *hormone response element* (*HRE*) and activates or represses specific genes (Fig. 13.1).

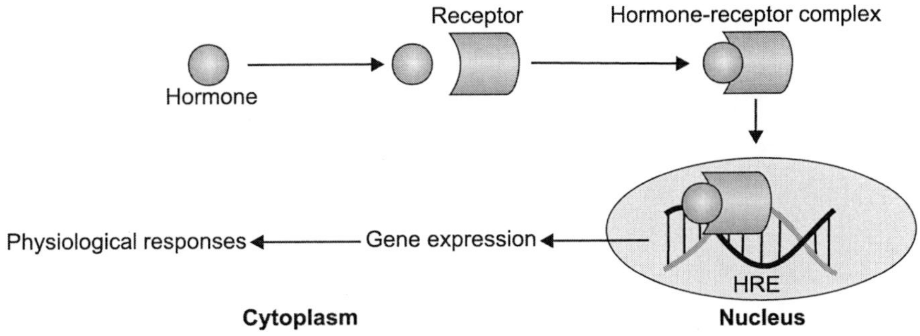

Fig. 13.1 Mode of action of group I hormones

- influence various steps in the information pathway, including
  - gene transcription
  - post-transcriptional processing
  - mRNA degradation
  - translation
  - protein degradation.

❑ Give a brief account of group II hormones.

The group II hormones

- are water-soluble.
- include catecholamines, polypeptides, proteins and glycoprotein hormones.
- bind to receptors on the plasma membrane of the target cell. The hormone–receptor interaction influences metabolic pathways *via* second messengers such as cyclic 3',5'-adenosine monophosphate (cAMP) and cyclic guanosine monophosphate (cGMP).
- are subclassified based on the second messengers that mediate their action.

❑ How are group II hormones subclassified?

Group II hormones are categorized into *four subgroups* based on the second messengers that mediate their action. These are:

- Group IIA hormones that use cAMP as the second messenger include peptide hormones such as adrenocorticotrophic hormone (ACTH), glucagon, and somatostatin.

- Group IIB hormones that use cGMP as the second messenger include atrial natriuretic factor (ANF), nitric oxide and sodium azide.
- Group IIC hormones that use Ca^{2+} or phosphatidylinositols as second messengers include cholecystokinin (CCK), gonadotrophin releasing hormone (GnRH), and thyrotrophin releasing hormone (TRH).
- Group IID hormones that act through a *kinase* or a *phosphatase* cascade include insulin, growth hormone (GH) and prolactin.

Discuss signal transduction by hormones that use cAMP as the second messenger.

- Group II A hormones use cAMP as the second messenger. Figure 13.2 shows a schematic representation of cAMP mediated signal transduction.

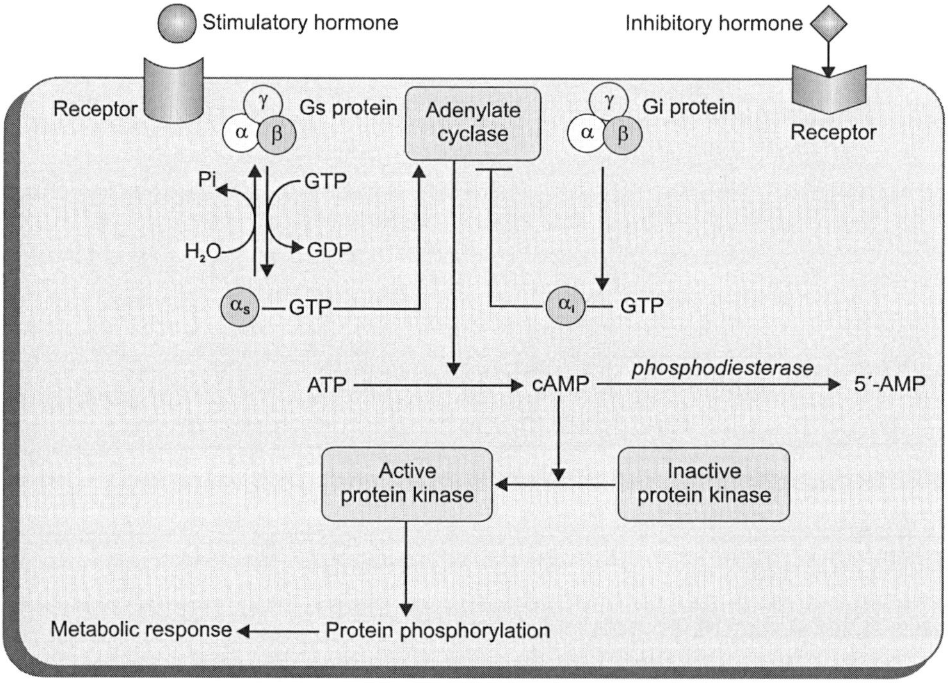

Fig. 13.2 cAMP-mediated signal transduction

- The hormone interacts with its receptor to either activate or inactivate *adenylate cyclase*, the enzyme responsible for cAMP synthesis.
- While ACTH, β-adrenergics and glucagon stimulate (*s*), the α_2-adrenergics, acetylcholine and somatostatin inhibit (*i*) cAMP production.
- cAMP binds to a heterotetrameric *protein kinase* that contains two regulatory (R) and two catalytic (C) subunits.
- Binding of cAMP activates the C subunits that catalyze phosphorylation of serine or threonine residues in proteins.
- cAMP mediated phosphorylation – dephosphorylation of proteins influences:
  - steroidogenesis

- metabolic pathways
- enzyme induction
- gene regulation
- cellular growth
- replication.

- The *adenylate cyclase* system is regulated as follows:
  - The *G-protein complex* comprises two GTP-dependent trimeric proteins — Gs and Gi containing α, β, and γ subunits. In the absence of the hormone, the G-protein complex is in the inactive GDP-bound form. Binding of the hormone to the receptor activates the G-protein complex. As a consequence, the α-subunit dissociates from the β-γ heterodimer, binds to, and activates adenylate cyclase or some other effector.
  - Methylated xanthine derivatives such as caffeine and theophylline that inhibit *phosphodiesterase*, the enzyme that degrades cAMP, increase intracellular cAMP levels and prolong the actions of these hormones.

Give a brief account of group II B hormones.

- Group II B hormones include
  - atrial natriuretic factor (ANF), an aldosterone inhibitor that induces natriuresis and vasodilation
  - nitric oxide and sodium azide, which are potent vasodilators and smooth muscle relaxants.
- These hormones use cGMP as the second messenger. They activate *guanylate cyclase* and increase cGMP levels, leading to activation of cGMP-dependent protein kinase and phosphorylation of smooth muscle proteins.

Discuss signal transduction by calcium and phosphoinositol system.

Signal transduction by calcium and phosphoinositol system involves the following steps:

- Group IIC hormones such as cholecystokinin (CCK), gonadotrophin releasing hormone (GnRH), and thyrotrophin releasing hormone (TRH) use Ca^{2+} or phosphatidylinositols as the second messengers.
- Binding of hormone to receptor activates *phospholipase C* that activates a calcium channel.
- *Phospholipase C* hydrolyses phosphoinositol 4,5-bisphosphate to form inositol triphosphate and 1,2-diacylglycerol.
- Inositol triphosphate releases Ca^{2+} stored in the mitochondria and sarcoplasmic reticulum.
- Ca^{2+} interacts with the calcium-binding protein *calmodulin* and regulates
  - enzymes of cyclic nucleotide metabolism
  - actin–myosin complex of smooth muscle
  - ion transport
  - cell motility
  - mitosis
  - endocytosis.
- Diacylglycerol activates protein *kinase C* in the presence of free Ca^{2+}.
- Activated protein *kinase C* phosphorylates specific substrates, leading to altered physiological responses.
- Agents that increase phosphatidylinositols include ACTH, angiotensin II, luteinizing hormone (LH), and cAMP.

Signal transduction by calcium and phosphoinositol system is summarized in Figure 13.3.

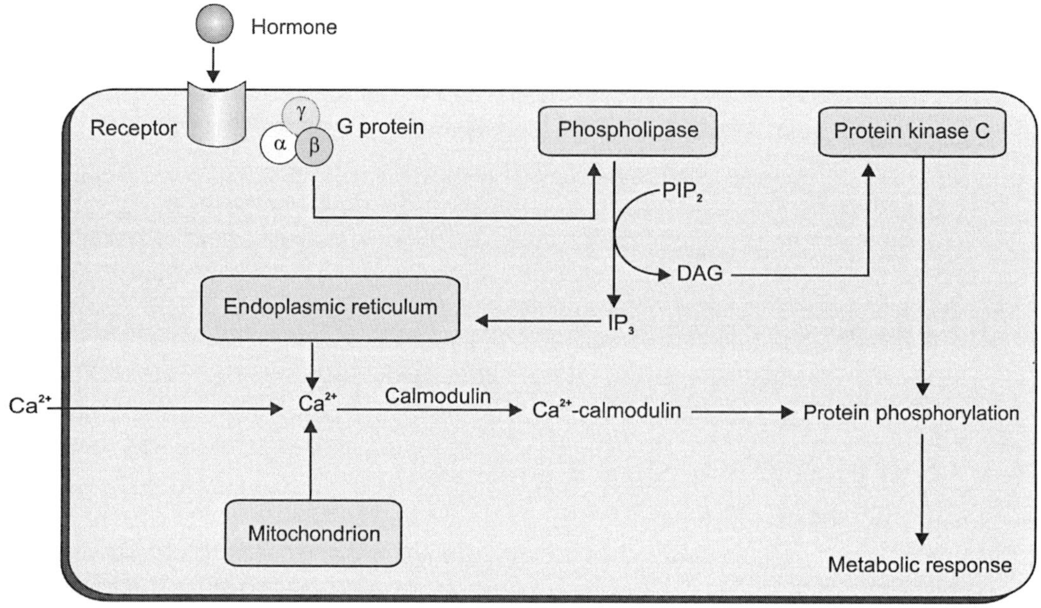

Fig. 13.3 Signal transduction by Ca^{2+} and phosphoinositol system (DAG = diacylglycerol; IP_3 = inositol; 1,4,5-triphosphate; PIP_2 = phosphoinositol 4,5-bisphosphate)

❑ Explain signal transduction through a kinase or a phosphatase cascade.

- Group II D hormones such as insulin, growth hormone (GH), and prolactin act through a *kinase* or a *phosphatase* cascade.
- Receptors for insulin, epidermal growth factor (EGF), and insulin-like growth factor-I (IGF-I) have intrinsic tyrosine *kinase* activity that can phosphorylate tyrosine residues in proteins, resulting in a cascade of events.
- Receptors for prolactin and growth hormone that lack intrinsic tyrosine *kinase* activity undergo dimerization on hormone binding, activate cytoplasmic protein *kinases* (JAKs or TYK) that phosphorylate
  - tyrosine residues on the receptor
  - cytosolic proteins termed signal transduction and activators of transcription (STATs) that associate with the phosphorylated receptor.
- Phosphorylated STAT protein
  - undergoes dimerization
  - translocates to the nucleus
  - binds to specific DNA elements and regulates transcription.

Chapter 14

Hormones of the Hypothalamus and the Pituitary

❑ What are trophic hormones?

- Hormones secreted by the anterior pituitary, which influence hormone secretion by other endocrine glands such as thyroid, adrenal cortex, etc., are termed *trophic hormones.*
- The trophic hormones are, in turn, controlled by the hypothalamus which synthesizes regulatory peptides that are transported by the hypothalamic portal system to the anterior pituitary.

❑ Enumerate the factors that regulate secretion of anterior pituitary hormones.

The factors regulating anterior pituitary hormone secretion include the following:
- Stress (physical, mental, emotional or hypoglycemia).
- Increased plasma levels of target cell hormones, which exert feedback inhibition.
- Circadian rhythm: This may be lost in disease states.
- Drugs: For example, L-dopa stimulates GH secretion.

❑ Give a brief account of the anterior pituitary hormones.

1. *Growth hormone* (*GH*) regulates growth and intermediary metabolism. GH secretion is stimulated by *growth hormone releasing hormone* (*GHRH*), and inhibited by *growth hormone release inhibiting hormone* (*GHRIH*) or *somatostatin*.
2. *Thyroid-stimulating hormone* (*TSH*) regulates thyroid function. TSH secretion is stimulated by *thyrotrophin releasing hormone* (*TRH*).
3. *Adrenocorticotrophic hormone* (*ACTH*) stimulates glucocorticoid synthesis, and secretion from the adrenal cortex. It is regulated by *corticotrophin-releasing hormone* (*CRH*) from the hypothalamus.
4. *Luteinizing hormone* (*LH*) plays a role in ovulation, progesterone synthesis, and testosterone production. It is under the control of hypothalamic *gonadotrophin-releasing hormone* (*GnRH*).
5. *Follicle-stimulating hormone* (*FSH*) regulates estrogen synthesis, oogenesis, and spermatogenesis. It is regulated by GnRH.
6. *Prolactin* acts on the mammary gland to induce lactation. It is regulated by *prolactin-release inhibiting hormone* (*PRIH*), and TSH.
7. *β-Lipotrophin* is a neurotransmitter and endorphin precursor that is regulated by CRH.

Give an account of the secretion and biological effects of growth hormone.

Secretion

- Growth hormone (GH) or *somatotrophin* is secreted by somatotrophs in the anterior pituitary.
- The human GH (hGH) is a single polypeptide chain containing 191 amino acid residues.
- Binding of GH to its receptor causes dimerization and phosphorylation of the receptor with activation of JAK2 tyrosine kinase and other signaling pathways.
- GH secretion
  - is low in infancy
  - increases during early childhood
  - reaches peak levels at puberty
  - decreases after the third decade
- Plasma GH is episodic and pulsatile, and shows diurnal variations with peak values seen during sleep.
- Stimuli for GH secretion include:
  - stress
  - fasting
  - exercise, arginine
  - hypoglycemia
  - glucagon
  - drugs such as L-dopa
- GH secretion is inhibited by hyperglycemia.

Biological effects

- *Growth promotion* by inducing hepatic synthesis of *insulin-like growth factors* (*IGF*) or *somatomedins*.
- *Carbohydrate metabolism* that
  - increases blood glucose levels by increasing hepatic glucose production and decreasing peripheral utilization
  - inhibits glucose transport and glycolysis.
- *Lipid metabolism* that mobilizes adipose tissue fat with increase in plasma free fatty acids which inhibit insulin release from the pancreas.
- *Protein metabolism* that
  - increases protein synthesis by stimulating amino acid uptake
  - increases amino acid transport into muscle.
- *Other effects*
  - retention of Na^+, K^+, and Cl^-
  - prolactin-like effects such as lactogenesis
  - stimulation of bone growth by promoting a positive calcium, phosphate, and magnesium balance
  - increase cartilage formation.

How is the secretion of growth hormone regulated?

- GH secretion is
  - stimulated by growth hormone-releasing hormone (GHRH) in a process that is calcium-dependent and mediated by cAMP, and
  - inhibited by *growth hormone-release inhibiting hormone* (*GHRIH*) or *somatostatin.*

    Somatostatin synthesized by the hypothalamus, gastrointestinal tract, and the D cells of the pancreatic islets decreases the response of GH to secretagogues.

- GH release is influenced by various factors that act via GHRH and/or somatostatin.
  - IGF-I inhibits GHRH release and stimulates somatostatin secretion. An increase in IGF-I/SM-C leads to inhibition of GH secretion and a decrease releases GH.
  - The neurotransmitter dopamine stimulates GHRH secretion, whereas acetylcholine and gamma-aminobutyric acid (GABA) inhibit somatostatin.
  - α-Adrenergic blockers such as phentolamine inhibit GH release, while β-adrenergic blockers stimulate GH secretion by inhibiting somatostatin secretion.
  - Thyroid hormones, VIP, cAMP, Ca^{2+}, and Na^{+} inhibit GH secretion by increasing somatostatin release.

❑ Discuss the clinical features, diagnosis, and management of growth hormone excess.

Excess growth hormone secretion causes *gigantism* in children and *acromegaly* in adults.

- ***Gigantism***
  - Occurs in childhood before fusion of the epiphyseal plates.
  - Affected children may reach a height of 7 – 9 feet.
  - Is characterized by postural defects.
  - Results in death due to infection, cardiac failure, or pituitary tumor.
- ***Acromegaly***
  - Occurs after epiphyseal fusion in middle age.
- ***Clinical features***
  - Overgrowth of bone and soft tissue.
  - Enlargement of hands, jaw, heart, tongue, larynx, liver, and kidney.
  - Protruding jaw and enlarged nose.
  - Excessive sebaceous gland secretion with oily skin.
  - Increase in BMR with increased sweating.
  - Headache, and muscle weakness.
  - Hypertension associated with low renin and aldosterone secretion.
  - Hirsutism and amenorrhea in women.
  - Impaired glucose tolerance and diabetes mellitus.
  - Death due to cardiovascular, cerebrovascular and respiratory diseases or malignancy.
- ***Diagnosis***
  - Measurement of plasma GH
  - *Glucose-suppression test:* An increase in plasma GH level not suppressed by high plasma glucose suggests autonomous GH secretion.

    The test involves oral administration of glucose (75 g in 300 mL) and estimation of plasma glucose and GH before the glucose load and at 30, 60, 90, and 120 minutes after glucose intake. In normal subjects, glucose suppresses plasma GH concentration, whereas in acromegaly, there is failure of suppression.
  - Increased *serum IGF-I* concentrations.
- ***Management***
  - Transsphenoidal surgery.
  - Conventional or heavy particle radiation if surgery has failed or is contraindicated.
  - Treatment with bromocriptine and octreotide, a long-acting somatostatin analog as an adjunct.

❑ Elucidate the different types of dwarfism due to GH deficiency.

- GH deficiency causes *dwarfism* in children.

  Three major types of dwarfism occur due to GH deficiency. They differ in the response to GH stimulation and plasma GH levels, as detailed below:

  1. *GH-deficient dwarfs*
     - have low levels of plasma IGF-I and IGF-II
     - show normal response to GH administration.
  2. *Pygmies*
     - have normal plasma levels of GH and IGF-II
     - have low levels of plasma IGF-I
     - the underlying cause may be a post-GH receptor defect.
  3. *Laron type dwarfs*
     - have high plasma GH
     - have low plasma IGF-I and IGF-II levels
     - lack functional GH receptors.

- ***Diagnosis***

  Low plasma GH during sleep and after excercise. Insulin stimulation test is performed to rule out GH deficiency. The test involves administration of insulin (0.15 U/kg body weight) to induce hypoglycemia. Blood samples are withdrawn before injection and at 30, 60, 90, and 120 minutes after the injection. Plasma GH level should be greater than 20 mU/L.

❑ Discuss the synthesis, biological effects, regulation, and clinical significance of thyrotrophin.

- ***Synthesis***
  - Thyroid stimulating hormone (TSH) or thyrotrophin is synthesized by the thyrotrophs of the anterior pituitary.
  - TSH is a glycoprotein containing an α-subunit, which it shares with LH, FSH, and human chorionic gonadotrophin (hCG), and a β-subunit that is unique to TSH. TSH secretion exhibits a circadian rhythm with levels rising in the evening, reaching a peak at 11 P.M., and declining to low levels by 11 A.M.
- ***Biological effects***
  - TSH stimulates the synthesis and secretion of thyroid hormones by binding to specific receptors in the plasma membrane of thyroid cells and activating *adenylate cyclase* with increase in cAMP levels.
  - TSH influences iodide uptake, organification, coupling and release of thyroid hormones from the thyroid gland.
  - Chronic effects on the thyroid: Increase in the size and number of thyroid cells, increased synthesis of proteins, phospholipids, and nucleic acids.
- ***Regulation***
  - Thyrotrophin releasing hormone (TRH) from the hypothalamus
    - increases the synthesis and secretion of TSH via a calcium–phosphatidylinositol mechanism.
    - increases the biological activity of TSH by increasing the rate of glycosylation.
    - stimulates prolactin release.

TRH is present in other regions of the brain besides the hypothalamus, as well as in the spinal cord, cerebrospinal fluid, gastrointestinal tract, and pancreatic islets. In addition to regulating TSH, it may function as a neurotransmitter.

- Somatostatin inhibits TSH release.
- Circulating thyroid hormones feedback inhibit TSH secretion.

- ***Clinical significance***
  - In *hyperthyroidism*, TSH release is suppressed due to feedback inhibition by high plasma thyroid hormone levels.
  - In *Graves' disease*, a serum factor called *long-acting thyroid stimulator* (*LATS*) with properties similar to TSH binds to the TSH receptor in the thyroid gland and activates *adenylate cyclase* with increase in cAMP levels.

    In *primary hypothyroidism*, TSH is released in response to low plasma thyroid levels with compensatory hypertrophy of the thyrotrophs and enlargement of the pituitary gland.

Elucidate the synthesis and regulation of ACTH secretion.

Synthesis

- ACTH is a single polypeptide containing 39 amino acid residues. It is synthesized by corticotrophs of the anterior pituitary as a large precursor molecule called *pro-opiomelanocortin* (*POMC*) of 285 amino acids.
- Cleavage of POMC yields ACTH and *β-lipotrophin*.

Regulation

- The release of ACTH is pulsatile and exhibits diurnal variations, with peak levels at 0800 hours and lowest levels at midnight.
- ACTH release is regulated by hypothalamic *corticotrophin releasing hormone* (*CRH*), which increases POMC *m*RNA in a cAMP-dependent process.
- ACTH secretion is stimulated by
  - stress due to surgery and hypoglycemia by increasing CRH release, and
  - immune mediators such as interleukins.
- ACTH secretion is inhibited by cortisol, which inhibits CRH secretion.

Enumerate the biological effects of ACTH.

ACTH binds to specific receptors on the plasma membrane of adrenal cortical cells, increases cAMP levels, and

- promotes synthesis and secretion of adrenal cortical steroids by stimulating the conversion of cholesterol to pregnenolone
- regulates cortisol secretion from the adrenal cortex and influences aldosterone secretion
- maintains the size and integrity of the adrenal cortex.

The biological effects of ACTH mediated via the adrenal cortical steroids include the following:

- increased gluconeogenesis
- increased adipose tissue lipolysis
- increased ketogenesis
- decreased protein synthesis (except in the liver)
- stimulation of salt and water reabsorption by the kidney.

Acting independently, ACTH exerts the following actions in tissues other than the adrenal cortex:

- pigmentation due to its *melanocyte-stimulating hormone* (MSH) activity
- glucose utilization and fatty acid release by the adipose tissue *in vitro*
- activation of *triglyceride lipase* in adipose tissue by cAMP stimulation
- stimulation of insulin release from pancreatic islet cells.

❑ What are gonadotrophic hormones?

The gonadotrophic hormones secreted by the gonadotrophs of the anterior pituitary include *follicle-stimulating hormone (FSH) and luteinizing hormone* (*LH*). These are glycoproteins and contain two subunits: an α-subunit that is identical, and a β-subunit unique to each hormone.

❑ Enumerate the biological actions of FSH.

FSH regulates the functions of the ovary and testis by binding to specific receptors in the plasma membrane of target cells, activation of adenylate cyclase, generation of cAMP, and stimulation of protein kinases.

Biological effects of FSH

In females

- follicular growth
- preparation of the follicle for ovulation
- increase in ovarian weight
- estrogen synthesis in the ovary during the first part of the menstrual cycle
- regulation of estradiol formation
- sensitization of the granulosa cells of the ovarian follicle by increasing LH receptor number.

In males

- growth of seminiferous tubules and testicles
- spermatogenesis
- synthesis of androgen-binding proteins essential for testosterone transport to seminiferous tubules and epididymis
- synthesis of estradiol in Sertoli cells.

❑ Enumerate the biological actions of LH.

LH regulates the functions of the ovary and testis by binding to specific receptors in the plasma membrane of target cells, activation of adenylate cyclase, generation of cAMP, and stimulation of protein kinases.

Biological effects of LH

In females

- stimulation of ovarian cells to synthesize estrogens, androgens, and androgen precursors
- formation and maintenance of corpus luteum
- stimulation of progesterone production by corpus luteum.

In males

- stimulation of testosterone secretion by Leydig cells
- spermatogenesis
- development and maintenance of accessory sex organs.

❑ How is the secretion of LH and FSH regulated?

- The synthesis and secretion of LH and FSH are stimulated by *gonadotrophin-releasing hormone* (*GnRH*) or *luteinizing hormone-releasing hormone* (*LHRH*), a decapeptide synthesized in the hypothalamus.
- GnRH secretion
  - is episodic and pulsatile
  - is increased by noradrenergic agonists and decreased by endogenous opioids
  - is high in the first three months after birth and declines until puberty
  - increases after puberty with increase in LH concentrations greater than FSH; after menopause, the FSH concentrations are higher than LH
  - is under positive feedback regulation by plasma estrogens with increase in LH secretion (*LH surge*) which stimulates ovulation and corpus luteum development; subsequently, high plasma estrogens and progesterone inhibit FSH and LH release
  - is inhibited by testosterone directly or after conversion to estradiol with decrease in the frequency of LH pulsations.
- *Inhibin* produced by the testes and ovaries inhibits FSH secretion.

❑ What is Kallman's syndrome?

Kallman's syndrome refers to congenital isolated gonadotrophin deficiency that occurs in postmenopausal women and in men with primary hypogonadism. LH and FSH levels are reduced due to defective synthesis or release of GnRH. Clinical symptoms include skeletal defects, mental retardation, cryptorchidism, and anosmia (lacking the sense of smell).

❑ Write notes on human chorionic gonadotrophin (hCG).

Human chorionic gonadotrophin (hCG), a gonadotrophic hormone,

- is produced by the human placenta
- is a glycoprotein with structural homology to LH
- maintains the corpus luteum during the first four to six weeks of pregnancy
- when determined in serum or urine, is useful as a pregnancy test
- is increased in malignancies such as chorionepithelioma and testicular tumors
- is produced ectopically by non-trophoblastic tumors such as hepatomas and lung carcinomas.

❑ Describe the secretion, biological effects, and regulation of prolactin.

Synthesis and secretion

- *Prolactin* (*PRL, lactogenic hormone, mammotrophin, luteotrophic hormone*), synthesized by lactotrophs of the anterior pituitary, is a single polypeptide with 198 amino acid residues and three intrachain disulphide bonds.
- Prolactin secretion
  - is high during sleep and stress
  - is higher in women than in men.
- High estrogen levels during pregnancy increase prolactin secretion from the pituitary and inhibit secretion in the breast.
- Low estrogen levels postpartum lead to commencement of lactation.
- After delivery, there is a rapid fall in prolactin levels.

Biological effects

- Initiation and maintenance of lactation.
- Weak GH-like effects such as
  - lipid mobilization
  - anabolic effects
  - anti-insulin action.
- Inhibition of the effects of gonadotrophins on ovarian steroidogenesis.
- *Prolactin receptor*s are present in the plasma membrane of mammary gland, gonads, liver, and adrenal glands.
- Hormone – receptor interaction increases synthesis of milk proteins such as casein.
- Prolactin receptor concentration is increased by
  - prolactin (an exception to the phenomenon of downregulation of receptors by hormones),
  - glucocorticoids
  - thyroid hormones.
- Prolactin receptor concentration is decreased by progesterone.

Regulation

- Prolactin secretion is inhibited by
  - dopamine from the hypothalamus by lowering cAMP levels. Increased plasma prolactin levels feedback inhibit hormone synthesis by enhancing dopamine synthesis.
  - *GnRH-associated peptide* (*GAP*), a neuropeptide that has both GnRH and *prolactin-release inhibiting hormone* (*PRIH*) activities.
- Prolactin secretion is stimulated by
  - TRH from the hypothalamus
  - *Vasoactive intestinal peptide* (*VIP*) during lactation.

❑ Write a short note on hyperprolactinemia.

Causes

- Pituitary disease with autonomous hormone synthesis by prolactin-secreting tumors (*prolactinomas*)
- Pituitary stalk section and surgery
- Hypothalamic disease
- Drugs such as
  - dopamine receptor blockers (phenothiazines, haloperidol)
  - dopamine depleting agents (methyldopa, reserpine)
- Ectopic production by non-pituitary tumors
- Primary hypothyroidism and estrogens, which lower dopamine-induced inhibition of prolactin secretion
- Chronic renal failure in which prolactin clearance from plasma is reduced

Clinical features

- Impotence, infertility, and gynecomastia in men with decrease in FSH, LH, and testosterone levels.
- Amenorrhea or oligomenorrhea, infertility and galactorrhea (breast discharge) in women with decreased secretion of LH and FSH.

Management: Administration of dopamine agonists such as bromocriptine or transsphenoidal resection.

❑ What are the hormones secreted by the posterior pituitary?

The posterior pituitary secretes
- *Oxytocin*, which regulates lactation and stimulates uterine contraction.
- *Vasopressin* or *antidiuretic hormone* (*ADH*), which stimulates water reabsorption from the renal tubules and collecting ducts.

Both the hormones are synthesized in the hypothalamus and transported together with the corresponding carrier proteins *neurophysins I and II* to the posterior pituitary gland.

❑ Elucidate the synthesis and secretion of vasopressin.

- *Arginine vasopressin* (*AVP*) is synthesized in the supraoptic nucleus of the hypothalamus and transported to the posterior pituitary gland.
- Vasopressin secretion is stimulated by
  - an increase in plasma osmolality sensed by osmoreceptors in the hypothalamus
  - a decrease in plasma volume sensed by stretch receptors in the left atrium
  - a fall in blood pressure that activates carotid and aortic baroreceptors
  - neurotransmitters and neuropeptides such as acetylcholine, histamine, bradykinin, angiotensin II, as well as stress
  - pharmacological agents such as morphine, nicotine cyclophosphamide, and vincristine.
- Vasopressin release is inhibited by
  - norepinephrine
  - ethanol
  - phenytoin.
  - gamma aminobutyric acid (GABA)
  - reserpine

❑ Enumerate the biological actions of vasopressin.

Vasopressin

- Increases blood pressure and peripheral vasoconstriction (in pharmacological amounts) mediated by the following sequence of events:
  1. binding of vasopressin to the V1 receptors
  2. activation of phospholipase C
  3. generation of inositol triphosphate (IP3) and diacylglycerol
  4. increase in intracellular Ca^{2+}
  5. activation of protein kinase C
- Stimulates reabsorption of water from the renal distal convoluted tubules (DCT) and collecting ducts by promoting permeability of these cells to water and concentrating the urine. This is mediated by the following sequence of events:
  1. binding of vasopressin to the V2 receptor
  2. activation of *adenylate cyclase* by the hormone – receptor complex
  3. generation of cAMP
  4. activation of a membrane-bound protein *kinase* by cAMP
  5. phosphorylation of membrane proteins
  6. increased permeability to water.

Discuss the causes, diagnosis, and management of diabetes insipidus.

Diabetes insipidus is characterized by

- Excretion of excessive amount of urine (*polyuria*).
- Compensatory excessive water intake (*polydipsia*) in response to thirst.
- Diluted urine with urine volume ranging between 2.5 and 6 L/d in mild cases, and between 16 – 24 L/d in severe cases.
- Increased plasma osmolality that stimulates thirst. If water is available, dehydration does not occur; whereas inadequate water intake can lead to severe dehydration with weakness, fever, mental disturbances, prostration, and death.

Causes

Diabetes insipidus may be of two types:

- *Central diabetes insipidus* (*cranial or neurogenic diabetes insipidus*) due to pituitary or hypothalamic disease results from the following causes:
  i. Tumors and infiltrative lesions of the pituitary or hypothalamus such as pituitary adenomas, craniopharyngiomas, sarcoidosis, etc.
  ii. Surgery of the pituitary or hypothalamus.
  iii. Severe head injuries such as fractures of the skull.
  iv. Idiopathic, with circulating antibodies to hypothalamic nuclei.
- *Nephrogenic diabetes insipidus* in which vasopressin secretion is normal but the kidneys fail to respond to the hormone due to
  i. Congenital nephrogenic diabetes insipidus, resulting from V2 receptor abnormality.
  ii. Chronic renal diseases.
  iii. Potassium deficiency as in primary aldosteronism.
  iv. Chronic hypercalcemia as in hyperparathyroidism.
  v. Systemic disorders such as multiple myeloma, sickle cell anemia, and sarcoidosis.
  vi. Pharmacological agents such as lithium.

Diagnosis

- Determination of plasma and urine osmolality.
- *Fluid deprivation test* to confirm diagnosis.

  Procedure

  i. The night before the test, fluids are allowed.
  ii. A light breakfast without fluid is given the next morning.
  iii. The body weight is recorded and fluid withheld for 8 hours.
  iv. The weight of the patient is taken every one hour and the test stopped if the body weight falls by >3%.
  v. After 8 hours, the patient is allowed fluid.
  vi. Desmopressin (1-desamino-D-arginine-vasopressin) is administered subcutaneously (1 μg) or intranasally (10 μg).
  vii. Urine samples are collected every one hour for 4 hours.

  Interpretation

  In normal subjects, deprivation of fluid concentrates the urine and plasma osmolality does not exceed 295 mmol/kg. In diabetes insipidus, there is impairment of the concentration mechanism and plasma osmolality exceeds 300 mmol/kg. Following desmopressin injection, the urine becomes concentrated in patients with cranial diabetes insipidus, but remains dilute in nephrogenic diabetes insipidus.

Management

- Desmopressin nasal spray for central diabetes insipidus.
- Administration of chlorpropamide, an oral hypoglycemic agent that enhances renal sensitivity to vasopressin in mild cases.
- Adequate fluid in patients with nephrogenic diabetes insipidus to avoid dehydration.
- Administration of thiazides to deplete sodium, increase reabsorption of sodium and water, and reduce polyuria.

❑ Explain the causes and clinical findings in syndrome of inappropriate ADH secretion (SIADH).

SIADH secretion is characterized by

- high plasma vasopressin (ADH) levels
- excretion of concentrated urine
- ECF hypoosmolality without edema.
- excessive water retention
- dilutional hyponatremia

The secretion of ADH is termed 'inappropriate' because it occurs in the presence of plasma hypoosmolality.

Clinical manifestations

- lethargy
- mental confusion
- weakness
- convulsions, and coma

Causes

- Central nervous system disorders such as:
  - skull fracture
  - tuberculous meningitis
  - subarachnoid hemorrhage
  - cerebral vascular thrombosis
  - subdural hematoma
  - acute encephalitis
- Drugs that stimulate release or increase the action of ADH such as
  - chlorpropamide
  - cyclophosphamide
  - tricyclic antidepressants.
  - vincristine
  - anesthetics
- Ectopic production of ADH by
  - small cell carcinoma of the lung
  - lymphosarcoma
  - carcinoma of the pancreas
  - Hodgkin's disease
  - non-neoplastic pulmonary diseases such as pneumonia and tuberculosis.

❑ Outline the diagnostic tests and management of SIADH.

Diagnosis

1. Concentrated urine with osmolality >300 mmol/kg.
2. Plasma osmolality <270 mmol/kg.
3. Low blood urea nitrogen (BUN), serum urate, creatinine, and albumin.
4. Plasma sodium <130 mmol/L.
5. *Water load test* to be carried out only when plasma sodium levels are >125 mmol/L.

   The test involves administration of 20 mL water/kg body weight up to 1500 mL and collection of one-hourly urine samples.

Normal response: At least 65% of the water is excreted in 4 h and 80% in 5 h with the lowest urinary osmolality (<100 mmol/kg) seen in the second hour. In SIADH, and in adrenal and renal insufficiency, there is an inability to excrete the water load.

Management

1. Restriction of fluid intake to 0.8 – 1.0 L/d.
2. Intravenous administration of 200 – 300 mL of 5% saline over several hours to raise plasma sodium in severe cases.

❑ Enumerate the biological actions of oxytocin.

Oxytocin, a nanopeptide with cysteine residues at positions 1 and 6 linked by a disulphide bond, has close structural homology with vasopressin.

Biological actions

- Contraction of the uterine smooth muscle in pharmacological amounts. The hormone is clinically used to induce labour.
- Promotion of milk ejection on suckling (*milk letdown response*).
- Stimulation of the contraction of myoepithelial cells surrounding the mammary alveoli causing milk release.
- Mild antidiuretic effects.
- Increased glucose oxidation and glucose utilization by adipose tissue at high concentrations.

Oxytocin receptors

- These are present on the membrane of uterine and mammary cells
- Their number is increased by estrogens, and decreased by progesterone.

❑ Discuss the causes, clinical features, diagnosis, and management of hypopituitarism.

◘ *Causes*

- Isolated hormone deficiencies
- Tumors of the pituitary or hypothalamus
- Surgery or radiation of the pituitary
- Postpartum infarction (*Sheehan's syndrome*)
- Inflammatory diseases (granulomatous hypophysitis, sarcoidosis)
- Anorexia nervosa associated with malnutrition and LHRH deficiency
- Emotional stress with GH deficiency and *psychosocial dwarfism*
- Severe illness, with deficiency of TSH and free T_4 (*sick euthyroid syndrome*).

◘ *Clinical manifestations*

- Growth retardation in children, and reduced muscle mass and tendency to hypoglycemia in adults due to GH deficiency.
- Gonadotrophin deficiency, characterized by
  a. delayed puberty in children
  b. oligomenorrhea, infertility, and atrophy of breasts and genitalia in women
  c. impotence, azoospermia, and testicular atrophy in men.
- Prolactin deficiency with failure of lactation.
- TSH deficiency leading to secondary hypothyroidism.

- ACTH deficiency resulting in
  a. secondary adrenocortical hypofunction
  b. cortisol deficiency
  c. water retention with dilutional hyponatremia and hypotension
  d. reduced skin pigmentation
  e. fasting hypoglycemia.
- ADH deficiency causing thirst and polyuria.

▫ *Diagnosis*

Target gland deficiency should be confirmed before evaluating pituitary hormone secretion. Following are some common diagnostic tools:

1. Hormone assays important in diagnosis include
   - LH and FSH
   - Testosterone levels in males and estradiol levels in females
   - Total or free T_4 and TSH
   - Prolactin
   - Cortisol concentration at 9.00 hours.
2. Combined hypothalamic pituitary stimulation test if the plasma levels of both the target gland and the trophic hormones are low.

   Protocol involves administration of insulin (0.15 U/kg body weight) to lower blood glucose concentration to <45 mg/dL (2.5 mmol/L) after an overnight fast (a lower dose of 0.05 – 0.1 U/kg body weight is given if hypopituitarism or adrenal cortical insufficiency is suspected). This is followed by injection of 200 μg of TRH and 100 μg of GnRH. Hormone assays (cortisol, GH, T_4, TSH, prolactin, LH, and FSH) are conducted in blood samples collected at 30, 60, 90, and 120 minutes after the injections. In normal subjects, insulin-induced hypoglycemia should elevate hormone levels.
3. Computerized tomography (CT) scanning or magnetic resonance imaging (MRI) of the pituitary.

▫ *Management*

1. Multiple hormone replacement in patients with panhypopituitarism.
2. Administration of levothyroxine if hypothyroidism is present.
3. Administration of estrogen – progestogen combinations in women and testosterone esters in men for hypogonadism.
4. Gonadotrophin administration in patients with pituitary disease.
5. Gonadorelin administration for hypothalamic disease, if fertility is desired.
6. GH administration in children with GH deficiency.
7. Desmopressin administration for diabetes insipidus.

Chapter 15

Thyroid Hormones

❑ Name the iodothyronine hormones synthesized by the thyroid gland.

The *iodothyronine hormones* synthesized by the thyroid gland are

- 3,5,3'-triiodothyronine (T_3)
- 3,5,3'5'-tetraiodothyronine (T_4, thyroxine).

Figure 15.1 shows the structure of the thyroid hormones and their precursors.

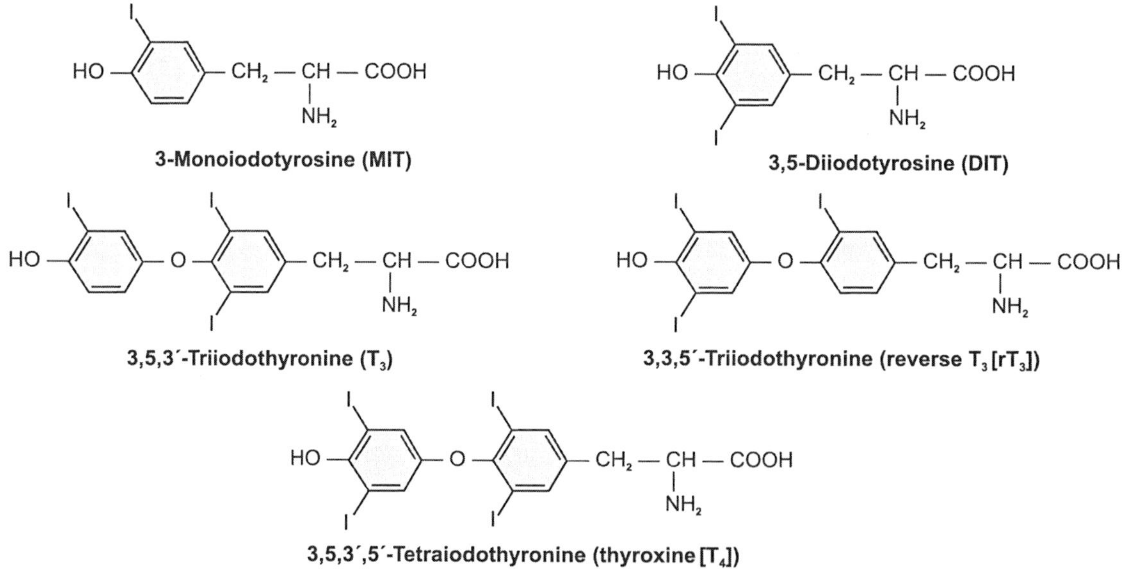

Fig. 15.1 Structures of the thyroid hormones and their precursors

❑ Outline the steps involved in thyroid hormone synthesis.

The thyroid hormones are synthesized on the tyrosine residues of *thyroglobulin*, a protein present in the follicular cells of the thyroid surrounding a cavity termed the 'colloid'. Thyroid hormone synthesis involves the following steps (Fig. 15.2):

1. ***Iodide (I^-) uptake*** from the blood by the follicular cells by an active transport mechanism. Iodide uptake is blocked by thiocyanate and perchlorate.

2. ***Oxidation of iodide*** by *thyroperoxidase*, which uses hydrogen peroxide as an oxidizing agent. This step is inhibited by thiourea.
3. ***Iodination of tyrosine residues*** in thyroglobulin (organification) by *thyroperoxidase* to form monoiodotyrosine (MIT) and diiodotyrosine (DIT). This step is inhibited by propylthiouracil and carbimazole.
4. ***Coupling*** of MIT to DIT to form T_3, and coupling of two residues of DIT to generate T_4.
5. ***Proteolysis and secretion.*** Thyroglobulin, the storage form of thyroid hormones, is taken up by the follicular cells followed by hydrolysis by lysosomal enzymes with release of the thyroid hormones into circulation. This step is stimulated by thyroid stimulating hormone (TSH) and inhibited by iodide.
6. ***Deiodination*** of MIT and DIT by a *deiodinase*. The iodine is reutilized. The major hormone synthesized by the thyroid gland is T_4, whereas T_3 and reverse T_3 are obtained by deiodination of T_4 in the liver, kidney, and muscle.

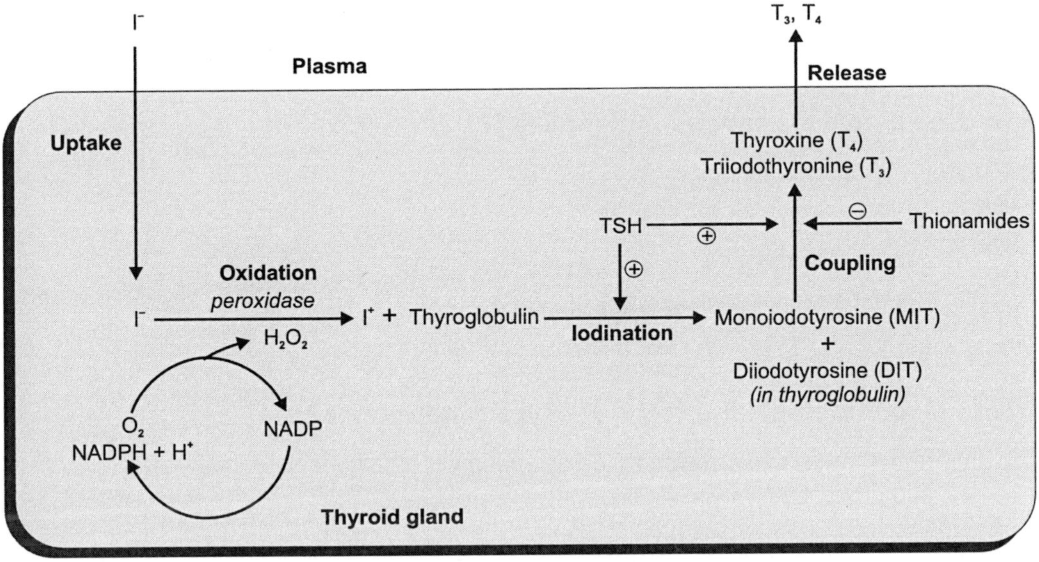

Fig. 15.2 Biosynthesis of thyroid hormones

❑ How is thyroid hormone secretion regulated?

- The adenohypophyseal hormone TSH stimulates the synthesis and secretion of thyroid hormones.
- TSH secretion is, in turn, stimulated by thyrotrophin-releasing hormone (TRH) from the hypothalamus.
- The secretion of both TRH and TSH is feedback inhibited by circulating levels of thyroid hormones.

❑ Give an account of goitrogens.

Goitrogens are antithyroid agents that inhibit thyroid function. The following agents function as goitrogens:

1. High concentrations of iodide inhibit iodination of tyrosine as well as release of thyroid hormones by suppressing TSH. Iodide is, therefore, clinically useful in the treatment of hyperthyroidism.

2. Monovalent anions (*e.g.*, thiocyanate, perchlorate), which competitively inhibit iodide uptake by the thyroid. Thiocyanate is present in body fluids and in plants of the genus *Brassica* such as cabbages. Ingestion of large quantities of monovalent anions can cause *goitre*.
3. Thiocarbamides/thionamides (*e.g.*, thiourea, propylthiouracil), which inhibit the oxidation of iodide, reduce the amount of DIT, and block coupling reactions.
4. Other agents such as
   a. *ouabain*, an inhibitor of Na^+K^+ATPase, which blocks iodide uptake
   b. *colchicine*, which inhibits endocytosis of colloid into follicular cells
   c. lithium salts, which inhibit thyroid hormone release.

❑ How are thyroid hormones transported in circulation?

More than 99% of thyroid hormones are transported in circulation, bound to
- thyroxine-binding globulin, TBG (predominant)
- prealbumin
- albumin.

The remaining (0.02% of T_4 and 0.3% of T_3) free unbound fraction is the biologically active form.

❑ How are thyroid hormones metabolized?

The thyroid hormones undergo
- deiodination
- deamination
- decarboxylation
- conjugation with glucuronate or sulphate and excretion in the urine.

During deiodination, T_4 is activated to T_3, the biologically active form.

❑ What are the biological effects of thyroid hormones?

The thyroid hormones exert a wide range of biological effects, including
- growth and development
- oxygen consumption and thermogenesis
- increase in basal metabolic rate (BMR)
- stimulation of gluconeogenesis
- lowering of plasma cholesterol levels
- protein anabolic effects in physiological amounts
- bone resorption and decalcification.

❑ Give a brief explanation of thyroid function tests.

The tests available for the assessment of thyroid function include the following:
1. Direct tests of thyroid function (uptake of radioactive iodine and technetium)
   a. *Radioactive iodine uptake* (*RAIU*) *test* that assesses the amount of iodide accumulated by the thyroid using radioiodine ^{131}I or ^{123}I. The uptake correlates inversely with the plasma iodide concentration and directly with the functional state of the thyroid.
   b. *Thyroid scintiscanning* that makes use of the isotope ^{99m}Tc to identify *'hot'* (active) or *'cold'* (inactive and malignant) nodules in the thyroid as well as in the differential diagnosis of hyperthyroidism.

2. Measurement of total hormone levels by radioimmunoassay (RIA). This however, reflects the fraction bound to the protein TBG, which is increased by estrogens and decreased by severe illness, androgens, and glucocorticoids.
3. Measurement of free hormone levels, which correlate better with the thyroid status.
4. Tests of thyroid hormone regulation by
   a. Measurement of TSH
   b. TRH stimulation test based on the principle that TRH administration (200 µg in 2 mL saline i.v.) induces increased secretion of TSH with increase in plasma TSH levels that reach a peak in 20 minutes and subsequently decline rapidly.
5. Miscellaneous tests
   a. Detection of antibodies to T_3 and T_4 is useful in autoimmune thyroid disease.
   b. Measurement of plasma TBG is useful in thyroid cancer.
   c. Detection of antithyroid peroxidase in Hashimoto's disease and Graves' disease.

❑ What are the causes of hyperthyroidism?

Hyperthyroidism is characterized by enlargement of the thyroid (*goitre*). The most common causes of hyperthyroidism are

1. Graves' disease, which
   a. may occur at any age
   b. is more common in females
   c. presents with exophthalmos (abnormal protrusion of the eyeballs)
   d. is characterized by the presence of thyroid antibodies and *thyroid stimulating immunoglobulins* (*TSI*) such as *long-acting thyroid stimulator* (*LATS*). TSI mimics TSH and causes excessive production of thyroid hormones and is insensitive to negative feedback by thyroid hormones.
2. Toxic nodular goitre, which
   a. is usually seen in older patients
   b. secrete thyroid hormones autonomously through nodules.
3. Other causes include
   a. excessive ingestion of thyroid hormones
   b. TSH-secreting tumors
   c. ectopic thyroid tissue such as thyroid tissue in an ovarian teratoma
   d. thyroiditis.

❑ Enumerate the clinical and metabolic features characteristic of hyperthyroidism.

The term *thyrotoxicosis* is used to refer to the clinical syndrome produced by hyperthyroidism.

The clinical symptoms in hyperthyroidism include

- goitre with or without exophthalmos
- increase in BMR and body temperature
- increased sweating and sensitivity to heat
- *cardiovascular symptoms* such as increased heart rate (*tachycardia*), cardiac arrhythmias
- weight loss
- diarrhea
- weakness, anxiety and insomnia
- warm and moist skin
- *Plummer's nail*: in this, the fingernail (ring finger) is separated from the nail bed.

The metabolic abnormalities include

- hypocholesterolemia
- hyperglycemia and glycosuria
- decreased glucose tolerance
- negative nitrogen balance.

As a result, plasma T_3 and T_4 are increased, basal TSH levels are suppressed, and there is a subnormal response of TSH to TRH stimulation.

Management of the disorder involves administration of antithyroid drugs, radioactive iodine, or β-blockers.

❑ How is hyperthyroidism diagnosed in the laboratory?

Hyperthyroidism can be confirmed by the following laboratory findings:

- Increase in total and free thyroid hormone levels, with increase in T_3 being greater and occurring earlier than T_4.
- Low plasma TSH levels.
- TRH Test: The response of TSH to TRH stimulation is subnormal.

❑ How is hyperthyroidism treated?

The treatment for hyperthyroidism involves

- antithyroid drugs (propylthiouracil, carbimazole)
- administration of radioactive iodine ^{131}I (contraindicated in pregnancy)
- administration of β-blockers (propranolol), which inhibit peripheral conversion of T_4 to T_3.

❑ Describe the pathogenesis of hypothyroidism.

- Hypothyroidism may be
  - *primary* due to thyroid disease
  - *secondary* due to pituitary or hypothalamic disease
- Severe hypothyroidism causes *cretinism* in children and *myxedema* in adults.
- The causes of primary hypothyroidism include
  - Autoimmune diseases (thyroiditis, Hashimoto's disease) in which thyroid tissue is progressively destroyed by high concentrations of circulating antithyroid antibodies.
  - Removal of thyroid gland for treatment of hyperthyroidism.
  - Iodine deficiency, which blocks thyroid hormone synthesis. As a result, TSH is not feedback inhibited and it stimulates thyroid hormone synthesis, leading to enlargement of thyroid gland. Intake of iodized salt protects against iodine-deficiency goitre in endemic areas such as the Himalayas and Andes.
  - Drugs such as propylthiouracil, lithium, iodides, and aminosalicyclic acid.
  - Inherited defects in iodination of tyrosine with failure of hormone biosynthesis.
- Secondary hypothyroidism, which results from anterior pituitary or hypothalamic disease, is characterized by inappropriately low plasma TSH levels.

❑ What are the clinical features and metabolic abnormalities of hypothyroidism?

The clinical symptoms of hypothyroidism include

1. decreased BMR, decreased body temperature and cold intolerance
2. weight gain
3. decreased growth
4. decreased appetite and constipation
5. fatigue, lethargy, somnolence, and decreased muscle tone

6. cardiovascular symptoms such as *bradycardia* (decreased heart rate), and decreased blood volume
7. decreased growth of hair, scaliness of skin, and hoarse voice.

The metabolic abnormalities in hypothyroidism include hypercholesterolemia and decreased glucose utilization.

❑ What are the characteristic features of cretinism?

Cretinism occurs in children with severe hypothyroidism. This is characterized by the following features:

- Short stature
- Coarse features
- Sparse hair
- Dry skin
- Impaired physical and mental development
- Delayed puberty and retarded sexual development.

❑ What are the characteristic features of myxedema?

Myxedema occurs in adults with severe hypothyroidism. This is characterized by the following features:

- Edematous appearance of the body
- Periorbital puffiness
- Dry, waxy, thickening of the skin due to infiltration of proteoglycans
- Dry hair and skin
- Hoarse voice
- Stiffness and cramps
- Carpal tunnel syndrome
- Slowing of intellectual and motor activity.

In severe untreated cases, a hypothermic, fatal stupor termed *myxedema coma* occurs.

❑ What are the laboratory findings in hypothyroidism?

1. Plasma T_4 is decreased.
2. Plasma TSH is
   a. increased in primary hypothyroidism due to decreased negative feedback
   b. decreased in secondary hypothyroidism.
3. TRH test
   a. In primary hypothyroidism, there is exaggerated rise in plasma TSH in response to TRH.
   b. In secondary hypothyroidism of pituitary origin, the response to TRH is reduced due to inability of the anterior pituitary to secrete TSH.
   c. In secondary hypothyroidism of hypothalamic origin, the response is delayed with higher concentrations of plasma TSH at 60 minutes than at 20 minutes.

❑ What is the treatment for hypothyroidism?

Levothyroxine is the drug of choice for hypothyroidism. In secondary hypothyroidism, levothyroxine is administered only after treatment of adrenal insufficiency.

Chapter 16

Adrenal Hormones

❑ What are the functional regions of the adrenal glands?

The adrenal glands are functionally divided into the *cortex* and the *medulla.*

❑ Name the various adrenal cortical hormones.

The adrenal cortex secretes the following hormones:

- *Mineralocorticoids,* secreted by the outer layer of the cortex (*zona glomerulosa*), are C_{21} steroids that maintain $Na^+ - K^+$ homeostasis. Aldosterone is the predominant mineralocorticoid.
- *Glucocorticoids* synthesized by middle (*zona fasciculata*) and inner (*zona reticularis*) layers, are C_{21} steroids that play a vital role in intermediary metabolism. The major glucocorticoids include *cortisol, corticosterone,* and *deoxycorticosterone.*
- *Androgens,* synthesized by the *zonae fasciculata* and *reticularis* are C_{19} steroids and include *dehydroepiandrosterone* (*DHEA*) and its sulphate, and smaller amounts of *androstenedione* and *testosterone.*

❑ Outline the steps in the biosynthesis of adrenal steroid hormones.

Cholesterol is the precursor of the adrenal cortical hormones. The major steps in steroid hormone biosynthesis include the following (Fig. 16.1):

1. Conversion of cholesterol to pregnenolone catalyzed by mitochondrial *cytochrome P450-dependent side chain cleavage* (*P450scc*). This involves hydroxylation of cholesterol at C_{22} and C_{20}, followed by removal of a 6-carbon fragment, isocaproaldehyde. This is the rate-limiting step in steroid hormone biosynthesis.
2. Conversion of pregnenolone to progesterone by the action of 3β-*hydroxysteroid dehydrogenase* (3β-*OHSD*) and $\Delta^{5,4}$-*isomerase* present in the smooth endoplasmic reticulum.
3. Progressive hydroxylation of progesterone at positions C_{21}, C_{11}, and C_{18}, to form aldosterone *via* the intermediates 11-deoxycorticosterone and corticosterone.
4. Formation of cortisol from pregnenolone, which involves oxidation, double bond isomerization, and hydroxylations at positions C_{17}, C_{21}, and C_{11}.
5. Removal of the side chain in pregnenolone by a *17, 20 lyase* to form androgens. DHEA, the major androgen produced by the adrenal cortex, is converted to its sulphate as well as testosterone via androstenedione.

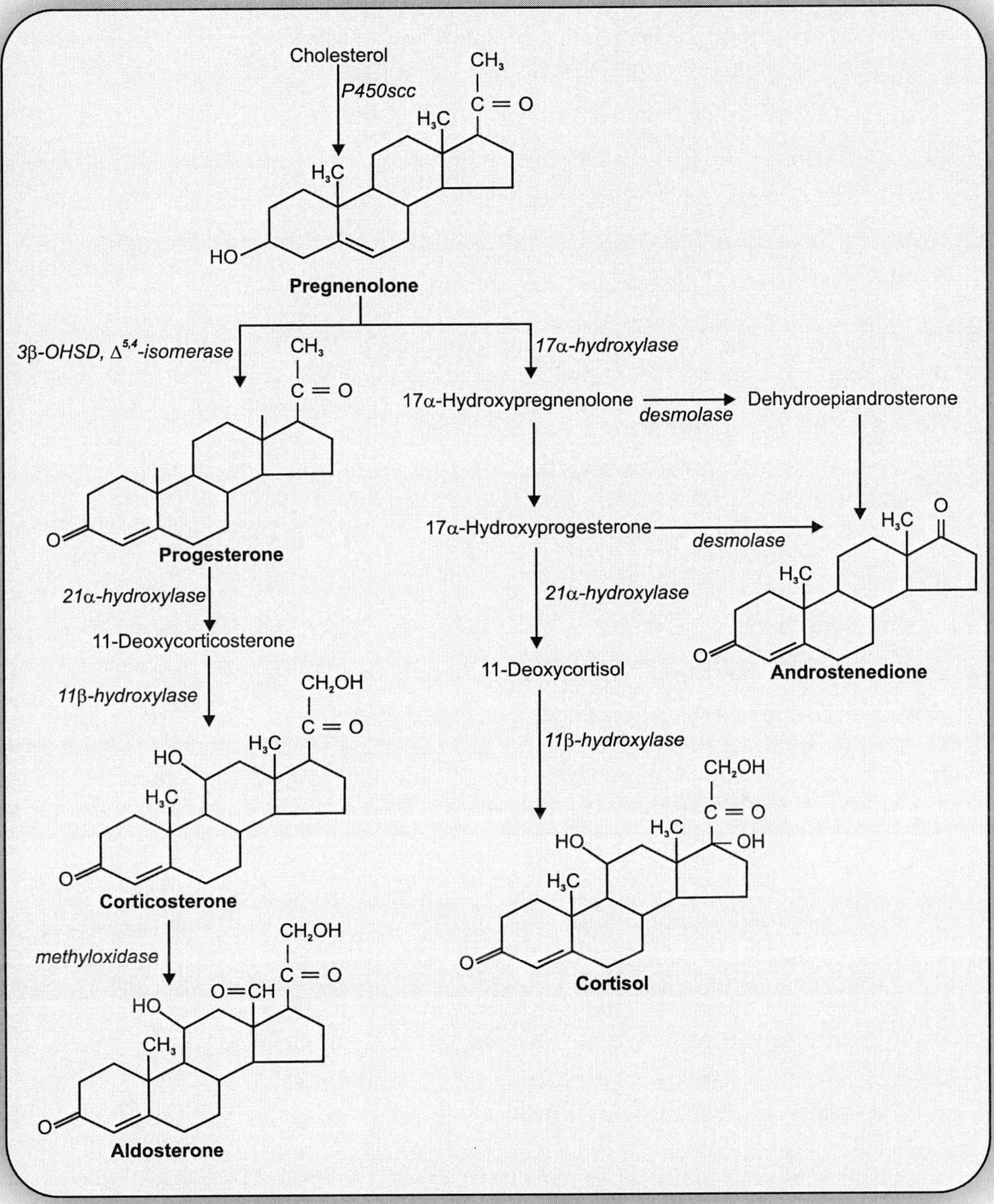

Fig. 16.1 Biosynthesis of adrenal steroid hormones

❑ How is glucocorticoid synthesis regulated?

The synthesis of glucocorticoids is regulated by the anterior pituitary and the hypothalamus

- *Corticotrophin-releasing hormone* (*CRH*) from the hypothalamus releases ACTH from the anterior pituitary which, in turn, stimulates the synthesis and secretion of glucocorticoids.
- The secretion of CRH and ACTH is increased by stress, and feedback inhibited by high cortisol levels.
- Both CRH and ACTH as well as cortisol exhibit *circadian variations* with high levels in the morning and gradually reaching a minimum during the first few hours of sleep.

❑ How are the adrenal steroid hormones transported in circulation?

- The adrenal steroid hormones circulate bound to plasma proteins.
- About 95% of cortisol is bound to plasma proteins, chiefly *corticosteroid-binding globulin* (*CBG*) or *transcortin*. The remaining 5% is found free in circulation and is the biologically active form.
- Androgens are bound to *sex hormone-binding globulin* (*SHBG*) and albumin in circulation.

❑ How are the cortical steroids metabolized?

- The catabolism of adrenal cortical steroids occurs primarily in the liver and to some extent in the skin, the kidney, and the adrenal gland.
- The cortical steroids undergo reduction, oxidation, hydroxylation, side chain cleavage, and conjugation.
- The conjugated metabolites are excreted mainly in the urine. The androgens are excreted as 17-keto compounds.

❑ Enumerate the biological effects of glucocorticoids.

Cortisol and cortisone exhibit the following biological effects:

- Increase blood glucose by inhibiting insulin secretion, and stimulating hepatic gluconeogenesis.
- Protein anabolic effect at physiological concentrations, and protein catabolic effect with negative nitrogen balance at higher concentrations.
- Increase fat mobilization from adipose tissue by activation of *lipase* with increase in plasma free fatty acids.
- Maintain ECF volume and normal blood pressure by inhibiting ADH secretion.
- Weak mineralocorticoid effects (renal tubular reabsorption of sodium and excretion of potassium).
- Anti-inflammatory effects: Glucocorticoids inhibit capillary permeability, and the production and action of inflammatory mediators such as lymphokines and prostaglandins and interferon.

❑ Briefly explain the principle of adrenal cortical function tests.

Adrenal cortical function can be assessed by the following tests:

- Measurement of plasma hormone levels
  - Plasma cortisol levels show diurnal variations with highest levels in the morning and lowest at night. Loss of diurnal variation is an early feature of adrenal hyperfunction.
  - Plasma aldosterone levels are influenced by posture, and by dietary Na^+ and K^+.
  - Low plasma ACTH indicates primary adrenal cortical tumor, while high levels occur due to an abnormality in the hypothalamic–pituitary axis or ectopic production.
- Urinary steroid levels
  - Urinary free cortisol is useful in the diagnosis of adrenocortical hyperactivity.

- Urinary aldosterone is useful in the diagnosis of hyperaldosteronism.
- Urinary 17-ketosteroids are increased in bilateral hyperplasia of the adrenal cortex, and in adrenal and testicular tumors, and are decreased in Addison's disease.

- Stimulation tests
  - Synacthen test: In the rapid test, *Synacthen* (250 µg), a synthetic ACTH that stimulates cortisol production, is administered by intramuscular injection. In normal persons, both the baseline and peak levels of cortisol are high ; whereas in adrenal failure, the response is blunted. In the long test, depot Synacthen (1 mg i.m.), which has longer duration of action, is given daily for 3 days and plasma cortisol determined every day and 24 h after the last dose. Plasma cortisol levels are low in primary adrenal failure and high in secondary adrenal failure.
  - *Stimulation tests* for mineralocorticoids are based on volume depletion by sodium restriction, diuretics, or upright posture. Aldosterone secretion is increased 3 – 5 days after severe sodium restriction.
- Suppression tests
  - Suppression tests for glucocorticoids use dexamethasone, a synthetic glucocorticoid that suppresses ACTH release. These tests are based on the fact that an increase in plasma cortisol feedback inhibits ACTH release.
  - Suppression tests for mineralocorticoids require increasing ECF volume by saline, oral salt load, or deoxycorticosterone which inhibit renin release with decrease in plasma and urinary aldosterone.
- Monitoring the pituitary–adrenal axis
  - Metyrapone test. It is based on the fact that administration of *metyrapone* [2-methyl-1,2 bis-(3'-pyridyl)-propan-1-one], an inhibitor of 11β-*hydroxylase* that catalyzes the formation of cortisol from 11-deoxycortisol, decreases cortisol levels with increased ACTH release by feedback stimulation.
  - Insulin hypoglycemia test is useful in investigating pituitary function because plasma cortisol increases in response to insulin-induced hypoglycemia.

❑ Enumerate the causes of Cushing's syndrome.

Cushing's syndrome is characterized by hypersecretion of cortisol resulting from

- Primary adrenal disease due to adrenal hyperplasia or cortisol-secreting tumors of the adrenal cortex with autonomous secretion of cortisol, which inhibits ACTH.
- Hypothalamic disease with inappropriate CRH release.
- *Cushing's disease* in which there is bilateral adrenal hyperplasia secondary to adenoma of the anterior pituitary gland. ACTH secretion is increased because the pituitary is insensitive to feedback inhibition by cortisol.
- Ectopic ACTH production by small cell carcinoma of the lung and carcinoid tumors that cause bilateral adrenal hyperplasia, with excessive production of cortisol that inhibits ACTH secretion by the pituitary.

❑ Elucidate the clinical and metabolic features of Cushing's syndrome.

- ***Clinical features***
  - Characteristic truncal obesity with *'moon face'*, and *'buffalo hump'*
  - Protuberant abdomen with *'purplish striae'* on the abdominal wall
  - Osteoporosis
  - Muscle wasting

- Reduced immunity
- Hypertension due to Na^+ and water retention
- Increased androgen secretion in women, leading to acne, hirsutism, and menstrual disturbances
- Psychiatric disturbances

- ***Metabolic abnormalities***
  - Impaired glucose tolerance with hyperglycemia and glycosuria
  - Increased protein catabolism associated with reduced protein synthesis
  - Proteinuria
  - Negative nitrogen balance
  - Hypokalemic alkalosis due to increased urinary loss of potassium.

❑ Explain the tests useful in the differential diagnosis of Cushing's syndrome.

The following tests are useful in the differential diagnosis of Cushing's syndrome:

- ***Plasma and urinary cortisol*:** Loss of diurnal variation in plasma cortisol levels with high levels in the late evening and increased urinary excretion of free cortisol are characteristic of Cushing's syndrome.
- ***Dexamethasone suppression test*:** Failure of suppression in the overnight test, and low plasma cortisol levels with suppression in the high dose test are characteristic of Cushing's disease. In ectopic ACTH production and adrenal tumors where the pituitary ACTH is already suppressed, even high doses of dexamethasone will have no effect.
- ***Plasma ACTH*:** It is low in adrenal tumors, moderately high in Cushing's disease, and very high in ectopic ACTH production.
- ***Insulin tolerance test*:** Plasma cortisol levels fail to increase in response to insulin-induced hypoglycemia.
- ***CRH test*:** While administration of CRH (100 μg i.v.) increases plasma ACTH and cortisol levels in Cushing's disease, no response is seen in ectopic ACTH secretion.

❑ What is hyperaldosteronism?

Hyperaldosteronism refers to increased mineralocorticoid activity. This is of two types:

- Primary aldosteronism *(Conn's syndrome)* in which hypersecretion of aldosterone is due to adrenal adenoma, bilateral adrenal hypertrophy, or adrenal carcinoma. The clinical and metabolic features include
  - high plasma aldosterone and low renin
  - failure of aldosterone suppression following salt load
  - sodium retention, potassium excretion, and hypokalemia without edema
  - hypertension, muscle weakness, and fatigue
  - polyuria and compensatory polydipsia
  - metabolic alkalosis with increase in plasma bicarbonate and H^+ excretion.
- Secondary aldosteronism is due to extra-adrenal causes such as hepatic cirrhosis, nephrotic syndrome, and congestive cardiac failure. This condition is characterized by an increase in plasma renin.

❑ What are the consequences of hypersecretion of androgens?

Hypersecretion of adrenal androgenic steroids occurs due to overproduction of dehydroepiandrosterone and androstenedione. Adrenal androgen excess results in

- adrenal virilism in adult women with masculinization, deepening of voice, cessation of menstruation, atrophy of the breasts, acne, and hirsutism
- excessive masculinization in men
- congenital adrenal hyperplasia commonly due to deficiency of C-21 hydroxylase involved in steroid hormone biosynthesis, with accumulation of 17α-hydroxyprogesterone. This causes virilization of female infants and cortisol deficiency with or without salt-losing tendency.

❑ What are the causes of adrenal cortical hypofunction?

Hypofunction of the adrenal cortex may be

- primary (Addison's disease), due to destruction of adrenal tissues by surgery, tuberculosis, autoimmune diseases, AIDS, or drugs such as rifampin.
- secondary, due to ACTH deficiency resulting from excess glucocorticoids or panhypopituitarism.

❑ What are the clinical and metabolic features of adrenal cortical insufficiency?

Adrenal cortical insufficiency is characterized by the following clinical and metabolic features:

- Weakness, fatigue, weight loss
- Gastrointestinal symptoms: Anorexia, nausea, and vomiting
- Deficiency of aldosterone, which causes
  - impaired Na^+ reabsorption
  - failure of K^+ and H^+ secretion in exchange for Na^+
  - hyperkalemia
  - loss of Na^+, Cl^-, and water in urine
  - mild acidosis
  - decrease in ECF volume, leading to hemoconcentration, decreased cardiac output, renal circulatory insufficiency, decrease in the glomerular filtration rate, and eventually uremia.
- Cortisol deficiency that causes
  - hypoglycemia
  - hypotension due to reduced sensitivity of arteriolar smooth muscle to catecholamines
  - hyponatremia
  - susceptibility to stress
- Increased pigmentation of skin and mucosa in primary adrenal cortical insufficiency due to high plasma ACTH, which has melanocyte stimulating activity, and absence of pigmentation in secondary adrenal insufficiency in which plasma ACTH is low.

❑ What is Addisonian crisis?

Addisonian crisis refers to clinical symptoms such as severe hypovolemia, shock, and hypoglycemia in patients with Addison's disease that is precipitated by stress such as infection, trauma, or surgery.

❑ Enumerate the laboratory findings useful in the diagnosis of adrenal cortical insufficiency.

- Plasma cortisol of < 50 nmol/L in a blood sample drawn at 0900 hours.
- Hyponatremia and hyperkalemia.
- Blunted or absent response to the short Synacthen stimulation test. In the long Synacthen test, low plasma cortisol levels are characteristic of primary adrenal disease, whereas an increase in plasma cortisol is seen in secondary adrenal failure.
- Failure of insulin-induced hypoglycemia to increase ACTH secretion from the anterior pituitary and plasma cortisol levels indicates pituitary dysfunction.

- Plasma ACTH is very high in primary adrenal failure, due to lack of negative feedback inhibition by cortisol, and low in secondary adrenal insufficiency.

❑ What are the causes of hypoaldosteronism?

Aldosterone deficiency with normal production of cortisol occurs due to the following:

- Hyporeninism
- Inherited defects in aldosterone synthesis
- Postural hypotension.
- Surgical removal of adenomas
- Heparin administration

The condition is characterized by hyperkalemia exacerbated by salt restriction.

❑ Name the various adrenal medullary hormones.

The hormones synthesized by the adrenal medulla are the catecholamines *epinephrine* and *norepinephrine* that mediate the "*fight* or *flight*" response.

❑ Outline the steps in the biosynthesis of catecholamines.

The biosynthesis of catecholamines takes place in the adrenal medulla and sympathetic neurons by the following sequence of reactions (Fig. 16.2):

- Conversion of tyrosine to 3,4-dihydroxyphenylalanine (dopa) by *tyrosine hydroxylase*. This is the rate-limiting step in the biosynthesis of adrenal medullary hormones.

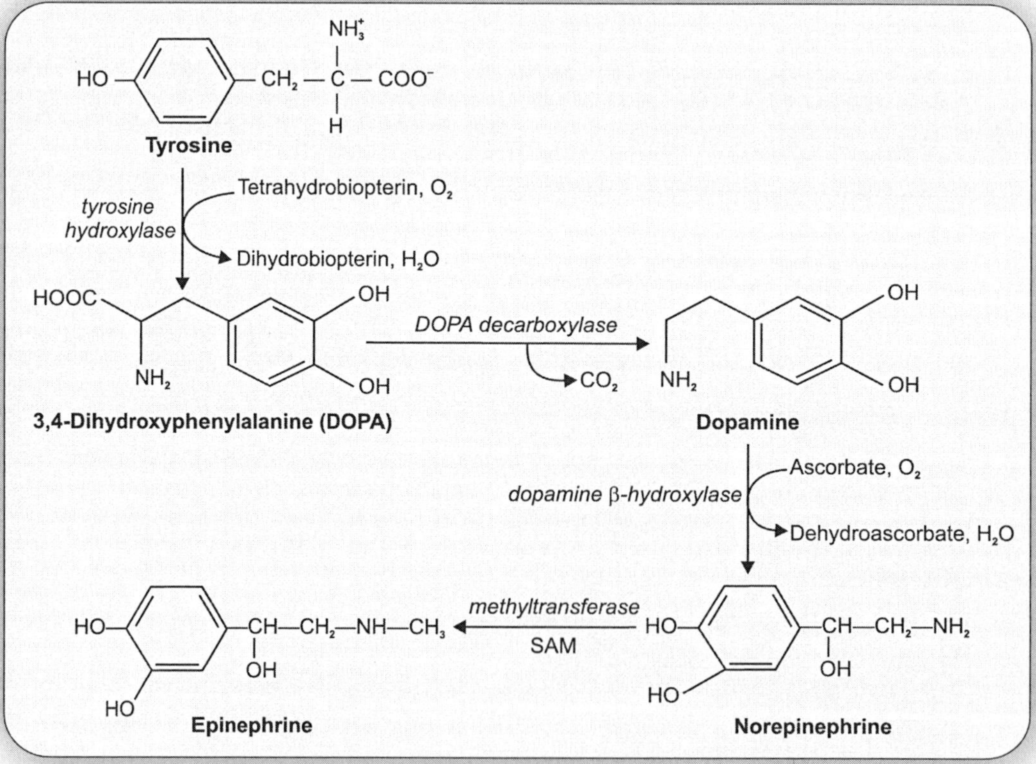

Fig. 16.2 Biosynthesis of catecholamines

- Conversion of dopa to dopamine catalyzed by *dopa decarboxylase* followed by entry of dopamine into vesicles called *chromaffin granules*. Inhibitors of *dopa decarboxylase* include α-methyldopa, an anti-hypertensive agent, and L-dopa, used in the treatment of Parkinson's disease.
- Conversion of dopamine to norepinephrine in the chromaffin granules by *dopamine β-hydroxylase*.
- Formation of epinephrine catalyzed by *phenylethanolamine N-methyltransferase* that uses S-adenosylmethionine as the methyl donor. This step is stimulated by glucocorticoids.

❑ How are catecholamines stored and secreted?

Catecholamines are stored in the chromaffin granules and secreted by calcium-dependent exocytosis. Secretion of catecholamines is stimulated by cholinergic and β-adrenergic agents and inhibited by α-adrenergic agents.

❑ Give a brief account of the metabolism of catecholamines.

The catecholamines are metabolised in the liver or in the target tissues by

- *methylation* catalysed by *catechol-O-methyltransferase* (*COMT*), a cytosolic enzyme that uses S-adenosylmethionine as a methyl group donor.
- *oxidative deamination* catalyzed by the mitochondrial enzyme *monoamine oxidase* (*MAO*) in the liver, stomach, intestine, and kidney. MAO inhibitors are useful as anti-hypertensives and antidepressants.
- *Conjugation:* The principal urinary metabolite 4-hydroxy-3-methoxy mandelic acid (*vanillylmandelic acid, VMA*) is excreted in the conjugated form.

❑ How is catecholamine metabolism regulated?

Catecholamine biosynthesis is increased by

- stress
- β-adrenergic stimulation
- administration of ACTH or corticosteroids
- feeding
- pituitary hyperactivity

Catecholamine biosynthesis is decreased by *disulfiram*, an inhibitor of *dopamine β-hydroxylase*.

❑ Enumerate the drugs that influence catecholamine levels.

1. Drugs that increase catecholamine levels are used as antidepressants. These include
   - *cocaine* and *amphetamines*, which increase the biological availability of catecholamines by inhibiting tissue reuptake.
2. Drugs that decrease catecholamine levels are used as anti-hypertensive agents and tranquillisers. These include
   - *α-methyldopa* and *L-dopa*, which inhibit catecholamine biosynthesis.
   - *reserpine* and *guanethidine*, which increase catecholamine catabolism.

❑ Elucidate the biological effects of catecholamines.

- *Cardiovascular effects*

  Epinephrine exhibits the following effects:
  - generalized vasodilator effect
  - vasoconstrictor effect in the arterioles of the skin, mucous membranes, and splanchnic viscera
  - stimulation of cardiac muscle contraction
  - increase in cardiac output

Norepinephrine has a generalized vasoconstrictor effect. Both the catecholamines increase blood pressure.

- *Effects on visceral smooth muscle*:
  Epinephrine causes relaxation of the smooth muscle of:
  - bronchioles: It is used as a bronchodilator in the treatment of asthma
  - urinary bladder, and
  - gastrointestinal tract.

  Epinephrine causes contraction of the sphincters of the stomach and bladder.
- *Effects on metabolism*:
  Epinephrine increases blood-glucose levels by
  - stimulation of glycogenolysis in the liver and muscle
  - increasing gluconeogenesis in the liver
  - decreasing peripheral utilization of glucose
  - inhibiting insulin release from the pancreas
  - releasing free fatty acids from adipose tissue into circulation and channeling the fatty acids to hepatic gluconeogenesis.
- *Effect on other hormones*:
  - Acting via the β-adrenergic receptors, epinephrine increases the secretion of glucagon, insulin, thyroxine, calcitonin, parathyroid hormone, renin, and gastrin.
  - Epinephrine inhibits insulin secretion from the pancreas by acting *via* the α-adrenergic receptors.

What are phaeochromocytomas?

Phaeochromocytomas are non-malignant tumors of the chromaffin cells that secrete high amounts of catecholamines.

The clinical and metabolic features include

- hypertension
- palpitations
- flushing
- sweating
- headache
- weight loss
- impaired glucose tolerance with hyperglycemia and sometimes glycosuria.

What are neuroblastomas?

Rare malignant tumors of nerve cells seen in infants and children as part of APUD (amine precursor uptake and decarboxylation) tumors are called *neuroblastomas*. These tumors may be found in the adrenal medulla or may be extra-adrenal. While most tumors secrete high amounts of catecholamines, some tumors secrete dopamine.

Chapter 17

Gonadal Hormones

❑ What are the types of cells present in the testes?

The testes contain three specialized cells:

1. The *spermatogonia*, in the seminiferous tubules where spermatogenesis occurs.
2. The *Sertoli cells* that produce androgen-binding protein and respond to *follicle stimulating hormone* (*FSH*) from the anterior pituitary.
3. The *Leydig cells* that secrete testosterone.

❑ Explain the steps in the biosynthesis of androgens and their regulation.

- *Sites of synthesis*: Testes, ovaries, adrenal cortex, and placenta.

  Testicular androgens are synthesized from cholesterol derived from
  1. Leydig cells in the testes, which synthesize cholesterol and store it as cholesterol ester.
  2. plasma by receptor-mediated uptake.
- *Rate-limiting step*: Cholesterol delivery to the inner mitochondrial membrane by the transport protein, *steroidogenic acute regulatory protein (STAR).*
- *Steps*:
  1. Conversion of cholesterol to *pregnenolone* by mitochondrial enzyme complex *cytochrome P450 side chain cleavage* (*P450 scc*), which catalyzes oxidative side chain cleavage of cholesterol. This step is regulated by LH and ACTH.
  2. Translocation of pregnenolone from Leydig cells to the endoplasmic reticulum where the remaining biosynthetic reactions occur.
  3. Formation of testosterone from pregnenolone by the concerted action of the enzymes such as 3β-*hydroxysteroid dehydrogenase* (3β-*OHSD*), $\Delta^{5,4}$ *isomerase*, 17α-*hydroxylase*, $C_{17\text{-}20}$ *lyase*, and 17β-*hydroxysteroid dehydrogenase* (17β-*OHSD*).
  4. In addition to testosterone, small quantities of dehydroepiandrosterone, androstanediol, *dihydrotestosterone* (*DHT*), and 17β-*estradiol* are also synthesized by the testes.
  5. Estradiol is derived mostly from aromatization of testosterone in peripheral tissues in males.
  6. In females, testosterone is mostly derived from the adrenals, and only one-third is synthesized in the ovaries.

Figure 17.1 gives an overview of the biosynthesis of sex hormones.

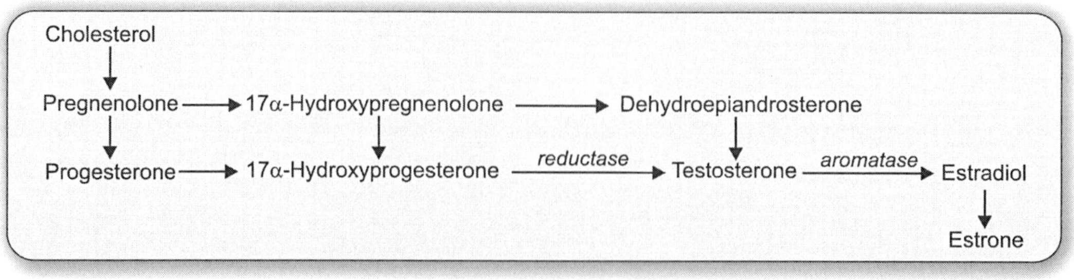

Fig. 17.1 Biosynthesis of sex hormones

Regulation

- The anterior pituitary hormone LH regulates the synthesis and secretion of testosterone by the following sequence of events:
  - Binding to receptors on the plasma membrane of Leydig cells
  - Activation of *adenylate cyclase*
  - Increase in cAMP levels
  - Increase in the rate of cholesterol side chain cleavage.
- *Regulation of LH secretion*
  - by GnRH from the hypothalamus.
  - feedback by testosterone that may be exerted at the level of the hypothalamus or the anterior pituitary.
- The regulatory effect of ACTH on the adrenals is similar to that of LH.
- FSH
  - sensitizes Leydig cells to the actions of LH by increasing the number of LH receptors.
  - promotes the synthesis of androgen-binding protein (ABP) by the Sertoli cells.
  - ABP binds testosterone and transports it from the Leydig cells to the site of spermatogenesis.
  - secretion is inhibited by *inhibin* produced by the Sertoli cells.

❑ How is testosterone transported?

About 97–99% of testosterone is transported in circulation bound to the following proteins:
- *sex hormone-binding globulin* (*SHBG*), or *testosterone–estrogen-binding globulin* (*TEBG*), a β-globulin synthesized in the liver, accounts for most of the testosterone binding. SHBG synthesis is increased by estrogens and decreased by androgens.
- plasma albumin.

The free fraction, which constitutes only 1–3% of total testosterone, is the biologically active hormone.

❑ How is testosterone metabolically activated and inactivated?

Metabolic activation

- Less than 10% of testosterone is activated to *dihydrotestosterone* (DHT) or *estradiol,* and 90% is converted to excretory metabolites.
- Testosterone is converted to its metabolically active form dihydrotestosterone by the enzyme *5α-reductase* in male accessory organs, brain, skin, and hair follicles. DHT is a potent androgen.
- Aromatization of testosterone forms estradiol in peripheral tissues such as adipose tissue.

The structures of testosterone and dihydrotestosterone are shown in Fig. 17.2.

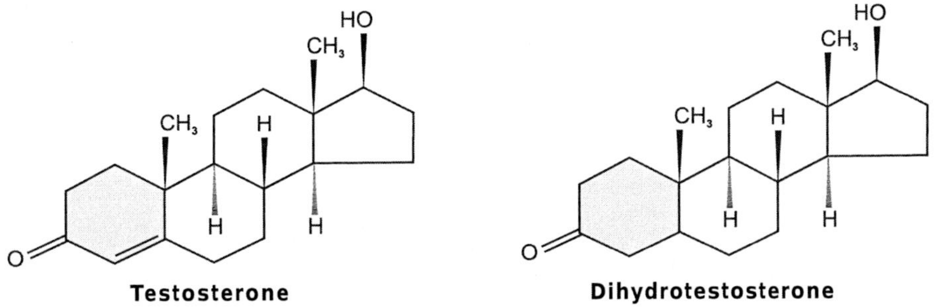

Fig. 17.2 Structures of testosterone and dihydrotestosterone

Metabolic inactivation

- Metabolic inactivation of testosterone takes place in the liver.
- Testosterone is converted to the 17-ketosteroids, androsterone and etiocholanolone, which undergo conjugation with glucuronate or sulphate to form water-soluble compounds that are excreted in the urine.

❑ Elucidate the biological actions of testosterone.

Testosterone and DHT exert the following *biological effects*:

- *Sexual differentiation during embryonic development, and development of male secondary sexual characteristics.* Testosterone is necessary for sexual differentiation and spermatogenesis, whereas DHT is essential for development of the male external genitalia. Both testosterone and DHT are required for the male phenotype.
- *Anabolic and growth promoting effects* such as growth of bone, increase in skeletal muscle mass, and redistribution of fat.

Mechanism of action:

- Free testosterone enters the cytoplasm of target cells through the plasma membrane where it is converted to DHT.
- Both testosterone and DHT bind to a specific intracellular receptor and the hormone–receptor complex binds to nuclear chromatin selectively activating transcription of specific genes.
- Testosterone increases the activity of *DNA polymerase* and *thymidine kinase*, and the synthesis of androgen-binding protein.

❑ Outline the causes and diagnostic tests for hypogonadism.

Hypogonadism refers to a clinical condition characterized by defective spermatogenesis and/or testosterone production.

Causes

Hypogonadism may be classified into two types:

- *Primary hypogonadism* due to testicular disease with defective seminiferous tubular function, decreased spermatogenesis, and infertility, or defective Leydig cell function with failure of spermatogenesis.

  Primary hypogonadism may arise due to

  - *Klinefelter's syndrome* with 47 XXY genotype, and enzyme defects such as 5α-*reductase*

- bilateral testicular torsion
- radiation, and cytotoxic drugs.

- *Secondary hypogonadism* may be due to
  - pituitary disorders panhypopituitarism, prolactinomas
  - hypothalamic disorders such as *Kallman syndrome.*

Diagnostic tests

- *Measurement of plasma testosterone*: Levels are low in both primary and secondary hypogonadism.
- *Measurement of LH and FSH*: Plasma LH is increased in Leydig cell failure, and plasma FSH in defective seminiferous tubular function. Decrease in both LH and FSH with oligospermia is suggestive of pituitary or hypothalamic disease.
- *hCG stimulus test*: Administration of human chorionic gonadotrophin (hCG) to test Leydig cell function.
  - Testosterone is estimated in blood drawn at 9.00 hours.
  - hCG 2000 IU is administered by intramuscular injection on day 1 and 3, and blood withdrawn on the fifth day for testosterone estimation.
  - An increase in plasma testosterone is seen in normal subjects, whereas in primary testicular failure, the response is decreased or absent.
  - The response may, however, be normal in secondary testicular failure.
- *GnRH test* is used in the differential diagnosis of hypogonadism.
  - The test involves injection of GnRH (100 µg i.v.) and measurement of plasma LH and FSH before, and at 30 and 60 minutes after GnRH injection.
  - In pituitary disease, the response of LH and FSH to GnRH may be normal, decreased, or absent; whereas in hypothalamic disease, the response may be delayed, normal, or decreased.

❑ Outline the causes of gynecomastia.

Gynecomastia refers to the development of breast in males. It results from an abnormality in the relative amount of estrogens and androgens.

Causes

- *Physiological gynecomastia* as in
  - neonates, exposed to maternal estrogens
  - puberty, when there is transient increase in oestrogen levels
  - old age, when testosterone secretion is low.
- *Pathological gynecomastia* due to
  - chronic liver disease, due to impaired hepatic conjugation and clearance of estrogens
  - Tumors in which estrogen levels are high
  - Klinefelter's syndrome in which androgen levels are low
  - Testicular feminization characterized by androgen insensitivity
  - Refeeding after starvation characterized by high LH secretion.
- *Drug-induced gynecomastia* by
  - estrogens
  - digoxin that binds to estrogen receptors
  - antiandrogens such as cyproterone.

❑ What are estrogens?

Estrogens are C18 steroids secreted by the ovaries, corpus luteum, and placenta. The estrogens include

- *β-Estradiol*, the most potent estrogen secreted by the ovarian follicle
- *Estrone*, a metabolic product of β-estradiol
- *Estriol*, a metabolic product of β-estradiol secreted by the placenta. The corpus luteum and the placenta also secrete estrogens. Estriol is the predominant estrogen.

The structures of estrogens are shown in Fig. 17.3.

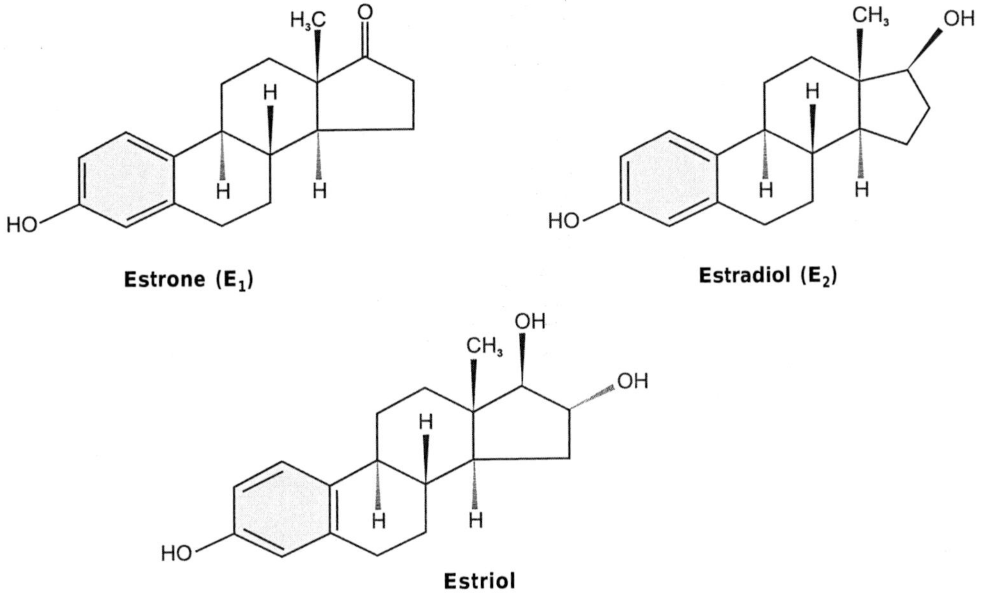

Fig. 17.3 Structures of estrogens

❑ Give a brief account of the biosynthesis of estrogen and its regulation.

- *Site of synthesis*: Ovary in non-pregnant females, the testes in males. Estrogen biosynthesis also occurs in the liver, muscle, adipose tissue, hypothalamus, and hair follicles.
- *Precursor molecule*: Testosterone
- *Conversion of testosterone to estradiol* is accomplished by a microsomal enzyme system called the *aromatase complex* in a process involving three successive hydroxylations and requiring three molecules each of NADPH and O_2.
- In males, peripheral aromatization of testosterone accounts for significant production of estrogens.
- Adrenal androgens are responsible for estrogen synthesis in postmenopausal women and for about 50% of estrogen synthesis during pregnancy.

Regulation

- LH stimulates theca cells to synthesize androgen precursors of estrogens.
- FSH
  - stimulates growth of the ovarian follicle
  - stimulates estrogen synthesis by granulosa cells

- stimulates development of LH receptors on both theca and granulosa cells
- is inhibited by inhibin.

- Secretion of LH and FSH
  - is regulated by GnRH
  - is inhibited by estrogens.

❑ How is estrogen transported in circulation?

- About 97–99% of estrogen is transported in circulation bound predominantly to SHBG, and to a minor extent, to albumin.
- The free fraction, which constitutes only 1–3% of total estrogen is the biologically active hormone. The concentration of the free hormone, increases during ovulation.

❑ How are estrogens metabolized?

Inactivation of estrogens occurs in the

- liver (predominant)
- extrahepatic tissues such as
  - placenta where an *estradiol-dependent transhydrogenase* interconverts estradiol and estrone.
  - the brain where *estradiol 2-hydroxylase* catalyzes hydroxylation of estrogens to form catecholestrogens that have no estrogenic activity. The catecholestrogens undergo O-methylation by *catecholamine O-methyltransferase* (*COMT*) to form methoxy derivatives that are excreted in the urine.

Estrogens and their metabolites are excreted as glucuronate and sulphate conjugates.

❑ What are the biological effects of estrogens?

- *Effects on the female reproductive system*
  - growth and development of uterus, vagina, accessory sex glands, breasts, and external genitalia.
  - proliferative effects on uterine and vaginal epithelium.
  - increase in the blood flow in the uterus.
  - stimulation of the growth of ductile elements in the breasts.
- *Effects on lipid metabolism*
  - increase in plasma VLDL and HDL.
  - HDL-enhancing effect may account for lower incidence of coronary heart disease in premenopausal women.
- *Effect on bone*
  - closure of bony epiphyses.
  - prevention of decalcification.
- *Other effects*
  - weak salt-retaining activity.
  - weak anabolic effects.
  - increased synthesis of renin substrate, plasma-binding proteins for steroids and thyroid hormones, some blood clotting factors.

Mechanism of action

- Estrogens bind to high affinity cytosolic receptors on target cells in the uterus, vagina, mammary gland, anterior pituitary, and hypothalamus.

- The hormone–receptor complex is translocated to the nucleus and binds to the *estrogen response element* (*ERE*) of DNA to increase transcription of specific genes.

❑ Discuss the transport, metabolism, and biological effects of progesterone.

Transport

- Progesterone is transported in circulation bound to *corticosteroid-binding globulin* (*CBG*) and *albumin*.
- The free fraction is the biologically active hormone.
- Plasma progesterone, which is 50–100 ng/dL (1.5–3 nmol/L) in the follicular phase increases in the luteal phase (1000–1500 ng/dL; 30–45 nmol/L) with peak values at the end of a full-term pregnancy (16,000 ng/dL or 480 nmol/L).

Metabolism

- Progesterone is metabolized to pregnanediol, conjugated with glucuronate or sulphate in the liver, and excreted in the urine.
- Pregnanetriol is a major urinary metabolite.

Biological effects

- induces mucus secretory activity in the uterine endometrium facilitating implantation of the ovum.
- is essential for completion of pregnancy.
- stimulates growth of the breasts.
- keeps the uterus quiescent during pregnancy.
- causes slight increase in temperature in the latter half of the menstrual cycle.
- *Mechanism of action*: Progesterone binds to specific intracellular receptors and the hormone–receptor complex binds to a specific DNA sequence termed the *progesterone response element (PRE)* influencing transcription of specific genes resulting in an altered metabolic response.
- Synthetic progestins that exert antiovulatory effect by feedback suppression of gonadotrophin secretion are useful as oral contraceptives.

❑ Elucidate the hormonal changes during the menstrual cycle.

Characteristic cyclic changes in the secretion of ovarian hormones and gonadotrophins occur during the menstrual cycle, which is divided into three phases:

- *Pre-ovulatory phase*
  - FSH stimulates development of the ovarian follicle.
  - Development of one of the follicles that is sensitive to FSH occurs.
  - Estrogen secretion is increased.
  - The uterine and vaginal epithelia undergo proliferative changes.
- *Ovulation*
  - Ripening of the dominant follicle occurs with increase in plasma estrogen.
  - *LH surge* occurs at midcycle followed by ovulation and rupture of the follicle.
- *Post-ovulatory phase* (*luteal or secretory phase*)
  - The ruptured follicle is transformed into a yellow structure called the *corpus luteum*. Progesterone and estradiol are produced by the corpus luteum.

- The corpus luteum regresses if the ovum is not fertilized accompanied by a fall in hormone secretion and menstruation ensues.
- Pregnancy occurs if the ovum is fertilized.

❑ What is menopause?

Menopause is the stage when all the ovarian follicles atrophy and menstruation ceases.

Menopause is accompanied by the following changes:

- Progressive loss of ovarian function.
- Estrogens are formed only from peripheral aromatization of adrenal androstenedione.
- Plasma estrogen levels are low.
- LH and FSH levels are high.
- Susceptibility to osteoporosis.
- Atrophy of urogenital epithelium and skin.
- Vasomotor instability manifested as hot flushes, nervousness, anxiety, irritability, and depression.

Estrogen therapy relieves symptoms in postmenopausal women.

❑ Discuss the causes, diagnosis, and management of amenorrhea.

Amenorrhea refers to the absence of menstruation.

Causes

Amenorrhea may result from

- Non-endocrinal causes such as pregnancy and menopause.
- Endocrinal causes such as
    1. *Primary ovarian failure* as in *Turner's syndrome* with 45XO karyotype, characterized by developmental abnormalities, delayed puberty, low plasma estrogens, and increase in FSH and LH.
    2. *Polycystic ovary syndrome* (*Stein-Leventhal syndrome*), characterized by large polycystic ovaries, high plasma LH levels, chronic anovulation, obesity, androgen excess, hirsutism, and infertility.
    3. *Hypogonadotrophic hypogonadism* resulting from pituitary or hypothalamic disease as in
        - panhypopituitarism following treatment of pituitary adenomas by surgery or radiation.
        - prolactinomas.
        - *Kallman syndrome,* due to defective GnRH synthesis or secretion, characterized by sexual immaturity and eunuchoid habitus.
        - deficiencies of GH, ACTH, thyroid hormone, and vasopressin.
        - excessive weight loss with anorexia nervosa.
        - malignancies.
        - end-stage kidney disease.

Diagnosis

- Increase in plasma FSH with decrease in estradiol suggests ovarian failure.
- Increase in plasma LH in the absence of pregnancy indicates polycystic ovary syndrome.
- Normal or low plasma FSH and LH levels with low plasma estradiol levels suggest pituitary or hypothalamic disorder.

Management

- Bromocriptine administration in amenorrhea due to a pituitary tumor.

- Cyclical estrogen and progesterone replacement in patients with ovarian, pituitary, or hypothalamic disease, when fertility is not required.
- Administration of FSH and LH in pituitary failure if fertility is required.
- Administration of clomiphene in hypothalamic disease to block estrogen receptors in the hypothalamus, and to stimulate GnRH with secretion of LH and FSH.

Describe the causes, diagnosis, and management of hirsutism and virilism.

- *Hirsutism* refers to an increase in body hair and an androgen type of hair distribution.
- *Virilism* refers to *masculinization,* characterized by excessive androgen secretion, enlarged clitoris, increased male pattern of hair growth, deepening of voice, and breast atrophy.
- Hirsutism and virilism arise due to increased androgen secretion, low plasma SHBG, and increase in free testosterone levels or enhanced sensitivity to androgens.

Causes

- Polycystic ovary syndrome
- Androgen-secreting tumors of the adrenal or ovary
- Congenital adrenal hyperplasia
- Cushing's syndrome
- Administration of androgens or progestrogens

Diagnosis

- Determination of total and free testosterone concentration, dehydroepiandrosterone, and SHBG.
- Measurement of adrenal androgens and *17-hydroxyprogesterone* (*17-OHP*).
- Very high testosterone levels suggest the presence of an androgen-secreting tumor.
- High 17-OHP levels in response to Synacthen injection in late-onset congenital hyperplasia.
- Elevated plasma dehydroepiandrosterone sulphate and androstenedione in adrenal tumors.

Treatment

- Removal of excess hair in mild cases.
- Administration of synthetic glucocorticoids that suppress adrenal androgen excretion, or anti-androgen drugs such as cyproterone in severe cases.

Outline the laboratory tests for the investigation of infertility.

The laboratory tests for investigation of infertility include the following:

- *In females*
  - Plasma progesterone level is a reliable indicator of ovulation. A value of < 10 nmol/L indicates failure in ovulation or luteal function.
  - Ovarian ultrasound examination to monitor follicular development and ovulation.
  - Evaluation of luteal function by histological examination of an endometrial biopsy specimen.
- *In males*
  - Semen analysis.
  - Evaluation of plasma FSH.
  - Low plasma testosterone with increase in LH indicates Leydig cell failure.
  - A decrease in both plasma LH and testosterone requires evaluation of pituitary and hypothalamic function.

- High plasma FSH indicates failure of seminiferous tubules, while low FSH with oligospermia pinpoints pituitary or hypothalamic disorder.
- High plasma prolactin indicates pituitary tumor.

❑ What are the functions of the placenta?

The placenta, which separates the maternal from fetal circulation, performs the following functions:

- Provides nourishment to the fetus
- Removes waste
- Synthesizes hormones necessary for pregnancy
- Functions as a barrier for the transport of large proteins and hormones. While lipid-soluble substances, gases (oxygen, carbon dioxide), ions (sodium and chloride) are passively transported, glucose, amino acids, and calcium are actively transported.

❑ Give an account of placental hormones.

The hormones synthesized by placenta include

- *Human chorionic gonadotrophin (hCG)*
  1. a glycoprotein heterodimer made up of two nonidentical subunits (an α-chain and a β-chain) synthesized by the syncytiotrophoblast cells.
  2. maintains the corpus luteum in the first weeks of pregnancy.
  3. α-subunit on glycation increases cyclic AMP, which stimulates progesterone synthesis, prevents menses, and maintains pregnancy.
  4. estimation in blood and urine is useful for
     - confirming pregnancy and establishing the accurate dates of pregnancy.
     - identification of ectopic and abnormal intrauterine pregnancies.
     - diagnosis and monitoring gestational trophoblastic neoplasms and hCG-producing tumours such as carcinoma of the lung.
- *Human placental lactogen (hPL)*
  1. also known as *human chorionic somatomammotrophin (HCS)*.
  2. a single polypeptide chain that has sequence homology with growth hormone.
  3. exerts biological effects such as
     - lactogenic, somatotrophic, erythropoietic, and aldosterone-stimulating effects.
     - metabolic effects: Inhibition of glucose uptake, increased lipolysis, nitrogen retention.
- *Placental steroid hormones* such as progesterone, estradiol, estriol, and estrone.

❑ What is amniotic fluid?

- Amniotic fluid is the fluid that protects the fetus against injury, maintains a constant temperature, and provides a medium for the movement of the foetus.
- During early gestation, the amniotic fluid is a complex dialysate of the maternal serum. As the fetus grows, the amniotic fluid undergoes changes in composition.
- While the osmolality and the sodium concentration are low, the levels of urea, uric acid, and creatinine increase.
- The prostaglandins PGE_1, E_2, F_{1a}, and F_{2a} are found in small amounts and levels of PGE_2, and $PGF_{2\alpha}$ reach high levels during labour.

❑ Enumerate the various biochemical changes during pregnancy.

- *Hematological changes*
  - increase in maternal blood volume
  - decrease in RBC count, and hemoglobin concentration.
  - increase in concentration of the clotting factors fibrinogen, and factors VII, VIII, IX, and X.
- *Plasma proteins*
  - low plasma albumin and total protein with a slight increase in plasma globulin.
  - increase in plasma thyroxine-binding globulin (TBG), corticosteroid-binding globulin (CBG), and sex hormone- binding globulin (SHBG) levels due to increased hepatic synthesis.
- *Plasma lipids*
  - increase in serum triglycerides, cholesterol, phospholipids, and free fatty acids.
- *Plasma enzymes*
  - decreased activity of plasma *cholinesterase*
  - increase in *alkaline phosphatase* (ALP) due to synthesis of a specific heat stable placental isoenzyme during the last trimester of pregnancy.
- *Kidney function*
  - increase in the glomerular filtration rate (GFR).
  - increase in plasma urea and creatinine.
  - proteinuria.
  - lactosuria.
  - glycosuria due to decreased renal threshold.
- *Hormones*

  Increased secretion of
  - progesterone
  - estrogen
  - prolactin after the 8th week of pregnancy
  - cortisol, aldosterone, and deoxycorticosterone
  - parathyroid hormone, vitamin 1, 25-$(OH)_2D_3$ with increase in intestinal calcium absorption

  Decrease in the secretion of LH and FSH.

❑ What is the common diagnostic test for pregnancy?

Pregnancy can be detected in the laboratory by measurement of the urinary concentration of hCG by sensitive immunoassays that give positive results within 10 days of conception.

❑ Write notes on the most common fetal abnormalities.

Fetal abnormalities can be investigated by analysis of amniotic fluid obtained after 14 weeks of gestation by amniocentesis or the maternal plasma.

The common fetal abnormalities include the following:

- *Neural tube defects* such as *spina bifida* in which concentrations of *alpha-fetoprotein (AFP)* in amniotic fluid and maternal plasma are abnormally high.
- *Down syndrome,* characterized by low maternal plasma AFP and high concentration of hCG.

- *Erythroblastosis fetalis* associated with hemolysis caused by Rh incompatibility. In this condition, IgG antibodies cross the placenta and destroy fetal erythrocytes, leading to fetal anemia, increased erythropoiesis, and liver damage. There is inability to conjugate the excess bilirubin resulting from hemolysis, which leads to deposition of bilirubin in the brain and destruction of brain cells (*kernicterus*). In severe cases, *hydrops fetalis* occurs eventually resulting in death.
- *Respiratory distress syndrome* (*hyaline membrane disease*), caused by lack of surfactant (a mixture of lecithin and small amount of sphingomyelin) that causes *focal atelectasis* with alveolar collapse and failure of oxygenation.

Chapter 18

Gastrointestinal and Pancreatic Hormones

❑ Give a brief account of the gastrointestinal hormones.

- The gastrointestinal tract produces a number of peptide hormones that perform specific functions.
- These peptides are also synthesized in the brain where they function as neurotransmitters and neuromodulators.
- The gastrointestinal hormones are grouped into two families on the basis of amino acid sequence and functions:
  - ***The gastrin family*** includes
    i. *gastrin*, which stimulates gastric acid and pepsin secretion.
    ii. *cholecystokinin* that causes contraction of the gallbladder and stimulates pancreatic enzyme secretion.
  - ***The secretin family*** comprises
    i. *secretin* that stimulates pancreatic bicarbonate secretion.
    ii. *gastric inhibitory peptide* (*GIP*) that inhibits gastric acid secretion, and increases glucose-mediated insulin release.
    iii. *vasoactive intestinal polypeptide* (*VIP*) has the following actions:
    Stimulation of pancreatic bicarbonate secretion, smooth muscle relaxation, and gastrointestinal motility.
    iv. *glucagon*, an insulin antagonist and ketogenic hormone.
    v. *glicentin* with glucagon-like immunoreactivity.
  - ***The neurocrine peptides*** do not belong to either family and include substance *P, somatostatin,* and *bombesin-like peptides.*

❑ Discuss the secretion and metabolic functions of glucagon.

- Glucagon is a single polypeptide containing 29 amino acid residues.
- It is synthesized in the A cells of the pancreatic islets as a large prohormone.
- Synthesis of glucagon is stimulated by hypoglycemia and fasting, and inhibited by insulin.
- The hormone has a short half-life (~5 minutes) and is inactivated in the liver by removal of the first two amino acids from the N-terminal end.

In the liver

Glucagon binds to specific receptors in the plasma membrane and activates adenylate cyclase with increase in cAMP.

The hormone

- increases glycogenolysis by activating *glycogen phosphorylase*
- decreases glycogenesis by inhibiting *glycogen synthase*
- stimulates gluconeogenesis by inducing the gluconeogenic enzymes *phosphoenolpyruvate carboxykinase (PEPCK), glucose-6-phosphatase* and *fructose-1,6-bisphosphatase*
- increases ketogenesis
- represses the enzymes *glucokinase, acetyl CoA carboxylase,* and *hydroxymethyl glutaryl CoA (HMG CoA) reductase.*

In the adipose tissue

- Glucagon increases cAMP levels
- Activates *hormone-sensitive lipase*
- Stimulates lipolysis. The fatty acids are either oxidized for energy or converted to ketone bodies.

❑ Elucidate the biological actions of somatostatin.

- Somatostatin is synthesized in the D cells of the pancreatic islets as a large prohormone that is processed into a molecule containing 14 amino acids.
- The hormone functions as a neurotransmitter in gastrointestinal tissues and in the central nervous system.

Biological effects

Somatostatin

- inhibits the release of other pancreatic hormones via a paracrine action
- prevents ketosis by inhibiting glucagon release
- inhibits transport of nutrients from the gastrointestinal tract into the circulation by
  - delaying gastric emptying
  - inhibiting secretion of gastrin and gastric HCl
  - inhibiting pancreatic exocrine secretion
  - decreasing splanchnic blood flow
  - slowing carbohydrate absorption.

❑ Give a brief description of the structure of insulin.

- Insulin is a heterodimeric polypeptide containing an A chain with 21 amino acids and a B chain with 30 amino acids, respectively (Fig. 18.1).
- The A chain contains one intrachain disulphide linkage between residues 6 and 11.
- There are two interchain disulphide bridges between A_7 and B_7, and A_{20} and B_{19}.

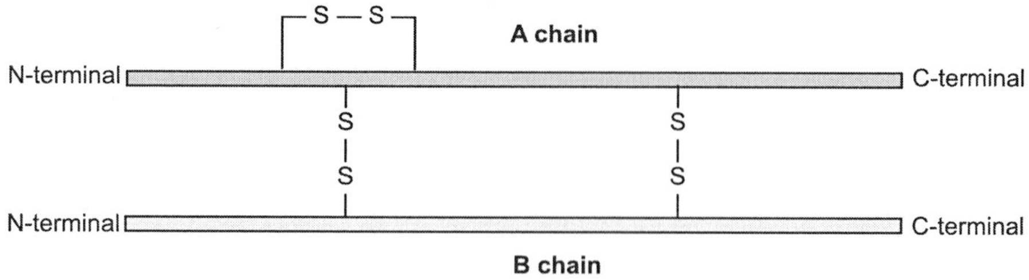

Fig. 18.1 Structure of insulin

❑ Explain the formation of insulin from proinsulin and the regulation of insulin secretion.

Synthesis

- Insulin is synthesized by the β cells of the pancreatic islets as a preprohormone.
- Preproinsulin is processed to form proinsulin by removal of the leader peptide, disulphide bond formation and folding.
- Proinsulin is transported to the Golgi apparatus where it is packaged into secretory granules followed by proteolysis to generate equimolar amounts of mature insulin and C-peptide.
- Proinsulin as well as insulin form hexamers with zinc.
- On receiving the stimulus for insulin release, the granules fuse with the plasma membrane and discharge their contents into the extracellular fluid by *emiocytosis*. A small amount of proinsulin is also released with insulin.

Regulation

- *Rate of secretion*: Approximately 40–50 units of insulin/day.
- *Stimulus for insulin secretion*: Hyperglycemia
- Insulin secretion is increased by
  - epinephrine acting *via* β-adrenergic receptors
  - growth hormone
  - cortisol
  - estrogens
  - placental lactogen
  - progestins
  - drugs such as sulphonylureas.
- Insulin secretion is decreased by
  - epinephrine acting via α-adrenergic receptors.

❑ Enumerate the biological effects of insulin.

Insulin exerts the following biological effects:

Membrane transport

- increases carrier-mediated facilitated diffusion of D-glucose, galactose, xylose, and arabinose by means of *glucose transporters* (*GLUT 1–4*).
- promotes entry of amino acids, K^+, Ca^{2+}, nucleosides, and inorganic phosphate.
- *In the liver*
  - not directly required for glucose entry.
  - indirectly increases glucose influx by inducing *glucokinase* that converts glucose to glucose-6-phosphate, thereby lowering free glucose concentration and facilitating glucose diffusion down a concentration gradient.

Carbohydrate metabolism

- enhances glucose utilization by
  - increasing hepatic glycolysis by increasing the activity and amount of *glucokinase* and *phosphofructokinase*.
  - decreasing glucose release into circulation by lowering the activity of hepatic *glucose-6-phosphatase*.

- promotes glucose entry into the skeletal muscle through the glucose transporter and enhances *hexokinase II* activity.
- stimulates glycogenesis in the liver and muscle by increasing the activities of
  - *glucokinase* and *hexokinase II* respectively.
  - *glycogen synthase* by
    i. stimulating *phosphodiesterase* with decrease in cAMP levels.
    ii. activating a *phosphatase* that dephosphorylates *glycogen synthase.*
- inhibits glycogenolysis by inactivating *glycogen phosphorylase.*
- inhibits gluconeogenesis by inhibiting transcription of key genes such as *phosphoenolpyruvate carboxykinase* (*PEPCK*).

Lipid metabolism

Insulin exerts an anabolic effect on lipid metabolism.

The hormone

- stimulates adipose tissue lipogenesis by providing acetyl CoA and NADPH for fatty acid synthesis, and glycerol for triglyceride synthesis.
- inhibits lipolysis in the adipose tissue and liver by activating a *phosphatase* that dephosphorylates *hormone-sensitive lipase*, thereby inhibiting its activity.
- decreases cAMP levels and inactivates *cAMP-dependent protein kinase.*
- decreases plasma free fatty acid levels.
- influences the metabolism of very low density lipoproteins (VLDL) and low density lipoproteins (LDL).

Protein metabolism

Insulin exerts an anabolic effect on protein metabolism.

The hormone

- stimulates protein synthesis
- inhibits protein catabolism
- promotes uptake of neutral amino acids into muscle.

Cell replication

- stimulates cell replication
- potentiates the effects of growth factors, hormones, prostaglandins, and cAMP analogs on cell cycle progression.

Explain the mechanism by which the insulin receptor mediates insulin action.

- Insulin binds to a glycoprotein receptor, a heterodimer ($\alpha_2\beta_2$) present on the plasma membrane of target cells.
- The extracellular α-subunit binds insulin, while the β-subunit is a transmembrane protein with *tyrosine kinase* activity and an autophosphorylation site involved in signal transduction and insulin action.
- The receptor undergoes autophosphorylation on insulin binding and in turn phosphorylates tyrosine residues on *insulin receptor substrate 1 (IRS-1)*. Phosphorylated IRS-1 binds to GTPases, protein kinases, and lipid kinases that mediate the metabolic actions of insulin.

- Receptor–hormone interaction activates several signaling pathways, including
  - *phosphoinositol-3-kinase* that generates novel inositol lipids which function as second messengers and in turn activate other signaling pathways such as the *p70S6 kinase.*
  - *GRB2,* which activates a cascade of threonine or serine kinases.
  - *p21RAS,* a GTPase that stimulates the *mitogen-activated protein (MAP) kinase pathway.*
- The transmembrane signals generated modulate intracellular activities including
  - glucose transport.
  - phosphorylation and dephosphorylation that alter enzyme activities. These effects are mediated by decreasing the activity of *cAMP-dependent protein kinase,* inhibition of *protein kinases,* or stimulation of *phosphoprotein phosphatases.*
  - cell growth and replication.
  - gene expression by regulating transcription and translation. Insulin influences the synthesis of over 100 different mRNAs and the activity of about 50 proteins in a wide range of tissues.

Figure 18.2 summarizes the signaling pathways of insulin.

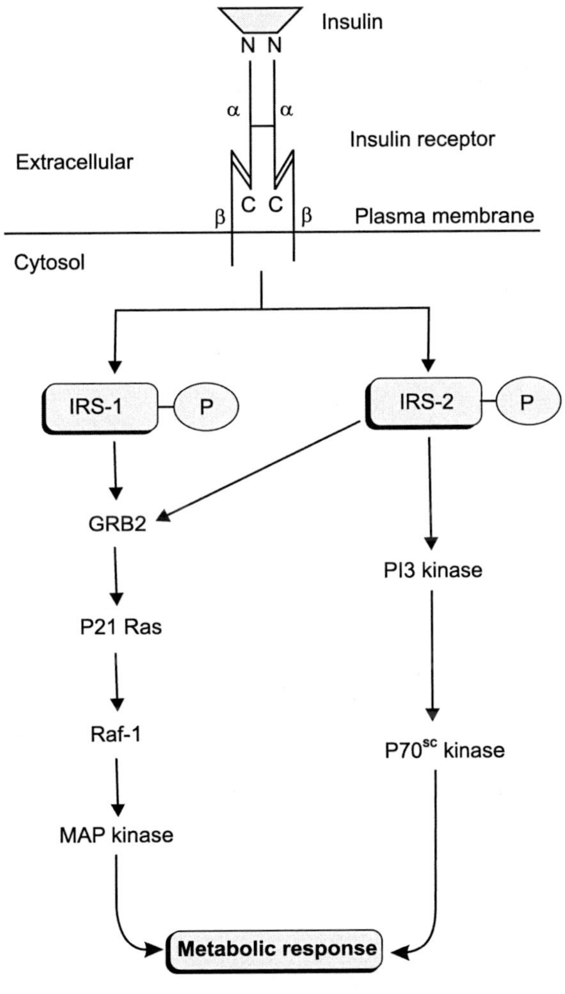

Fig. 18.2 Insulin signaling pathways

Chapter 19

Elements of Nutrition

❑ Define respiratory quotient (RQ).

Respiratory quotient is the ratio of the volume of CO_2 eliminated to the volume of oxygen utilized in oxidation.

❑ Give the RQ values for carbohydrates, fats, and proteins.

- The RQ value for carbohydrates is 1.
- Fats have a lower RQ than carbohydrates because the oxygen content of fats is less than the CO_2 content, and more oxygen from outside is required for oxidation. The RQ value of fat is 0.7.
- The RQ of proteins is variable and is approximately 0.8.

❑ Define basal metabolic rate (BMR).

BMR is the lowest level of energy production that is compatible with life.

❑ Enumerate the factors that influence BMR.

The factors that influence BMR include the following:

- *Surface area*: Although the BMRs of different individuals, expressed in terms of the surface area (m^2), are constant, smaller individuals have higher BMR per unit of surface area compared to larger individuals.
- *Age*: BMR is low in newborns, reaches a maximum at age 5, and declines steadily. A small rise occurs before puberty.
- *Sex*: BMR in women is lower than that in men.
- *Climate*: BMR is lower in warm climates; it decreases in summer and increases in winter.
- *Race*: BMR is lower in Orientals. High BMR values have been reported in Eskimos.
- *Nutritional state*: BMR is low in starvation and malnutrition.
- *Disease*: BMR is increased in fever, leukemia, polycythemia, cardiac failure, and hypertension due to increased cellular activity.
- *Hormones*: Thyroxine has a role in maintaining BMR. BMR is increased in hyperthyroidism and decreased in hypothyroidism.

❑ How is the daily caloric requirement estimated?

- Daily caloric requirement can be estimated with the help of the Harris – Benedict equation.

- The basal energy expenditure (BEE) is as follows

  $BEE_{women} = 655 + (9.5 \times W) + (1.8 \times H) - (4.7 \times A)$ kcal/d

  $BEE_{men} = 66 + (13.7 \times W) + (5 \times H) - (6.8 \times A)$ kcal/d,

  where W is the weight in kg, H is the height in cm, A is age in years.

What is meant by nitrogen balance?

- Nitrogen balance is a state in which the total daily loss of nitrogen equals the daily nitrogen intake.
- In a normal healthy adult, essential amino acids and the amino acid nitrogen lost from the body are replaced by dietary protein.
- The normal daily intake of nitrogen is 10–15 g in dietary protein (1 g of nitrogen is derived from 6 g of protein).
- Loss of nitrogen occurs through the following routes:
  - Urine accounts for excretion of 90–95% of nitrogen with urea constituting 80–85%, and smaller amounts as urate and creatinine.
  - Feces in the form of:
    a. undigested and unabsorbed nitrogenous dietary material.
    b. end products of nitrogenous metabolism from shed intestinal cells or digestive enzymes,
    c. intestinal bacterial nitrogen.
  - Saliva
  - Sweat
  - Desquammated skin
  - Hair and nails.

Define positive and negative nitrogen balance.

- *Positive nitrogen balance* is a state in which the nitrogen loss is less than the intake. It is seen in the following conditions:
  - During growth
  - Pregnancy
  - Convalescence
  - Excess secretion of anabolic hormones such as androgens, estrogens and growth hormone.
- *Negative nitrogen balance* is a state in which the nitrogen loss is greater than the intake. It occurs due to the following causes:
  - Inadequate protein and energy intake
  - Decreased protein digestion (chronic pancreatic disease)
  - Decreased amino acid absorption
    i. after surgery
    ii. intestinal disease
    iii. inherited transport defects
  - Increased protein catabolism as in
    i. endocrine hyperactivity (thyrotoxicosis, Cushing's syndrome)
    ii. starvation
    iii. severe trauma
    iv. infections.

- Protein loss
  i. during menstruation
  ii. following burns
  iii. nephrotic syndrome
  iv. protein-losing gastroenteropathies as in malignancies
  v. hemorrhage.

❑ What are essential amino acids?

Human beings are incapable of synthesizing 9 of the 20 amino acids required for protein biosynthesis. These amino acids, which must be supplied in the diet, are termed *essential amino acids*. These include the following:

Leucine	Methionine	Histidine*
Isoleucine	Phenylalanine	Arginine*
Valine	Tryptophan	Lysine
Threonine		

* Histidine and arginine are semiessential.

❑ Write a note on protein quality.

- Protein quality refers to the content of the essential amino acids present in the diet.
- Proteins that lack any of the essential amino acids have less biological value.
- The quality of a protein depends on:
  - the content and proportion of amino acids
  - content of essential amino acids
  - easy digestibility
  - efficiency of absorption.
- Based on quality, proteins are classified into two:
  - *First class proteins*, which include animal proteins that contain all the essential amino acids in correct proportions.
  - *Second class proteins*, which include vegetable proteins, lack some essential amino acids. For example, cereal proteins are poor in lysine, methionine, and tryptophan and are less digestible.

❑ When does protein energy malnutrition occur?

Protein energy malnutrition (PEM) occurs under conditions where the diet is inadequate to meet the needs of the body for protein, energy or both.

The disorder is prevalent in developing countries

- It mainly affects infants and pre-school children.
- Mild PEM is seen in elderly subjects and during pregnancy and lactation.

❑ Enumerate the causes of PEM.

The causes of PEM include

- inadequate food
- inadequate protein intake

- malabsorption
- increased metabolic demands in disease conditions
- increased nutrient losses
- infections
- maternal malnutrition.

❑ How is PEM classified?

Protein energy malnutrition is categorized as

- ***Mild PEM*** in which
  - the weight loss is ≤ 10%
  - the weight for height is 80–89 (normal 90–110)
  - the height for age is 90–94 (normal 95–105)
  - there is no disruption of physiological functions or increase in morbidity.
- ***Moderate PEM*** in which
  - the weight loss is > 10% and ≤ 20%
  - the weight for height is 70–79 and the height for age is 85–89
  - there is significant loss of body protein
  - there is impairment in biological functions and increased morbidity.
- ***Severe PEM*** in which
  - the decrease in body weight is > 20%
  - the weight for height is less than 70
  - the height for age is less than 85
  - there is severe impairment of physiological functions
  - morbidity is very high.

Severe PEM is seen in *kwashiorkor* and *marasmus*.

❑ Give an account of the prevalence, clinical symptoms, and biochemical findings in kwashiorkor.

Prevalence

'*Kwashiorkor*' is a term that was first used by the Ga tribe in Ghana to describe "the sickness the older child gets when the next baby is born".

The disease

- is prevalent in less developed areas of tropical America, Central and South Africa and India;
- occurs in pre-school children who are fed a diet rich in carbohydrate but poor in protein;
- is precipitated by outbreaks of malaria, measles, and gastroenteritis.

Clinical manifestations

- Failure of growth: The weight of the child is 60–80% of the expected weight for the age.
- Failure to thrive.
- Edema due to low plasma albumin and potassium, and high plasma sodium. The edema
  - is soft, pitting, and painless,
  - occurs in the lower limbs,
  - is associated with protruding abdomen and full, well-rounded face (*moon face*).

- Skin lesions include
  - erythema
  - dryness
  - hyperkeratosis
  - desquammation
  - hyperpigmentation of the buttocks, backs of thighs, and axillae (*crazy pavement dermatoses*).
- Sparse, soft, thin, dry, and brittle hair.
- Gastrointestinal symptoms: Loss of appetite, vomiting, diarrhea, steatorrhea, hepatomegaly, and fatty liver.
- Anemia due to iron and folate deficiency.
- Vitamin A deficiency, leading to xerophthalmia and keratomalacia.
- Riboflavin deficiency presenting as glossitis and angular stomatitis.
- Mental changes: Apathy and general unhappiness.

Diarrhea, respiratory and skin infections can eventually result in death.

Biochemical findings

- *Proteins*
  - Depletion of visceral protein due to low amino acid supply
  - Decreased hepatic synthesis of albumin
  - Low plasma albumin levels (0.7–2.2 g/dL) have the following consequences:
    i. decreased intravascular oncotic pressure
    ii. fluid accumulation in extravascular tissues
    iii. decreased cardiac output and GFR
    iv. retention of sodium and water
    v. entry of water into the extravascular space, resulting in edema
- *Lipids*
  - Decreased total cholesterol
  - Increased hepatic fatty acid synthesis
  - Decreased synthesis of very low density lipoproteins (VLDL), leading to fatty liver and hepatomegaly
  - Low fasting blood glucose
  - Decreased activities of *amylase, lipase,* and *cholinesterase*
- *Electrolytes and water*
  - Hypokalemia due to diarrhea
  - Low plasma magnesium
  - Water retention due to low plasma albumin
- Plasma glucocorticoids lower than in marasmus.

❑ Discuss the predisposing factors, clinical features, and biochemical findings in marasmus.

Predisposing factors

Marasmus (Greek 'to waste') is a PEM commonly seen in weaned infants, characterized by deficiency of both protein and energy.

The predisposing factors for marasmus include
- rapid succession of pregnancies
- early abrupt weaning
- inadequate artificial feeding of infants with dilute milk or milk products
- recurring infections treated with starvation.

Clinical features

- Growth retardation: The weight of the child is less than 60% of the expected weight for the age.
- Absence of edema.
- Gross wasting: Generalized muscular wasting with absence of subcutaneous fat, shrunken abdomen, and atrophy of mucous membranes gives the child a "skin and bones" appearance.
- Skin: dry, thin, and wrinkled with loss of elasticity.
- Hair: sparse, thin, and dry with loss of sheen.
- Intolerance to food, constipation, diarrhea and weakness.
- Bradycardia (decrease in heart rate), hypotension, and hypothermia.
- Palpable lymph nodes.
- Mental retardation, apathy, irritability, and fretfulness.
- Death occurs due to acute gastroenteritis, dehydration, and respiratory and systemic infections.

Biochemical findings

- Increased muscle protein breakdown leads to:
  - muscle wasting
  - release of amino acids that are used for gluconeogenesis and plasma protein synthesis.
- Plasma albumin: 2.5–3 g/dL
- Visceral proteins are preserved.
- Decreased fasting blood glucose levels
- High plasma glucocorticoid levels.

Enumerate the differences between kwashiorkor and marasmus.

The major difference between marasmus and kwashiorkor is the adaptation to PEM. The adaptive response of marasmus to PEM is better than that shown by kwashiorkor. The difference is due to the higher glucocorticoid levels in marasmus.

In addition, the two types of PEM differ in clinical and biochemical features.

Occurrence

While kwashiorkor occurs in pre-school children, marasmus is seen in weaned infants.

Clinical features

- Weight: 60–80% of expected weight for age in kwashiorkor and < 60% in marasmus.
- Edema and subcutaneous fat are present in kwashiorkor but not in marasmus.
- Facial appearance: Moon face in kwashiorkor, old man's face in marasmus.
- Abdomen: Protruding in kwashiorkor, shrunken in marasmus.

Biochemical findings

- *Proteins*
  1. visceral proteins are depleted in kwashiorkor, but preserved in marasmus. Muscle proteins are broken down in marasmus.
  2. plasma albumin low in kwashiorkor, normal in marasmus.

- *Lipids*
  - low serum triglycerides, cholesterol, phospholipids, LDL, VLDL with elevated FFA associated with fatty liver in kwashiorkor.
  - no changes in lipid profile in marasmus.
- *Plasma cortisol*: Increased in marasmus.

❑ What is marasmic kwashiorkor?

- Marasmic kwashiorkor is a condition in which clinical features of both marasmus and kwashiorkor are seen.
- It is characterized by
  - edema
  - weight of the child is less than 60%
  - muscle wasting
  - reduced subcutaneous fat.

❑ Outline the principles of management of PEM.

Treatment of PEM involves

- compensation of fluid and electrolyte loss due to diarrhoea and vomiting,
- adminstration of antibiotics for infections,
- blood transfusions if anemia is severe,
- administration of glucose for hypoglycaemia,
- administration of vitamin A,
- improvement of nutrition by providing easily digestible foods with high protein and energy and supplementation of vitamins and minerals.

❑ Give an account of anorexia nervosa.

Anorexia nervosa is an eating disorder seen predominantly in young women who develop a fear of becoming fat. It is characterized by the following:

- Restriction of food intake.
- Markedly decreased body weight, undetectable body fat.
- Amenorrhea due to low basal levels of luteinizing hormone (LH) and follicle stimulating hormone (FSH).
- Cardiovascular changes such as bradycardia, hypotension, and arrhythmias.
- Skin changes such as hirsutism, dryness, and carotenemia.
- Hypothermia, parotid gland enlargement.
- Hypoalbuminemia and edema.
- Anemia, leukopenia, hypokalemia.

❑ Elucidate the characteristic features of bulimia.

- *Bulimia* is an eating disorder seen predominantly in young women who develop a fear of becoming fat.
- It is characterized by episodes of binge eating, followed by self-induced vomiting.
- The weight is near normal and amenorrhea is infrequent.

- Cardiovascular and skin changes are uncommon unlike anorexia nervosa.
- Metabolic alkalosis and hypokalemia are seen.

❑ Briefly explain the causes and consequences of obesity.

- *Obesity* refers to a condition in which the body weight is 20% or more than the ideal body weight.
- The energy intake exceeds the energy expenditure and the excess energy is stored in the adipose tissue as fat.
- *Causes* of obesity may be
  - *Primary* such as overeating, genetic, environmental and social factors.
  - *Secondary* to diseases such as
    - i. hypothyroidism due to decreased energy requirements
    - ii. Cushing's disease
    - iii. insulinoma due to increased energy intake as a result of hypoglycemia
- *Consequences*
  - Impaired glucose tolerance and fasting hyperlipidemia.
  - Predisposition to osteoarthritis, thromboembolism, cholelithiasis, hypoventilation, hyperventilation, and hypoxemia.
  - Morbidity and mortality due to cardiovascular complications.
- *Treatment*
  - Weight reduction by energy restriction.
  - Jejunoileal bypass surgery, or gastroplasty to delay gastric emptying, in patients who have failed weight loss regimens and who are 50–100% of ideal body weight.

Section 5

Homeostasis

- Sodium, Potassium, and Water Homeostasis
- Hydrogen Ion Homeostasis
- Blood Glucose Homeostasis, Diabetes and Hypoglycemia

Chapter 20

Sodium, Potassium, and Water Homeostasis

❑ Explain the distribution of water in the body.

In a healthy human adult, the total body water is about 42 litres which is distributed as follows:

1. *Intracellular fluid (ICF)*, which constitutes two-third of the total body water.
2. *Extracellular fluid (ECF)*, which constitutes one-third of the total body water. The ECF is present in several subcompartments:
   - *Intravascular volume* of plasma (one-third of ECF)
   - *Interstitial fluid* (two-third of ECF)
3. *Specialized fluids*
   - *Lymph* (1.5 L)
   - *Transcellular fluid* (1.5 L), which includes cerebrospinal fluid, synovial fluid, pleural and peritoneal fluids, aqueous humor and alimentary secretions.

The distribution of water between the different body compartments depends on the concentration gradient of solutes on either side of the membrane (osmotic gradient).

❑ Explain the distribution of electrolytes in the body.

The chief electrolytes in the body are Na^+, K^+, Cl^- and HCO_3^-.

- Sodium is the predominant cation in the ECF and the normal concentration is 135–145 mmol/L. The active transport of sodium across cell membranes is catalyzed by *Na^+K^+ ATPase.*
- Potassium is the chief intracellular cation. About 90% of the total body potassium is free, while the remaining 10% is bound to erythrocytes, bone, and brain tissue.
- Chloride is the predominant anion in the ECF. Erythrocytes are rich in chloride.
- Magnesium and phosphate are predominantly intracellular.
- Bicarbonate is present in equal proportions in the ECF and ICF.

❑ Define osmolality and plasma oncotic pressure.

The osmotic activity of solutes is expressed as osmolality in mmol/kg of solvent. ECF and ICF are isoosmotic (have equal osmolalities) except in the kidney. The normal plasma osmolality is 275–295 mmol/kg. Sodium and its associated anions contribute to 90% of plasma osmolality. Glucose, urea, and protein also contribute to ECF osmolality.

The higher concentration of proteins in the plasma compared to the interstitial fluid causes a difference in osmolality between these two compartments. This is known as *plasma colloid osmotic pressure* or *plasma oncotic pressure.* The oncotic pressure opposes the hydrostatic pressure, which tends to force fluid from the intravascular to the extravascular space.

❑ Give an account of the intake and excretion of sodium and water.

Source of body water and sodium

- *Water*: Diet (drinking water, beverages, water content of foods) and oxidation of foods.
- *Sodium*: Diet (food, drinks, and seasoning).

Daily intake

The daily water and sodium intake, which depend on the amount lost from the body, are approximately 1.5–2.0 litre and 60–150 mmol, respectively.

Loss

Water and sodium are lost through

- *Urine*: About 500 mL/day of water is required to excrete waste products. This is termed *obligatory requirement*. Urinary excretion beyond the obligatory requirement is known as *facultative water excretion*. Approximately 1.5–2 liters of water and 100 mmol of sodium are lost per day in the urine.
- *Feces*: 100 mL of water and 15 mmol sodium are lost in the feces every day with a considerable increase in diarrhea and vomiting.
- *Lungs*: About 400 mL of water per day is removed through the expired air. This is increased in fever with hyperventilation.
- *Skin*: About 500 mL of water is lost per day without obvious sweating. The amount of sodium and water lost by sweating increases in hot climates, and in fever.

❑ Explain the regulation of water balance.

Water balance in the body is regulated by

- the *thirst center* located in the third ventricle in the hypothalamus, which is stimulated by ECF hyperosmolality and hypovolaemia, and inhibited by ECF hypoosmolality or hypervolaemia
- *vasopressin* or *antidiuretic hormone* (*ADH*), the anterior pituitary hormone that enhances water reabsorption in response to an increase in ECF osmolality or hypovolaemia. Vasopressin secretion is regulated by the hypothalamus, which responds to impulses from stretch receptors in the left atrium and baroreceptors in the aortic arch and carotid sinus.

❑ Enumerate the factors that regulate sodium balance.

The factors that regulate sodium balance include the following:

- *Aldosterone*, a hormone secreted by the adrenal cortex that stimulates Na^+ reabsorption by the renal tubular cells and increases ECF volume. Aldosterone secretion is controlled by the *renin – angiotensin system.*
- *Atrial natriuretic factor*, a peptide secreted from the right atrial wall that
  - increases sodium excretion (*natriuresis*)
  - increases the glomerular filtration rate (GFR)
  - lowers renin activity
  - inhibits aldosterone, angiotensin II, and norepinephrine.

❑ How does the renin–angiotensin system regulate aldosterone secretion?

Renin is a proteolytic enzyme secreted by the juxtaglomerular apparatus that cleaves *angiotensinogen* to release the decapeptide *angiotensin I.* Angiotensin I is converted to the biologically active octapeptide *angiotensin* II by a converting enzyme present in the lungs (Fig. 20.1).

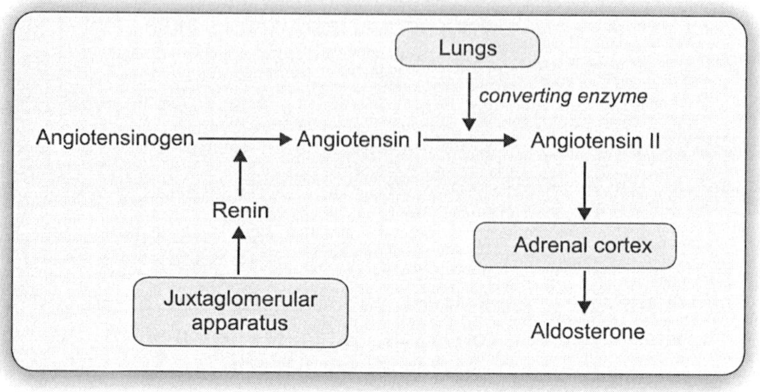

Fig. 20.1 The renin–angiotensin aldosterone system

Angiotensin II stimulates
- vasoconstriction with increase in blood pressure,
- aldosterone biosynthesis and reabsorption of sodium and water from the renal tubules,
- catecholamine release from adrenal medulla,
- thirst.

❑ Outline the causes, and consequences of water depletion, and the compensatory mechanisms.

□ ***Causes***

Water depletion without loss of equivalent amount of sodium occurs under the following conditions:

1. *Decreased intake* as in
   - elderly debilitated persons
   - unconscious patients
   - severe dysphagia
   - postoperative patients, when oral fluid intake has stopped.
2. *Increased loss* due to
   - sweating, during fever
   - hyperventilation
   - infantile gastroenteritis
   - diabetes insipidus due to ADH deficiency
   - diabetes mellitus due to osmotic diuresis
   - nephritis
   - acute renal failure.

□ ***Consequences*** of water depletion include
- increase in plasma sodium and osmolality

- decreased renal blood flow, which stimulates aldosterone secretion with increased reabsorption of sodium that aggravates hypernatremia
- increase in ECF osmolality, resulting in diffusion of water from the cells to the ECF.

  The homeostatic mechanisms that tend to compensate include
  1. stimulation of the thirst center with increased intake of water
  2. ADH secretion with increased water reabsorption (except in diabetes insipidus).

❑ Enumerate the clinical and biochemical findings and management of water depletion.

Clinical symptoms are apparent only after 2 liters of water is lost. These include
- apathy
- dehydration of muscle and nerve cells, leading to weakness and confusion
- thirst
- decreased salivation
- dryness of the tongue and mouth
- oliguria.

Loss of 20% of total body water can result in death.

Biochemical findings
- ↓ ECF volume
- hemoconcentration
- plasma proteins: Normal or slight ↑
- ↓ urine sodium
- ↑ ECF osmolality
- ↑ plasma sodium
- ↓ urine volume
- mild uremia

Management

Water administration (oral or 5% glucose i.v.) until urine volume exceeds 1500 mL.

❑ Elucidate the causes and compensatory mechanisms in sodium depletion.

Sodium depletion occurs when there is loss of both sodium and water followed by replacement of water without sodium.

Causes
- *Diminished intake*: After prolonged starvation in anorexia nervosa or chronic alcoholism.
- Overuse of diuretics
- Infusion of fluid with less sodium content to patients with isoosmolar fluid depletion
- Loss through skin (excessive sweating, cystic fibrosis, burns, exudative skin lesions)
- Loss through the gut (fistulae, vomiting and diarrhea, steatorrhea, intestinal obstruction)
- Adrenal cortical insufficiency (Addison's disease)
- Diabetic ketoacidosis
- Chronic renal failure
- Sick cell syndrome (part of acute or chronic illnesses due to a failure of the sodium pump).

Homeostatic mechanisms involve the following sequence of events:
- Inhibition of ADH secretion
- Decrease in ECF volume
- Reabsorption of sodium
- Loss of water in the urine
- Stimulation of aldosterone secretion
- Restoration of plasma osmolality.

❑ Discuss the clinical symptoms, biochemical features and management of sodium depletion.

Clinical symptoms that appear after a sodium equivalent of about 4 L of isotonic saline is lost include

- tiredness, apathy
- weight loss
- hypotension
- oliguria
- muscle weakness
- peripheral circulatory insufficiency
- tachycardia
- coma, and death.

Biochemical findings

- ↓ ECF volume
- ↓ Plasma sodium
- ↓ Urine volume
- ↓ ECF osmolality
- ↑ Plasma proteins
- Urine sodium: ↑ in Addison's; ↓ in other conditions.

Management: Intravenous infusion of saline.

❑ Enumerate the causes and homeostatic mechanisms in water and sodium depletion in isoosmolar amounts.

Causes

- Intestinal fistulae and small intestinal obstruction
- Failure of renal tubular function
- Removal of ECF as in edema, ascites, and paralytic ileus.

Homeostatic mechanisms

Decrease in intravascular volume has the following consequences:

- Decrease in renal blood flow
- Stimulation of aldosterone secretion
- Reabsorption of sodium
- Increase in plasma osmolality
- Stimulation of ADH secretion
- Water reabsorption with distribution of water to all the compartments, and expansion of cells.

In renal failure, homeostatic mechanisms do not operate.

❑ Outline the clinical features, biochemical findings and management of water and sodium depletion.

Clinical features

- Weakness
- Postural hypotension
- Apathy
- Oliguria

Biochemical findings

- ↓ ECF volume
- ↓ Urine volume
- Uremia
- ↑ Plasma proteins
- ↓ Urine sodium

Management: Replacement of sodium and water in isotonic amounts.

❑ Enumerate the causes and homeostatic mechanisms of water excess.

Causes

Accumulation of water without an equivalent amount of sodium occurs as a result of:

- *Increased intake* due to:
  - administration of large quantities of salt-free fluids
  - excessive consumption of water.
- *Decreased excretion* as in:
  - syndrome of inappropriate ADH (SIADH) secretion
  - ectopic ADH production by malignant tumors such as bronchogenic carcinomas
  - chronic ADH administration
  - intravenous administration of oxytocin, an ADH agonist that has antidiuretic effects
  - severe renal failure.

Homeostatic mechanisms

The excess water is distributed osmotically to all the fluid compartments.

The increase in ECF volume

- increases renal blood flow
- decreases sodium reabsorption
- inhibits ADH secretion
- inhibits aldosterone secretion
- decreases plasma sodium and osmolality
- causes excretion of dilute urine.

❑ Outline the clinical and biochemical features, and management of water excess.

Clinical features

- Anorexia
- Confusion
- Coma
- Edema
- Lethargy
- Fits
- Cardiac failure

Biochemical features

- ↑ ECF volume
- ↓ Plasma sodium
- ↑ Urine volume
- ↓ ECF osmolality
- ↓ Plasma proteins
- ↑ Urine sodium

Management

- Restriction of water intake
- Infusion of hypertonic saline if water intoxication occurs.

❑ Outline the causes and consequences of sodium excess.

Causes

Accumulation of sodium, not accompanied by an equivalent amount of water, occurs due to

- increased intake of sea water,
- excess administration of hypertonic sodium bicarbonate after cardiac arrest,
- *decreased excretion* as in
  i. renal failure due to low GFR
  ii. Cushing's syndrome and Conn's syndrome, due to increased tubular reabsorption.

Homeostatic mechanisms

Increase in ECF osmolality has the following consequences:

- Diffusion of water from ICF to ECF
- Increase in ECF volume
- Inhibition of aldosterone secretion
- Urinary excretion of sodium
- Stimulation of ADH secretion with reabsorption of water.

❑ Elucidate the clinical features, biochemical findings, and management of sodium excess.

Clinical features

- ECF hyperosmolality causes thirst, mental confusion, coma, and death due to cerebral dehydration.
- Hypervolemia can cause hypertension, weight gain, and edema.

Biochemical findings

- ↑ ECF volume
- ↑ Plasma sodium
- ↓ Urine volume
- ↑ ECF osmolality
- ↓ Plasma proteins
- Urine sodium: ↓ in Conn's; ↑ in others.

Management

- Treatment of the underlying cause
- Restriction of sodium intake
- Diuretic therapy for promoting sodium excretion
- Dialysis if kidney function is impaired.

❑ Discuss the causes, homeostatic mechanisms, biochemical findings, and management of water and sodium excess.

□ ***Causes***

Accumulation of water and sodium in isotonic amounts is seen in secondary aldosteronism associated with

- chronic liver disease, nephrotic syndrome, and kwashiorkor associated with hypoalbuminemia and edema,
- cardiac failure presenting with edema,
- essential, malignant or renal hypertension without edema.

□ ***Homeostatic mechanisms***

- High plasma aldosterone levels cause reabsorption of sodium and water.
- In hypoalbuminemia and heart failure,
  i. the retained water passes into the interstitial fluid leading to edema.
  ii. the increase in interstitial fluid volume stimulates ADH secretion with retention of sodium-free water.
  iii. mild hyponatremia is seen despite excess total body sodium.

□ ***Biochemical findings***

- Hemodilution
- ↓ Plasma protein
- Low urinary sodium.
- ↓ Plasma urea
- Normonatremia or mild hyponatremia

□ ***Management*** involves treatment of the underlying cause.

❑ Enumerate the causes of hyponatremia.

Hyponatremia may be

- *dilutional hyponatremia* resulting from water excess due to
  - increased water intake
  - inappropriate administration of fluid with low sodium content
  - Addison's disease
  - syndrome of inappropriate ADH secretion
  - congestive cardiac failure
  - nephrotic syndrome.
- *appropriate hyponatremia* associated with high plasma osmolality due to:
  - increased plasma concentrations of urea, glucose, alcohol, and other solutes,
  - infusion of mannitol or amino acids.
- *pseudohyponatremia* associated with hyperproteinemia and hyperlipidemia and normal plasma sodium concentrations.

❑ Outline the causes of hypernatremia.

Hypernatremia occurs due to

- decreased water intake
- increased water excretion as in diabetes mellitus and diabetes insipidus.
- increased dietary intake (rare)
- drugs such as metronidazole
- increased sodium reabsorption as in Cushing's syndrome and Conn's syndrome.

❑ How is potassium balance maintained?

Potassium balance is maintained by

1. ***The kidneys***
   - Potassium is filtered by the glomeruli, reabsorbed in the proximal convoluted tubules (PCT), and actively secreted by the distal tubular cells in exchange for sodium or hydrogen ions.
   - Potassium secretion is stimulated by the adrenal cortical hormone, aldosterone.
   - Urinary potassium excretion depends on:
     i. the availability of sodium for exchange with potassium
     ii. the relative amounts of hydrogen and potassium ions in the distal tubular cells
     iii. plasma aldosterone.
   - In acidosis, hydrogen ions are excreted and potassium is reabsorbed, whereas in alkalosis, potassium is excreted and hydrogen ions are conserved.
2. ***The gastrointestinal tract***
   - Potassium from the diet and gastric juice is reabsorbed; in the small intestine and secreted in the colon and rectum in exchange for sodium.
   - Small amount of potassium (< 10 mmol/d) is excreted in the feces.
   - In chronic diarrhea and intestinal fistulae, fecal excretion is increased.
3. ***Transcellular movement***
   - Potassium ions exchange with hydrogen or sodium ions.
   - Na^+ K^+-ATPase maintains high intracellular concentrations of potassium by catalyzing the exchange of three sodium ions from the cells for two extracellular potassium ions, creating an electrochemical gradient across the cell membrane.
   - Insulin stimulates the intracellular uptake of potassium.

❑ Enumerate the causes of hypokalemia.

Hypokalemia may occur due to

- ***Decreased intake*** as in
  - dietary deficiency of potassium (rare)
  - starvation
  - severe illness
  - postoperatively, after gastrointestinal surgery
  - kwashiorkor.
- ***Increased urinary excretion*** due to
  - Fanconi syndrome
  - hyperaldosteronism
  - Bartter's syndrome in which there is increased secretion of renin and aldosterone
  - Cushing's syndrome
  - ectopic ACTH production
  - substances with aldosterone-like activity such as liquorice and tobacco
  - drugs that potentiate the action of aldosterone (*e.g.*, carbenoxolone)
  - thiazide diuretics that enhance Na^+- K^+ exchange and potassium excretion.
  - renal tubular acidosis in which there is decreased availability of H^+ for exchange with K^+.
- ***Loss from the gut*** as in
  - diarrhea
  - vomiting
  - steatorrhea
  - intestinal fistula
- ***Redistribution from ECF to ICF***
  - in alkalosis, where potassium enters cells is associated with increased urinary loss of potassium.
  - after insulin treatment.

❑ Outline the clinical features and management of hypokalemia.

Clinical features

- *Neuromuscular manifestations*: Weakness, lethargy, hypotonia, depression, confusion, cramps, tetany.
- *Cardiac manifestations*: Arrhythmias, hypotension, ECG changes, ST depression, flattened T waves, prolonged P-R interval, prominent U waves.
- *Renal manifestations*: Inability to concentrate urine, polyuria
- Alkalosis.

Management

- Oral administration of potassium chloride.
- Intravenous infusion of potassium if oral administration is not possible.

❑ Discuss the causes, clinical symptoms, and management of hyperkalemia.

- ***Causes***

  Hyperkalemia may result from
  - Excessive parenteral administration for treatment of hypokalemia.
  - Decreased urinary excretion as in
    - i. adrenocortical insufficiency (Addison's disease)
    - ii. congenital adrenal hyperplasia

iii. diuretics with K^+-retaining effect (*e.g.*, spironolactone)
iv. antihypertensive drugs such as captopril that block aldosterone secretion
v. severe renal failure due to decreased availability of sodium for exchange with potassium.

- Redistribution from ICF to ECF as in
  i. severe tissue damage
  ii. hypoxia with failure of the sodium pump
  iii. crush injury, burns, and postoperative conditions
  iv. acidosis associated with increased urinary excretion of potassium.

□ ***Clinical features***

- *Neuromuscular manifestations*: ↓ membrane potential, ↑ repolarization
- *Cardiac manifestations*: ↓ Cardiac action potential, bradycardia, cardiac arrest, ventricular fibrillation, *ECG changes* (high T waves, loss of P wave, broad QRS complex).

□ ***Management***

- Dietary control
- Correction of acidosis or water and sodium depletion
- Intravenous administration of calcium gluconate to reverse the cardiac changes.
- Insulin administration to promote potassium entry into cells. Insulin-induced hypoglycemia can be prevented by glucose administration.
- Bicarbonate infusion to promote entry of potassium into cells.
- K^+ binding ion-exchange resins that increase potassium excretion by exchanging potassium for sodium or calcium ions.
- Dialysis to remove excess potassium.

Chapter 21

Hydrogen Ion Homeostasis

❑ Define acid, base, and pH.

- An *acid* is a proton donor that is capable of dissociating into hydrogen ions. While strong acids such as hydrochloric acid are completely dissociable and generate many hydrogen ions, weak acids such as carbonic acid produce relatively lesser number of hydrogen ions.
- A *base* is a proton acceptor, *i.e.*, an acceptor of hydrogen ions. Sodium hydroxide is a strong base.
- *pH* is the negative logarithm of hydrogen ion concentration and is a measure of hydrogen ion activity. It is expressed as

$$pH = -\log [H^+]$$

❑ What is a buffer?

- A *buffer* resists a change in pH by replacing a strong acid (base) with a weaker acid (base), thereby decreasing the number of free hydrogen ions.
- A buffer contains a weak acid and its conjugate base. For example,

$$\underset{\text{Strong acid}}{H^+Cl^-} + \underset{\text{Buffer}}{NaHCO_3} \rightleftharpoons \underset{\text{Weak acid}}{H_2CO_3} + \underset{\text{Neutral salt}}{NaCl}$$

❑ What does the Henderson-Hasselbalch equation express?

The *Henderson – Hasselbalch equation* expresses pH as a function of the ratio of a conjugate base (bicarbonate) to a weak acid (carbonic acid).

$$pH = pK' + \log \frac{[HCO_3^-]}{[H_2CO_3]}$$

The value of pK' is approximately 6.1. Since plasma carbonic acid is in equilibrium with dissolved carbon dioxide, it can be determined by multiplying the partial pressure of carbon dioxide (P_{CO_2}), expressed in kilopascals, by the solubility constant (S) for CO_2 (0.23). Substituting

$$pH = 6.1 + \log \frac{[HCO_3^-]}{P_{CO_2} \times 0.23}$$

❑ What is the normal blood pH?

- The normal blood pH ranges from 7.35 to 7.45, with a mean of 7.4. Venous blood is more acidic (pH 7.37) than arterial blood (pH 7.4).
- Maintenance of normal blood pH depends on the ratio of bicarbonate to PCO_2.

❑ Briefly explain the sources and fate of hydrogen ions.

▫ *Sources of hydrogen ions*

1. *Dietary phosphoproteins and phospholipids,* which are converted to phosphoric acid.
2. *Metabolism* of
   - sulphur-containing amino acids, cysteine, and methionine to sulphate,
   - amino nitrogen to urea
   - formation of lactate and ketone bodies with release of equimolar amounts of hydrogen ions.
3. *Pathological conditions*:
   - lactic acidosis
   - diabetic ketoacidosis.

▫ *Fate of hydrogen ions*

1. Incorporation into water during oxidative phosphorylation
2. Removal from the system by the kidneys and intestines coupled to bicarbonate generation.

❑ Name the various lines of defence for the maintenance of hydrogen ion homeostasis.

Hydrogen ion homeostasis is maintained by the following:

1. *Buffers*: Bicarbonate–carbonic acid buffer and hemoglobin are the major buffer systems that regulate blood pH.
2. *Erythrocytes,* by generating bicarbonate from carbon dioxide.
3. *Lungs and respiratory center* that determine carbon dioxide tension in the alveolar air and regulate carbonic acid.
4. *The kidneys* regulate acid–base balance by bicarbonate reabsorption and bicarbonate generation.
5. Other factors
   - *Gastrointestinal secretions*
   - *Bone*
   - *Cellular buffering,* by exchange of Na^+, K^+, and H^+ across cell membranes.

❑ Name the physiologically important buffers.

The physiologically important buffers include

- Bicarbonate–carbonic acid buffer system that accounts for more than 60% of the total buffering capacity.
- Hemoglobin, the most important buffer in the erythrocytes that accounts for approximately 40% of the total buffering capacity.
- Proteins and phosphate, which have minor blood buffering potential due to low plasma concentrations.

❑ Explain the role of erythrocytes in hydrogen ion homeostasis.

Erythrocytes generate bicarbonate from carbon dioxide and maintain the blood pH (Fig. 21.1). This occurs by the following sequence of steps:

1. Diffusion of carbon dioxide from tissues through plasma into erythrocytes.
2. Formation of carbonic acid from carbon dioxide and water by the action of *carbonate dehydratase* (*CD; carbonic anhydrase*).
3. Dissociation of carbonic acid (H_2CO_3) to form hydrogen ion (H^+) and bicarbonate ion (HCO_3^-).
4. Buffering of H^+ by hemoglobin to form HHb
5. Diffusion of HCO_3^- into the ECF
6. Diffusion of chloride (Cl^-) into the erythrocyte to maintain electrochemical neutrality. This is termed *chloride shift*.

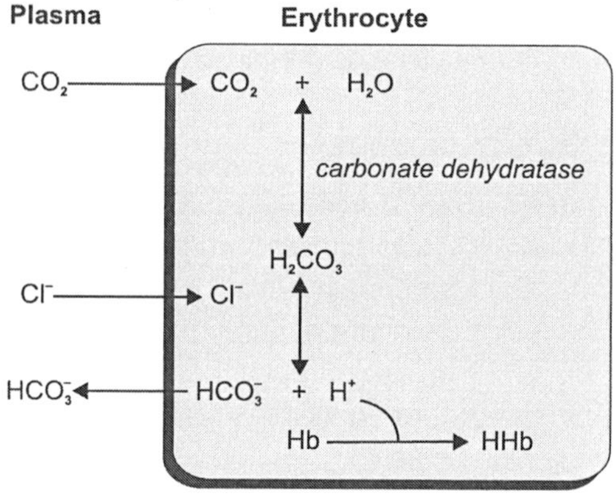

Fig. 21.1 Generation of bicarbonate in the erythrocytes

❑ How do the lungs and respiratory center handle changes in blood pH?

Role of lungs

- Lung ventilation determines carbon dioxide tension in the alveolar air and regulates carbonic acid in blood by the following sequence of events:
  - oxygen transport from the lungs to the tissues by hemoglobin
  - oxygen utilization by tissues for metabolism, with the release of carbon dioxide
  - diffusion of carbon dioxide from the cells into ECF and elimination through the lungs.

Role of respiratory center

- The respiratory center present in the brainstem and in the carotid and aortic bodies contains chemoreceptors that are sensitive to changes in pH and PCO_2.
- A decrease in pH stimulates the respiratory center, leading to increase in ventilation, decrease in PCO_2 and the $[HCO_3^-]/[CO_2]$ ratio, restoring the pH to normal.
- An increase in pH decreases the respiratory rate with increase in PCO_2 and decrease in pH.
- Respiratory compensation is rapid but not complete.

❑ What is the role of bone in hydrogen ion homeostasis?

Bone regulates acid–base balance in prolonged acidosis by releasing Ca^{2+} and HPO_4^{2-} into plasma facilitating urinary buffering of H^+. Although effective, this eventually causes demineralization of bone.

❑ Briefly explain the regulation of acid-base balance by gastrointestinal secretions.

- In the stomach, bicarbonate ions diffuse from parietal cells into the ECF with simultaneous secretion of H^+ and Cl^- into the lumen.
- In the intestinal cells, HCO_3^- is secreted into the lumen and H^+ is pumped into the ECF.

- The gain of bicarbonate by the ECF due to gastric HCl secretion is corrected by intestinal secretion of HCO_3^-.

❑ Explain how hydrogen ion homeostasis is maintained by cellular exchange of ions.

- Exchange of Na^+, K^+, and H^+ across cell membranes maintains hydrogen ion homeostasis.
- In acidosis, H^+ that enters cells in exchange for Na^+ is buffered by proteins and organic phosphates.
- In alkalosis, H^+ leaves the cells in exchange for Na^+ and/or K^+.

❑ Discuss the role of the kidneys in maintaining hydrogen ion homeostasis.

The kidneys acidify urine by bicarbonate reabsorption and generation.

Bicarbonate reabsorption

Bicarbonate filtered by the glomerulus is reabsorbed by the tubules and very little is excreted.

Bicarbonate reabsorption occurs predominantly in the proximal convoluted tubule (PCT) and to a minor extent in the proximal part of the distal convoluted tubule (DCT) by the following sequence of events (Fig. 21.2):

1. The filtered bicarbonate combines with H^+ secreted by the tubules to form H_2CO_3.
2. H_2CO_3 dissociates into carbon dioxide and water. The reaction is catalyzed by *carbonate dehydratase* in the PCT and is spontaneous in the DCT.
3. Carbon dioxide diffuses from the lumen into the tubular cells down a concentration gradient.
4. Carbon dioxide reacts with water to form H_2CO_3.
5. H_2CO_3 dissociates into H^+ and HCO_3^-.
6. HCO_3^- diffuses into the ECF, while H^+ is secreted into the tubular lumen in exchange for Na^+ which is reabsorbed along with bicarbonate into the ECF.
   - In this process, filtered bicarbonate is completely reabsorbed and there is no net gain of bicarbonate or a net change in hydrogen ion balance.
   - This process maintains a steady state but does not mitigate acidosis.

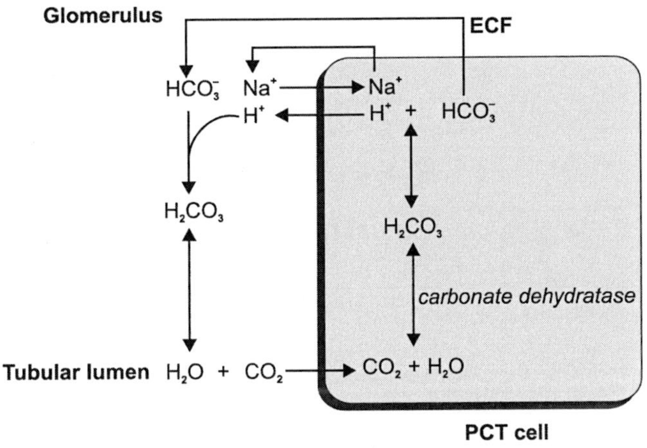

Fig. 21.2 Bicarbonate reabsorption by the renal tubules

Bicarbonate generation

This occurs in the DCT coupled to a net loss of hydrogen ions in urine and involves two urinary buffers—the phosphate and ammonia buffers.

- ***The phosphate buffer***
  - Phosphate in the form of monohydrogen phosphate (HPO_4^{2-}) is filtered by the glomerulus into the tubular lumen.
  - In the tubular cells, H_2CO_3 dissociates into HCO_3^- and H^+.
  - HCO_3^- ions diffuse into the ECF.

- H^+ is secreted into the lumen where it combines with HPO_4^{2-} to form dihydrogen phosphate ($H_2PO_4^-$), which is excreted in the urine. This is known as the *titrable acidity* of urine.
- About 30 – 40 mM of H^+ is excreted in the form of titrable acid every day in the urine.

Figure 21.3 shows bicarbonate generation coupled to excretion of titrable acidity.

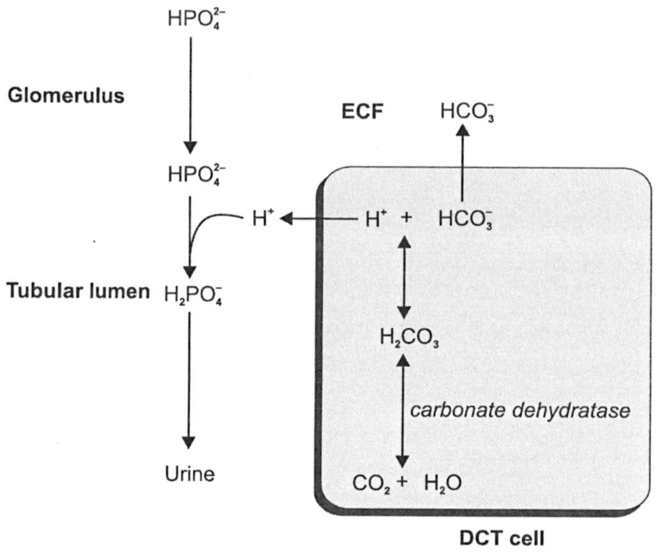

Fig. 21.3 Bicarbonate generation coupled to excretion of titrable acidity

- ***The ammonia buffer***
  - In the renal tubule cells, glutamine undergoes deamination to form glutamate and ammonium ion in a reaction catalyzed by *glutaminase*.

$$\text{Glutamine} + H_2O \longrightarrow \text{Glutamate} + NH_4^+$$

  - Glutamate is converted to 2-oxoglutarate and enters gluconeogenesis.
  - The ammonium ion dissociates to form NH_3 and H^+.
  - NH_3 diffuses from the tubular cells into the lumen where it combines with H^+ and is excreted in the urine as NH_4Cl.
  - CO_2 combines with water in the tubule cell, to form H_2CO_3 in a reaction catalyzed by *carbonate dehydratase*.
  - H_2CO_3 dissociates into H^+ and bicarbonate.
  - While H^+ is secreted into the lumen in exchange for Na^+, bicarbonate diffuses into the ECF.

Figure 21.4 gives a schematic representation of bicarbonate generation coupled to ammonia production.

❑ What are the types of acid – base imbalance?

- Maintenance of normal pH depends on the ratio of HCO_3^- to PCO_2.
- Acid–base disorders are classified into two types:
  - *Acidosis*, which occurs when there is a fall in the HCO_3^-/PCO_2 ratio and therefore the blood pH. The pH is less than 7.36 and in severe cases, less than 6.8.
  - *Alkalosis*, which occurs due to a rise in the ratio and the blood pH. The pH is greater than 7.44 and in severe cases, greater than 7.8.

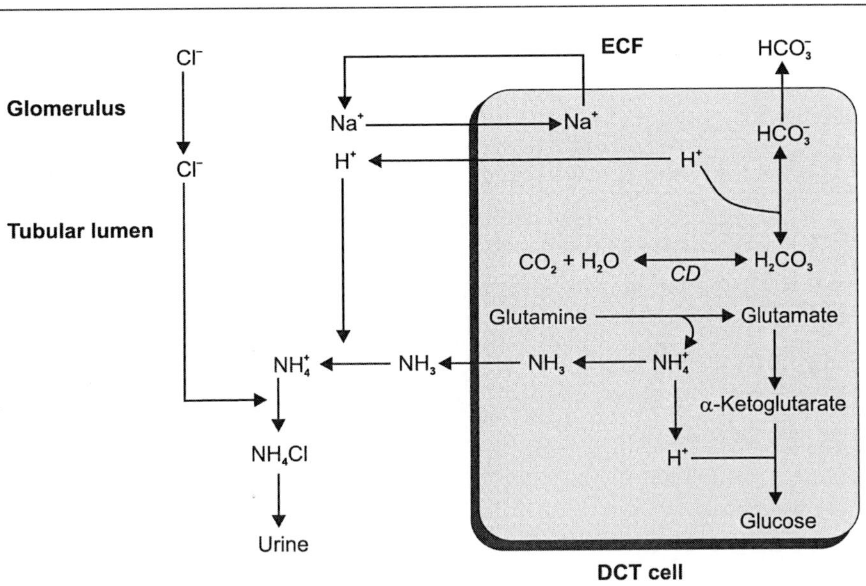

Fig. 21.4 Bicarbonate generation coupled to ammonia production

- Both acidosis and alkalosis are further categorized into
  1. *respiratory*, in which there is a primary abnormality in P_{CO_2}
  2. *non-respiratory* (*metabolic*), in which there is a primary abnormality in bicarbonate concentration.

❑ Outline the causes and compensatory mechanisms in metabolic acidosis.

Primary defect: Bicarbonate deficit.

Causes

Metabolic acidosis is categorized based on the anion gap into two types: *Increased anion gap acidosis* and *Normal anion gap acidosis*. Anion gap refers to the difference between the total concentration of measured cations [Na^+ and K^+] and measured anions [Cl^- and HCO_3^-]. This constitutes the unmeasured anions.

Increased anion gap acidosis

The increase in unmeasured anions occurs due to the following conditions:

- *Overproduction of acids,* as in lactic acidosis and ketoacidosis.
  - Plasma bicarbonate is decreased due to buffering.
  - Plasma chloride levels are normal.
- *Gain of acid* resulting from the metabolism of exogenous substances such as ammonium chloride, ethanol, methanol, and salicylates.

  There is a decrease in plasma bicarbonate and plasma cations, with retention of other anions such as chloride.

Normal anion gap acidosis

This arises due to the following conditions:

- *Bicarbonate loss* (*e.g.*, severe diarrhea) that exceeds bicarbonate regeneration by the kidney. Anion gap is unaffected because an equivalent amount of sodium is lost.

- *Impaired H^+ secretion* as in
  - Generalized renal failure in which reabsorption and generation of HCO_3^- is impaired and a mixture of anions and cations is lost in urine.
  - *Renal tubular acidosis* (*RTA*) characterized by H^+ retention and failure of HCO_3^- reabsorption and generation. There is consequent reabsorption of Na^+ and Cl^- with secretion of K^+, leading to *hyperchloremic acidosis* with *hypokalemia.*
  - Administration of *acetazolamide*, a *carbonate dehydratase* inhibitor, which impairs H^+ secretion and HCO_3^- formation in erythrocytes and renal tubular cells, leading to *hyperchloremic acidosis* with *hypokalemia.*

Compensatory mechanisms

Erythrocytes: Stimulation of *carbonate dehydratase* reaction. When plasma HCO_3^- falls, HCO_3^- enters from red cells in exchange for Cl^-.

Respiratory: Stimulation of the *respiratory center,* leading to hyperventilation with compensatory decrease in PCO_2.

Renal: Stimulation of *carbonate dehydratase,* leading to increased excretion of NH_4^+ and titrable acid. Increased reabsorption of bicarbonate. Urine becomes acidic.

Cellular exchange: H^+ enters cells in exchange for Na^+.

Discuss the biochemical findings and management of metabolic acidosis.

Biochemical findings

Blood
- ↓ pH
- ↓ [HCO_3^-]
- ↓ PCO_2
- [Cl^-] ↑ in renal tubular acidosis, and acetazolamide therapy; normal in other conditions.

Urine
- ↑ NH_4^+
- ↑ titrable acidity
- ↓ HCO_3^-

Management
- Treatment of the underlying cause.
- Bicarbonate administration in small amounts to buffer hydrogen ions when the arterial pH is less than 7.0.

Enumerate the causes and compensatory mechanisms in respiratory acidosis.

Primary defect: Carbon dioxide retention with consequent increase in PCO_2.

Causes

Respiratory acidosis can result from the following conditions:
- *Respiratory failure* as in
  - brochopneumonia
  - status asthmaticus
  - chronic obstructive airways disease

- *Pulmonary diseases* such as
  - respiratory distress syndrome
  - pulmonary fibrosis
- *Extrapulmonary diseases* such as
  - cardiac arrest
  - motor neuron disease
  - poliomyelitis
  - tetanus
- *Drugs and toxins* such as
  - anesthetics
  - morphine
  - opiate
  - neurotoxins

Compensatory mechanisms

Compensation is more effective in chronic than in acute cases.

- *Erythrocytes*
  Increase in carbonate dehydratase mechanism with compensatory generation of bicarbonate.
- *Respiratory*
  - *In acute respiratory failure*: Stimulation of the respiratory center by PCO_2, leading to hyperventilation to remove carbon dioxide.
  - *In chronic respiratory failure*: Respiratory compensation is ineffective because the respiratory center loses sensitivity to increased carbon dioxide tension.
  - *In severe cases*: Carbon dioxide narcosis with cerebral edema, coma, and muscular twitching occurs.
- *Renal*
  - *In acute renal failure*: Increase in bicarbonate generation. However, pH is not corrected due to insufficient time for renal compensation.
  - *In chronic renal failure*: Increase in bicarbonate reabsorption and generation, excretion of acidic urine, restoration of pH to normal.
- *Cellular exchange*: H^+ enters cells in exchange for Na^+.

❑ Outline the changes in blood and urine in respiratory acidosis and the management.

Blood	*Acute*	*Chronic*	***Urine***
pH	↓	Slight ↓ or Normal	↑ Titrable acidity
$[HCO_3^-]$	slight ↑	↑	↑ NH_4^+
PCO_2	↑	↑	↓ HCO_3^-

(more effective in chronic than in acute cases)

Management

- Improvement of alveolar ventilation by physiotherapy, bronchodilators, and antibiotics.
- Supply of oxygen at high concentrations for patients with acute respiratory failure.

❑ Discuss the causes, compensatory mechanisms, laboratory findings and management of metabolic alkalosis.

Primary defect: Bicarbonate excess accompanied either by a decrease in plasma chloride or an increase in plasma Na^+.

Causes

Metabolic alkalosis may result from the following:

- *Alkali administration*
  - Overmedication with antacids such as sodium bicarbonate for the treatment of indigestion or peptic ulcer.
  - Inappropriate intravenous bicarbonate infusion in acidosis.
- *Loss of hydrogen ions.* This is due to
  - removal of gastric acid by aspiration
  - vomiting in patients with pyloric stenosis due to loss of gastric HCl
  - administration of diuretics by causing hypovolaemia stimulates renal reabsorption of sodium accompanied by increase in bicarbonate reabsorption and excretion of K^+ and H^+.
  - Cushing's syndrome: Excess mineralocorticoid increases H^+ excretion secondary to enhanced sodium reabsorption.

Compensatory mechanisms

- *Erythrocytes*: Inhibition of *carbonate dehydratase* mechanism.
- *Respiratory*: Depression of the respiratory center with hypoventilation, carbon dioxide retention, increase in PCO_2, and hypoxemia. This stimulates respiration, making respiratory compensation ineffective.
- *Renal*: Inhibition of H^+ and HCO_3^- formation by *carbonate dehydratase,* with decreased H^+ secretion and increased urinary loss of bicarbonate.

Laboratory findings

Blood	**Urine**
↑ pH	↓ NH_4^+
↑ $[HCO_3^-]$	↓ Titrable acidity
↑ PCO_2 (compensated)	↑ HCO_3^-

Management

- Treatment of the underlying cause.
- Infusion of isotonic sodium chloride solution to correct hypovolemia and hypochloremia and to allow excretion of bicarbonate.
- Administration of potassium chloride to prevent hypochloremic alkalosis.

❑ Enumerate the causes and compensatory mechanisms in respiratory alkalosis.

Primary defeact: Carbon dioxide deficit due to stimulation of the respiratory center, leading to hyperventilation and a fall in PCO_2.

Causes

- *Hyperventilation*
  - Mechanical overventilation
  - Hysterical voluntary hyperventilation
- *Hypoxia*
  - High altitudes
  - Severe anemia
- *Others*
  - Fever
  - Anxiety
- *Pulmonary diseases*
  - Pulmonary embolism
  - Pulmonary edema
- *Stimulation of respiratory center*
  - Salicylates
  - Encephalitis
  - Brain stem lesions

Compensatory mechanisms

- *Erythrocytes*: Decrease in bicarbonate generation by *carbonate dehydratase.*
- *Respiratory*: Depression of respiratory center with carbon dioxide retention. This is however, ineffective.
- *Renal*: Decrease in bicarbonate generation by *carbonate dehydratase.* Increase in bicarbonate excretion. Excretion of alkaline urine.
- *Cellular exchange*: Exchange of Na^+ for cellular K^+ and H^+ mitigates extracellular alkalosis. However, this causes intracellular alkalosis and depletion of K^+, which is excreted in urine.

❑ Briefly summarize the laboratory findings and management of respiratory alkalosis.

Laboratory findings

Blood	Acute	Chronic	Urine
pH	↑	Normal or slight ↑	↓ NH_4^+
$[HCO_3^-]$	low normal	↓	↓ Titrable acidity
PCO_2	low normal	↓	↑ HCO_3^-

(more effective in chronic cases than in acute)

Management

The affected subject is made to breathe into a paper bag to correct the decrease in PCO_2 (temporary).

❑ What are the conditions in which hyperchloremic acidosis occurs?

Hyperchloremic acidosis occurs in the following conditions:

1. Transplantation of the ureter into colon (*ureterosigmoidostomy*) following cystectomy for bladder cancer. When urine which contains chloride enters the ileum or colon, bicarbonate is excreted in exchange for chloride leading to hyperchloremic acidosis.
2. *Renal tubular acidosis (RTA),* characterized by
   - H^+ retention
   - failure of HCO_3^- reabsorption and generation
   - reabsorption of Na^+ and Cl^- (as a consequence)
   - secretion of K^+
   - *hyperchloremic acidosis* with *hypokalemia.*
3. Administration of acetazolamide, a carbonate dehydratase inhibitor, which impairs H^+ secretion and HCO_3^- formation in erythrocytes and renal tubular cells, leading to *hyperchloremic acidosis* with *hypokalemia.*

❑ What is renal tubular acidosis?

- *Renal tubular acidosis (RTA)* is characterized by
  - H^+ retention
  - failure of HCO_3^- reabsorption and generation
  - reabsorption of Na^+ and Cl^- (as a consequence)
  - secretion of K^+
  - *hyperchloremic acidosis* with *hypokalemia.*

- There are two types of RTA
  - *RTA type I* or *classical RTA,* characterized by a distal tubular defect in H^+ secretion with inability to acidify the urine.
  - *RTA type II,* in which there is a defect in the *carbonate dehydratase* mechanism leading to impairment in bicarbonate reabsorption in the PCT. The ability to acidify urine is, however, retained.

Explain the acid–base imbalance in vomiting due to pyloric stenosis.

- The acid-base imbalance in vomiting in patients with pyloric stenosis is metabolic alkalosis due to loss of gastric HCl.
- Loss of H^+ (from HCl) induces *carbonate dehydratase* in gastric mucosal cells, leading to an increase in plasma bicarbonate.
- Loss of Cl^- (from HCl) reduces the availability for renal reabsorption with Na^+.
- The increase in Na^+ due to volume depletion enhances reabsorption of HCO_3^-, leading to *hypochloremic alkalosis.*
- In addition, there is also hypokalemia and mild uremia.

What is paradoxic aciduria?

- *Paradoxic aciduria* refers to the excretion of acidic urine inspite of alkalosis.
- Impairment of urinary bicarbonate excretion aggravates alkalosis.
- This occurs due to two factors:
  - Hypovolemia that decreases GFR and therefore, bicarbonate excretion.
  - Hypochloremia that decreases chloride availability for reabsorption with Na^+. As a consequence,
    i. Na^+ becomes more available for exchange with H^+ and K^+.
    ii. K^+ is lost in urine depleting cellular K^+.
    iii. H^+ enters the cells in exchange for K^+, leading to *extracellular alkalosis* and *intracellular acidosis* aggravating *hypokalemia* due to alkalosis.

In potassium-depleted individuals as in patients taking thiazide diuretics,
- H^+ is excreted in exchange for Na^+, which is reabsorbed
- the urine is inappropriately acidic despite alkalosis because the kidneys tend to *conserve cations* at the expense of acid–base balance. This is termed *paradoxic aciduria.*

What is the acid–base imbalance in salicylate poisoning?

Salicylates
- stimulate the respiratory center and cause *respiratory alkalosis*
- uncouple oxidative phosphorylation with consequent *lactic acidosis.*

Plasma bicarbonate is low in both the cases.

pH is
- high if respiratory alkalosis predominates
- low if metabolic acidosis predominates
- normal if both the effects are equal.

Chapter 22

Blood Glucose Homeostasis, Diabetes Mellitus and Hypoglycemia

❑ What is the normal range for plasma glucose?

Fasting: 3.6–5.8 mmol/L (65–105 mg/dL)
2 hours postprandial: < 7.8 mmol/L (< 140 mg/dL)

❑ Why is maintaining blood glucose levels within the normal range of utmost importance?

Maintenance of normal blood glucose level is important due to the following reasons:

- The brain uses glucose for energy except during starvation when ketone bodies are used.
- The brain is dependent on normal levels of blood glucose because it
  - lacks reserve sources of energy.
  - cannot take up glucose from ECF at low concentrations.
  - is incapable of synthesizing glucose.
- The brain cells are susceptible to both hypoglycemia and hyperglycemia. While hypoglycemia impairs cerebral function, ECF hyperosmolality due to hyperglycemia causes loss of water from brain cells into ECF, resulting in cerebral dehydration, confusion, and coma.

❑ Enumerate the factors that regulate blood glucose levels.

The major factors that regulate blood glucose are as follows:

- ***The tissues***
  - Liver.
  - Extrahepatic tissues — adipose tissue, skeletal muscle, and the kidney
  - In the well-fed state,
    i. the brain and muscle use glucose preferentially.
    ii. the liver converts excess glucose to glycogen, and converts the remaining into triacylglycerols, which are stored in the adipose tissue.
  - In the fasting state,
    i. the liver releases glucose into circulation.
    ii. the adipose tissue fat is mobilized, and ketone bodies are used as energy source.
- ***Hormones***

  Blood glucose homeostasis is maintained by the concerted action of
  - the hypoglycemic hormone insulin.
  - hyperglycemic hormones: Glucagon, epinephrine, growth hormone, glucocorticoids, ACTH, and thyroxine.

❑ Discuss the central role of the liver in maintaining blood glucose homeostasis.

The liver plays a central role in regulating blood glucose in the following manner:

- Supplies glucose to other tissues including the brain.
- Buffers the hyperglycemic effect of a high carbohydrate meal because of following factors:
  - the liver cells are freely permeable to glucose.
  - liver *glucokinase*
    1. is not subject to feedback inhibition by glucose-6-phosphate
    2. has high K*m* for glucose.
    3. is primarily involved in glucose uptake at higher blood glucose concentrations.
- Synthesizes glucose when blood glucose levels are normal.
- In the well-fed state
  - increases glycogenesis.
  - converts excess glucose into triacylglycerols, which are transported by very low density lipoproteins (VLDL) and stored in the adipose tissue.
- During fasting
  - increases hepatic glycogenolysis and gluconeogenesis, thereby releasing glucose into circulation.
  - synthesizes ketone bodies used by extrahepatic tissues as a source of energy.

❑ Elucidate the role of extrahepatic tissues in maintaining blood glucose homeostasis.

The extrahepatic tissues involved in blood glucose homeostasis include:

- ***The adipose tissue***
  - Stores energy in the form of triacylglycerol.
  - Requires insulin for glucose uptake.
  - In the well-fed state, stores triacylglycerol (after transport from the liver as VLDL, hydrolysis by lipoprotein lipase and re-esterification).
  - in the fasting state, hydrolyses triacylglycerols to glycerol and fatty acids, which are transported to the liver. While glycerol enters gluconeogenesis, fatty acids are converted to ketone bodies and used as energy source in extrahepatic tissues.
- ***The skeletal muscle***
  - Requires insulin for glucose entry.
  - Takes up glucose and stores it as glycogen for use as energy source during muscle contraction.
  - Forms lactate during muscle contraction which is transported to the liver and used for gluconeogenesis.
  - In the resting state and during carbohydrate shortage, oxidises ketone bodies and free fatty acids, sparing glucose from oxidation. The spared glucose inhibits adipose tissue lipolysis. This is known as the *glucose fatty acid cycle*.
- ***The kidney***
  - Glucose is filtered by the glomeruli and completely reabsorbed by the tubules. The *tubular maximum for glucose* or *TmG* is 375 mg/min.
  - The *renal threshold* for glucose is 10 mmol/L (180 mg/dL). Glucose spills over into the urine (*glycosuria*) when blood glucose level exceeds the renal threshold.
  - Gluconeogenesis occurs in the renal cortex, which is significant during fasting, to maintain hydrogen ion homeostasis.

❑ Elucidate the role of hormones in maintenance of blood glucose homeostasis.

Blood glucose homeostasis is maintained by the concerted action of

1. the hypoglycemic hormone insulin
2. hyperglycemic hormones: glucagon, epinephrine, growth hormone, glucocorticoids, ACTH, and thyroxine.

Insulin

- Is produced by the β-cells of the pancreatic islets.
- Is secreted in response to a rise in blood glucose levels.
- Facilitates glucose entry into the adipose tissue and muscle (glucose entry into liver is not insulin dependent)
- Exerts hypoglycemic effect by:
  - inducing the synthesis of the glycolytic enzymes-*glucokinase, phosphofructokinase* and *pyruvate kinase;*
  - repressing the synthesis of the gluconeogenic enzymes-*glucose-6-phosphatase*, *fructose 1,6-bisphosphatase*, *PEP carboxykinase*, and *pyruvate carboxylase*; and
  - promoting glycogen synthesis by stimulating *glycogen synthase,* and inhibiting *glycogen phosphorylase* activities.

Glucagon

- Is produced by the α-cells of the islets of Langerhans.
- Is stimulated by hypoglycemia and fasting, and inhibited by insulin.
- Exerts hyperglycemic effect by increasing
  - hepatic glycogenolysis by activating *glycogen phosphorylase* and inactivating *glycogen synthase.*
  - gluconeogenesis
  - lipolysis
  - ketogenesis.

Epinephrine

- Is synthesized by the adrenal medulla.
- Is stimulated by stress.
- Promotes glycogenolysis in the muscle and liver with increase in blood glucose concentration.
- Influences the secretion of other hormones that regulate blood glucose levels.
- Suppresses insulin secretion.
- Stimulates glucagon secretion.

Glucocorticoids

- Are synthesized by the adrenal cortex.
- Are stimulated by hypoglycemia and stress.
- Promote gluconeogenesis by enhancing protein catabolism and providing amino acids.
- Inhibit glucose utilization in extrahepatic tissues by providing free fatty acids as energy source.

Thyroxine

- Increases intestinal absorption of carbohydrates.
- Promotes tissue oxidation and increases the need for insulin.

Growth hormone (GH)

- Is secreted by the adenohypophysis.
- Is stimulated by hypoglycaemia, stress and sleep.
- Decreases peripheral utilization of glucose by mobilizing lipids and providing fatty acids as fuel.
- Increases hepatic gluconeogenesis.
- Has a protein anabolic effect.
- Increases insulin secretion from pancreas (*insulinotrophic effect*), resulting in β-cell exhaustion.

Adrenocorticotrophic hormone (ACTH)

- Is secreted by adenohypophysis
- Is stimulated by hypoglycemia and stress.
- Increases the synthesis and secretion of adrenal cortical hormones.
- Acting *via* adrenal cortical hormones
  - increases gluconeogenesis
  - inhibits glucose uptake by extrahepatic tissues
  - stimulates fat mobilization.

❑ How is glucose metabolized in the well-fed state?

In the well-fed state

- The increase in blood glucose stimulates secretion of insulin, which promotes glucose uptake into muscle and adipose tissue.
- Brain and muscle utilize glucose preferentially due to the presence of *hexokinase* that has a high affinity for glucose.
- A small amount of glucose is stored in the muscle in the form of glycogen.
- Glucose enters the liver via the GLUT2 transporter by a process that does not require insulin.
- In the liver, glucose can be converted to
  - glycogen and stored.
  - triglycerides that are transported to the adipose tissue and stored.

❑ How is glucose metabolized in the fasting state?

Unavailability of glucose during starvation leads the following:

- Increase in hepatic glycogenolysis and gluconeogenesis.
- Mobilization of adipose tissue fat with release of free fatty acids (FFA) and glycerol that are transported to the liver.
- In the liver
  - glycerol is used for gluconeogenesis
  - FFA undergo β-oxidation to form acetyl CoA which is converted to ketone bodies.
- Ketone bodies are used as energy source by extrahepatic tissues including the brain and muscle.
- Muscle protein breakdown releases amino acids, which also enter the gluconeogenic pathway.
- Glucose synthesised in the liver by gluconeogenesis is released into circulation to maintain blood glucose levels.

❑ List the conditions associated with hyperglycemia.

Hyperglycemia refers to an increase in blood glucose above normal levels.

It is seen in the following conditions:

- Diabetes mellitus
- Hyperactivity of the thyroid, pituitary, and adrenals.
- Pancreatic diseases such as pancreatitis and pancreatic carcinoma.
- Infectious diseases.
- Intracranial diseases such as meningitis, encephalitis.
- Anesthesia and asphyxia (transient).
- Emotional stress.

❑ What is diabetes mellitus?

- Diabetes mellitus is a clinical syndrome caused by an absolute or relative deficiency of insulin.
- In diabetes mellitus, the fasting venous plasma glucose concentration is greater than 7.8 mmol/L (140 mg/dL) or a concentration $\geq$ 11.1 mmol/L ($\geq$ 200 mg/dL) 2 h after a carbohydrate meal or 2 h after oral ingestion of 75 g of glucose.

❑ Discuss the classification of diabetes mellitus.

Based on the American Diabetes Association as well as the World Health Organisation (WHO), diabetes mellitus is classified into four main subtypes.

Type 1 diabetes mellitus

- Formerly known as *insulin-dependent diabetes mellitus (IDDM), juvenile-onset diabetes.*
- Includes immune mediated and idiopathic forms of β-cell function that lead to absolute insulin deficiency.
- Generally presents in childhood and adolescence.
- About 25% of the cases may occur at any age.
- Patients are liable to develop *ketoacidosis*.
- Individuals with the histocompatibility antigens *HLA DR3* and *DR4* are at high risk. *HLA DQβ1*0302* is a susceptibility gene associated with high risk for diabetes, while HLA Dβ1*0602 decreases risk.
- Pathogenesis involves the following sequence of events: Genetic predisposition → viral infection (*Coxsackie* B_4) → inflammatory response in the pancreas ("*insulitis*") → transformation of β-cell surface from "*self*" to "*non-self*" *(autoimmunity)* → activation of the immune system → destruction of β-cells → diabetes mellitus.

Type 2 diabetes mellitus

- Formerly known as *non-insulin-dependent diabetes mellitus (NIDDM), adult-onset diabetes.*
- More common, less severe.
- Although occurs in subjects over 40 years of age, it is also seen in young people. Patients are not prone to diabetic ketoacidosis.
- Caused by:
  - decreased insulin secretion.
  - relative insulin deficiency due to insulin resistance in target tissues due to a decrease in insulin receptors or post-receptor defects.
- Combination of genes and environmental factors contribute to susceptibility.

- Candidate genes identified in disease pathogenesis include *glucokinase* and *transmembrane glucose transferase.*
- High risk factors for the disease include diet, life-style, obesity, and stress.

Type 3 diabetes mellitus

Includes
- genetic defects of β-cell function, insulin action.
- diseases of exocrine pancreas (chronic pancreatitis, cystic fibrosis, hemochromatosis).

Type 4 diabetes mellitus, is induced by pregnancy (*gestational diabetes*).

❑ Elucidate the metabolic abnormalities in diabetes mellitus.

The metabolic abnormalities in diabetes mellitus result from insulin deficiency. Insulin lack causes the following alterations in the metabolic profile:

Carbohydrate metabolism

- Decreased glucose uptake by tissues due to decreased glucose transport across cell membrane.
- Reduced glycolytic activity as a result of decrease in the insulin: glucagon ratio which lowers hepatic concentration of *fructose 2,6-bisphosphate* with inhibition of *phosphofructokinase*.
- Stimulation of gluconeogenesis.
- Depletion of glycogen stores with glucose release from the liver into circulation. This leads to intracellular glucose deficit and extracellular glucose excess.
- If blood glucose levels exceed the renal threshold,
  - glucose spills over into the urine (*glycosuria*) associated with loss of water and electrolytes (*osmotic diuresis*).
  - urinary output is increased (*polyuria*) with compensatory increase in thirst (*polydipsia*).
  - in extreme cases, nausea and vomiting occur leading to
    - i. dehydration,
    - ii. decrease in blood volume and renal blood flow, and
    - iii. anuria and coma.

Lipid metabolism

- Decreased lipogenesis.
- Increased adipose tissue lipolysis with release of free fatty acids (FFA) into circulation.
- FFA are used for hepatic triglyceride synthesis.
- Decreased activity of lipoprotein lipase.
- *Hypertriglyceridemia*, characterized by increase in plasma very low density lipoproteins (VLDL) and decrease in high density lipoprotein (HDL).
- Cholesterol synthesis is increased.
- FFA are channelled to ketone body formation.
- Ketone body production exceeds rate of utilization, leading to increased plasma levels (*ketonaemia*) with spillover into urine (*ketonuria*), leading to *ketoacidosis* in severe cases.

Protein metabolism

- Increased protein catabolism with release of amino acids, which are used for gluconeogenesis.
- Increased amino acid catabolism with increase in urea nitrogen.
- Muscle wasting in advanced cases.

❑ Enumerate the diagnostic tests for diabetes mellitus.

1. ***Urine glucose***
   Glucose appears in urine when blood glucose is >10 mmol/L (>180 mg/dL).
2. ***Blood glucose***: A fasting venous plasma glucose concentration of ≥7.8 mmol/L (140 mg/dL) or a random plasma glucose of ≥11.1 mmol/L (200 mg/dL) on more than one occasion confirms diagnosis.
3. ***Oral glucose tolerance test (OGTT)***:
   - Not required in frank diabetes.
   - Indicated
     - in mild diabetes when fasting and random glucose levels are normal or slightly increased.
     - when clinical symptoms of diabetes mellitus are evident with normal blood glucose levels.
     - acromegaly.
   - *Method*
     - After an overnight fast, blood is withdrawn for fasting plasma glucose estimation.
     - Urine specimens are collected.
     - Seventy-five grams of glucose dissolved in 300 mL of water is given and blood withdrawn every half an hour for 2½ h.
     - Two one-hourly urine samples are collected.
   - *Normal response*: Plasma glucose levels reach a peak ½ h or 1 h after glucose intake with return to normal fasting levels in 1½ to 2 h. Glucose is absent in the urine samples.
4. ***Glycosylated hemoglobin (HbA1c)***
   - Reflects the average plasma glucose concentration during the preceding 6–8 weeks.
   - A 2–3-fold increase (normal: 6% of total hemoglobin) is seen in diabetes mellitus.

❑ Outline the principles of management of diabetes mellitus.

Management of *diabetes mellitus* involves the following:

- Alleviation of clinical symptoms.
- Prevention of metabolic complications.
- Insulin administration in type 1.
- Administration of fluids and electrolytes to correct osmotic and electrolyte imbalances.
- Treatment with the immunosuppressive drug cyclosporin A to induce remission in type 1.

Management of type 2

- Weight reduction.
- Dietary control
  - carbohydrates:
    i. comprise 40–60% of total energy intake.
    ii. intake of complex carbohydrates that are absorbed slowly.
    iii. avoidance of simple sugars like glucose and sucrose that are rapidly absorbed.
  - proteins comprise 15% of total energy intake.
  - fat
    i. comprises 30–35% of the total energy intake.
    ii. restriction of total fat and saturated fat to prevent coronary heart disease.
    iii. intake of polyunsaturated fats, such as sunflower oil, and soya bean oil and monounsaturated fats.
  - high intake of dietary fiber, which causes satiety and lowers plasma glucose and lipid levels.

- Administration of insulin during periods of stress.
- Oral hypoglycemic agents in patients not responding to diet therapy. These include the following:
  - *sulphonylureas* (tolbutamide and chlorpropamide)
    - i. are readily absorbed in the liver.
    - ii. increase insulin output by β-cells.
    - iii. enhance insulin action on peripheral tissues via insulin receptors or through post-receptor mechanisms.
    - iv. side effects: hypoglycemia from overdosage, hyponatremia, and sometimes, cardiovascular disease.
  - *biguanides s*uch as metformin and phenformin
    - i. inhibit intestinal glucose absorption, TCA cycle, and gluconeogenesis.
    - ii. side effect: lactic acidosis.

❑ What are the metabolic complications of diabetes mellitus? Enumerate the causes of the complications.

Metabolic complications in *diabetes mellitus* may be

Acute complications
- Diabetic ketoacidosis
- Hyperosmolal non-ketotic coma

Long-standing complications
- Vascular disease
- Diabetic nephropathy
- Diabetic retinopathy
- Diabetic neuropathy
- Diabetic foot

Causes of complications in diabetes mellitus

The factors implicated in causing various complications in diabetes mellitus are as follows:

1. *Sorbitol*, formed by the reduction of glucose in the polyol pathway that decreases myoinositol and *Na^+K^+ ATPase* activity, and causes abnormalities in phosphoinositide metabolism.
2. *Glycation* of proteins forms *advanced glycation end products (AGE)* that bind to receptors on macrophages and endothelial cells with release of cytokines, endothelin-1, and tissue factor. Glycation of proteins has the following consequences:
   - glycated hemoglobin (**HbA1c**) formed by binding of glucose non-enzymatically to N-terminal amino group of β-chain of HbA alters the function of Hb.
   - glycation of LDL increases its half-life since it is not recognized by the normal LDL receptor.
   - glycated HDL turns over more rapidly than native HDL.
   - glycated collagen becomes more resistant to degradation by collagenase.
   - glycation of structural proteins in the arterial wall of the glomerular basement membrane contributes to renal complications.
   - glycation of lens proteins leads to cataracts.
3. *Increased blood flow* or *hyperperfusion* may contribute to diabetic complications by increasing the filtration of tissue toxins.

❑ Discuss the causes, clinical and biochemical features, and management of diabetic ketoacidosis.

Causes

- Cessation of insulin therapy
- Vomiting
- Trauma
- Infection
- Emotional stress
- Myocardial infarction.

Clinical and biochemical features

- *Hyperglycemia* [*blood glucose level*: 22.2 – 38.8 mmol/L (400 – 700 mg/dL) or higher] that leads to glycosuria, osmotic diuresis, loss of water and electrolytes. Vomiting can aggravate fluid electrolyte loss leading to dehydration.
- Decrease in extracellular fluid (ECF) volume leads to decreased renal blood flow with reduced glomerular filtration rate (GFR) and uremia.
- Increase in ECF osmolality depletes intracellular fluid (ICF) resulting in cellular dehydration. Dehydration of cerebral cells leads to confusion and coma.
- Increased ketone body formation accompanied by release of hydrogen ions and anion gap metabolic acidosis. A deep sighing breathing (*Kussmaul respiration*) with acetone odor in the breath is a characteristic finding.
- Electrolyte changes:
  - Hyponatremia secondary to hyperglycemia or hyperlipidemia.
  - Hyperkalemia because of decreased entry into cells: but total body potassium is decreased due to osmotic diuresis.
  - Plasma bicarbonate is decreased due to buffering ketone bodies.

Management

- Administration of insulin with isoosmolal saline results in glucose entry into cells followed by sodium-free water along the osmotic gradient.
- Potassium replacement to correct hypokalemia, which occurs due to re-entry of potassium into cells after treatment.
- Bicarbonate administration if the acidosis is severe.

❑ Give an account of hyperosmolal non-ketotic coma.

- Hyperosmolal non-ketotic coma occurs only in type 2 elderly diabetics in whom water intake is inadequate.
- It is precipitated by:
  - dialysis
  - high intake of carbohydrates and proteins
  - administration of osmotic agents such as mannitol and urea.
- Insulin is adequate to prevent excessive lipolysis and generate ketone bodies but not sufficient to prevent hyperglycemia.
- Plasma glucose level: >500 mmol/L (900 mg/dL) leads to
  - osmotic diuresis, which causes decrease in GFR with increase in plasma urea, creatinine, and protein.
  - severe dehydration. Plasma sodium is increased due to loss of water and excess of sodium.
  - very high plasma osmolality.
  - ketosis is absent.
  - acidosis is minimal, plasma bicarbonate is slightly above normal due to decreased hydrogen ion excretion.
  - coma occurs due to cerebral cellular dehydration.

Management

- Rehydration.
- Administration of insulin by intravenous infusion together with hypotonic saline. Heparin to prevent hyperviscosity and thrombosis.
- After treatment, management with diet alone, or with diet and oral hypoglycemic agents.

Outline the long-term complications of diabetes mellitus.

The long-term complications of diabetes mellitus include:

- ***Vascular disease***
  Atherosclerosis, that is accelerated due to
  - non-enzymatic glycation of lipoproteins.
  - LDL oxidation.
  - cerebrovascular and peripheral vascular insufficiency.
- ***Diabetic retinopathy*** progressing to blindness. Hyperglycemia increases blood flow to the retina causing dilation of capillaries, endothelial cell proliferation, and production of vascular endothelial growth factor (VEGF), which stimulates angiogenesis and increased vascular permeability. This leads to retinal leakage and exudation.
- ***Diabetic nephropathy*** associated with proteinuria, nephrotic syndrome, progressive renal failure, and hypertension.
  Diabetic nephropathy may be
  - diffuse with widening of the basement membrane.
  - nodular characterized by
    i. deposition of periodic acid Schiff (PAS)-positive material in the glomerular tufts (*Kimmelsteil-Wilson lesion*).
    ii. increase in the renal threshold.
    iii. hyperglycemia without glycosuria.
- ***Diabetic neuropathy*** is a relatively early complication seen in 30% of diabetics. It occurs secondary to metabolic abnormalities and may involve motor, sensory, and autonomic nerves.
- ***Diabetic foot***: Ulcers of the feet and lower extremities occur secondary to diabetic neuropathy. Decreased blood supply and microbial infections contribute to foot ulcers.

Define hypoglycemia. What are the clinical symptoms of hypoglycemia?

Hypoglycemia refers to a clinical condition in which the plasma glucose level is less than 2.5 mmol/L (45 mg/dL).

Clinical manifestations

Adrenergic symptoms:
- sweating
- tremor
- anxiety
- tachycardia
- hunger

Central nervous system (CNS) symptoms:
- dizziness
- reduced mental acuity
- visual abnormalities
- convulsions
- loss of consciousness

Prolonged hypoglycemia leads to coma, cerebral damage, and death.

❑ How is hypoglycemia classified?

There are two main classes of hypoglycemia:

1. *Postprandial or reactive hypoglycemia*, in which clinical manifestations appear within 5–6 hours after a meal. It may be precipitated by ingestion of certain types of food or drugs.
2. *Fasting hypoglycemia*, in which symptoms are manifest in the night or in the early morning. It is precipitated by fasting or strenuous exercise.

❑ Enumerate the causes of reactive hypoglycemia.

The causes of reactive hypoglycaemia are as follows:

1. *Alimentary hyperinsulinism* following meals in patients after gastrectomy, vagotomy, and gastrojejunostomy due to rapid passage of food into the intestine and rapid absorption of glucose, which stimulates excessive insulin secretion.
2. *Drug-induced hypoglycemia* due to insulin overdosage, sulphonylureas, salicylates, and antihistamines, and overdoses of paracetamol.
3. *Alcohol*, which causes hypoglycaemia 2–10 hours after excessive ingestion.
4. *Inherited metabolic disorders* such as hereditary fructose intolerance and galactosemia.
5. *Leucine-sensitivity hypoglycemia* precipitated by casein that contains high amount of leucine, a stimulant of insulin secretion. Leucine also precipitates a condition termed *nesidioblastosis* in infants with a developmental abnormality of the pancreatic ducts and islet cells.

❑ Outline the causes of fasting hypoglycemia.

Fasting hypoglycemia results due to decreased hepatic glucose production and increased utilization by peripheral tissues.

- ***Decreased glucose production*** occurs due to the following reasons:
  1. *Enzyme deficiencies* such as glucose-6-phosphatase deficiency in von Gierke's disease.
  2. *Deficiency of substrate* such as in severe malnutrition and wasting.
  3. *Liver and kidney diseases.*
  4. Deficiency of hormones antagonistic to insulin such as in adrenal insufficiency.
  5. *Alcohol* by increasing the NADH/NAD^+ ratio during hepatic oxidation with consequent inhibition of gluconeogenesis, decreased hepatic glucose output and hypoglycaemia.
- ***Overutilization of glucose*** occurs under the following conditions:
  1. *Insulinomas*, insulin-secreting primary tumors of the β-cells of the pancreatic islets in which plasma insulin levels are high despite low fasting plasma glucose levels.
  2. *Insulin autoimmunity* due to
     a. antibodies to insulin that release excessive insulin from the pancreas.
     b. antibodies to insulin receptors that cause insulin resistance.
  3. *Non-pancreatic tumors* such as carcinomas of the liver due to increased glucose uptake by the tumor, reduced hepatic output, or to *insulin-like growth factors* (*IGFs*).
  4. *Unavailability of fatty acids for oxidation* as in carnitine deficiency or due to deficiency of enzymes involved in fat oxidation. There is consequent overutilization of glucose as an energy source.

❑ How is hypoglycemia diagnosed in the laboratory?

Determination of plasma insulin and C-peptide is useful in the differential diagnosis of hypoglycemia. Plasma insulin and C-peptide levels are appropriately low in hypoglycaemia due to

- alcohol
- non-pancreatic tumors
- hypopituitarism
- adrenal failure.

Inappropriately, high plasma insulin with increase in C-peptide concentration is seen in

- insulinomas
- after administration of sulphonylurea which stimulates insulin secretion from pancreas.

Inappropriately high plasma insulin with decrease in C-peptide concentration is seen after

- exogenous insulin administration.

Insulin suppression test confirms diagnosis. This involves the following steps:

1. Insertion of an indwelling intravenous cannula after an overnight fast. The cannula is kept patent with sodium citrate solution or heparinized saline.
2. Collection of fasting blood sample for estimation of plasma glucose and C-peptide.
3. Intravenous injection of insulin (0.15 U/kg body weight).
4. Collection of blood samples every half an hour for 2½ h.

Normal response: Plasma C-peptide is less than 50% of the initial value. Failure of plasma C-peptide to fall confirms the diagnosis of insulinoma. Further investigations are required if hypopituitarism or adrenal cortical failure is suspected.

❑ Outline the treatment for hypoglycemia.

- Intravenous administration of 10 – 20 mL of 10% glucose in children and 50% glucose in adults.
- Removal of insulinoma by surgery.
- Combination of diazoxide and chlorothiazide if surgery is contraindicated.

Section 6

Function Tests and Plasma Proteins

- Liver Function
- Kidney Function
- Gastric Function
- Pancreatic and Intestinal Function
- Plasma Proteins and Blood Coagulation

Chapter 23

Liver Function

❑ Briefly describe the ultrastructure of the liver.

- The liver is made up of *lobes*.
- Each lobe contains a number of *lobules*, which are the functional units of the liver.
- Each lobule is made up of hepatic cells.
- The hepatic cells are separated from each other by vascular channels known as *sinusoids*.
- The sinusoids are lined by phagocytic cells called *Kupffer cells*, which form part of the reticuloendothelial system.
- *Bile canaliculi* present between hepatic cells empty into the bile ducts.

❑ What are the functions of the liver?

The liver performs several functions such as

- *Metabolism*
  - Glycogen metabolism
  - Gluconeogenesis
  - Urea cycle
  - Amino acid metabolism
  - Fatty acid oxidation
  - Cholesterol synthesis
  - Hormone synthesis.
- *Synthesis* of plasma proteins such as albumin, coagulation factors, lipoproteins, and primary bile acids.
- *Storage* of glycogen, iron, and vitamins A and B_{12}.
- *Secretion* of bile that plays a key role in the digestion and absorption of lipids.
- *Detoxication* of drugs, alcohol, bilirubin, and xenobiotics.

❑ Elucidate the composition and functions of bile.

- About 1 litre of bile is secreted by the hepatic parenchymal cells into the bile canaliculi from where it passes through the intrahepatic ducts and enters the intestine through the common bile duct.
- The gallbladder absorbs water and electrolytes, and concentrates the bile.
- Presence of partially digested food in the intestine stimulates gallbladder contraction by cholecystokinin and emptying of bile into the duodenum.

Composition

The components of bile include

- *Bile acids* present as bile salts.
  a. The *primary bile acids* are *cholic acid* and *chenodeoxycholic acid.*
  b. The *secondary bile acids include deoxycholic acid* and *lithocholic acid.*
  c. The bile acids are conjugated by the liver to form water-soluble bile salts—sodium and potassium *glycocholate* and *taurocholate.*
  d. The bile salts are reabsorbed from the intestines into the portal blood, transported to the liver, and reexcreted into the bile. This is known as *enterohepatic circulation.*
- *Bile pigments bilirubin, biliverdin* and small amounts of *coproporphyrin.*
- Cholesterol
- Lecithin
- Electrolytes
- Fatty acids
- Alkaline phosphatase.

Functions

- *Absorption of fat*: Bile salts help in fat absorption by
  - lowering the surface tension.
  - emulsification of fats in the intestine.
  - formation of micelles with cholesterol and phosphatidylcholine.
  - activation of lipases in the intestine.
- *Neutralization of acidic chyme* from the stomach.
- *Excretion* of bile acids, bile pigments, cholesterol, drugs, toxins, and inorganic substances such as mercury, zinc, and copper.

❑ Enumerate the various tests of liver function.

Liver function can be evaluated by assessing the excretory, metabolic, synthetic, and detoxication functions of the liver as well as by assay of liver marker enzymes.

- ***Excretory function*** can be assessed by measurement of plasma bilirubin.
  - *Unconjugated bilirubin* is increased due to excessive hemolysis or defective conjugation.
  - *Conjugated bilirubin* is increased when the excretory functions of the liver are impaired.
- ***Metabolic function*** can be assessed by the ability of the liver to metabolize galactose. The test involves oral administration of 40 g of galactose (or 0.5 g galactose/kg body weight i.v.) and estimation of blood galactose over the next two hours. In normal subjects, blood galactose levels decline progressively and reach fasting levels by 2 hours. Galactose utilization is decreased in cirrhosis and hepatitis.
- ***Synthetic function*** of the liver can be assessed by estimating plasma proteins and prothrombin time.
  - Plasma albumin is decreased with increase in globulin and reversal of A/G ratio in liver diseases with consequent edema.
  - *Prothrombin time* is prolonged in both liver disease and obstruction to bile flow (*cholestasis*) because prothrombin synthesis requires functioning liver cell mass and vitamin K.
- ***Detoxication function tests***
  The liver is involved in detoxification of xenobiotics by the action of *monooxygenases* followed by conjugation to form polar, water-soluble compounds that can be excreted in the bile or urine.

The tests that are used to assess the detoxicating function of the liver include the following:
- *Hippuric acid synthesis test*
- *Blood ammonia*
- *Aminopyrine breath test*

- ***Determination of plasma enzymes***

The following plasma enzymes are useful to assess liver function:
- *Aminotransferases*
- *Alkaline phosphatase*
- *Gamma-glutamyl transferase*

❑ Briefly explain the principle of tests that are used to assess the detoxification capacity of the liver.

1. *Hippuric acid synthesis test* is based on detoxification of benzoic acid by conjugation with glycine to form hippuric acid, which is excreted in the urine. The test involves administration of benzoic acid as sodium benzoate either orally (6 g in 200 mL of water) or intravenously (1.77 g in 20 mL of water) and estimation of the amount of hippuric acid excreted. In normal persons, at least 3 g of hippuric acid should be excreted 4 h after oral ingestion or 1.15 g 2 h after intravenous injection. In liver cell damage, excretion of hippuric acid is decreased.
2. *Blood Ammonia*: In liver cell damage, due to advanced cirrhosis and hepatic coma, urea synthesis from ammonia is affected and blood ammonia levels are increased.
3. *Aminopyrine breath test* involves measurement of $^{14}CO_2$ formed from [^{14}C] methyl-labeled aminopyrine in the liver by N-demethylation. In liver diseases such as cirrhosis and hepatitis, the amount of $^{14}CO_2$ formed is decreased.

❑ What are the plasma enzymes that are useful as indicators of liver function?

The following plasma enzymes are of diagnostic value in liver diseases:

1. *Aminotransferases*: Aspartate aminotransferase (AST) and alanine aminotransferase (ALT) are both found in high concentrations in the liver. ALT is predominantly cytosolic, while AST is found both in the cytoplasm and the mitochondria.
   - Plasma AST and ALT are increased in parenchymal liver cell damage.
   - Increase in ALT is greater than AST in viral and toxic hepatitis where the cytoplasm sustains major damage.
   - The increase in AST is greater than ALT in cirrhosis where both the cytoplasm and mitochondrial membranes are damaged.
   - A sustained increase in plasma transaminases for more than six months indicates chronic liver disease.
2. *Alkaline phosphatase (ALP)*: The hepatic isoenzyme originates from the lining of the bile canaliculi and sinusoidal surfaces of the hepatocytes.
   - The presence of the enzyme is a sensitive indicator of cholestasis, biliary obstruction, and liver infiltration.
   - High values are seen in posthepatic obstruction, primary biliary cirrhosis.
   - Progressive increase is characteristic of malignancy.
3. *Gamma-glutamyl transferase (GGT)*:
   - The enzyme is predominantly located in the endoplasmic reticulum of the cells of the hepatobiliary tract.
   - It is increased in cholestasis, especially in posthepatic obstruction.
   - Disadvantage: Non-specific and can be induced by drugs and alcohol.

❑ How is jaundice classified?

- Jaundice is characterized by increase in plasma bilirubin levels associated with yellow pigmentation of the skin and sclera.
- The current classification based on the nature of altered bilirubin metabolism categorizes jaundice into two broad classes:
  - *Retention jaundice* characterized by increased amounts of unconjugated bilirubin in circulation.
  - *Regurgitation jaundice,* characterized by an increase in plasma conjugated bilirubin.

❑ Enumerate the causes of retention jaundice.

Retention jaundice is characterized by elevated plasma *unconjugated bilirubin*. Being protein-bound and water insoluble, unconjugated bilirubin is not excreted in the urine. This type of jaundice is therefore referred to as *acholuric jaundice.*

Causes

- *Prehepatic (hemolytic) jaundice*: Excessive hemolysis with overproduction of bilirubin occurs in
  - congenital erythrocyte abnormalities (*e.g.*, hemoglobinopathies, sickle cell anemia)
  - *glucose-6-phosphate dehydrogenase* (G6PD) deficiency,
  - ineffective erythropoiesis (*e.g.*, pernicious anaemia),
  - incompatible blood transfusions,
  - infections by hemolytic organisms such as *Plasmodium* species,
  - drugs (*e.g.*, sulphonamides, salicylates) that displace bilirubin from plasma albumin.
- *Defective hepatic uptake* of bilirubin due to:
  - sepsis,
  - prolonged fasting,
  - drugs such as flavaspidic acid, used in the treatment of tapeworm infestation that compete with bilirubin for binding to ligandin.
- *Decreased conjugation* due to
  - decrease in functioning liver cell mass as in chronic hepatitis and cirrhosis.
  - defect in conjugation as in familial unconjugated hyperbilirubinemias such as:
    a. *Crigler-Najjar syndrome* in which there is complete absence (type I) or deficiency (type II) of *UDP-glucuronyl transferase* activity. Plasma bilirubin levels are < 20 mg/dL (345 μmol/L) in type I but do not exceed 20 mg/dL (345 μmol/L) in type II. Phenobarbital therapy is effective in type II but not in type I.
    b. *Gilbert's syndrome,* a common form of familial jaundice in which plasma bilirubin levels are between 1.2 and 2.5 mg/dL (20–40 μmol/L) due to defective hepatic uptake and conjugation.
    c. *Acquired deficiency of UDP-glucuronyl transferase* due to
       i. presence of pregnane-3β-20α-diol, an inhibitor of *glucuronyl transferase* in breast milk (*breast milk jaundice*).
       ii. hypothyroidism, which delays maturation of the enzyme.
       iii. drugs such as novobiocin that inhibit enzyme activity.
- *Neonatal jaundice (physiological jaundice)*, characterized by
  - excessive hemolysis combined with an immature hepatic system for the uptake, conjugation, and secretion of bilirubin.

- decreased activity of *UDP-glucuronyl transferase*
- increase in plasma unconjugated bilirubin (30 mg/dL (517 μmol/L) that can penetrate the blood-brain barrier, leading to hyperbilirubinemic toxic encephalopathy (*kernicterus*) eventually leading to brain damage and death.

The condition can be treated by phenobarbital administration or by exposure to visible light, which promotes hepatic excretion of unconjugated bilirubin.

❑ Enumerate the causes of regurgitation jaundice.

Regurgitation jaundice is characterized by an increase in plasma *conjugated bilirubin* and appearance of bilirubin in urine (*choluric jaundice*). As a result, the amount of bilirubin entering the intestine is reduced with decreased formation of urobilin and pale stools.

Causes

- *Impaired hepatic excretion of bilirubin* as in
  - extensive liver cell damage as in hepatitis and cirrhosis with impairment in uptake, conjugation and excretion of bilirubin. Plasma bilirubin does not exceed 20 mg/dL (345 μmol/L). Plasma ALP activity is normal, while AST and ALT are raised. Cholestasis is present and prothrombin time is prolonged.
  - *familial conjugated-hyperbilrubinaemia* such as *Dubin-Johnson syndrome* in which bilirubin conjugation is normal but hepatic secretion of conjugated bilirubin into the bile is defective. The condition is associated with accumulation of a brown pigment with staining properties of *lipofuscin* and abnormal coproporphyrin excretion.
- *Cholestasis* (obstruction to bile flow) that may be
  1. *Intrahepatic cholestasis* due to impaired bile secretion from hepatocytes into the bile canaliculi caused by
     - drugs (chlorpromazine, oral contraceptives, methyltestosterone)
     - alcohol
     - primary biliary cirrhosis
     - inflammation of the biliary tract (cholangitis)
     - carcinoma of the bile duct
     - cystic fibrosis.
  2. *Extrahepatic cholestasis* (*posthepatic or obstructive jaundice*) due to
     - congenital atresia of the main bile ducts
     - bile duct stricture after surgery
     - gallstones in the common bile duct or the Ampulla of Vater
     - carcinoma of the head of the pancreas.

❑ What are the characteristics of cholestasis?

Obstruction to bile flow referred to as *cholestasis* is characterized by:

- Increase in plasma ALP activity
- High plasma bilirubin levels of 50 mg/dL (862 μmol/L)
- Prothrombin deficiency and hemorrhage due to malabsorption of vitamin K
- Hypercholesterolemia due to decreased cholesterol excretion in bile.

❑ Summarize the alterations in bilirubin metabolism in retention and regurgitation jaundice.

Table 23.1 summarizes the laboratory findings in jaundice.

Table 23.1 Laboratory findings in retention and regurgitation jaundice

Parameter	*Retention jaundice*	*Regurgitation jaundice*
Plasma bilirubin		
Total	↑	↑
Unconjugated	↑	—
Conjugated	—	↑
Urine bilirubin	Absent because unconjugated bilirubin bound to albumin is water insoluble and not filtered by the glomerulus	Present because conjugated bilirubin is water soluble
Urine urobilinogen	↑	Normal in viral hepatitis and cirrhosis. Absent in extrahepatic cholestasis because bilirubin does not enter the intestine
Fecal urobilinogen	↑	Low in viral hepatitis and cirrhosis. Absent in extrahepatic cholestasis
Bile salts	—	Present in cholestasis

❑ How is prothrombin time useful in the differential diagnosis of jaundice?

- Prothrombin time is prolonged in both hepatocellular disease and in cholestasis.
- In severe liver disease, prothrombin time fails to return to normal even after parenteral administration of vitamin K; whereas in cholestasis, where vitamin K absorption is defective, parenteral administration of vitamin K corrects the abnormality.

❑ Elucidate the causes and biochemical findings in acute hepatitis.

Causes

- Hepatotoxins such as carbon tetrachloride, benzene derivatives, acetaminophen, and mushroom poisoning, which induce necrosis and fatty infiltration.
- Drugs such as isoniazid, halothane, phenytoin, and methyldopa that induce changes similar to viral hepatitis or cholestasis.
- Alcohol. Hepatitis caused by alcohol is accompanied by markedly elevated plasma GGT activity, uric acid, and triglyceride levels.
- Viruses such as hepatitis A, B, C, D, and E viruses as well as cytomegalovirus, Epstein-Barr virus, and Coxsackie virus

Table 23.2 lists the mode of transmission of hepatitis viruses and the diagnostic tests.

Biochemical findings

- *Bile pigments*
  - Plasma bilirubin levels do not exceed 20 mg/dL (345 μmol/L).
  - Increase in urinary bilirubin and urobilinogen in the pre-icteric phase.
  - Decreased excretion of urobilinogen in the urine and the feces in the icteric stage.
  - Pale stools.

Table 23.2 Hepatitis viruses : transmission and diagnosis

Virus	*Transmission*	*Diagnosis*
Hepatitis A	Fecal–oral route, food and water-borne infection	IgM anti-HAV in serum
Hepatitis B	Percutaneous, sexual, or perinatal	Viral surface antigen (HbsAg) and internal component of the virus (HbcAg) in serum
Hepatitis C	Percutaneous transmission and intravenous drug use	Anti-HCV in serum
Hepatitis D	Percutaneous in HBsAg-positive intravenous drug users or by transfusion in hemophiliacs	Anti-HDV in serum
Hepatitis E	Water-borne epidemic	Anti-HEV IgM and IgG in serum

- *Plasma enzymes*
  - Increase in serum transaminase activities
  - Normal or moderately increased plasma ALP
  - Prolonged prothrombin time in severe liver damage.

❑ Write notes on chronic hepatitis.

Chronic hepatitis refers to a group of disorders characterized by chronic hepatic inflammation.

Histological types

- *Chronic persistent hepatitis,* a benign, asymptomatic condition that does not progress to cirrhosis. There is a slight increase in plasma transaminases.
- *Chronic lobular hepatitis,* characterized by foci of necrosis and inflammation in the liver lobule and increase in plasma transaminases that progresses to cirrhosis.
- *Chronic active hepatitis,* characterized by hepatic necrosis, lobular inflammation, and fibrosis that progresses to cirrhosis and eventually liver failure and death. The biochemical findings include
  - increase in plasma IgG levels
  - low plasma albumin
  - hyperbilirubinemia and jaundice
  - increase in plasma transaminases with AST activity greater than ALT as the disease progresses.

❑ Discuss the causes, consequences, biochemical findings, and management of cirrhosis.

Cirrhosis is the end result of many inflammatory and metabolic diseases of the liver. It is characterized by fibrosis, and disorganization of hepatic architecture.

Causes

- Alcoholism (most common cause)
- Chronic active hepatitis
- Wilson's disease
- Toxins
- Viral hepatitis
- Hemochromatosis
- Drugs

Clinical manifestations

- Anorexia
- Vomiting
- Fatigue
- Jaundice
- Nausea
- Diarrhea
- Fever

Consequences

- Distortion of normal hepatic architecture due to fibrosis.
- Disruption of hepatic blood supply by regeneration of nodules of hepatocytes with increased pressure in the portal vein and portal hypertension.
- Loss of functioning liver cell mass, leading to jaundice.
- Accumulation of fluid in the peritoneal cavity (*ascites*), *hepatic encephalopathy*, and *edema*.
- Glucose intolerance and central hyperventilation resulting in respiratory alkalosis.
- Hepatocellular failure and primary hepatocellular carcinoma in advanced cases.
- Renal failure in end-stage alcoholic cirrhosis.
- Hormonal disturbances leading to feminization of males, gynecomastia, decreased body hair, impotence, and testicular atrophy.

Biochemical findings

- *Bile pigments*
  - normal or slightly increased plasma bilirubin.
  - absence of bilirubin in the urine.
  - excessive urinary excretion of urobilinogen.
- *Plasma enzymes*
  - increase in plasma AST is greater than that of ALT.
  - moderate increase in plasma ALP and GGT.
- Decreased synthesis of prothrombin and other vitamin K dependent clotting factors, leading to coagulopathy and prolonged prothrombin time.
- *Plasma proteins*
  - decreased plasma albumin that contributes to edema.
  - hyperglobulinemia.
  - increased synthesis of IgG and IgA.
  - characteristic serum protein electrophoretic pattern of β-γ fusion.
- *Procollagen type III peptide* measurement is useful in monitoring fibrosis.
- *Minerals*
  - hypomagnesemia and hypophosphatemia due to dietary deficiency and increased urinary excretion.
  - dilutional hyponatremia in patients with ascites.
  - hypokalemia due to increased urinary loss.
  - prerenal azotemia.

Management

- Management of hepatic encephalopathy controlling gastrointestinal bleeding, restricting dietary protein, and by use of laxatives or enemas.
- Neomycin treatment to control toxin production by intestinal bacteria.

- Maintenance of calorie intake, and fluid and electrolyte balance.
- Management of ascites by sodium restriction.
- Avoidance of alcohol.

❑ Discuss the composition, pathogenesis, and management of gallstones.

Gallstones are crystalline structures formed by concretion or accretion of bile constituents that may be *cholesterol stones, mixed stones,* or *pigment stones.*

Cholesterol and mixed stones

- Account for 80% of total stones.
- More prevalent in western countries.
- Are made up of cholesterol monohydrate (> 70%), calcium salts, bile acids, bile pigments, proteins, fatty acids, and phospholipids.
- Occur due to supersaturation of bile with cholesterol as a result of
  a. obesity and high caloric diets.
  b. increased activity of HMG-CoA reductase, or decreased activity of *7α-hydroxylase* with decreased formation of bile acids from cholesterol.
  c. decreased secretion of bile salts and phospholipids.
  d. hypolipidemic drugs such as clofibrate that increase cholesterol secretion in the bile.

Pigment stones

- Are more common in Oriental countries, especially in Asia.
- Are composed of calcium bilirubinate and small amount of cholesterol (<10%).
- Occur in association with
  a. biliary tree infections,
  b. chronic hemolytic states (*e.g.,* hereditary spherocytosis) in which increased hemoglobin breakdown results in overproduction of bilirubin and increased biliary secretion.

Consequences of gallstones

- Acute or chronic cholecystitis.
- Bacterial infection due to a block in the cystic duct.
- Obstructive jaundice due to obstruction of the common bile duct.

Treatment of gallstones

- *Cholecystectomy* (surgical removal of the gallbladder).
- Oral administration of the bile acids, chenodeoxycholic acid and urodeoxycholic acid, to dissolve gallstones by reducing the relative saturation of bile with cholesterol.
- Shock-wave lithotripsy to shatter gallstones into tiny fragments and enable passage through the cystic duct.

Chapter 24

Kidney Function

❑ Briefly explain the structure of a nephron.

- *Nephron* is the functional unit of the kidney.
- Each nephron is made of a *renal corpuscle* and a *renal tubule*.
- The *renal corpuscle* contains a thin-walled sac-like structure termed the *Bowman's capsule* into which the afferent arteriole from the renal artery forms a capillary network called the *glomerulus*.
- The *renal tubule* consists of the *proximal convoluted tubule* (*PCT*), which dips into the renal medulla to form the *loop of Henle*. The tubule reenters the renal cortex to form the *distal convoluted tubule* (*DCT*). Several DCTs merge to form a *collecting duct*. Several collecting ducts join and open into the *renal pelvis*. The *juxtaglomerular apparatus* present between the DCT and the afferent arteriole elaborates *renin*.

❑ What are the functions of the kidney?

The kidneys perform the following biological functions:

- Excretion of waste products
- Acid–base homeostasis
- Fluid electrolyte homeostasis
- Metabolism
- Elaboration of hormones such as
  - *renin* produced in the juxtaglomerular apparatus, which forms *angiotensin II*, a powerful vasoconstrictor, that increases blood pressure and causes contraction of smooth muscle.
  - *erythropoietin* that stimulates erythropoiesis in bone marrow.
  - *1,25-dihydroxycholecalciferol*: 1-hydroxylation of vitamin D occurs in the kidney to form 1,25-dihydroxycholecalciferol that has a key role in calcium homeostasis.
  - *Prostaglandins* PGA_2, PGE_2, and $PGF_{2\alpha}$ produced in the renal medulla induce relaxation of smooth muscles, vasodilatation, decrease in renal blood pressure, increase in renal blood flow, and excretion of sodium.
  - *Bradykinin* and *kallidin*, peptide hormones synthesized by the kidney are potent vasodilators that influence reabsorption of sodium and water. Kinins also play a role in blood coagulation, increase renal blood flow, and exhibit antihypertensive effect. Bradykinin stimulates the synthesis of renal prostaglandins.
- Receptor for hormones such as parathyroid hormone, and ADH.
- Site of inactivation of hormones such as insulin, glucagon, etc.

❑ Briefly explain the formation of urine.

Urine formation is a composite of three functions:

1. *Glomerular filtration*:
   - Of the ~1 litre of blood that flows through the kidneys per minute, one-tenth is filtered by the glomerulus. The *glomerular filtration rate* (*GFR*) is 120 mL per minute.
   - Glomerular filtration depends on
     - renal blood flow
     - shape, charge, and size of molecules filtered: anionic derivatives, globular molecules, and molecules of molecular weight >68,000 are retained.
   - The glomerular filtrate is a protein- and cell-free filtrate of whole blood but contains essential metabolites such as glucose and amino acids as well as waste products, urea, uric acid, and creatinine.
2. *Tubular reabsorption*
   - This is an active process.
   - *Threshold substances* are substances that are completely reabsorbed when their plasma levels are normal and appear in the urine when the levels exceed the normal range.
   - *High threshold substances* are those that are reabsorbed efficiently (*e.g.*, glucose, amino acids).
   - *Low threshold substances* are those that are not reabsorbed or reabsorbed only slightly (*e.g.*, urea, uric acid).
   - Under normal circumstances, glucose is completely reabsorbed by the tubules.
   - Water reabsorption
     - more than 99% reabsorbed by counter current multiplication and counter current exchange.
     - reabsorption from the distal convoluted tubule (DCT) and collecting ducts is facilitated by the pituitary hormone ADH.
   - Sodium is reabsorbed by both symport and antiport systems. All the potassium that is filtered is reabsorbed in the proximal convoluted tubule (PCT).
   - Urate is reabsorbed in the PCT by both passive and active transport.
3. *Tubular secretion*
   - Substances actively secreted by the tubular cells include
     - endogenous substances (*e.g.*, hippuric acid, bile salts, urate),
     - exogenous substances (*e.g.*, p-aminohippurate, penicillin, diuretics, salicylates).
   - Creatinine is secreted when plasma levels increase above normal level.
   - Potassium is secreted in the DCT in exchange for Na^+ reabsorbed.
   - Aldosterone favors sodium reabsorption and excretion of potassium. In Addison's disease, sodium is excreted and potassium is retained; whereas in primary aldosteronism, there is excessive excretion of potassium.

❑ How is glucose reabsorbed?

- Normally, glucose is completely reabsorbed.
- The maximum rate at which glucose is reabsorbed, referred to as the *tubular maximum for glucose* (*TmG*), is 350 mg per minute.

- When blood glucose exceeds the *renal threshold* for glucose, which is 10 mmol/L (180 mg/dL), glucose spills over into the urine (*glycosuria*).
- Water is excreted as a solvent for glucose (*osmotic diuresis*).

❑ Explain the mechanism of water reabsorption by the renal tubules.

More than 99% of water entering the tubules is reabsorbed.

◻ *In the PCT*
- About 80% of water is reabsorbed.
- Water is reabsorbed by passive transport as a solvent for actively reabsorbed sodium chloride and glucose (*obligatory reabsorption*).
- Tubular filtrate is isoosmotic with plasma.

◻ *In the descending limb*
- The tubular filtrate is *hyperosmotic.*
- Water without solute is lost. This is termed *counter current multiplication.*
- *Counter current multiplication*
  a. functions in the absence of *ADH,*
  b. is an active process that creates high medullary osmolality in the renal medulla and produces a dilute urine.

◻ *In the ascending limb*
- Sodium chloride is lost without water.
- Tubular filtrate is *hypoosmotic.*

◻ *In the DCT and the collecting duct*
- Water without solute is reabsorbed along the osmotic gradient created by *counter current multiplication.* This is known as *counter current exchange.*
- *Counter current exchange*
  a. is facilitated by antidiuretic hormone (ADH) (*facultative reabsorption*).
  b. concentrates the urine.

Factors influencing ADH secretion include
- Osmotic pressure of the blood: Osmoreceptors in the hypothalamus that are sensitive to changes in the osmotic pressure of the blood influence the synthesis and secretion of ADH.
- Blood volume: Volume receptors in the carotid sinus and left atrium influence ADH secretion.
- An increase in osmotic pressure or a decrease in blood volume stimulates ADH secretion.

❑ How is sodium reabsorbed?

Sodium reabsorption in the PCT occurs by two types of *co-transport* systems.
- *Symport,* which involves co-transport of sodium with other substances such as glucose, amino acids, and phosphate, is catalyzed by Na^+K^+ *- ATPase* present in the basolateral plasma membrane of the epithelial cell. This creates a sodium gradient that accomplishes active transport of sodium out of the cell while potassium enters the cell.
- *Antiport* involves exchange of sodium for another cation such as in the Na^+ - H^+ *antiport system.*

❑ Enumerate some defects in renal tubular reabsorption.

- *Renal glycosuria* in which glucose appears in urine despite normal plasma glucose levels. It occurs during pregnancy, and in Fanconi syndrome.

- *Vitamin D-resistant rickets* in which phosphate reabsorption is defective.
- *De Toni Fanconi syndrome* associated with glycosuria, phosphaturia, and renal tubular acidosis.
- *Renal amino aciduria* as in cystinosis and Fanconi syndrome.

❑ What are the tests that are useful in assessing kidney function?

Kidney function tests are categorized into

- ***Glomerular function tests*** that include
  - *measurement of clearance* of substances filtered by the glomeruli but neither reabsorbed nor secreted by the tubules, such as inulin, urea, and creatinine.
  - *measurement* of plasma urea, creatinine, and β_2-microglobulin.
- ***Tubular function tests*** that include
  - *measurement of tubular secretory mass*
  - *concentration test*
  - *dilution test*
  - *acid excretion test.*

Define clearance

Clearance of a substance is the volume of plasma completely cleared of the substance in a minute. It is expressed as follows:

$$\text{Clearance} = \frac{\text{mg of substance excreted per minute}}{\text{mg of substance per mL of plasma}} = \frac{UV}{P},$$

where U and P are the concentration of the substance in urine and plasma, respectively, and V is the volume of urine excreted per minute.

❑ What is GFR? How is clearance used as a measure of GFR?

- GFR is the volume (in ml) of plasma filtered at the glomeruli every minute. It is 120 ml/minute.
- The clearance of substances that are filtered at the glomeruli but neither reabsorbed nor secreted by the tubules is used as a measure of GFR.
- These include exogenous substances (*e.g.*, inulin, EDTA) and endogenous substances (*e.g.*, urea creatinine).
- *Inulin clearance*, although close to the GFR, is not preferred because it requires inulin infusion into circulation.
- *^{15}Cr-labeled EDTA clearance*: The GFR is calculated from the rate of fall of plasma radioactivity as the isotope is cleared.
- *Urea clearance*, although useful in assessing renal failure, is not an approximate measure of GFR because
  - 40% of urea is passively reabsorbed by the tubules.
  - it is influenced by urine output, protein diet, protein catabolism, and protein loss by hemorrhage.
- *Creatinine clearance* is 120 mL/minute, which is close to the GFR. However, measurement of creatinine clearance is not reliable because
  - creatinine is actively secreted by the tubules.
  - the test is incapable of differentiating prerenal, renal, and postrenal causes of renal failure.

❑ Explain how plasma urea, creatinine, and β_2-microglobulin levels are useful in assessing renal function.

▫ *Urea*

- The end product of protein catabolism.
- Is synthesized in the liver.
- Is filtered by the glomerulus.
- Is passively reabsorbed by the tubules at low rates of urine flow.
- Levels in plasma useful as an index of glomerular function.
- Disadvantage: Extrarenal factors such as dietary protein, protein catabolism, and protein loss due to hemorrhage, dehydration, and liver disease can affect plasma urea concentration.

▫ *Creatinine*

- Is derived from breakdown of tissue creatine.
- Plasma creatinine
  a. is related to muscle mass.
  b. is not significantly influenced by high protein diet and hypercatabolic state.
  c. increase indicates impairment of renal function.

▫ *β_2-microglobulin*

- Small peptide found on the surface of mast cells.
- Forms part of class I antigens of the major histocompatibility complex.
- Plasma concentration
  a. is a reliable index of GFR because it is completely filtered by the glomeruli and reabsorbed and catabolized by the cells of the PCT.
  b. is not influenced by diet or muscle mass.
  c. is increased in inflammatory diseases and malignancies.

❑ Give an account of tubular function tests.

▫ *Tubular secretory mass*

The mass of functioning tubular tissue can be assessed by measuring the tubular maximum for PAH (TmPAH).

$$Tm = UV\text{-}P \times GFR$$

For PAH, the Tm is 76 mg/minute.

▫ *Concentration test*

- The concentrating ability of the kidney can be assessed by measuring the specific gravity of urine after overnight water deprivation.
- This depends on the ability of the posterior pituitary to elaborate ADH in response to water deprivation and on the renal tubular response to ADH.
- The test is useful in the differential diagnosis of psychogenic polydipsia and diabetes insipidus.
- One day before the test, fluid is deprived after breakfast, but food is allowed.
- In addition to the early morning specimen, urine samples are collected at 60 and 120 minutes.
- In kidney disease, the maximum specific gravity is <1.022 in any of the three samples. However, the ADH test is preferred to test the concentrating ability.

▫ *Dilution (water load) test*

- In renal disease, the kidneys lose the ability to dilute urine.
- When the specific gravity does not fall to 1.003 or below after a water load, it indicates impairment of renal function. This test is however not routinely performed.

- ***Acid excretion test***
  - This test measures acid excretion by the kidneys following administration of 0.1 g of ammonium chloride/kg body weight after an overnight fast.
  - Urine samples are collected before and at 2, 4, and 6h following ammonium chloride administration.
  - In renal failure, there is impairment in acid excretion (normal excretion: 30–90 mEq ammonia/minute with a urinary pH of 5.3).

❑ Elucidate the characteristics of urine.

- ***Volume***
  - Normal urine output: 600 – 2500 mL per day, with an average of about 1200 mL per day.
  - It is influenced by fluid intake, external temperature, diet, and beverages. Coffee, tea, and alcohol have a diuretic effect.
  - *Polyuria* (increased excretion of urine) is seen in
    a. chronic renal insufficiency due to tubular damage or osmotic diuresis
    b. diabetes insipidus
    c. diabetes mellitus due to osmotic diuresis.
  - *Nocturia* (increased urine excretion at night) occurs during early stages of renal disease followed by *oliguria* (decrease in urine volume).
  - *Anuria* (complete cessation of urination) is seen in the terminal stages of kidney disease.
- ***Odor***
  - Normal odor of freshly voided urine is aromatic.
  - Odor is altered by
    a. diet
    b. ketoacidosis in which there is acetone odor
    c. infections:
      - urea-splitting organisms impart an ammoniacal odor.
      - Gram-negative organisms give an unpleasant odor.
  - maple syrup urine disease imparts a characteristic odor of *maple syrup* or *burnt sugar* to the urine.
- ***Color***
  - Normal amber color is mainly due to the pigment *urochrome*. Urobilin, uroerythrin, and hematoporphyrin also contribute to the color to a minor extent.
  - Urine color is altered in the following conditions:
    a. greenish yellow due to presence of bile pigments in jaundice.
    b. red when blood or hemoglobin is present.
    c. dark brown color due to presence of methemoglobin.
    d. port wine color due to porphyrins.
    e. darkens on standing due to presence of homogentisic acid in alkaptonuria.
- ***pH***
  - Acidic with an average pH of 6.0 (range 4.6 – 8.0).
  - Titrable acidity is 250 – 700 mL of 0.1N acid, 25 mEq H^+.
  - Acidity is increased by
    a. high protein diet due to increase in sulphate and phosphate
    b. fever
    c. acidosis.

- Urine becomes alkaline
  a. on standing due to ammonia formation
  b. in alkalosis
  c. after meals (*postprandial alkaline tide*).
- ***Specific gravity***
  - Normal range: 1.003–1.030.
  - is low in polyuria due to diabetes insipidus.
  - is high in polyuria due to diabetes mellitus.

❑ What are the normal constituents of urine?

Inorganic constituents

1. **Ions** such as chloride, sodium, potassium, and ammonia.
2. **Phosphates** that include alkaline phosphates (sodium and potassium), and alkaline earth phosphates (calcium and magnesium). The excretion of phosphate is increased in bone diseases and hyperparathyroidism and decreased in pregnancy, renal disease, diarrhea, and hypoparathyroidism.
3. **Sulphates** including *inorganic* (completely oxidized) *sulphur*, *ethereal* (*conjugated*) *sulphur*, and *neutral* (incompletely oxidized) *sulphur*.

Organic constituents

1. ***Urea***
   - is the end product of protein metabolism.
   - normal daily excretion: 30 g
   - excretion is increased by high protein diet, fever, diabetes mellitus, and adrenal cortical hyperactivity.
   - excretion is decreased in terminal stages of liver diseases and acidosis.
2. ***Uric acid***
   - is the end product of purine metabolism.
   - normal daily excretion: 0.7 g
   - excretion is increased in leukemia, severe liver disease, and gout.
3. ***Creatinine***
   - is the anhydride of creatine and a waste product.
   - normal daily excretion: 1–2 g.
   - excretion is related to functioning muscle mass and is more in men than in women.
4. ***Creatine***
   - normally very little is excreted.
   - excretion occurs in children and is greater in women than in men.
   - excretion is increased in pregnancy, starvation, diabetes mellitus, fever, hyperthyroidism, and muscular dystrophy.
5. ***Amino acids***
   - normally very little is excreted (150–200 mg/day).
   - increased excretion (aminoaciduria) is abnormal.
6. ***Hippuric acid***
   - is the product of the conjugation of benzoic acid (present in fruits) with glycine in the liver.
   - excretion is used as a liver function test.
   - normal daily excretion: 0.7 g.
7. ***Other substances*** such as urobilinogen, water-soluble vitamins, hormones, and enzymes.

❑ Enumerate the abnormal constituents of urine.

The abnormal constituents of urine include the following:

1. ***Proteins***
   - Normal daily excretion: <150 mg.
   - Increased excretion of protein (*proteinuria*) occurs due to
     - increased plasma concentrations
     - increased glomerular permeability
     - reduced tubular reabsorption
     - increased tubular secretion.
   - The proteins excreted in urine include plasma proteins, hemoglobin, myoglobin, *Bence Jones protein*, and mucus.
2. ***Amino acids***
3. ***Reducing sugars*** such as glucose, fructose, lactose, galactose, pentoses, and homogentisic acid.
4. ***Ketone bodies***
   - Normal daily excretion: 3 – 15 mg
   - Increased excretion (*ketonuria*) is seen in starvation and diabetes mellitus.
5. ***Bile salts and bile pigments*** seen in jaundice.
6. ***Porphyrins***
7. ***Blood*:** Presence of blood in urine (*hematuria*) is seen in nephritis and in urinary tract lesions.

❑ Enumerate the causes of albuminuria.

Albuminuria may be

- *Physiological albuminuria,* which occurs in
  - adolescents
  - last trimester of pregnancy
  - severe stress
  - postural (in the upright position) due to increased hydrostatic pressure in the renal veins.
- *Pathological albuminuria,* which is subclassified into
  - prerenal albuminuria resulting from impaired renal circulation as in
    i. dehydration
    ii. congestive heart disease
    iii. intraabdominal pressure due to tumors, ascites, etc.
  - renal albuminuria seen in renal diseases such as nephritis, nephrotic syndrome, etc.
  - postrenal albuminuria that occurs due to urinary tract lesions.

❑ When do hemoglobin and myoglobin appear in urine?

- **Hemoglobinuria** occurs due to intravascular hemolysis as in
  - sickle cell anemia
  - thalassemia
  - hemolytic anemia
  - paroxysmal (induced by cold and stress)
  - exposure to chemicals (sulfonamides, arsenicals).
- **Myoglobinuria** occurs due to breakdown of muscle as in
  - crush injuries

- burns
- electric shock
- malignant hyperthermia
- following infarction due to release from myocardium
- paroxysmal after exercise.

❑ What are Bence Jones protein?

Bence Jones protein are γ-globulins that precipitate between 40 and 60°C, redissolve at 100°C, and reprecipitate on subsequent cooling. They are seen in the urine of multiple myeloma patients.

❑ Give an account of the types of aminoaciduria.

Excessive urinary excretion of amino acids is termed *aminoaciduria.*

There are two types of aminoaciduria:

- ***Overflow aminoaciduria*** in which aminoaciduria occurs due to high plasma concentrations that exceed the tubular reabsorbtion capacity. This is seen in
  - inborn errors of metabolism such as phenylketonuria, alkaptonuria, etc.
  - secondary to cirrhosis, chronic liver diseases, acute hepatic necrosis, and wasting diseases.
- ***Renal aminoaciduria*** results from decreased renal tubular reabsorption of amino acids. Plasma amino acid levels are however normal. This is seen in
  - inborn errors in amino acid transport system (*e.g.*, cystinuria, Hartnup disease).
  - generalized tubular damage which occurs in
    i. Fanconi syndrome characterized by glycoaminophosphaturia.
    ii. Wilson's disease, in which copper deposition causes renal tubular damage.
    iii. cystinosis, in which cystine crystals are deposited in the kidney.
    iv. heavy metal poisoning (lead, cadmium, mercury).

❑ What are the reducing substances found in urine?

The reducing substances found as abnormal constituents of urine include

- ***Glucose,*** which appears in urine with or without hyperglycemia. Glycosuria without hyperglycenia is termed renal glycosuria.
- ***Lactose*** found in the urine of women during pregnancy, lactation, and weaning.
- ***Galactose*** excreted in urine in galactosemia, in which the conversion of galactose to glucose is impaired.
- ***Fructose*** appears in urine
  - after ingestion of fructose-rich foods such as fruits, honey, etc.
  - in liver disease.
  - in essential fructosuria due to fructokinase deficiency.
  - in hereditary fructose intolerance, characterized by *fructose-1-phosphate aldolase* deficiency.
- ***Pentoses*** appear in urine
  - after ingestion of large amounts of cherries, prunes, and plums.
  - in essential pentosuria.
- ***Homogentisic acid*** is excreted in alkaptonuria, an inborn error of tyrosine metabolism.

❑ What are the types of glycosuria?

- Glucose appears in urine when plasma glucose levels exceed the renal threshold (*glycosuria*).
- Glycosuria without hyperglycemia is due to defective renal tubular reabsorption and is called *renal glycosuria.*
- Glycosuria with hyperglycemia occurs in the following conditions:
  - diabetes mellitus
  - endocrinal hyperactivity (hyperthyroidism, Cushing's syndrome, etc.)
  - pancreatic diseases
  - infections
  - anesthesia
  - asphyxia
  - stress due to increased epinephrine secretion
  - surgical removal of the pancreas
  - alloxan-induced diabetes
  - administration of phlorizin, an inhibitor of the renal tubular reabsorption of glucose.
- *Renal glycosuria* is seen in
  - late pregnancy
  - Wilson's disease
  - Fanconi syndrome.

❑ What are the types of glomerulonephritis?

Glomerulonephritis results from an inflammatory response to an immunological reaction in the basement membrane and may be acute or chronic.

- ***Acute glomerulonephritis***
  - is of sudden onset.
  - occurs after bacterial (hemolytic streptococcal) or viral infections of the throat.
  - is characterized by
    - i. impaired glomerular function
    - ii. oliguria
    - iii. hematuria that imparts a smoky appearance to urine
    - iv. proteinuria
    - v. increase in plasma urea and creatinine (*azotaemia*)
    - vi. hypertension
    - vii. edema due to sodium and water retention, and ECF dilution.
  - changes are reversible in 80% of cases, some cases develop anuria and acute renal failure while others progress to chronic glomerulonephritis and terminal renal failure.
- ***Chronic glomerulonephritis***
  1. Is characterized by oliguria, hematuria, and albuminuria.
  2. With advancing disease, renal failure occurs with atrophy of nephrons, polyuria, decreased albumin excretion, increase in blood urea, and tendency towards acidosis.

❑ Discuss the causes, biochemical findings, and management of acute renal failure (ARF).

Acute renal failure (ARF) is characterized by a rapid deterioration in renal function with retention of salt, water, urea, creatinine, and hydrogen ions with hypertension and edema. ARF can induce permanent renal damage, leading to chronic renal failure.

ARF may be classified into the following types:

- ***Prerenal ARF*** that occurs due to:
  - renal hypoperfusion as in hemorrhage, burns, congestive cardiac failure.
  - use of non-steroid anti-inflammatory drugs (NSAIDs).

  Inadequate renal perfusion leads to the following sequence of events:

  Decrease in GFR → hypovolemia → stimulation of the renin angiotensin–aldosterone system → ADH secretion → excretion of concentrated urine → high plasma levels of sodium and urea.

- ***Intrinsic renal failure*** that results from
  - renal ischemia.
  - nephrotoxic drugs such as cephalosporins.

  Intrinsic renal failure leads to renal tubular necrosis, loss of the ability to concentrate urine, and proteinuria.

- ***Postrenal ARF*** that occurs as a result of urinary tract obstruction due to prostatic enlargement, calculi, or neoplasms.

Biochemical findings

- Increase in blood urea nitrogen (BUN) and serum creatinine.
- Hyperkalemia due to decreased excretion, movement from the cell to the ECF, and buffering of hydrogen ions.
- Hyperphosphatemia due to phosphate retention and leakage from the cell into the interstitial fluid. This can inhibit 1α-hydroxylation of 25-hydroxycholecalciferol to calcitriol.
- Hypermagnesemia due to decreased urinary excretion.
- Hyponatremia due to increased generation of water from oxidation, inappropriate fluid infusion, or decreased excretion.
- Low plasma calcitriol that causes resistance to the action of parathyroid hormone, leading to hypocalcemia.
- Low pH, decreased urinary excretion of hydrogen ions, and anion gap metabolic acidosis.
- Low plasma bicarbonate due to buffering.

Management

- *Prerenal ARF*: Expansion of ECF volume and improvement of renal perfusion.
- *Intrinsic renal failure*:
  a. renal vasodilatation by diuretics
  b. prevention of progression to acute tubular necrosis by:
     - adequate carbohydrate intake and dietary control of sodium, water, and protein.
     - prevention of hyperkalemia, acidosis, and infection.
  c. dialysis when plasma urea, creatinine, or potassium levels are high, in fluid overload or severe acidosis.

Postrenal ARF: Treatment of urinary obstruction and drainage of urine.

❑ Discuss the metabolic abnormalities and management of chronic renal failure (CRF).

Chronic renal failure (CRF) is the end result of chronic conditions such as diabetic nephropathy, hypertension, glomerulonephritis, and polycystic kidney disease. Uremia occurs due to impairment of excretory, metabolic, and endocrine functions of the kidney.

Metabolic abnormalities

- Protein malnutrition, inadequate energy intake.
- Hypertriglyceridemia.
- Inability to concentrate urine, leading to polyuria, nocturia, and hypovolemia.
- Inability to dilute urine at a later stage.
- Sodium retention with hypertension and edema.
- Hyperchloremic acidosis due to impaired acid excretion, bicarbonate reabsorption, and chloride retention.
- Hypocalcemia with hyperphosphatemia leading to secondary hyperparathyroidism, bone resorption, metastatic calcification in blood vessels, and osteosclerosis.
- Hyperkalemia occurs at a later stage due to sudden impairment of renal function or to inappropriate use of potassium-sparing diuretics.
- Hypermagnesemia.
- Pulmonary edema as a result of increased capillary permeability.
- Decreased erythropoiesis with normochromic, normocytic anemia, and bleeding tendency.
- Reduced synthesis of testosterone and estrogen, thyroid dysfunction, hyperinsulinemia, and glucose intolerance.

Management

- Symptomatic treatment followed by maintenance dialysis or renal transplantation. Administration of diuretics to promote sodium excretion.
- Oral bicarbonate to control acidosis.
- Oral ion-exchange resins to treat hyperkalemia.
- Oral administration of aluminium or magnesium salts to treat hyperphosphatemia. Restriction of dietary protein.

❑ Outline the causes, consequences and management of nephrotic syndrome.

Nephrotic syndrome is characterized by the triad of albuminuria, hypoalbuminemia, and edema.

◘ *Causes*

- *Most common causes:* Membranous glomerulonephritis and minimal change glomerulonephritis in which proteinuria is highly selective and high molecular weight proteins are retained. The condition responds to steroids.
- *Other causes:* Other forms of glomerulonephritis, focal glomerulosclerosis, diabetic nephropathy, malignancy, infections such as malaria and hepatitis B, and drugs such as penicillamine and NSAIDs.

◘ *Consequences*

- Increase in glomerular permeability causes
  i. albuminuria with consequent hypoalbuminemia, movement of water from plasma into extravascular space leading to hypovolemia, stimulation of renin–angiotensin–aldosterone system, reabsorption of sodium and water, and edema.
  ii. loss of immunoglobulins leading to susceptibility to infection.
  iii. serum electrophoretic pattern shows a characteristic decrease in albumin with increase in α_2-globulin.
  iv. loss of antithrombin III, an inhibitor of clotting factors, that leads to a *hypercoagulable state* and fetal thrombosis.
  v. low plasma levels of transferrin and hormone-binding proteins.

- Increased apolipoprotein synthesis leading to hypercholesterolemia, hypertriglyceridemia, and accelerated atherosclerosis.
- Hypocalcemia due to decreased protein binding or increased renal excretion of vitamin D metabolites.

▫ ***Management***
- Treatment of edema by high protein, low salt diet and diuretics such as spironolactone.
- Antibiotics to prevent infection.
- Anticoagulants to prevent thrombosis.

❑ How are renal stones formed?

Renal stones (*nephrolithiasis* or *calculi*) form in the urinary tract due to an imbalance between conservation of water and excretion of insoluble substances by the kidney. Crystal formation occurs when the urine is supersaturated with substances with low solubility.

❑ Enumerate the factors that predispose to renal stone formation.

Predisposing factors for stone formation include
- high concentrations of one or more normal constituents of the glomerular filtrate.
- alterations in urinary pH due to bacterial infection.
- obstruction of urinary flow.
- absence of inhibitors of crystal growth.

❑ Give an account of the different types of renal stones

The substances present in stones include calcium oxalate and calcium phosphate (80% of the total), struvite or $MgNH_4PO_4$ (15%), uric acid (5%), and cystine (1%).

Calcium stones

▫ *Calcium oxalate stones*
- are very hard, fusiform, and are found in the ureters.
- are caused by
  a. Hypercalciuria with or without hypercalcemia.
  b. Hyperuricosuria due to stimulation of calcium oxalate crystals by uric acid.
  c. Hyperoxaluria, which favors formation of calcium oxalate even when calcium excretion is normal.

▫ *Calcium phosphate stones*
- are soft, white '*staghorn*' calculi in the renal pelvis that precipitate under alkaline conditions.
- are seen during chronic renal infection with *Proteus vulgaris*, a urease producing organism.

▫ *Treatment* of calcium stones involves reducing urinary calcium by
- decreasing calcium and oxalate intake.
- decreasing the intestinal absorption of calcium by administration of oral phosphate.
- high fluid intake.
- management of hypercalcemia or urinary tract infection.
- administration of thiazide diuretics which reduce urinary calcium excretion.

Struvite stones

- are common in women and result from urinary tract infection with *Proteus vulgaris*.
- enlarge, assume a "staghorn" appearance, fill the renal pelvis, thereby causing obstruction.
- are radio opaque.
- can be managed by antibiotics that suppress infection and prevent recurrence.

Uric acid stones

- are small, yellowish-brown, radiotransluscent, and more common in men.
- occur in acidic urine.
- are seen in gout, diarrhea, inflammatory bowel disease, and myeloproliferative disorders.
- treatment involves
  - increasing fluid intake
  - alkalinization of urine
  - administration of allopurinol
  - restriction of dietary purine intake.

Cystine stones

- are rare.
- are found in homocystinuria.
- are more soluble in alkaline urine.
- treatment involves
  - increasing urinary volume
  - alkalinization of urine
  - administration of penicillamine that binds cystine.

Chapter 25

Gastric Function

❑ Elucidate the composition and functions of gastric juice.

Gastric juice contains:

1. *Hydrochloric acid (HCl)* secreted by parietal (oxyntic) cells. The functions of gastric HCl include:
   - bactericidal action
   - activation of pepsinogen to pepsin
   - denaturation of proteins.
2. *Intrinsic factor*, also secreted by parietal cells. Intrinsic factor is required for vitamin B_{12} absorption in the ileum.
3. *Pepsinogen*, secreted by chief (peptic) cells. Pepsinogen is activated by HCl and by auto-catalysis by pepsin. Pepsin is an endopeptidase that hydrolyses denatured proteins to peptides. The enzyme specifically cleaves peptide bonds of aromatic amino acids or dicarboxylic amino acids. Pepsinogens in gastric juice are separable into 7 fractions by immunochemical techniques.
4. *Mucus*, secreted by the gastric mucosal epithelium and gastric glands. Gastric mucus is a glycoprotein with four subunits linked by disulphide bonds and the polysaccharide blood group substances A, B, or H. It functions as a lubricant and protects the mucosa from injury.
5. *Gastrin*, a hormone secreted by the gastric antral mucosa, is a potent stimulant of gastric acid secretion.
6. *Rennin*, which converts caseinogen to casein.
7. *Lipase*, which together with lingual lipase initiates lipid digestion by hydrolysing triacylglycerols to free fatty acids and 1,2-diacylglycerol.
8. *Potassium chloride* (KCl), 0.007M.

❑ Explain the mechanism of gastric acid secretion.

- Gastric HCl is formed from secretion of H^+ and chloride (Cl^-) into the lumen by active transport (Fig. 25.1).
- While chloride is derived from plasma, H^+ is derived from CO_2 that arises from blood and cellular metabolism. CO_2 combines with H_2O to form carbonic acid (H_2CO_3) by the action of *carbonic anhydrase* that subsequently dissociates into HCO_3^- and H^+.
- The secretion of H^+ is catalyzed by H^+-K^+-ATPase located in the plasma membrane of the parietal cell by the movement of K^+ into the cell in exchange for H^+, in a process requiring ATP hydrolysis.
- Bicarbonate moves into the plasma in exchange for chloride to maintain electrochemical neutrality and water moves from plasma into the gastric lumen to maintain osmolality.

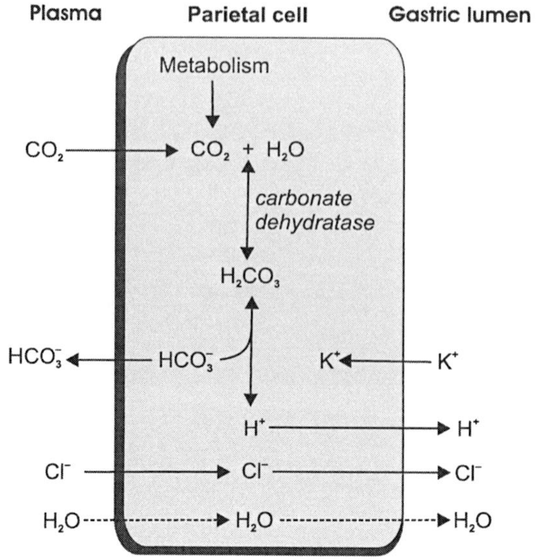

Fig. 25.1 The mechanism of gastric acid secretion

❑ How is the secretion of gastric HCl regulated?

The secretion of gastric HCl is stimulated by the following:

- Vagus via acetylcholine, which interacts with cholinergic receptors on the surface of the parietal cell causing Ca^{2+} release.
- Gastrin by interacting with receptors on parietal cell, causing changes in Ca^{2+} flux. Synthesis of gastrin is stimulated by peptides, amino acids, vagus nerve and inhibited by low pH, secretin, and somatostatin.
- Histamine by combining with the *H2-receptor* stimulates adenylate cyclase with formation of cAMP, which mediates H^+ secretion. H2-receptor antagonists such as cimetidine and ranitidine inhibit gastric acid secretion.

The secretion of gastric HCl is inhibited by the gastrointestinal hormones secretin, pancreozymin, gastric inhibitory peptide, vasoactive intestinal peptide (VIP), somatostatin, prostaglandins, neurotensin, peptide YY and urogastrone.

❑ Define achlorhydria, hyperchlorhydria hypochlorhydria and achylia gastrica.

1. *Achlorhydria* refers to the absence of HCl. Physiological achlorhydria is seen in ageing. Pathological achlorhydria is seen in tuberculosis, chronic gastritis, terminal stages of malignancy, and in gastric carcinoma.
2. *Hyperchlorhydria* refers to an increase in free acidity. The combined acidity may be normal. Hyperchlorhydria is common in duodenal ulcer and Zollinger–Ellison syndrome.
3. *Hypochlorhydria* refers to a decrease in the concentration of free acid.
4. *Achylia gastrica*, refers to the absence of both acid and pepsin, and is seen in pernicious anemia.

❑ Discuss the biochemical analysis of gastric juice.

After an overnight fast, the resting contents are collected and examined for the following:

1. An increase in volume of gastric jucice (normal 20–50 mL), which results from
   - hypersecretion of gastric juice as in duodenal ulcer.

- retention of gastric contents or delayed emptying of stomach associated with achlorhydria and presence of organic acids. This is seen in gastric carcinoma.
- regurgitation of bile from the duodenum.

2. Presence of starch indicates delayed emptying of stomach.
3. Presence of bile due to regurgitation from the duodenum.
4. Presence of blood. While fresh red blood can be due to trauma from insertion of the nasogastric tube, brown or reddish brown blood is pathological and is due to the formation of *acid hematin* by the action of HCl on red blood cells. Blood is found in resting contents in gastric ulcer and gastric carcinoma. In gastric carcinoma, the blood gets mixed with food due to delayed emptying of the stomach to give a *coffee grounds* appearance.
5. pH (normal 1–2), and concentrations of free and combined acid. An increase in the concentration of combined acid occurs due to the presence of organic acids such as lactic acid, and butyric acid formed by microbial action on food remaining in the stomach. Achlorhydria (absence of HCl) favors microbial fermentation. In gastric cancer, large amounts of lactic acid are found.

Biochemical analysis of resting contents is no longer carried out routinely and tests of gastric function following stimulation are more useful and reliable. These include fractional test meal as well as histamine, alcohol, insulin, and pentagastrin tests.

❑ Briefly explain the tests currently used to assess gastric function.

Endoscopy and contrast radiography are of diagnostic value in gastric disorders and biochemical tests are not of much significance except to evaluate gastric acid secretion.

Gastric function tests measure acid secretion following stimulation. Earlier tests used food as a stimulus for acid secretion which has been replaced by potent stimulants such as histamine, insulin, alcohol, and pentagastrin. The pentagastrin test and plasma gastrin levels are however more useful than the other tests of gastric function.

Pentagastrin test

Pentagastrin is the stimulant of choice because it is free from side effects. The test is used in the diagnosis of Zollinger–Ellison syndrome and pernicious anemia, and to evaluate acid secretion following gastric surgery.

Method: The stomach contents are aspirated after an overnight fast followed by two 15-minute specimens for basal secretion. After subcutaneous injection of pentagastrin (6 μg/kg body weight), specimens are collected every 15 minutes for one hour.

Findings: The normal basal secretion is 1.0–2.5 mmol/h and maximal secretion is 20–40 mmol/h. In Zollinger–Ellison syndrome, the basal secretion is increased, whereas in duodenal ulcer, the maximal secretion is high.

Plasma gastrin

Measurement of plasma gastrin by radioimmunoassay is also useful in evaluating gastric function. The normal plasma gastrin levels range from 10–50 pmol/L. Hypergastrinemia is seen in the following conditions:

- Zollinger–Ellison syndrome (>100 pmol/L)
- pernicious anemia
- chronic gastritis
- gastric carcinoma
- in patients receiving H2-receptor antagonists such as ranitidine and cimetidine.

Chapter 26

Pancreatic and Intestinal Function

❑ Elucidate the composition and functions of pancreatic exocrine secretion.

Pancreatic exocrine secretion is alkaline and consists of the following:

- ***The aqueous component*** containing:
  1. bicarbonate which neutralizes gastric HCl.
  2. cations: Na^+ (148 meq/L), K^+ (4 meq/L), Ca^{2+} (6 meq/L).
  3. anions: Cl^- and HCO_3^- (80 meq/L), HPO_4^{2-} (1 meq/L).
- ***The enzymic component*** containing proenzymes and enzymes required for the digestion of
  1. proteins (trypsinogen, chymotrypsinogen, procarboxypeptidases, and proelastase).
  2. carbohydrates (α-amylase).
  3. lipids (prophospholipase, lipase, colipase).
  4. nucleic acids (ribonuclease and deoxyribonuclease).

❑ Explain the mechanism of pancreatic bicarbonate secretion.

Sodium bicarbonate ($NaHCO_3$) secretion by ductules lining the pancreatic acinar cells involves the following mechanism (Fig. 26.1):

1. Diffusion of carbon dioxide from the blood into the ductule cell.
2. Combination of carbon dioxide with water to form carbonic acid by the action of *carbonate dehydratase*.
3. Dissociation of carbonic acid into HCO_3^- and H^+ ions.
4. Active transport of HCO_3^- through the lumenal border of the cell into the lumen of the duct.
5. Diffusion (or active transport) of Na^+ in exchange for H^+ ions.
6. Osmosis of water into the pancreatic duct due to osmotic gradient created by the movement of Na^+ and HCO_3^- from the blood to the lumen.

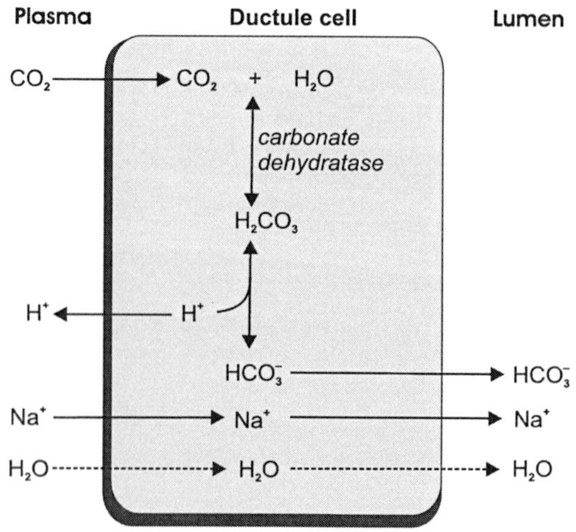

Fig. 26.1 Mechanism of pancreatic bicarbonate secretion

❑ How is pancreatic secretion regulated?

Pancreatic secretion is regulated by:

- *Vagus*, which releases acetylcholine that stimulates enzyme secretion by mobilizing Ca^{2+} stores.
- *Hormones*
  - *Secretin*, released into circulation by the presence of acid in the duodenum stimulates aqueous component of pancreatic juice by a cyclic AMP mediated process.
  - *Cholecystokinin (CCK)*, released due to the presence of lipids and polypeptides in the intestine stimulates pancreatic enzyme secretion via mobilization of Ca^{2+} stores.

❑ Briefly explain the tests used currently to assess pancreatic function.

Pancreatic function tests may be classified into direct and indirect tests.

- Direct tests involve quantitative estimation of fluid, bicarbonate, and enzymes. These include
  1. ***Secretin–Cholecystokinin test:*** This involves the following steps:
     a. Aspiration of gastric and duodenal juices after an overnight fast. The resting contents are discarded.
     b. Intravenous administration of secretin after an overnight fast.
     c. Collection of three 10-minute duodenal samples.
     d. Intravenous administration of cholecystokinin and collection of duodenal contents.
     e. Determination of the volume of duodenal secretions, the bicarbonate content, and the activities of amylase and trypsin.

     Normal findings:
     - Volume of aspirate: 150 mL/h.
     - Peak concentration of bicarbonate: 90 mmol/L.
     - Peak activities of amylase and trypsin: 270 IU/mL and 30 IU/mL, respectively. Both the bicarbonate content and the enzyme activities are decreased in chronic pancreatic insufficiency.
  2. ***Lundh test:*** This involves
     a. Administration of a test meal of corn oil, skimmed milk powder, and dextrose to stimulate pancreatic secretion.
     b. Aspiration of duodenal secretion for 2 hours.
     c. Assay of trypsin activity: Enzyme activity is low in chronic pancreatic insufficiency.
- Indirect tests that measure plasma enzymes such as amylase (limited value) and tubeless tests such as fluorescein dilaurate and the PABA tests.
  1. ***Fluorescein dilaurate test:*** This involves
     a. Oral administration of fluorescein dilaurate (0.5 mmol). This is hydrolysed in the gut and the fluorescein released is conjugated in the liver and excreted in the urine as glucuronide.
     b. Administration of an equivalent amount of free fluorescein the following day as control.
     c. Measurement of fluorescein excreted in the urine sample collected for 10 hours.

     Results: The test/control index is greater than 30% if pancreatic function is normal and the index is less than 20% in pancreatic insufficiency.
  2. ***PABA test:*** This is a sensitive and specific test that involves
     a. Oral ingestion of a synthetic peptide labeled with ^{14}C-p-aminobenzoic acid (^{14}C-PABA).
     b. Release of PABA from the peptide by chymotrypsin, followed by absorption in the intestine and excretion in the urine.
     c. Measurement of ^{14}C-labeled PABA excreted in the urine.

     PABA excretion is significantly reduced in chronic pancreatitis.

❑ What is the composition of intestinal secretion?

The intestinal secretion contains enzymes involved in the digestion of:

- protein (*aminopeptidase, exopeptidase, dipeptidase,* and *enteropeptidase*)
- carbohydrates (*disaccharidases, maltase, sucrase, isomaltase, lactase,* and *trehalase*)
- lipids (*lipase, phosphatidylcholinase, phospholipase*)
- nucleic acids (*nucleotidase, nucleosidase,* and *nucleases*)
- *phosphatase.*

❑ What do tests of intestinal function evaluate?

Tests of intestinal function evaluate the absorption of carbohydrates, amino acids, and lipids as well as overgrowth of bacteria and terminal ileal function.

❑ Elucidate the tests to assess intestinal absorption of carbohydrates.

- *Xylose absorption test*: Xylose, a pentose, is absorbed without digestion, is not metabolized, and is excreted as such in the urine. Measurement of urinary xylose is therefore useful in assessing intestinal absorption. The test involves
  1. Oral administration of xylose (5 g) in water after an overnight fast and emptying the bladder.
  2. Ingestion of about 500 mL of water over the next 2 hours.
  3. Determination of plasma xylose after 1 hour. Normal value: >1.3 mmol/L.
  4. Measurement of xylose in a 5-hour urine sample. Normal excretion: >7.0 mmol/5 hours. Xylose excretion is decreased in intestinal disease, bacterial overgrowth of the small intestine, renal failure, and delayed gastric emptying.

  Alternatively, 2.5 g of 3-O-methyl-D-glucose can be administered simultaneously with xylose and the plasma concentrations of the two sugars measured. The normal molar [xylose]/[3-O-methyl-D-glucose] ratio of 1:3 is lowered in mucosal disease.
- *Disaccharidase deficiency* is generally due to lactase deficiency that may be congenital or acquired. Hydrogen, a by-product of the fermentation of lactose by colonic bacteria, is measured in expired air.

❑ What are the forms of fecal fat ?

Under normal circumstances, most of the dietary fat is absorbed in the small intestine and very little is excreted in the feces. Three forms of fat are present in the feces:

1. Neutral fat (unsplit fat), which is mainly present in food in the form of triglycerides of fatty acids.
2. Free fatty acids (split fat) that arise from breakdown of triglycerides by intestinal lipases.
3. Combined fatty acids formed from fatty acids in the intestine are alkali and alkaline earth salts of fatty acids (soaps).

The normal fat content of feces is approximately 25% of dry weight, and the ratio of split to unsplit fat is 3 : 1.

❑ What is steatorrhea?

Steatorrhea refers to an increase in the amount of fat in the feces. This is seen when there is malabsorption of fat due to pancreatic disease, intestinal disease, or deficiency of bile salts. The stools are foul smelling, bulky, pale, and greasy.

❑ What are the tests for assessing fecal fat?

Fecal fat test. The test involves oral administration of a colored non-absorbable marker such as carmine and collection of feces for three consecutive days after the marker appears in the feces. Feces collection is terminated after a second marker administered 72 or 120 hours following administration of the first marker appears in the stool. The patient is required to consume 90 – 100 g fat/day for two days before and during the collection period. In addition to being unpleasant, the test is also of limited diagnostic value.

^{14}C-Triolein breath test. This test is an alternative for the fecal fat test. The test is based on oral administration of ^{14}C-labeled triglyceride (as 10 μCi ^{14}C-triolein in 60 g fat meal) and quantitation of ^{14}C labeled carbon dioxide in the breath. After an overnight fast, a basal sample of expired carbon dioxide (1 mmol) is collected. Although this is a sensitive test for fat malabsorption, it is not useful in the differential diagnosis of malabsorption.

❑ What are the tests to monitor bacterial overgrowth in the intestine?

Bacterial overgrowth in the small intestine occurs due to stasis of gut contents after surgery, jejunal diverticulosis, or hypomotility. Bacteria can deconjugate bile salts, thereby blocking the formation of mixed micelles leading to fat malabsorption.

^{14}C-Xylose Breath test is a reliable test for detecting bacterial overgrowth in the small intestine. This involves the following steps:

1. Collection of a basal sample of expired CO_2 (1 mmol) after an overnight fast.
2. Oral administration of xylose (1 g containing 10 μCi ^{14}C-xylose). Xylose is metabolized by intestinal bacteria to form radiolabeled CO_2, which can be measured in the expired air.
3. Collection of samples (1 mmol) of expired CO_2 every one hour for 7 hours.
4. Quantification of radioactivity of CO_2 in the samples. The stable isotope ^{13}C can be used to avoid radioactivity.

❑ Explain the Schilling test for terminal ileal function.

Terminal ileal function can be tested by the Schilling test for vitamin B_{12} absorption. The test involves oral administration of a small dose of radiolabeled B_{12} and measurement of its urinary excretion. In malabsorption due to pernicious anemia, administration of intrinsic factor together with labeled B_{12} restores absorption, whereas in intestinal disease, malabsorption persists.

❑ Discuss the causes, clinical features, biochemical findings, and management of acute pancreatitis.

▫ *Causes*

- Alcoholism
- Abdominal trauma
- Peptic ulcer
- Drugs: sulphonamides, estrogens, tetracycline
- Gallstones
- Infections (mumps, viral hepatitis)
- Hypertriglyceridemia, hypercalcemia

▫ *Clinical features*

- *Gastrointestinal manifestations:* Acute abdominal pain, nausea, vomiting, acute inflammation of the pancreas.
- *Cardiovascular manifestations:* Tachycardia, hypotension, hemorrhage.

- *Dermatological manifestations:* Erythematous skin nodules, blue discoloration in the periumbilical area (*Cullen's sign*), and blue-red-purple or green-brown discoloration of the flanks due to tissue catabolism of hemoglobin (*Turner's sign*).

Biochemical findings

1. Lipaemic plasma.
2. Plasma enzymes:
   - ↑ amylase that returns to normal in 48–72 hours.
   - ↑ lipase levels that remain elevated for 7–14 days.
   - ↑ aspartate transaminase activity in pancreatitis due to gallstones.
3. Leucocytosis, methaemalbuminemia (in severe cases).

Management

- Intravenous infusion to replace fluid, electrolyte, and protein lost.
- Analgesics for pain.
- Antibiotics if there is infection.

Outline the causes, clinical features, diagnosis, and management of chronic pancreatitis.

Causes

- Alcohol (most important cause): Alcohol causes precipitation of proteins in the ducts, leading to ductal obstruction, acinar cell atrophy, and fibrosis.
- Acute pancreatitis
- Hypercalcemia
- Hypertriglyceridemia
- Hemochromatosis
- Cystic fibrosis

Clinical features

- Weight loss
- Impaired glucose tolerance
- Symptoms of malabsorption
- Abdominal pain
- Steatorrhea

Diagnosis

Biochemical tests are not valuable in diagnosis, and plasma amylase and lipase are generally normal. Diagnosis can be made by abdominal X-ray, ultrasound imaging, CT scanning, and endoscopic retrocholangiopancreatography (ERCP).

Management

- Management of the underlying cause.
- Analgesics to relieve pain.
- Addition of pancreatic extracts for treatment of malabsorption.

Discuss the pathogenesis, clinical consequences, diagnosis, and management of cystic fibrosis.

Cystic fibrosis (CF) is a multisystem disease that presents

- in neonates, with a failure to pass the first feces containing bile, intestinal debris, and mucus (*meconium ileus*).
- in early childhood, with respiratory infections.
- in the adult.

- ***Inheritance:*** Autosomal recessive disorder
- ***Pathogenesis:*** Cystic fibrosis arises due to mutations in a gene located on chromosome 7 encoding the *cystic fibrosis transmembrane regulator (CFTR)* protein that regulates transmembrane chloride transport. The most common mutation is termed ΔF_{508} which refers to deletion of three base pairs, resulting in the absence of phenylalanine at position 508 in the amino acid sequence of CFTR.
- ***Clinical consequences***

  Absence of the CFTR chloride channel leads to the following consequences:
  - Exocrine pancreatic insufficiency with impaired secretion of sodium, bicarbonate, and water resulting in increased viscosity (*mucoviscoidosis*), obstruction of pancreatic ducts, pancreatic fibrosis, and destruction of pancreatic tissue.
  - Chronic airways infection that affects mucus secretion in the bronchi with recurrent respiratory infections, bronchiectasis, and chronic lung disease.
  - Malabsorption, cirrhosis of the liver, and cholelithiasis due to defective secretion of water and chloride.
  - Abnormal sweat gland function due to excessive excretion of Na^+ and Cl^- in sweat.
  - Abnormal urogenital function.
- ***Diagnosis***
  - Clinical symptoms.
  - Measurement of pilocarpine-induced sweat electrolyte concentrations. Na^+ and Cl^- (>70 mmol/L) is diagnostic.
  - Neonatal screening test: ↑ plasma immunoreactive trypsin.
  - Prenatal screening for ΔF_{508} mutations.
- ***Management***
  - Prevention of respiratory infections by physiotherapy and antibiotics.
  - Maintenance of proper nutrition and addition of pancreatic enzymes to the diet.

❑ Enumerate the causes of malabsorption.

The causes of malabsorption include the following:

1. ***Defective digestion*** as in:
   - patients after subtotal gastrectomy due to rapid gastric emptying.
   - pancreatic enzyme deficiency in chronic pancreatitis, cystic fibrosis, and carcinoma of the pancreas.
2. ***Bile salt deficiency*** with impairment of micelle formation, malabsorption of fat, and steatorrhea due to:
   - parenchymal liver disease
   - cholestasis
   - bacterial overgrowth in the small intestine due to stagnation of intestinal contents as in small intestinal diverticula and fistula
   - drugs such as neomycin and cholestyramine that cause sequestration or precipitation of bile salts.
3. ***Inadequate intestinal absorptive surface*** on account of surgery of the small intestine, jejunoileal bypass, or gastroileostomy.
4. ***Intestinal mucosal disease*** such as
   - parasitic infections
   - inflammation (Crohn's disease)

- *gluten-sensitive enteropathy* (coeliac disease), characterized by sensitivity to the wheat protein *α-gliadin*, with villous atrophy and reduction in the absorptive area of the intestine.
- tropical sprue.

5. ***Specific defects in absorption***
   - lactase deficiency that may be congenital, or acquired associated with severe diarrhea or colicky abdominal pain.
   - *sucrase* and *isomaltase* deficiency.
   - protein-losing gastroenteropathy.
   - Hartnup disease, in which there is defective intestinal transport of neutral amino acids.
   - cystinuria, in which there is a defect in the intestinal transport of cystine, ornithine, arginine, and lysine.
6. ***Endocrine and metabolic disorders*** such as
   - diabetes mellitus
   - hypothyroidism
   - adrenal insufficiency
   - Zollinger–Ellison syndrome
   - carcinoid syndrome in which there is excessive synthesis of serotonin by argentaffin cell tumors in the small intestine, associated with release of the gut peptide *Substance P* and increased intestinal motility.

❑ Outline the consequences of malabsorption.

Prolonged malabsorption can lead to the following consequences:

- Impaired fat absorption which causes
  - vitamin D deficiency, leading to hypocalcemia, hypophosphatemia, secondary hyperparathyroidism with consequent rickets or osteomalacia.
  - vitamin K deficiency with reduced synthesis of prothrombin and other clotting factors leading to hemorrhage.
- Malabsorption of amino acids and peptides which can cause
  - muscle and tissue wasting, and osteoporosis
  - hypoalbuminemia and consequent edema
  - low immunoglobulin levels leading to recurrent infections.
- Deficiency of iron, folate, and B_{12} leading to anemia.

❑ What are the clinical and biochemical findings in malabsorption?

Following are the clinical and biochemical features of malabsorption:

- Weight loss
- *Gastrointestinal manifestations:* Steatorrhea, abdominal discomfort, distension, flatulence with or without diarrhea
- *Hematological findings:* Anemia, bleeding tendency with prolonged prothrombin time
- Edema
- Recurrent infections
- Hypocalcemia causing tetany
- Hypophosphatemia
- Increase in alkaline phosphatase activity
- Osteomalacia or rickets.

Chapter 27

Plasma Proteins and Blood Coagulation

❑ What are the formed elements in blood?

The formed elements in blood include the *red blood cells (erythrocytes)*, *white blood cells (leukocytes)*, and *platelets (thrombocytes)*.

❑ What is the difference between serum and plasma?

- When blood is allowed to stand, measures taken to prevent clotting and it is centrifuged, the straw colored fluid that forms the top layer is known as *plasma*. Plasma contains a number of proteins that have important physiological functions. Plasma represents blood minus the cells.
- *Serum* is the clear yellow fluid that separates from blood, which is allowed to clot. Serum contains blood minus the cells and fibrinogen.

❑ What are the major plasma proteins?

The major plasma proteins are:

- *fibrinogen* involved in blood coagulation.
- *albumin* that maintains plasma oncotic pressure and has transport functions.
- *globulins* include
  i. α- and β-globulins that function as transport proteins.
  ii. γ-globulins that comprise cryoglobulins and immunoglobulins that have a role in defense against infection.

Fibrinogen is removed during clotting and serum contains only the albumin and globulin fractions.

❑ Enumerate the methods available for separation of serum proteins.

- Serum proteins can be separated by electrophoresis based on the differences in charge, size, and shape. Five distinct bands are obtained on cellulose acetate or agar gel electrophoresis: albumin, α_1, α_2, β, and γ globulins (Fig. 27.1).
- Isoelectric focusing allows visualization of more proteins including prealbumin, C-reactive protein, and α-fetoprotein.
- Immunoblotting, on nitrocellulose filter or nylon membrane, allows detection of specific proteins.
- The other methods for quantitation of serum proteins include nephelometry, turbidimetry, immunodiffusion, radioimmunoassay, fluorescence immunoassay, and enzyme immunoassay.

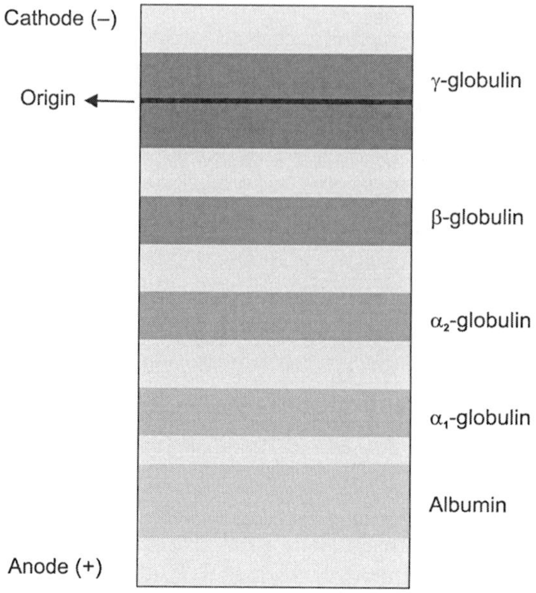

Fig. 27.1 Normal serum protein electrophoretic pattern

❑ Briefly explain the metabolism of plasma proteins.

- *Synthesis*
  - Chiefly in the liver and to a minor extent in extrahepatic tissues.
  - Hepatic synthesis is regulated by
    - i. oncotic pressure in the hepatic ECF
    - ii. humoral factors
    - iii. amino acid availability.
- *Catabolism*
  - Occurs in most cells of the body.
  - Amino acids released are utilized by tissues.
  - Those with carbohydrate side chains (*e.g.*, transferrin) are degraded in the liver.
  - Immunoglobulins and complement proteins bind to surface receptors and are phagocytosed and degraded by macrophages.
  - Products of protein degradation are filtered by the glomeruli, reabsorbed by the tubules, and catabolized further.

❑ Enumerate the functions of plasma proteins.

1. *Osmotic balance*: Albumin maintains osmotic balance between blood and the tissue spaces. An imbalance in the fluid exchange mechanism can cause accumulation of excess fluid in tissue spaces, which is known as *edema*.
2. *Buffering function*: The amphoteric nature of plasma proteins is responsible for buffering action and maintenance of pH.
3. *Protein reserve*: For metabolic needs during starvation, and parenteral administration in hypoproteinemia.
4. *Defence against infection*: The γ-globulins (immunoglobulins) play an important role.

5. *Anti-inflammatory effects*: Complement and coagulation proteins, C-reactive protein, α_1-antitrypsin, ceruloplasmin, and haptoglobin exert anti-inflammatory effects.
6. *Blood coagulation*: Fibrinogen and other clotting factors are required for blood coagulation.
7. *Transport* of substances insoluble in plasma such as hormones, metals, and drugs.

❑ Enumerate the functions of albumin.

The functions of albumin include the following:

1. *Osmotic regulation*: Albumin regulates the movement of water between intravascular and extravascular fluids.
2. *Transport* of substances sparingly soluble in water such as bilirubin, free fatty acids, tryptophan, steroid hormones, metal ions (Ca^{2+} and Cu^{2+}), and drugs (*e.g.*, aspirin, sulphonamides).
3. *Source of amino acids* for peripheral tissues.
4. *Antioxidant activity* by binding free fatty acids, bilirubin, divalent cations, and HOCl.
5. *Buffering action* in non-physiological conditions by binding free ions and charged molecules.
6. *Anti-inflammatory effects* by inhibiting the production of leukotrienes and actin.

❑ Elucidate the causes and consequences of hypoalbuminemia.

Causes

Hypoalbuminemia arises due to one or a combination of the following factors:

- Expanded vascular pool decreases plasma albumin levels due to
  - infusion of protein-free fluid (*dilutional hypoalbuminemia*).
  - redistribution of fluid from intravascular space to extravascular space as in hypoxemia, septicemia, and in the recumbent position.
- Decreased synthesis due to
  - prolonged malnutrition as in kwashiorkor.
  - chronic liver disease.
  - malabsorption.
- Increased catabolism as in
  - fever
  - sepsis
  - trauma.
- Increased loss through
  - the skin, as in burns and exudative skin lesions.
  - the kidney, as in nephrotic syndrome with edema.
  - the gut, as in gastrointestinal cancers and ulcerative colitis (*protein-losing gastroenteropathy*).
- Other causes such as
  - pregnancy due to hemodilution, inadequate protein intake, or increased protein utilization by the fetus.
  - malignancy, due to protein-losing enteropathy, reduced synthesis, liver metastasis, or enhanced uptake by the tumor.

Consequences of hypoalbuminemia

- Edema, due to loss of intravascular volume to the extravascular space.
- Reduced binding of drugs and other substances that are normally bound to albumin lead to:
  - increase in free drug concentrations and drug toxicity.
  - hyperbilirubinemia leading to kernicterus in neonates.
  - low plasma protein-bound calcium levels.

❑ List the various α-globulins.

There are two subfractions of α-globulins— α_1 and α_2, based on electrophoretic mobility. The α-globulins include several transport proteins such as

1. ***α_1-Acid glycoprotein (AAG)***
   - is a glycoprotein synthesized by the liver and to a lesser extent by granulocytes and monocytes.
   - binds lipophilic substances such as progesterone and drugs such as chlorpromazine, propranolol, and benzodiazepines.
   - is significantly increased in inflammation.
   - is decreased in nephrotic syndrome and protein-losing enteropathy.
2. ***α-Fetoprotein (AFP)***
   - is an oncofetal protein present in normal human fetal plasma and amniotic fluid and in very low concentrations in adults.
   - plasma levels are high in hepatocellular carcinoma.
3. ***Prealbumin and retinol-binding protein (RBP)***
   - exists as a complex with retinol-binding protein and transports thyroid hormones and retinol.
   - is a negative acute phase protein and an indicator of the protein status.
   - plasma levels are low in inflammation, malignancy, and in protein-wasting diseases.
   - plasma levels are elevated in Hodgkin's disease.
   - plasma RBP is increased in chronic renal disease and reduced in malnutrition.
4. ***Ceruloplasmin***
   - is an α_2-globulin.
   - transports copper.
   - has ferrooxidase activity and promotes incorporation of iron into transferrin.
   - is an acute phase protein.
   - is increased in infections, myocardial infarction, rheumatoid arthritis, viral hepatitis, collagen disorders, and hyperthyroidism.
   - is low in Wilson's disease associated with increase in plasma copper.
5. ***Haptoglobins (Hp)***
   - are α_2-globulins that form high molecular weight complexes with hemoglobin (Hb) and are released into plasma during intravascular hemolysis. The complex is not excreted by the kidney and prevents renal damage by Hb.
   - free Hp as well as Hp–Hb complex plays an important role in the control of inflammation.
   - plasma levels are high in acute inflammation, tissue necrosis, and in protein-losing states such as nephrotic syndrome.
   - plasma levels are low in hemolytic conditions, chronic liver disease, and metastatic disease.
6. ***Protease inhibitors*** such as α_1-antitrypsin and α_2-macroglobulin.

❑ Discuss the functions and clinical significance of α_1-antitrypsin and α_2-macroglobulin.

α_1-Antitrypsin (AAT)

- is an α_1-globulin that inhibits
  i. serine proteinases.

ii. collagenase and elastase released by the alveolar macrophages and polymorphonuclear leukocytes, thereby clearing the debris inhaled by the lungs.
iii. plasmin, thrombin, urokinase, kallikrein, and other enzymes in the clotting cascade.

- is an acute phase protein, and plasma levels are increased during inflammation and necrosis.
- shows genetic polymorphism with over 30 different alleles.
- deficiency causes:
  i. neonatal hepatitis that progresses to cirrhosis.
  ii. progressive proteolytic damage to the elastic tissues of the lungs, leading to emphysema and chronic obstructive pulmonary disease (COPD).
- deficiency is commonly due to the PiZZ variant.

α_2-Macroglobulin (AMG)

Functions

- Inhibits serine, cysteine, aspartate, and metalloproteases.
- Inhibits enzymes in the coagulation, fibrinolytic, and complement pathways.
- Transports insulin, growth hormone, and Zn^{2+}.
- Plays a role in immunological and inflammatory reactions.

Clinical significance

Plasma levels are

- Low in acute phase response, pancreatitis, advanced carcinoma of the prostate.
- High in nephrotic syndrome.

❑ Discuss the functions and clinical significance of the β-globulins.

The β-globulins include the following:

Lipoproteins that transport cholesterol, triglycerides, and phospholipids (Chapter 7).

Transferrin *(siderophilin)*

- binds two atoms of Fe^{3+} per molecule.
- transports iron to tissues that synthesize heme.
- plasma levels are high in iron deficiency, pregnancy, and by oral contraceptive use.

Hemopexin

- forms a complex with heme thereby preventing heme from being excreted in urine. The complex is removed from circulation by the liver and the iron reutilized.
- plasma levels are elevated in hemolytic anemia.

C-reactive protein (CRP)

- binds to bacterial polysaccharides or phospholipids released from damaged tissues and activates the complement pathway.
- responds to inflammation by inducing opsonization, phagocytosis, and lysis of invading organisms.
- plasma levels are increased in acute infections, after myocardial infarction, inflammation, surgery, and malignancies.

β_2-Microglobulin (BMG)

- is synthesized by nucleated cells and is found in all body fluids.
- is shed into circulation by lymphocytes and tumor cells.
- is identical to the smaller subunit of the HLA-histocompatibility complex and is therefore useful in assessing graft rejection.

- is filtered at the glomeruli and completely reabsorbed and degraded in the proximal convoluted tubule (PCT). Plasma levels are high in inflammation, renal failure, and malignancy of the lymphoid system.

❑ What are cryoglobulins?

Cryoglobulins

- are immunoglobulins that precipitate on cooling serum.
- are present in the form of immunoglobulin – anti-immunoglobulin complexes in inflammatory conditions such as rheumatoid arthritis.
- may be monoclonal or polyclonal.

❑ What are immunoglobulins?

The immunoglobulins

- are γ-globulins that function as antibodies.
- are synthesized by B lymphocytes in response to antigens.
- bind to specific antigens to form an antigen–antibody complex that is degraded by the complement system or by the macrophages.

❑ Explain the structure of immunoglobulins.

- The immunoglobulins are Y-shaped molecules that contain four polypeptide chains—two identical heavy (H) chains and two identical light (L) chains (Fig. 27.2).
- The H and L chains are linked by disulphide bonds.
- The H chains are of 5 types: γ, α, μ, δ, and ε, which determine the immunoglobulin classes IgG, IgA, IgM, IgD, and IgE respectively.

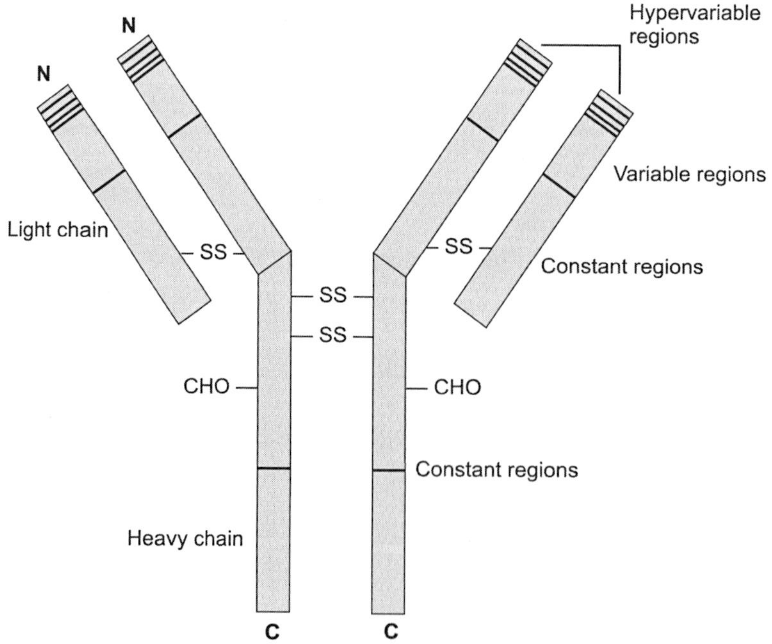

Fig. 27.2 Structure of an immunoglobulin molecule

- There are two types of light chains—κ and λ.
- The H and L chains of individual immunoglobulin molecules exhibit variations in the N-terminal amino acid sequence. This region termed the *variable region* or *V-region* is responsible for antigen specificity.
- The rest of the molecule that is invariant known as the *constant region* or *C-region* contains effector sites responsible for interaction of the molecule with cells and complement.

❑ Give an account of the different classes of immunoglobulins.

There are 5 classes of immunoglobulins. These include

- *IgG*
  - accounts for ~75 % of plasma immunoglobulins.
  - is a low molecular weight monomer found in high concentration in all extracellular fluids.
  - only immunoglobulin that can cross the placenta.
  - is capable of diffusion into the interstitial fluid and protection against tissue infection.
  - plasma levels are increased in chronic hepatitis.
- *IgA*
  - is a dimer.
  - is found in
    a. plasma cells that underlie the mucosa of the gastrointestinal and respiratory tracts, the skin, and exocrine glands.
    b. colostrum, sweat, and tears.
  - protects neonates from gastrointestinal infection.
- *IgM*
  - is a high molecular weight pentamer.
  - is found in the vascular space.
  - is the first immunoglobulin secreted by B lymphocytes stimulated to form plasma cells and the first line of defense against invading organisms.
  - high levels before birth indicate intrauterine infections such as congenital toxoplasmosis, cytomegalovirus, herpes simplex virus, and syphilis.
- *IgE*
  - is a monomer.
  - binds to mast cells and basophils causing degranulation of these cells with release of histamine, prostaglandins, and kinins. These substances induce smooth muscle contraction and vascular permeability.
  - plasma levels are increased in allergic conditions such as asthma and hay fever.
- ***IgD***, constitutes the membrane receptor of the B lymphocyte together with IgM.

❑ What are the causes of hypogammaglobulinemia?

Hypogammaglobulinemia may be

- *Transient* as in infants due to the time required for endogenous synthesis.
- *Primary* as in
  - *X-linked agammaglobulinemia* (*Bruton's disease*), characterized by complete absence of immunoglobulins and consequent recurrent bacterial infections.

- *dysgammaglobulinemias* with defect in one or two immunoglobulins, chiefly IgA associated with gastrointestinal, respiratory, or renal infections.

- *Secondary hypogammaglobulinemia* due to
  - malignancies of the hematopoietic and lymphoreticular system such as chronic lymphatic leukemia, multiple myeloma, and Hodgkin's disease.
  - protein-losing states such as nephrotic syndrome in which IgG is lost.
  - cytotoxic drug treatment.

❑ What are the causes of hypergammaglobulinemia?

Hypergammaglobulinemia may be

- *polyclonal* involving a diffuse increase in all the fractions that occurs due to
  a. acute and chronic infections.
  b. autoimmune diseases such as systemic lupus erythematosus (SLE).
  c. chronic liver disease.
- *monoclonal* involving a single immunoglobulin fraction. Usually, only select B cell clones are switched on.

❑ What are paraproteins?

- Paraproteins are immunoglobulins that are formed by a single clone of cells.
- These can be detected by the appearance of a discrete band on serum electrophoresis in the γ-region.
- Sometimes, paraproteins also appear in the α_2 or β region.
- Paraproteins are reliable tumor markers, especially of B-cell tumors.

❑ Enumerate the causes and consequences of paraproteinemias.

Causes

1. *Multiple myeloma* characterized by
   - proliferation of plasma cells that secrete IgG.
   - bone pain due to pressure from the proliferating cells, osteoporosis, and fractures.
   - presence of low molecular weight L chains or their fragments in urine such as *Bence Jones Protein (BJP)*.
   - increase in plasma IgG, β_2-microglobulin, and nonprotein nitrogen.
2. *Waldenstrom's macroglobulinemia,* a B-cell tumor associated with hyperviscosity that leads to slugging of red blood cells in capillaries and formation of thrombus. The paraprotein is IgM.
3. *B-cell lymphomas and chronic lymphatic leukemias* in which small lymphocytes produce IgM.

Consequences

1. Proliferation of a single clone of cells causes suppression of immunoglobulin synthesis from other clones, thereby increasing susceptibility to infection (*immune paresis*).
2. *Hyperviscosity syndrome* that impairs blood flow, leading to thrombosis and peripheral gangrene.
3. Normochromic, normocytic anemia, hemorrhages, and increased erythrocyte sedimentation rate (ESR) and adherence of erythrocytes to one another (*rouleaux formation*).
4. *Hyperproteinemia.*
5. *Pseudohyponatremia* due to space-occupying effect of the paraprotein.
6. Damage to renal tubular cells by *Bence Jones protein* resulting in obstruction, amyloid deposition in tissues, and '*myeloma kidney*'.

❑ What are benign paraproteinemias?

Benign paraproteinemias are characterized by plasma paraprotein concentration of <10 g/L and absence of Bence Jones protein in urine. They usually occur in the elderly and do not manifest clinical features of B-cell tumors.

❑ Elucidate the role of proteins and platelets in blood coagulation.

- Clotting factors are present in the zymogen form, and are activated by an enzymatic *cascade.*
- Accessory factors that enhance the rate of activation of zymogens include
  - acidic phospholipids from platelets or damaged tissues.
  - Ca^{2+}.
  - activated form of a clotting factor.
- Platelets are required
  - for formation of platelet plug.
  - as a source of clotting factors (*e.g.*, Factor XIII, phospholipids).
  - as a surface for binding of clotting factors.

❑ Discuss the pathways of blood coagulation.

Blood coagulation can be separated into:

- *The intrinsic pathway* that involves factors normally present in circulation and is initiated by surface contact.
- *The extrinsic pathway,* which is rapid, and is initiated by tissue factor abundant in the lungs, brain, saliva, and placenta.
- *The common pathway,* which is the convergence of the intrinsic and extrinsic pathways that eventually results in clot formation.

The intrinsic pathway involves the following stages:

- *Contact Stage* involves:
  - Conversion of *prekallikrein* to *kallikrein* when the complex of factor XII, prekallikrein, and high molecular weight kininogen (HMwK) contacts a surface such as glass *in vitro* or collagen *in vivo.*
  - Conversion of factor XII to XIIa by kallikrein.
  - Factor XIIa activates more kallikrein, which generates additional factor XIIa in a process called *reciprocal activation.*
  - Activation of factor IX by factor XIIa and high molecular weight kininogen as accessory factor.
- *Activation of factor IX* by factor IXa formed in the contact stage and by factor VIIa of the extrinsic pathway with phospholipids and Ca^{2+} as accessory factors.
- *Activation of factor X* by factor IXa with Ca^{2+}, phospholipids, and factor VIIIa as accessory factors. In patients with classic *hemophilia (hemophilia A),* factor VIII is absent.

The extrinsic pathway

- *Activation of factor VII* bound to tissue factor by either thrombin, factor Xa, or factor XIIa.
- *Activation of factor X* by a complex of tissue factor and factor VIIa. This step is inhibited by the circulating lipoprotein *extrinsic pathway inhibitor (EPI)* or *lipoprotein-associated coagulation inhibitor (LACI).*

The common pathway involves

- *Conversion of prothrombin (factor II) to thrombin (factor IIa)* by factor Xa formed from both the intrinsic and extrinsic pathways with phospholipids, Ca^{2+}, and factor Va as accessory factors. This involves cleavage of two peptide bonds in prothrombin to generate thrombin.
- *Conversion of fibrinogen (factor I) to fibrin (factor Ia)* by active thrombin. The fibrin monomers aggregate to form the soft clot that is converted to a hard clot by factor XIIIa, a *transglutaminase* that forms crosslinks between the fibrin monomers.

Figure 27.3 summarizes the pathway of blood congulation.

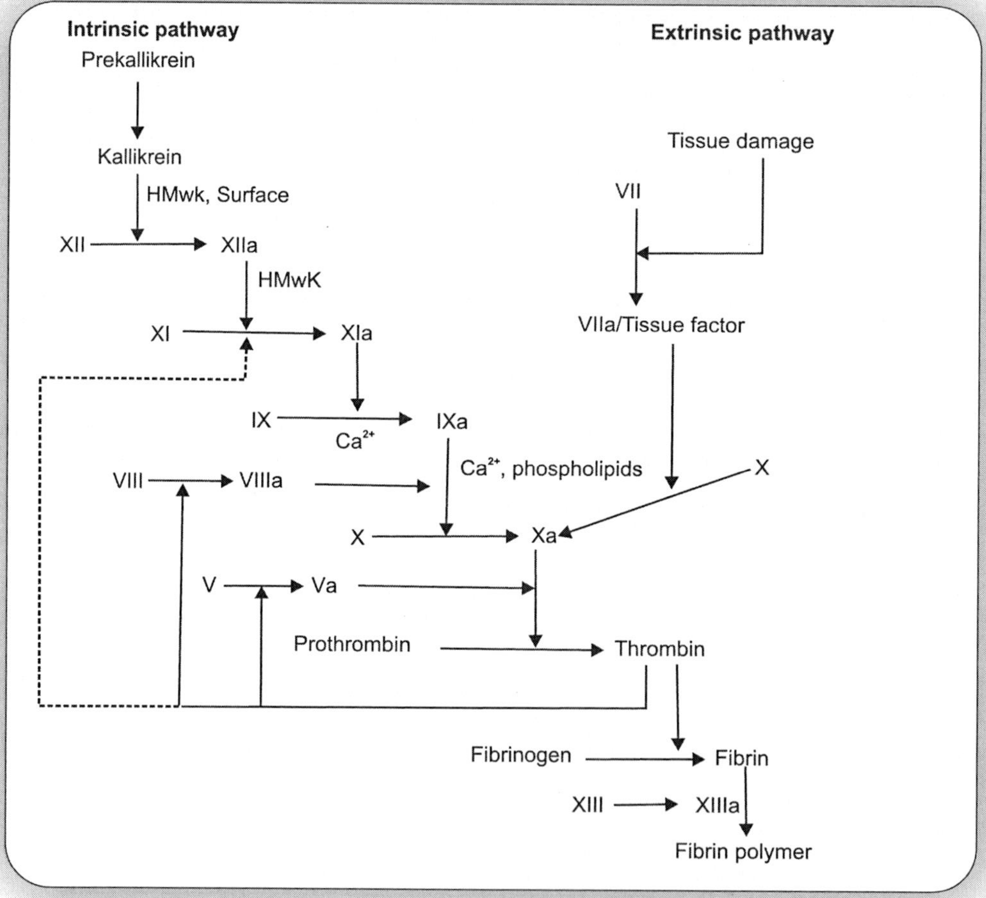

Fig. 27.3 Blood coagulation pathways

❑ Elucidate the role of vitamin K in blood coagulation.

- Vitamin K is responsible for the formation of *γ-carboxyglutamate* (*Gla*) residues at the N-terminal ends of prothrombin and factors VII, IX, and X.
- The *Gla* residues are required for binding to Ca^{2+}.

- *Gla* residues are formed by γ-carboxylation of glutamate residues as a post-translational modification by a *vitamin K-dependent carboxylase.*
- Vitamin K hydroquinone, the cofactor for the carboxylase, is oxidized to the inactive quinone form by *epoxide reductase.*
- Regeneration of the hydroquinone involves an NAD(P)H dependent reaction.
- *Dicumarol,* an antimetabolite of vitamin K, inhibits *epoxide reductase* and regeneration of the hydroquinone.

Figure 27.4 summarizes the vitamin K cycle.

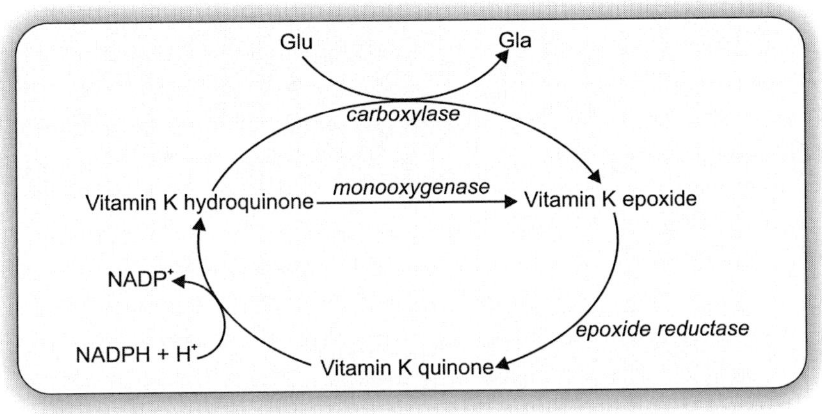

Fig. 27.4 The vitamin K cycle

What are the components of the fibrinolytic system?

- Fibrinolysis is the lysis of fibrin clots that occurs few days after clot formation.
- The fibrinolytic system comprises
  - plasminogen
  - plasminogen activators
  - plasmin
  - α_2-antiplasmin.
- Plasminogen is activated to plasmin by hydrolysis of a peptide bond between Arg and Val by plasminogen activators.
- *Plasminogen activators:*
  - are serine proteases.
  - examples include:
    a. *urokinase* synthesized in the kidney, and found in the urine.
    b. *streptokinase* from β-hemolytic *Streptococcus.*
    c. complex of factor XIIa, prekallikrein, and HMwK of the intrinsic pathway.
    d. inhibitor of plasminogen activation: α_2-*antiplasmin.*

The fibrinolytic system is shown in Fig. 27.5.

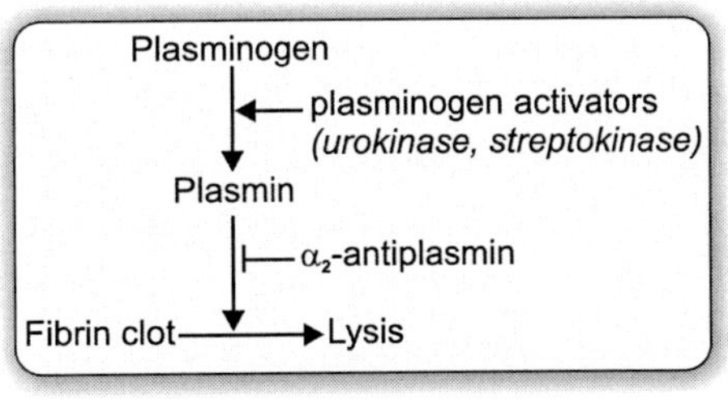

Fig. 27.5 The fibrinolytic system

How is blood clotting regulated?

Blood coagulation is regulated by the following factors:

1. Dilution of activated clotting factors by the large volume and flow of blood.

2. Degradation of clotting factors by the liver.
3. Inhibition of the clotting factors IXa, Xa, XIa, and thrombin by antithrombin III.
4. Inactivation of factors Va and VIIIa by activation of *protein C* induced by binding of thrombin to thrombomodulin.
5. Degradation of the clot by the fibrinolytic system.

❑ Give an account of anticoagulants used for therapeutic purposes.

The anticoagulants used for therapeutic purposes include the following:

- ***Heparin***
  - is a polyanionic glycoprotein rich in sulphated acidic polysaccharides.
  - is found in high concentrations in the mast cells that line the endothelium of blood vessels.
  - binds to antithrombin III to form a complex that inhibits clotting factors such as IIa, IXa, Xa, XIa, and XIIa.
  - used *in vivo:*
    a. during blood transfusions
    b. during surgery
    c. to prevent venous thrombosis
    d. in patients with prosthetic implants.
  - Complication: Causes hemorrhage that can be reversed by administration of *protamine* that inactivates heparin.
- ***Coumarins***
  - *e.g.,* warfarin
  - function as vitamin K antagonists by inhibiting γ-carboxylation of glutamate residues in prothrombin and factors VII, IX, and X.
  - prevent clot formation on prosthetic implants.
  - cause hemorrhage that can be reversed by administration of vitamin K.

Section 7

Molecular Biology and Genetic Engineering

- Chromatin Organization
- DNA Replication and Repair
- Transcription and Post-transcriptional Processing
- Translation and Regulation of Gene Expression
- Recombinant DNA Technology

Chapter 28

Chromatin Organization

❑ Discuss the structure and organization of the *E.coli* chromosome.

- The *E.coli* chromosome
  - is present in a region of the cell termed the *nucleoid.*
  - contains DNA, RNA, and protein.
  - is attached to the plasma membrane.
- *E.coli* DNA
  - is a single, circular, double-stranded molecule containing 4.6 mi bp.
  - encodes 1000–5000 proteins.
  - is organized into 50–100 domains attached to a protein-membrane scaffold. Some of the domains are supercoiled, while others lack supercoils.
- DNA-binding proteins required for packaging DNA into the nucleoid include
  - *Protein HU*, a small dimeric protein analogous to integration host factor (IHF) that stimulates DNA replication.
  - *HNS (protein H1)*, a monomeric neutral, histone-like protein.
- In addition, RNA polymerase, *m*RNA, and site-specific DNA-binding proteins such as IHF are also necessary for organization of the DNA domains.

❑ Explain the structure and functions of plasmids.

- Plasmids are extrachromosomal replicons present in most prokaryotic systems and in some eukaryotes.
- Most plasmids are double-stranded, closed, circular DNA.
- Exceptions include
  1. single-stranded, circular plasmids in *Clostridium* species.
  2. linear, double-stranded plasmids in *Streptomyces* species.
  3. RNA plasmids in viroids and yeast.
- Size: 1×10^6 to $>200 \times 10^6$ daltons; 2000–100,000 bp.
- Genes: 1 to >100 genes.
- Plasmids are autonomous and encode proteins essential for replication and maintenance in the host.

- *Functions*

  Most plasmids encode proteins that confer phenotypic traits on their host cells such as:
  1. resistance to antibiotics,
  2. degradation of aromatic compounds,
  3. production of antibiotics, hemolysin, and hydrogen sulphide.

❑ Give a brief outline of the hierarchies of organization of eukaryotic chromatin.

In eukaryotes, each chromosome is a single, linear DNA duplex that is packaged in the form of a nucleoprotein complex called *chromatin*, which contains more than 50% protein.

The eukaryotic chromatin contains several hierarchies of organization:

- The first level of packaging is the *nucleosome* in which DNA winds around a histone core to produce *'beads on string'* appearance. This is the least condensed form of chromatin.
- The second level of packaging is the *30 nm fiber* or *solenoid* in which 6 nucleosomes are coiled into a left-handed helical fiber. This is a more condensed structure and is seen in interphase chromatin. Packing of nucleosomes into the 30 nm fiber requires histone H1.

Figure 28.1 illustrates the packaging of DNA into nucleosomes and 30 nm fiber.

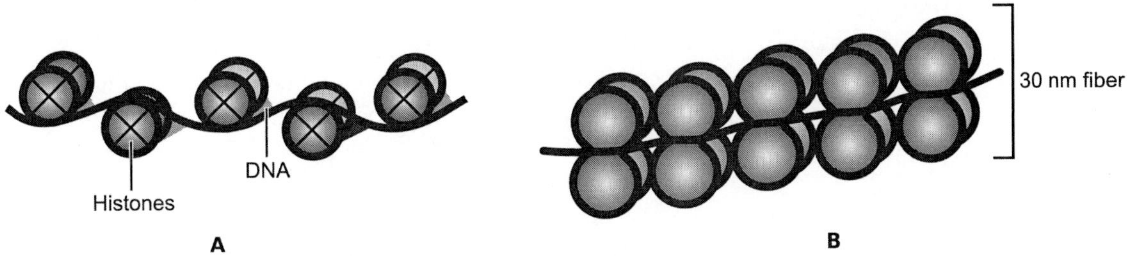

Fig. 28.1 Packaging of DNA into nucleosomes and 30 nm fiber. (A) Packaging of DNA with histones to form nucleosomes (*beads on string*), and (B) Packaging of nucleosomes into 30 nm fiber

- *Higher-order chromatin structure* involves organization of the 30 nm fiber into loops, scaffolds, and domains.
  - The looped solenoid, a highly compact structure is seen during mitosis and meiosis.
  - Each loop of the solenoid contains 20–100 kb DNA attached to a *nuclear scaffold* via AT-rich sequences termed *matrix-associated regions (MARs)* or *scaffold-associated regions (SARs)*.
  - The scaffold contains proteins such as histone H1 and topoisomerase II.
  - The DNA is compacted further to form coils upon coils. Each chromatid contains 10 coils. Two sister chromatids are joined together at the centromere in a metaphase chromosome.
  - A disk-like structure is formed in some eukaryotes by association of 18 loops of the solenoid.
  - Hundreds of disks stack together to form a condensed chromosome.

❑ Give an account of histones.

- Histones are basic proteins rich in the basic amino acids arginine and lysine.
- Histones are categorized into five types based on the amount of lysine and arginine:
  - H1, which is lysine rich and useful in packaging chromatin.
  - H2A and H2B, which contain equal amounts of lysine and arginine.
  - H3 and H4, which are arginine rich.

- *Functions*
  - Histones being positively charged form salt bridges with negatively charged phosphate groups on DNA and stabilize nucleosomes.
  - Confer stability to DNA.
  - Protect DNA against attack by DNase.
  - Function as general repressors and regulate gene expression: Replication and transcription occur only when histones are removed.
- *Chemical modifications*
  - The N-terminal extension (*tail*) regions of histones undergo modifications such as methylation, acetylation, and phosphorylation.
  - Histone acetylation and phosphorylation decrease the positive charge of the histone tail, which reduces the affinity for the negatively charged phosphate groups of DNA.
  - Acetylation of lysine residues renders chromatin transcriptionally active, whereas deacetylation represses transcription.

❑ What are nonhistone proteins?

Nonhistone proteins are a heterogeneous group of proteins that include

- *Enzymes* in DNA and histone metabolism, replication, recombination, and transcriptional regulation.
- *Scaffold proteins* involved in higher-order chromatin structure.
- *SMC (structural maintenance of chromosomes) proteins* that include cohesins and condensins.
- *High-mobility group (HMG) proteins* involved in structural organization and gene regulation.
- *Protamines* required for DNA packaging into sperm head.
- *Topoisomerases.*

❑ Discuss the structural organization of the nucleosome.

- The nucleosome is the first level of packaging of eukaryotic chromosomes.
- In the nucleosomes, DNA is wound around a histone core approximately two times to produce *'beads on string'* appearance (Fig. 28.2).
- Each nucleosome consists of
  - ***core histones*** containing two molecules each of H2A, H2B, H3, and H4. The histones protect DNA against attack by *micrococcal nuclease* and are not separated even when chromosomes duplicate.

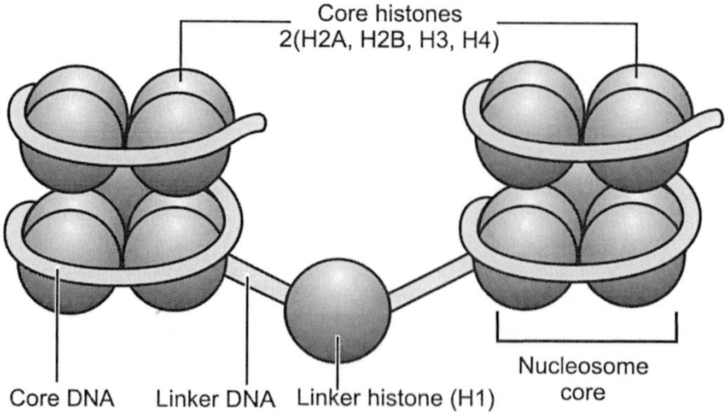

Fig. 28.2 Structure of nucleosome

- ***core DNA*** is the 146 bp length of DNA associated with the core histones.
- ***linker DNA*** is the DNA between adjacent nucleosomes that varies from 8 to 114 bp.
- ***linker histone (histone H1)*** is a large molecule that binds to linker DNA and stabilizes DNA between the nucleosome cores. It has an important role in condensation of chromatin to higher order structure. In some cell types, H5 is present instead of H1.

❑ What are telomeres?

- Telomeres are the tips of linear eukaryotic chromosomes that provide stability to the chromosome and ensure survival.
- Telomeres are protected against degradation by nucleases by palindromic hairpin loops or by proteins that bind to the 5' ends.
- Telomeric DNA are extremely short, highly repetitive TG-rich sequences of the form 5(T/A) *x* Gy3', where *x* is between 1 and 4, and *y* is greater than one.
- Telomere length varies between species, with ciliates containing only 36 bp, whereas mice have up to 50 kb. The telomere sequence TTAGGG extends for 10,000 nucleotides in humans.
- Telomeres form novel folded structures such as the *G quartet* containing non-Watson–Crick base pairing, which prevent enzymatic degradation.

❑ Define euchromatin and heterochromatin.

- *Euchromatin* refers to lightly stained regions of the chromatin that are transcriptionally active. Due to less condensed packaging, euchromatin is more sensitive to digestion by *Dnase I*, an enzyme that cuts DNA not protected by bound protein.
- *Heterochromatin* refers to darkly stained regions of the chromatin that are transcriptionally inactive. It is present in a highly condensed form due to closely packed regions of 30 nm fiber. Heterochromatin can be recognized by methylation of cytosine residues in the sequence 5′-CG-3′ in mammals.

❑ What are CpG islands?

- *CpG islands* are islands of unmethylated CpG found near the 5', end of eukaryotic genes.
- These are 1–2 kb in length and found ubiquitously in the genome.
- The human genome contains ~45,000 CpG islands of which 50% is associated with housekeeping genes and the rest with cell-type specific genes.
- Methylation of cytosine residues in these islands renders the chromatin transcriptionally inactive in mammals.

❑ Define the term genome. What are the sequence components in the genome?

- *Genome* is the total genetic complement of a cell.
- The DNA sequence components in cellular genomes can be divided into
  - *unique sequence DNA* represented only once or in few copies.
  - *repetitive DNA* represented in many copies.

❑ Explain the organization of the bacterial genome.

- The bacterial genome is comprised mainly of unique sequence DNA that contains genes and regulatory elements.

- The *E. coli* genome contains very little repetitive DNA accounting for <1% of the genome. The repetitive sequences include
  - *enterobacterial repeat internal consensus (ERIC)*
  - *repeated extragenic palindrome (REP).*

❑ Discuss the organization of the eukaryotic genome.

The sequence components in eukaryotes include the following:

- ***Unique sequence DNA***
  1. present in one or few copies
  2. contains most of the genes.
- ***Repetitive DNA***
  - varies from 5% in microbial eukaryotes to 40–50% in mammals and upto 80% in flowering plants.
  - two classes: moderately repetitive and highly repetitive.
  - ***Moderately repetitive DNA*** present as
    a. *tandem gene clusters* that have no spaces between individual units and contains the rRNA and histone genes.
    b. *dispersed repetitive DNA* found scattered throughout the genome and includes two classes: *SINES* (short interspersed nuclear elements) such as *Alu element* in humans and *B1 element* in mice and *LINES* (long interspersed nuclear elements) such as the *L1 element.*

    The *Alu* and *L1* elements that constitute 10% of the human genome are found in introns and flanking regions of genes. Although their exact function is not known, they may represent selfish DNA that may be involved in gene regulation.
  - ***Highly repetitive DNA*** includes
    a. *Satellite DNA* which are sequences of 2–30 bp present around centromeres and telomeres. Variations in repeat length (polymorphisms) occur between species.
    b. *Minisatellite DNA* that are small clusters of 9–70 bp present in telomeric and subtelomeric regions. Polymorphisms termed *variable number of tandem repeats (VNTR)* are used as markers of biological identity and form the basis of the *DNA fingerprinting* technique.
    c. *Microsatellite DNA* are short repeat units of 1–4 bp widely distributed throughout the genome of multicellular organisms. The human genome contains 50,000–100,000 $(dC\text{-}dA)_n$ microsatellites. Microsatellites exhibit high degree of polymorphism and are useful as genetic markers.

❑ Define a gene.

A gene can be defined as a segment of DNA that encodes a polypeptide or an RNA molecule.

A gene contains coding sequences called *exons* and noncoding intervening sequences termed *introns.*

❑ How does a bacterial gene differ from a eukaryotic gene?

Bacteria

- Bacterial genes are generally small and contain only exons.
- In bacteria, a single transcription unit contains:
  - several genes, each gene encoding a polypeptide.

- short internal noncoding regions (internal UTRs) separating the genes.
- noncoding regions in the 5' and 3' end referred to as 5' *untranslated region* (5' *UTR*) or *leader sequence* and 3' *untranslated region* (3' *UTR*) or *trailer sequence*, respectively. While the 5' UTR regulates ribosome binding, the 3' UTR maintains mRNA stability.

Figure 28.3 shows a schematic representation of the organization of a bacterial gene.

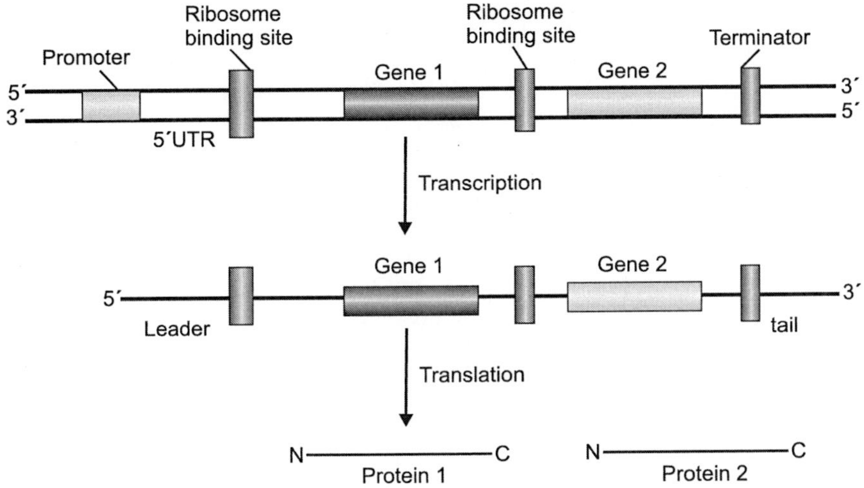

Fig. 28.3 Organization of a bacterial gene

Eukaryotes

- The genes in higher eukaryotes are generally large and contain both exons and introns. The large size of the genes is due to the presence of many large introns.
- In eukaryotes, the transcription unit contains a single gene containing a central coding region and flanking noncoding regions. The mature transcript may encode a single polypeptide or several polypeptides if alternative splicing occurs.

Figure 28.4 shows a schematic representation of the organization of a eukaryotic gene.

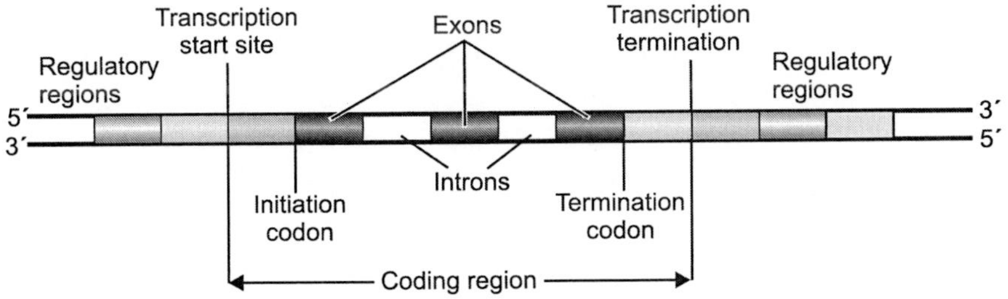

Fig. 28.4 Organization of a eukaryotic gene

Chapter 29

DNA Replication and Repair

❑ What is the central dogma of molecular biology?

- The *central dogma of molecular biology* states that genetic information flows from DNA to RNA to protein by the processes of
  - *Replication* that involves formation of daughter DNA molecules using a parental DNA as a template.
  - *Transcription* in which the genetic information in DNA is transcribed to form messenger RNA (mRNA).
  - *Translation* of mRNA into the amino acid sequence of proteins.
- In retroviruses, which have RNA genomes, genetic information flows in the reverse direction, *i.e.*, from RNA to DNA.

Figure 29.1 depicts the flow of genetic information.

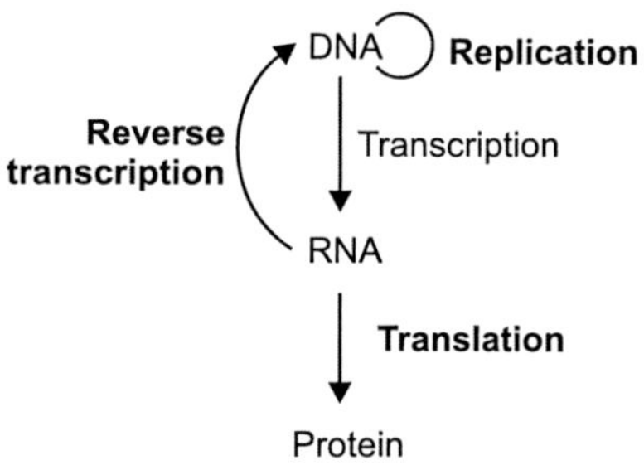

Fig. 29.1 The flow of genetic information

❑ Define the terms replicon, replication fork, and origin of replication.

- *Replicon* is the unit of DNA in which individual acts of replication occur. While the bacterial chromosome contains a single replicon, the eukaryotic chromosome has a large number of replicons.
- *Replication fork*, also known as the *growing point*, is the point at which replication occurs. Based on whether one or more replication forks start from the origin, replication may be *unidirectional* or *bi-directional.*
- The sites at which replication begins are called *origins of replication*. These sites are generally AT-rich to facilitate unwinding. Proteins and enzymes required for replication assemble at the origins.

❑ Elucidate the role of helicases and single-strand binding proteins in replication.

Helicases

- Separate double-stranded DNA to single-stranded DNA during replication using the energy derived from ATP hydrolysis.

- Unwind the DNA duplex just ahead of the replication fork at a rate of 1000 bp/s. Two DNA helicases unwind DNA at a replication fork moving in opposite directions, one on the leading strand template and another on the lagging strand template.
- *Functions*
  - DNA unwinding during replication: *Helicases* in conjunction with *topoisomerase* relieve torsional strain.
  - Role in homologous recombination, nucleotide excision repair, transcription termination, and conjugation.

Single-strand binding proteins (SSB)

- Also known as *helix-destabilizing proteins.*
- Bind tightly in a cooperative manner to single-stranded DNA and makes it available as a template for replication.
- Stabilize DNA in the single strand state and prevent base pairing.

❑ Explain the mechanism of action of topoisomerases.

Topoisomerases

- relax the torsional strain generated during DNA unwinding by helicases by causing a transient break in DNA followed by resealing. Topoisomerases bind covalently to DNA and cleave a phosphodiester bond that is reformed after the enzyme dissociates from the DNA.
- remove knots and catalyze catenation (linking together of double-stranded, circular DNA) and decatenation.
- type I topoisomerases cleave only one strand and catenate/decatenate substrates containing a nick, whereas type II enzymes cleave both strands of DNA and catenate/decatenate covalently closed circles.

Figure 29.2 shows the mechanism of action of type II topoisomerases.

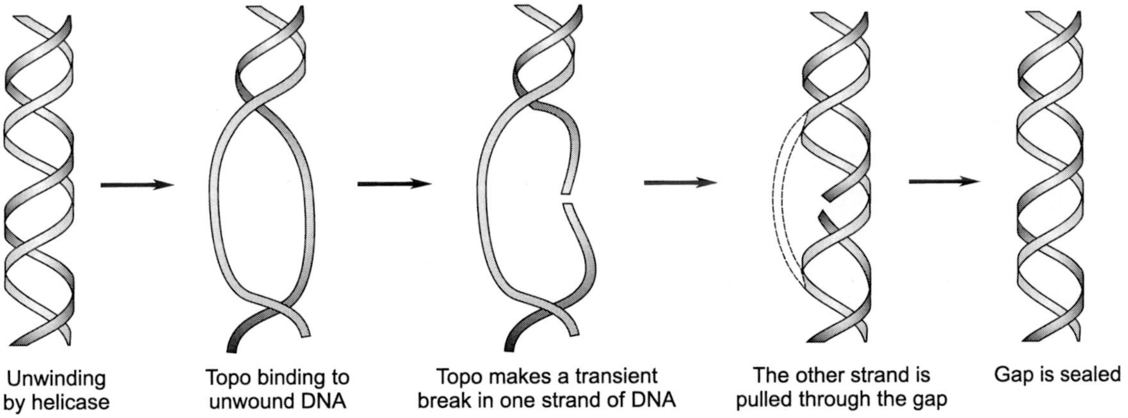

Fig. 29.2 The mechanism of topoisomerase action

❑ Explain the polymerization and exonuclease activities of DNA polymerases.

DNA-dependent DNA polymerases catalyze DNA synthesis on a DNA template during replication, following DNA damage, recombination, and after removal of primer from the lagging strand.

- ***5' → 3' Polymerization***
  - DNA polymerases catalyze DNA synthesis by successive addition of dNTPs to the free 3′-OH of a growing DNA chain by the following reaction:

$$(dNMP)_n + dNTP \longrightarrow (dNMP)_{n+1} + PP_i$$

  - The mechanism involves nucleophilic attack of the free 3'-OH group on the α-phosphorous atom of the incoming dNTP with release of PPi and formation of a phosphodiester bond. PPi is hydrolyzed by *pyrophosphatase* pulling the reaction to completion *in vivo.*
  - The enzyme requires
    i. a primer containing a free 3'-OH group base paired to the template.
    ii. all four dNTPs.
    iii. Mg^{2+} ions.
  - The enzyme has low processivity and cannot synthesize long DNA.
- ***5' → 3' exonuclease activity***
  This is essential for
  - removal of RNA primers
  - repair of damaged DNA
  - excision of a nonpaired stretch of DNA or a segment containing chemically modified or mutated nucleotides.
- ***3' → 5' exonuclease***
  - It serves proofreading function and ensures high fidelity of DNA replication (only 1 error for every 10^9 nucleotides or 1 wrong nucleotide/10,000 cells).
  - Proofreading function removes
    i. a primer terminus not base paired with the template
    ii. incorrect nucleotides added to the growing chain
    iii. frayed end of a primer caused by partial melting.

❑ What is meant by processivity?

- *Processivity* refers to the number of nucleotides added to the nascent chain before the polymerase dissociates from the template.
- Processivity of DNA polymerase can be increased by an accessory protein functioning as a sliding clamp that holds the polymerase firmly while it moves on the DNA and releases the enzyme at a region of double-stranded DNA. Assembly of the clamp requires a clamp loader.

❑ Give an account of *E.coli* DNA polymerases.

There are 5 DNA polymerases in *E.coli* of which pol I, II, and III have been extensively studied.

DNA Pol I

- Is also known as the *Kornberg enzyme.*
- Is a single polypeptide encoded by the *polA* gene
- contains 1 or 2 atoms of Zn at the active site.
- enzyme activities:
  - 5' → 3' polymerase
  - both 5' → 3' and 3' → 5' exonuclease activities.

- proteolytic cleavage yields a large C-terminal *Klenow fragment* with polymerase and 3' → 5' exonuclease activities, and a small fragment with 5' → 3' exonuclease activity.
- *Functions*
  - gap filling and exonuclease activity during replication, repair, and recombination.
  - nick translation.

DNA Pol II

- minor component during normal growth.
- induced by the SOS response.
- enzyme activities:
  - 5' → 3' polymerase
  - has 3' → 5' exonuclease activity, lacks 5' → 3' exonuclease activity.
- *Functions*
  - proofreading
  - DNA repair.

DNA Pol III

- most active and has the highest rate of chain elongation and processivity among the *E.coli* polymerases.
- encoded by the *dnaE* gene.
- enzyme activities:
  - 5' → 3' polymerase
  - 3' → 5' exonuclease activities.
- enzyme functions in the form of a large complex called *DNA polymerase III** with 13 subunits (Fig. 29.3). Addition of four β-subunits that function as a sliding clamp increases processivity. The γ-complex functions as a clamp loader.

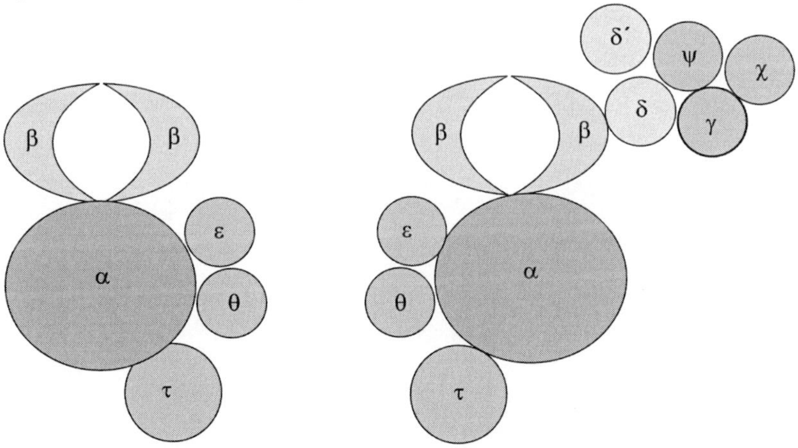

Fig 29.3 Structure of DNA pol III

❑ What is nick translation?

Nick translation involves degradation of a DNA (RNA) segment by the 5' → 3' exonuclease activity of DNA pol I, followed by gap filling by polymerase activity and nick sealing by *ligase*. This is useful in radiolabeling nucleic acid strands.

❑ Give an account of eukaryotic DNA polymerases.

There are five DNA polymerases in eukaryotes:

- ***Pol α***
  - is present in the nucleus.
  - has polymerase activity, lacks proofreading function.
  - is involved in the synthesis of short primers that are extended by pol δ.
  - provides the template for *replication factor C (RFC)*, an accessory factor that loads polymerase (similar to *E.coli* γ-complex).
- ***Pol β***
  - is present in the nucleus as a dimer.
  - has lowest fidelity and processivity.
  - functions in DNA repair.
- ***Pol γ***
  - is involved in mitochondrial replication.
- ***Pol ε***
  - functions in DNA repair.
- ***Pol δ***
  - is the main enzyme in eukaryotic replication.
  - synthesizes both leading and lagging strands.
  - has proofreading function.
  - binds *proliferating cell nuclear antigen (PCNA)*, an accessory factor that functions as a sliding clamp, and increases processivity (similar to β subunit of *E.coli* pol III).

❑ Explain the mechanism of action of DNA ligase.

- DNA ligase catalyzes the formation of a phosphodiester bond between two DNA fragments.
- The cofactor for the enzyme is NAD^+ in *E.coli* and ATP in eukaryotes.
- *Reaction mechanism* (Fig. 29.4)
  - the enzyme reacts with NAD^+ (ATP) to form a covalent adenylyl enzyme intermediate (E-AMP) and nicotinamide mononucleotide (NMN)/PPi.

$$E + NAD^+ (ATP) \rightleftharpoons E\text{-}AMP + NMN\ (PPi)$$

  - the adenylyl group is transferred from E-AMP to the 5'-phosphate of DNA to form a pyrophosphate bond.
  - nucleophilic attack of the 5'-phosphate by the 3'-OH group of the adjacent nucleotide forms a phosphodiester bond.

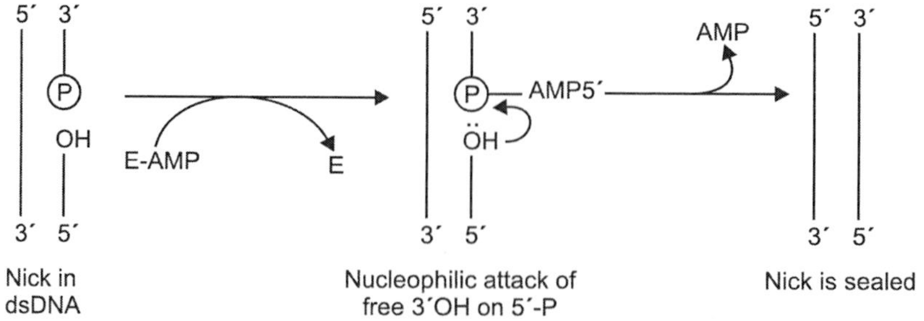

Fig. 29.4 The mechanism of action of DNA ligase

- *Functions*
  - repairs single-strand nicks in duplex DNA.
  - links the ends of linear double-stranded DNA to form circles.
  - joins the ends of Okazaki fragments.

Discuss the mechanism of replication in bacteria.

- Replication in bacteria is bidirectional.
- Two replication forks move in opposite directions from a *replication bubble* or *replication eye,* which becomes larger and assumes a θ-shaped structure.
- *Stages*: Replication involves 3 stages—initiation, elongation, and termination.
- ***Initiation*** involves the following sequence of steps:
  - Replication is initiated at *oriC,* a DNA sequence of 245 bp containing four 9 bp sequences (nonamers) and three AT-rich 13 bp sequences (13-mers).
  - Binding of many molecules of DnaA protein to the nonomers of *oriC* in a co-operative manner with melting of the 13-mers.
  - Loading of DnaB by DnaC to the unwound region of DNA activated by DnaA. Release of DnaC from the DnaB–DnaC complex after loading.
  - Unwinding of DNA bidirectionally by DnaB, a *helicase* creating two replication forks.
  - Stabilization of the two strands separated by *helicase* by SSB, and the torsional stress induced by unwinding relieved by *DNA gyrase.*

Figure 29.5 illustrates replication initiation at *oriC.*

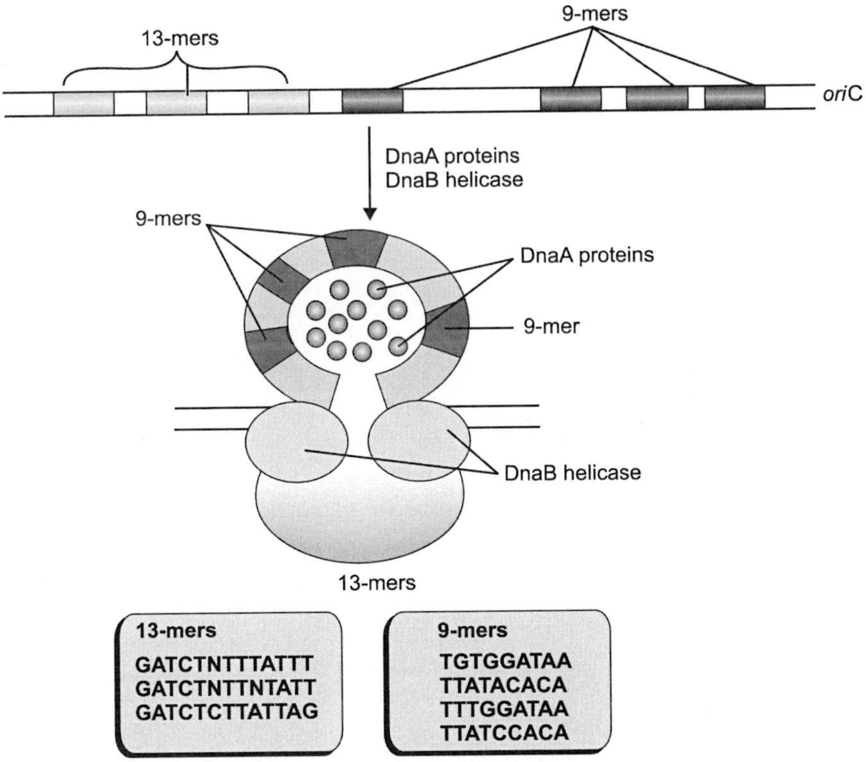

Fig. 29.5 Replication initiation at *oriC*

Elongation

- The two daughter strands are elongated simultaneously in the 5′ → 3′ direction.
  - the strand that is replicated continuously by DNA pol in the direction of the replication fork is the *leading strand*.
  - the strand replicated discontinuously in short fragments of 1000–2000 bp termed *Okazaki fragments* in a direction opposite to the movement of the replication fork is the *lagging strand*.
- *RNA priming* is required for the synthesis of both leading and lagging strands.
  - RNA primer or initiator RNA (iRNA) is 4–12 nucleotides in length base paired to the template and provides a free 3'-OH group for initiating DNA synthesis by polymerase.
- *Extension* of the RNA primer by DNA pol III which functions as a dimer with one core complex of the holoenzyme moving continuously along the leading strand template, while the other cycles from one Okazaki fragment to the other on the lagging strand.
  - As the leading strand is being synthesized, a corresponding portion of the lagging strand is looped through a β-clamp enabling coordinate synthesis of both strands.
  - Both the core complex and β-clamp dissociate after synthesis of the Okazaki fragment and again reassociate to initiate the next Okazaki fragment.
- *Primer removal and gap filling* by DNA pol I. The adjacent Okazaki fragments are linked together by the action of DNA ligase.

Figure 29.6 illustrates the mechanism of replication elongation in bacteria.

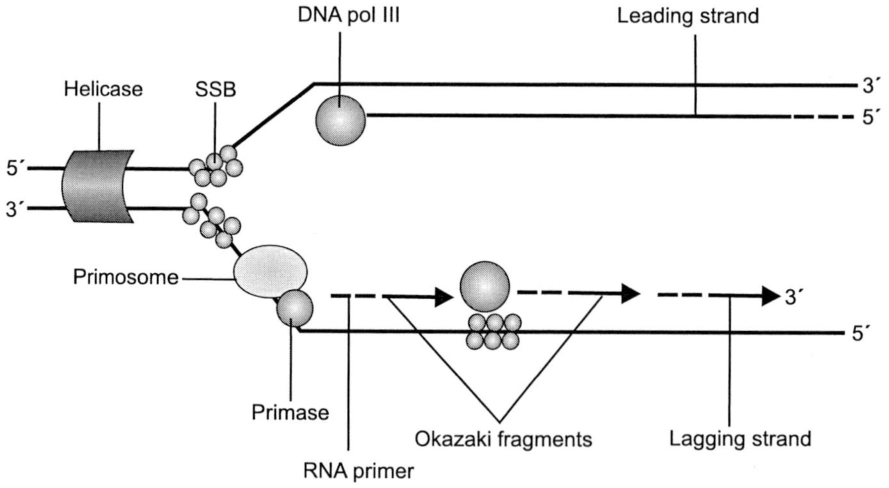

Fig. 29.6 Mechanism of replication elongation in bacteria

Termination

- The two replication forks meet at a specific sequence of 20 bp called the *terminator region (ter)* that binds a protein factor called *Tus* (for terminus utilization substance).
- The Tus–Ter complex arrests a replication fork in one direction and the other fork halts when it encounters the arrested fork.
- Two topologically interlinked circular chromosomes called *catenanes* generated by replication are separated by a *type II topoisomerase* and segregated into daughter cells at cell division.

❑ Briefly outline the salient features of eukaryotic replication.

- In eukaryotes, replication occurs in the S phase of the cell cycle.
- Replication is bidirectional occurring at multiple origins simultaneously.
- The replication origins are present in clusters called *replication units*. In humans, there are about 100 *ori* of replication of over 1000 bp each.
- Each replicon consists of a replication bubble with two replication forks moving in opposite directions. Replication continues until the replication bubbles merge together.
- The mechanism is similar to that seen in prokaryotes.
- There are five different DNA polymerases—α, β, γ, δ, and ε that catalyze replication and repair.
  - *Pol α* loaded by RFC to the initiation complex at the origin synthesizes an RNA primer of 10 bp followed by 20–30 bp of DNA called *initiator DNA (iDNA)*. The iDNA is then extended by polymerase δ/ε. This is known as *pol switch*.
  - *Pol δ* which synthesizes both leading and lagging strands binds PCNA, a cyclin that functions as a sliding clamp.
  - *Pol ε* is involved in lagging strand synthesis.
- The Okazaki fragments in mammals are removed by *RNase H1*, which makes an endonucleolytic cut, and *FEN1* that cleaves the primer.
- The newly synthesized DNA is packaged into nucleosomes by proteins termed *chromatin assembly factors*.

❑ What is the end-replication problem and how is it solved?

- In eukaryotes, replicating the ends of the lagging strand becomes a problem because there is no template available for the RNA primer of the last Okazaki fragment to bind.
- As a result, there is an overhang at the telomeres with gradual shortening of the chromosomes after each round of replication.
- Telomere shortening is considered to be responsible for ageing.
- *Telomerase*, an RNA-protein enzyme complex that carries an RNA template, recognizes the G-rich tip of a telomere sequence and extends it in the 5' → 3' direction before replication.
- Consequently, the 3' end of the telomere is longer than the 5' end and forms a *T loop* which protects against nuclease attack.
- Extension of the parental strands by telomerase is followed by
  - strand separation
  - synthesis of new strands
  - removal of primers
  - gap filling.
- The newly synthesized strands that are formed are not shorter than the parental strands.
- Telomerase
  - is present in microorganisms, plants, animals, and in germ-line cells of humans.
  - is absent in normal human somatic cells but present in cancer cells.
  - may be responsible for the immortalization of cancer cells.
  - is a potential target for anticancer agents.

The mechanism of telomerase-catalysed end replication is illustrated in Fig. 29.7.

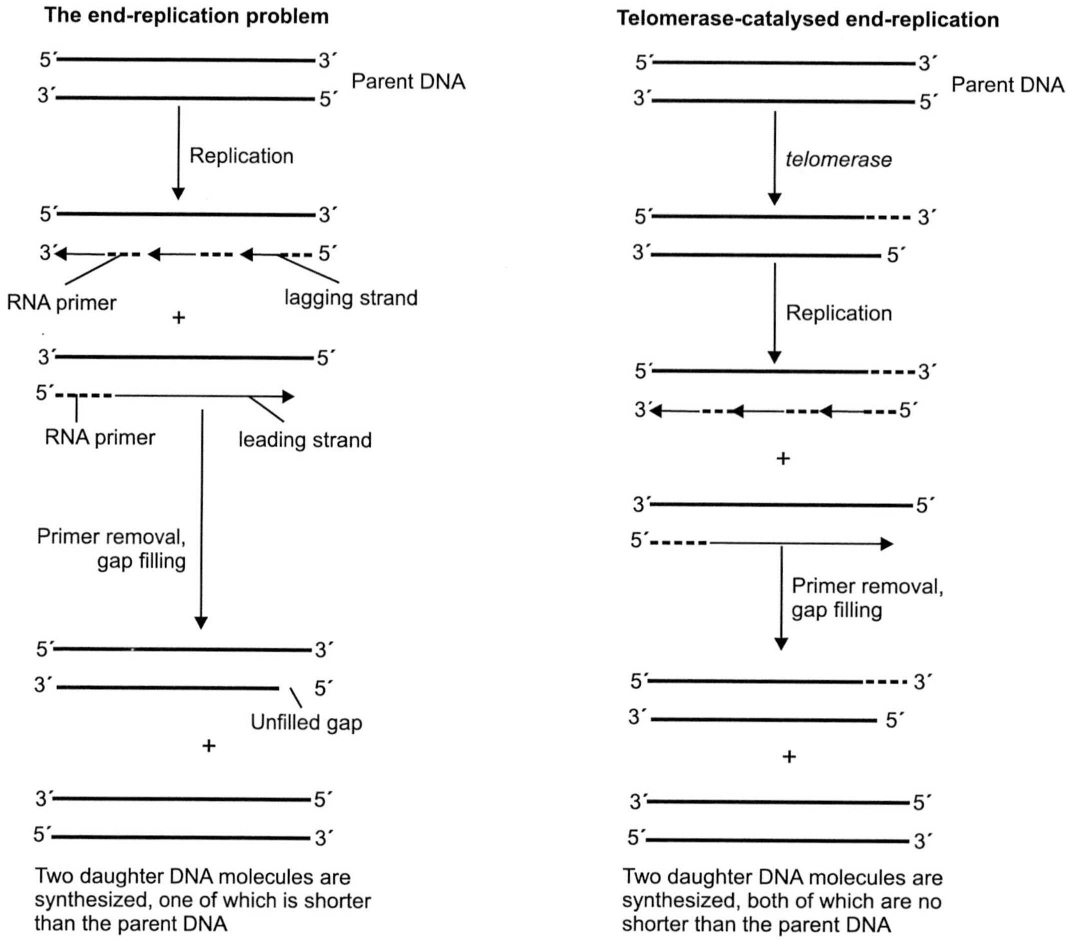

Fig. 29.7 Mechanism of telomerase-catalyzed end replication

❑ Give an account of the inhibitors of replication.

1. *Antimetabolites* such as 5-fluorouracil and methotrexate inhibit the formation of dNTPs, the substrates for replication.
2. *Substrate analogs* are incorporated into DNA after phosphorylation. These include *azothymidine* (*AZT, zidovudine*) used in the treatment of human immunodeficiency virus (HIV) infection, and cytosine arabinoside (*araC, cytarabine*), an antileukemia drug.
3. *Direct inhibitors of DNA* include
   - *intercalators* containing aromatic rings that insert between adjacent base pairs in DNA. These include the anticancer agents such as daunorubicin, produced by strains of *Streptomyces*.
   - *DNA-damaging drugs* including
     - *alkylating agents* that are strong electrophiles that bind to DNA. Examples include methylmethane sulfonate (MMS) and ethylnitrosourea.

- *Platinum-coordination complexes* such as cisplatin that interact with the N^7 of guanine in DNA to form crosslinks between adjacent guanines.
- *Bleomycins* produced by *Streptomyces verticillus* cause DNA damage by interacting with O_2 and Fe^{2+}. These are used as cancer chemotherapeutic drugs.

4. *Inhibitors of replicative enzymes* such as nalidixic acid and fluoroquinolones used in the treatment of urinary tract infections. These inhibit bacterial topoisomerase *DNA gyrase* and prevent it from relieving positive supercoils during replication.

❑ What are the types of DNA damage?

The types of damage that can occur to DNA include the following:

- Single- and double-strand breaks
- Cross-linking between:
  - bases in the same strand
  - bases in the opposite strands
  - DNA and protein.
- Base changes such as:
  - insertions
  - deletions
  - covalent modification.
- Substitutions, which may be:
  - *transitions* (replacement of one purine/pyrimidine by another purine/pyrimidine)
  - *transversions* (replacement of a purine by a pyrimidine or *vice versa*).
- Oxidative damage by reactive oxygen species.

❑ How is DNA damage caused?

Damage to DNA may be

Spontaneous such as

- deamination of cytosine to uracil that base pairs with A instead of G.
- depurination (apurination).
- depyrimidination.

Induced by

- *Radiation*
  - ionizing radiation (X-rays and γ-rays) which causes depurination and hydrolysis of phosphodiester bonds.
  - UV radiation from sunlight forms a thymine–thymine dimer between successive T residues, leading to replication errors.
- *Chemicals*
  - alkylating agents such as methylmethane sulfonate (MMS) which bind DNA.
  - non-alkylating agents such as nitrous acid, which causes deamination of cytosine, adenine, and guanine to uracil, hypoxanthine, and xanthine, respectively.
  - intercalators such as daunorubicin.

- base analogs such as bromodeoxyuridine (BrdU), an analog of T which is incorporated into DNA instead of T. It mispairs with G instead of A, resulting in AT to GC transition.

❑ Explain how thymine dimers are repaired by photoreactivation.

- Photoreactivation repairs UV-induced cyclobutane thymine–thymine dimers. This is accomplished by the enzyme *DNA photolyase*, which contains two chromophores capable of absorbing light:
  - N^5, N^{10}-methenyltetrahydrofolyl polyglutamate (MTHFpolyGlu)
  - $FADH^-$.
- The enzyme uses the energy derived from absorbed light to reverse the damage.

❑ Give an account of the types of excision repair.

Excision repair is of two types: base excision and nucleotide excision repair.

- ***Base excision repair***
  - Deamination of cytosine, adenine, and guanine forms uracil, hypoxanthine, and xanthine, respectively.
  - These bases are removed by *DNA glycosylases* to generate apurinic or apyrimidinic sites.
  - An endonuclease cleaves the abasic sugar and phosphate.
  - The gap is repaired by *DNA pol I* and sealed by *ligase*.

  Figure 29.8 illustrates the mechanism of base excision repair mechanism.

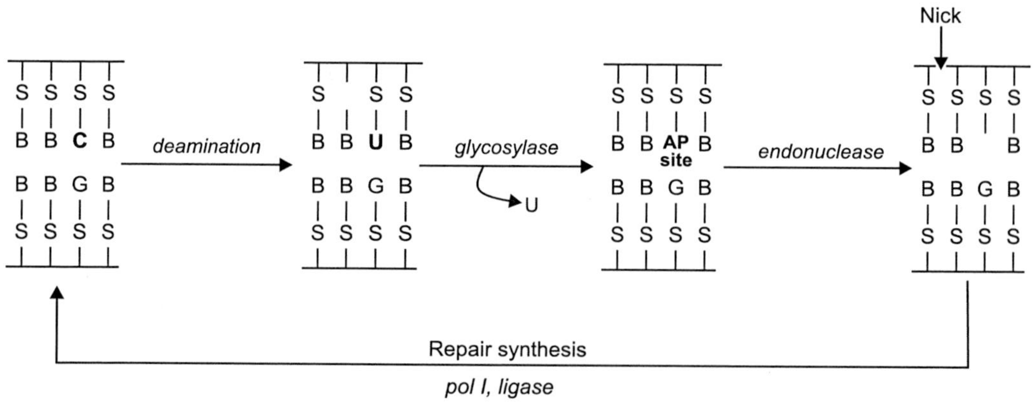

Fig 29.8 Base excision repair mechanism

- ***Nucleotide-excision repair*** (Fig. 29.9)
  - Repairs lesions such as thymine-thymine dimers, 6,4-photoproducts, and base adducts.
  - Upto 30 b of damaged DNA is replaced by excision of a short stretch of nucleotides on either side of the lesion.
  - Two phosphodiester bonds on the strand containing the defect are hydrolysed by an excision nuclease termed *excinuclease* present in both *E.coli* and humans. The gap is filled by *DNA pol I* in *E.coli* and pol δ/ε in humans.
  - The nick is sealed by *DNA ligase*.

- In *xeroderma pigmentosum* (XP), there is an inherited defect in the repair of thymine dimers on DNA. As a result, there is increased sensitivity to the UV rays of the sun, which leads to skin cancers and premature death.

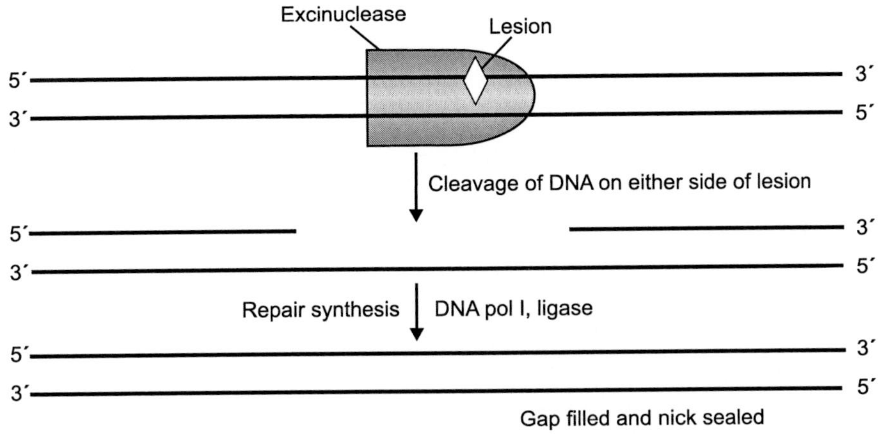

Fig. 29.9 Nucleotide excision repair mechanism

❑ Elucidate the mechanism by which DNA mismatches are repaired.

- The DNA mismatch repair (MMR) system corrects errors made during DNA replication to ensure a high degree of fidelity of replication.
- The mismatches may arise due to
  - mispairing
  - insertion of extra unpaired bases
  - deletion as a result of polymerase slippage.
- Before the mismatch is repaired, the parental (template) strand is methylated on the N^6 position of all adenine residues within the 5′GATC3′ sequences by *Dam methylase* in order to differentiate it from the newly synthesized strand.
- Repair is thus targeted to the transiently unmethylated newly synthesized strand.
- This system repairs up to 100 b from a hemimethylated GATC sequence.

Repair mechanism

- In *E.coli*, the proteins required to recognize the mutation and catalyze nicking include MutS, MutH, and MutL.
- MutS recognizes the mismatch and binds to it.
- MutH attached to MutS by MutL cleaves the unmethylated DNA strand by site-specific endonuclease activity, marking the strand for repair.
- An exonuclease degrades the unmethylated strand containing the mutation from the GATC through the mutation to remove the damaged DNA.
- If the mutation is present between two GATC sites, the DNA between these two sites is removed, the gap filled by polymerase, and nick sealed by ligase.
- Several thousands of bases may be replaced in order to correct one mismatch.
- In humans, defective mismatch repair in hereditary nonpolyposis colon cancer (HNPCC) is associated with microsatellite instability.
- hMSH2 and hMLH1 are human analogs of *E.coli* MutS and MutL.

Figure 29.10 illustrates the mechanism of DNA mismatch repair.

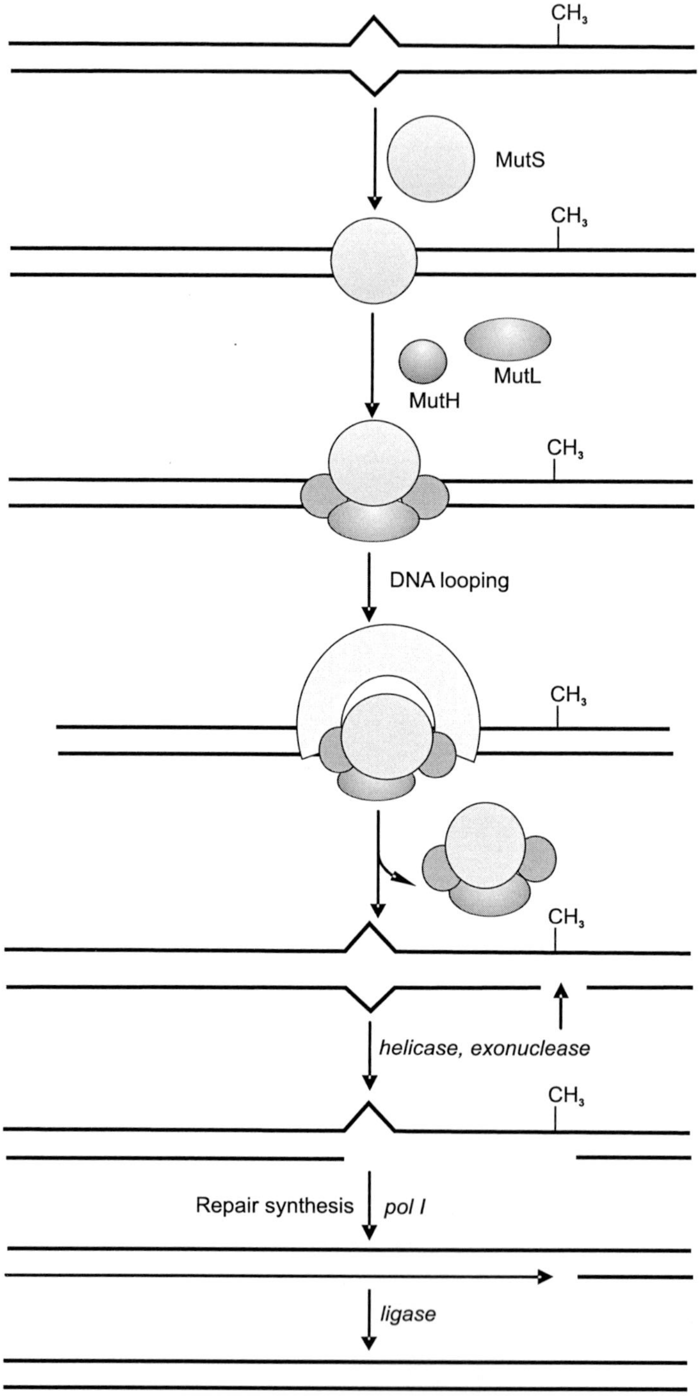

Fig. 29.10 DNA mismatch repair mechanism

Chapter 30

Transcription and Post-transcriptional Processing

❑ What is transcription?

- Transcription refers to the synthesis of RNA from a DNA template catalyzed by the enzyme *RNA polymerase*.

$$(NMP)_n + NTP \longrightarrow (NMP)_{n+1} + PPi$$

- The enzyme requires Mg^{2+} and Zn^{2+} for activity, all the four ribonucleoside 5'-triphosphates (UTP, ATP, CTP, GTP), and DNA template.
- RNA synthesis occurs in the 5' → 3' direction.
- The RNA synthesized is complementary to one of the strands of DNA termed the *template strand* and is identical to the other strand called the *coding strand* (Fig. 30.1).
- Transcription is initiated by binding of *RNA polymerase* to a specific region of DNA called the *promoter* and stops at a *terminator* sequence.

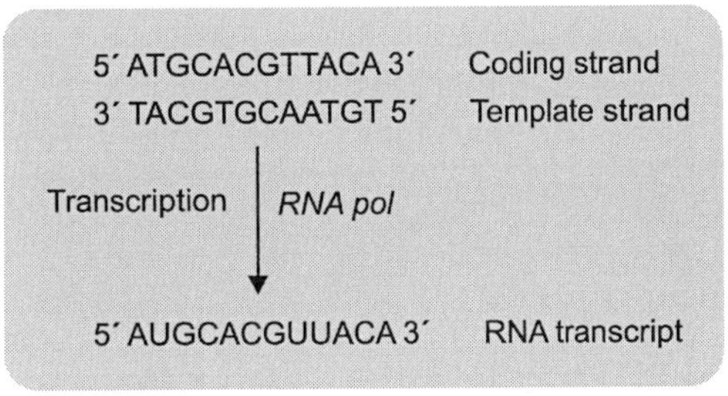

Fig. 30.1 Synthesis of RNA from DNA

- The DNA sequence extending from the promoter to the terminator is termed a *transcription unit*.
- Transcription begins at a specific site in DNA called the *start site (point)*.
- DNA sequences before the start point are referred to as *upstream* sequences and are denoted by a negative (–) sign; whereas sequences after the start point are described as *downstream* sequences and are indicated by a positive (+) sign.
- The product of transcription is known as a *primary transcript*.
- Except prokaryotic mRNA, all the other primary transcripts undergo post-transcriptional modifications to yield mature products.

❑ What is reverse transcription?

- Reverse transcription is the synthesis of DNA from RNA catalyzed by *reverse transcriptase* found in *retroviruses*.
- The retroviruses contain three genes:
  - *gag* that encodes proteins that form the core of the virus particle,
  - *pol* that encodes *reverse transcriptase* and *integrase*,
  - *env* that encodes viral envelope proteins.

- These genes are flanked by long terminal repeats (LTR) that help in viral integration into the host genome (Fig. 30.2).
- The RNA genome in retroviruses replicates via a double-stranded DNA intermediate that gets integrated into the host cell genome.

Fig. 30.2 The retroviral genome

Briefly describe the structure of *E.coli* RNA polymerase.

- In *E.coli*, a single RNA polymerase catalyzes the synthesis of all three types of RNA.
- The holoenzyme contains 5 polypeptide subunits 2α, 1β, 1β', 1ω, and 1σ that can be separated into the core enzyme ($\alpha_2\beta\beta'\omega$) and the σ factor. The σ factor is required for initiation after which it is released and the core enzyme undertakes elongation. The β and the β' subunits constitute the catalytic center.
- *α-subunit*: There are two identical α subunits, each of which is encoded by the *rpo*A gene. The α subunit is essential for core protein assembly.
- *β-subunit*: This is encoded by the gene *rpo*B. The antituberculosis drug rifampicin binds to the β subunit and inhibits transcription initiation, whereas the streptolydigins block transcription elongation.
- *β'-subunit*: This subunit, which is encoded by the *rpo*C gene, binds two Zn^{2+} ions that may be required for catalytic activity. Heparin inhibits transcription *in vitro* by binding to the β' subunit.
- *σ-factor*: σ^{70} is the most common sigma factor in *E.coli* that is responsible for transcription initiation by promoter recognition. Other σ factors that are expressed under altered environmental conditions such as σ^{32} during heat shock and σ^{N} during nitrogen starvation have been recognized.

Elucidate the salient features of the bacterial promoter.

- The *E.coli* σ^{70} promoter sequence ranges between 40 and 60 bp.
- RNA polymerase binds to the promoter around –55 to +20.
- Four features are conserved in bacterial promoters:

  TTGACA.......16–18bp........**TA**TAAT......5–8bp..........Start point

  - *The start point* is usually a purine (A or G) and is the central base in the sequence CG(A)T.
  - *The –10 sequence* present ~10 nucleotides upstream of the start site contains a 6 bp AT-rich consensus sequence TATAAT with the following percentage of conservation:

    $$T_{80}\ A_{95}\ T_{45}\ A_{60}\ A_{50}\ T_{96}$$

  - *The –35 sequence* present ~35 nucleotides upstream of the start site has the consensus sequence TTGACA and functions as a *recognition site* for RNA polymerase and enhances interaction with the σ factor.
  - *Distance between the –10 and –35 sequences* is 6–18 bp.

Discuss the stages in bacterial transcription.

The stages in transcription include template recognition, initiation, elongation, and termination (Fig. 30.3).

Template Recognition

- RNA polymerase recognizes and binds to the –35 and –10 sequences on the promoter.
- The σ factor increases the affinity for the promoter.
- The initial binary complex of the enzyme and the promoter DNA is termed the *closed complex*.
- RNA pol unwinds the DNA over a short distance of about 17 bp, converting the closed complex into an *open complex*.
- DNA unwinding forms a *transcription bubble* and makes the template strand available for transcription.

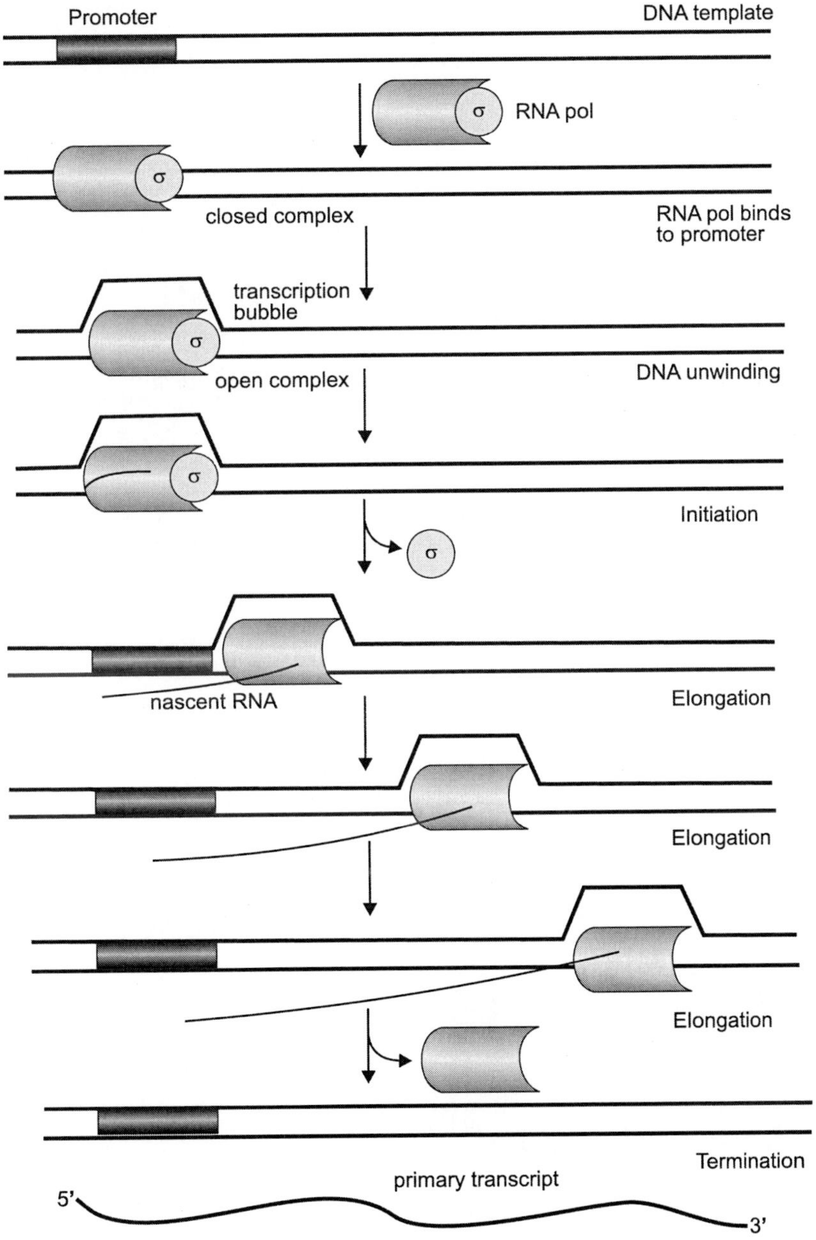

Fig. 30.3 Mechanism of transcription in prokaryotes

- ***Initiation***
  - Involves formation of a phosphodiester bond between the first nucleotide (usually GTP or ATP) and the second nucleotide (usually UTP or CTP), thereby creating a *ternary complex* of RNA pol-DNA-nascent RNA.
  - The 5'-triphosphate group on the first nucleotide is not hydrolysed but remains intact throughout transcription.
  - RNA pol makes short transcripts of up to 9 b, releases them, and reinitiates RNA synthesis by a process termed *abortive initiation.*
  - The mean time for promoter clearance by RNA polymerase is 1–2 s.
  - Following initiation, the σ factor is released from the holoenzyme and the core enzyme undertakes elongation.
- ***Elongation***
  - RNA pol moves along the DNA, unwinding short regions for elongation in front of the transcription bubble and rewinds DNA behind it.
  - Nucleotides are successively added to the growing RNA chain.
  - The 3' end of RNA forms a transient hybrid of 8–12 bp with the template DNA.
  - RNA pol is released from elongation arrest by Gre A and Gre B.
- ***Termination***
  - Transcription continues until RNA pol encounters the *terminator* sequence.
  - Termination involves the following events:
    a. release of transcript
    b. dissociation of the RNA–DNA hybrid
    c. reformation of the DNA duplex.
  - Two types of termination can occur in prokaryotes: *factor independent* or *intrinsic termination* and ρ-dependent termination (Fig. 30.3).

❑ Explain the mechanism of transcription termination in prokaryotes.

Two types of termination can occur in prokaryotes.

Factor independent (intrinsic termination) is based on two structural features at the end of the RNA transcript (Fig. 30.4).

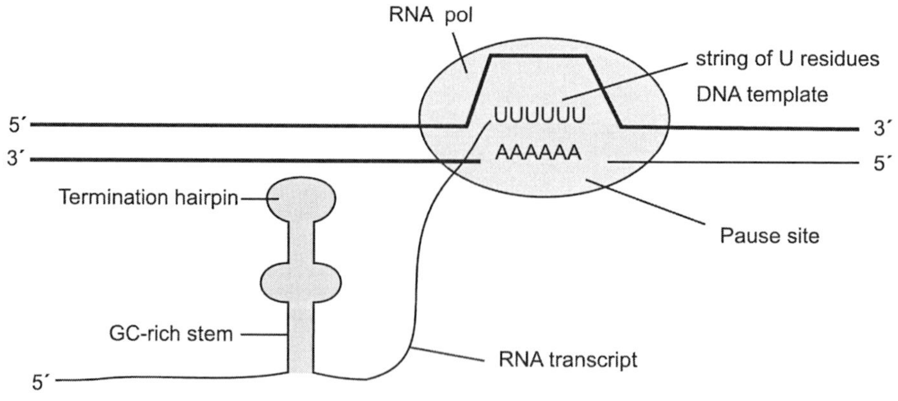

Fig. 30.4 Mechanism of intrinsic termination of transcription

- A self-complementary hairpin structure 15–20 nucleotides before the end of the RNA. The base of the hairpin, which is GC-rich, prevents RNA binding and disrupts the RNA–DNA hybrid.
- A string of U residues that base pair only weakly with the corresponding A residues in the template DNA strand. As a result, RNA dissociates from the RNA–DNA hybrid (Fig. 30.4).

ρ-dependent termination

- A hexamer with 5' → 3' helicase activity and RNA-dependent ATPase activity binds to nascent RNA upstream of the terminator.
- At the termination sequence, ρ factor unwinds the RNA–DNA duplex and dissociates the RNA transcript from DNA terminating transcription.

❑ Give an account of eukaryotic RNA polymerases.

- Eukaryotes contain three different RNA polymerases–RNA polymerase I, II, and III that differ in their sensitivity to the fungal toxin, *α-amanitin*.
- All the three enzymes are large complex proteins containing 2 large subunits and about 10 smaller subunits. The subunits show homology to α, β, and β' subunits of *E.coli* RNA pol.
- All the three enzymes require *transcription factors* to initiate transcription.
- *RNA polymerase I (RNA pol I)* present in the nucleolus transcribes all the rRNA except 5S rRNA. It is insensitive to α-amanitin.
- *RNA polymerase II (RNA pol II)* present in the nucleoplasm and chromatin transcribes the genes for mRNA and some small nuclear RNA (snRNA). It is highly sensitive to α-amanitin.
- *RNA polymerase III (RNA pol III)* located in the nucleoplasm transcribes the genes for tRNA, 5SrRNA, U6 snRNA, and some other small RNAs. It is moderately sensitive to α-amanitin.

❑ Give a brief account of the sequences found in the RNA pol II promoter.

- RNA pol II located in the nucleoplasm catalyzes the transcription of protein-coding (mRNA) genes and some snRNA genes.
- The RNA pol II promoter contains the following sequences (Fig. 30.5):
  - *Minimal (core) promoter elements* include the
    a. *TATA box* located 25–35 bp upstream of the transcription start site with the consensus sequence 5'-TATA(A/T)A(A/T)-3' which positions RNA pol II for correct transcription initiation.
    b. *Initiator (INR) element* present around the transcription start site has the consensus sequence 5'-C/TC/TANT/AC/TC/T-3'.

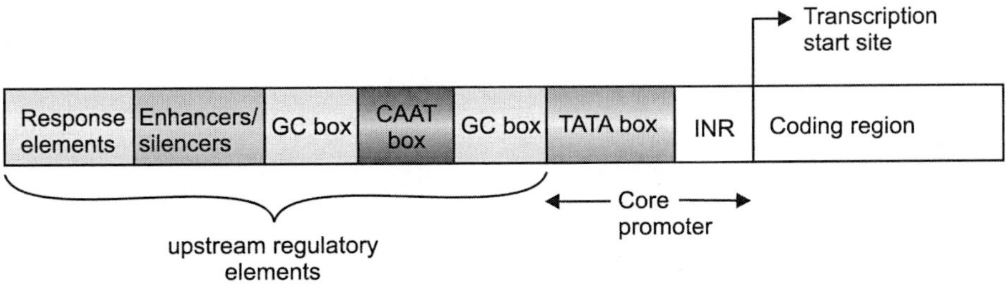

Fig. 30.5 RNA pol II promoter

- *Upstream regulatory elements* (URE) located several hundred bp upstream of the core promoter determine the *frequency* of transcription. These include
  a. *GC boxes* that bind Sp1 and Sp2.
  b. *CAAT box* that binds the CAAT binding factor (CTF).
  c. *enhancers* and *silencers* that increase or decrease the rate of transcription initiation, respectively. These are found either upstream or downstream from the transcription start site and can exert their effects even when located thousands of bp away from the transcription unit.
  d. *response elements* identify particular groups of genes and include the *heat shock response element (HSE), glucocorticoid response element (GRE)*, and *metallothionein response element (MRE)*.

❑ Discuss the mechanism of eukaryotic transcription with RNA pol II as an illustrative example.

Transcription by RNA pol II involves the formation of a *preinitiation complex* followed by initiation, elongation, and termination stages of transcription.

Formation of preinitiation complex (PIC)

Transcription initiation requires the formation of PIC consisting of RNA pol II and the general TFs TFIIA, B, D, E, F, H, and J.

The steps involved in the formation of PIC include the following (Fig. 30.6):

1. Binding of TFIID to the promoter. TFIID is made up of 2 components– TATA-binding protein (TBP) and about 11 TBP-associated factors (TAFs). TBP recognizes and binds to the TATA box.
2. Binding of TFIIA and TFIIB.
3. Binding of RNA pol II together with TFIIF followed by binding of TFIIE, H, and J. TFIIF targets RNA pol II to the promoter by decreasing RNA pol binding to nonspecific DNA sites. TFIIF and H are required for promoter clearance. TFIIH helps RNA pol II to gain access to the template strand.
4. After synthesizing short lengths of RNA at the promoter, TFIIH phosphorylates the C-terminal domain (CTD) of RNA pol II due to which the enzyme undergoes a conformational change that facilitates promoter clearance. The general TFs are cleared and RNA pol II continues transcription (Fig. 30.6).

Transcription initiation

Although the process is similar to that seen in bacteria, the following are required:

1. *Transcriptional activator proteins* that bind to enhancer sequences on DNA. These proteins position RNA pol and the general TFs by direct action on the transcription machinery and by histone acetylation of the chromatin around the promoter catalyzed by *histone acetyltransferases (HATs)*.
2. *Mediator*, a protein complex that mediates interaction between the activator protein, RNA pol, and general TFs.
3. *Local recruitment of chromatin* to facilitate the assembly of transcription initiation proteins onto DNA.

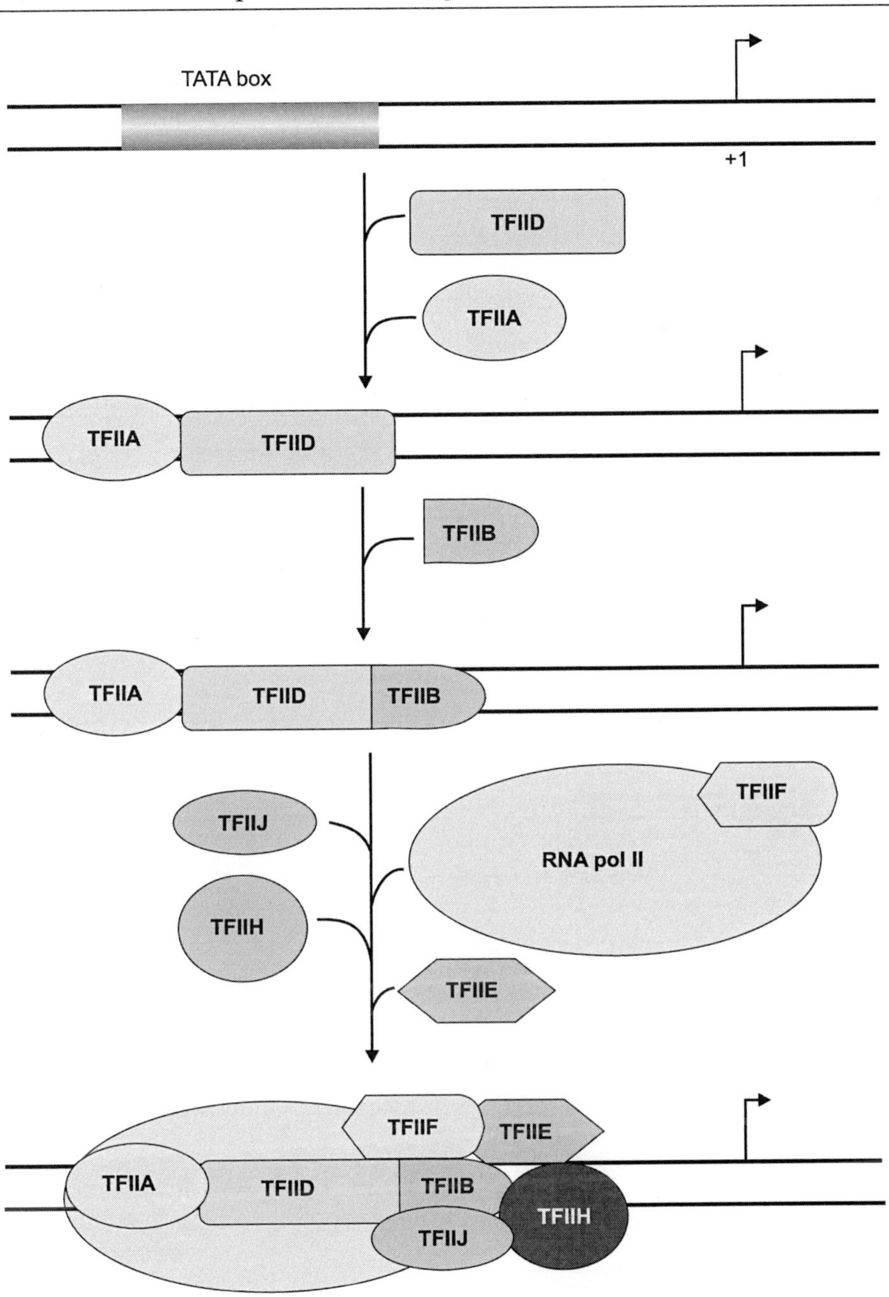

Fig. 30.6 Formation of RNA pol II preinitiation complex

Elongation

The requirements and mechanism are similar to that in prokaryotes.

Termination

Transcription continues for up to several 100 bp beyond the polyadenylation site.

❑ Explain the post-transcriptional processing of prokaryotic rRNA.

- Prokaryotic rRNAs arise from a long precursor 30S *prerRNA* that contains one copy each of 5S, 16S, and 23S rRNAs. The spacer region between 16S and 23S encodes tRNA.
- During and following transcription, stem-loop structures are formed by base pairing between complementary sequences.
- Processing involves specific endonucleolytic cleavages by RNase III, RNase P, E, and F.
- RNase III cuts within the stem-loop structures to generate pre16S and pre23S molecules.
- 16S and 23SrRNA undergo methylation during ribosomal assembly to form N^6, N^6-dimethyladenine and $O^{2'}$-methylribose residues that protect the adjacent phosphodiester bonds from degradation by intracellular RNases.
- Secondary processing trims the 5' and 3' ends of prerRNAs by RNases M5, M16, and M23 to generate mature rRNAs after association of the preRNA with the ribosomes.

Figure 30.7 illustrates post-transcriptional processing of prokaryotic rRNA.

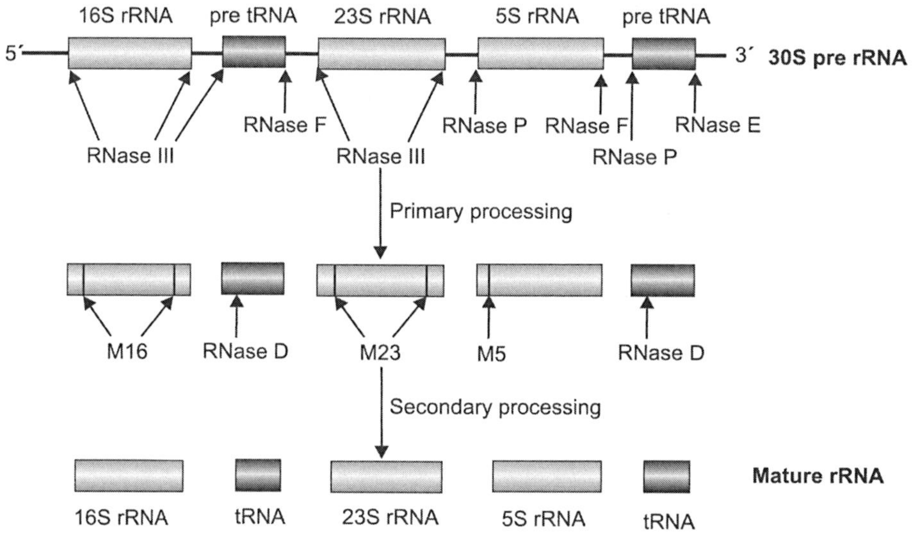

Fig. 30.7 Post-transcriptional processing of prokaryotic rRNA

❑ Explain the post-transcriptional processing of eukaryotic rRNA.

- rRNA is generated from a single, long 45S precursor molecule that contains 18S, 5.8S, and 28S rRNA (Fig. 30.8).
- The 45S precursor molecule undergoes modifications with the help of guide RNAs. These include
  - folding and complexation with proteins.
  - methylation to generate 2'-O-methylribose.
  - isomerization of uridine nucleotides to pseudouridine.
  - extensive cleavage to release the 20S rRNA and 32S rRNA precursors that are trimmed further.
  - further methylation.
- The 5.8S rRNA base pairs with the 28S rRNA followed by self assembly with ribosomal proteins to form the 60S subunit.

- The 18S rRNA associates with constituent peptides to generate the 40S subunit.

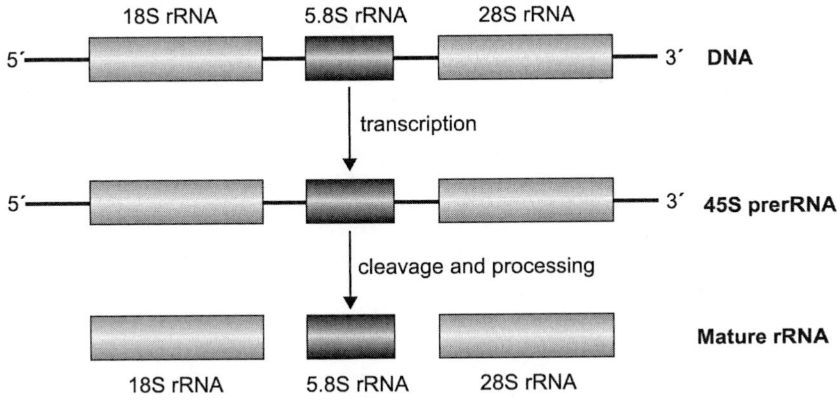

Fig. 30.8 Post-transcriptional processing of eukaryotic rRNA

Outline the post-transcriptional processing of prokaryotic tRNA.

- In *E.coli*, the tRNA genes are mostly distributed in clusters while some are found within the rRNA operons.
- tRNAs are formed from larger precursor molecules known as *pretRNA* that contain 1–7 different tRNAs as well as additional sequences at the 5' and the 3' ends referred to as *leader* and *trailer* regions and spacer regions.
- Processing involves
  - folding into the characteristic tRNA structure with stems and loops.
  - removal of the flanking sequence at the 3' end by the endonuclease RNase E or F leaving a precursor with 9 extra nucleotides, followed by successive removal of 7 of these nucleotides by *Rnase D*.
  - endonucleolytic cleavage by *RNase P* to generate the mature 5' end. RNase P is an endonuclease and the RNA component functions as an endonuclease. RNase P is an example of a catalytic RNA or *ribozyme*.
  - removal of the remaining two nucleotides from the 3' end by *RNase D* to generate the mature 3' end.
  - base modifications such as methylation, deamination, or reduction.

Elucidate the post-transcriptional processing of eukaryotic tRNA.

- tRNA is generated from a long precursor pretRNA molecule containing a 5' leader sequence, an intron containing 14 nucleotides, and two extra 3' nucleotides.
- The primary transcript folds into the characteristic tRNA structure with stems and loops.
- Processing involves
  - removal of the 5'-leader sequence and the two 3'-nucleotides.
  - addition of 5'-CCA terminus to the 3' end by the enzyme *tRNA nucleotidyltransferase* to generate the mature 3' end.
  - removal of the intron.
  - splicing of exons.
  - base modifications as in prokaryotes.

❑ What is a ribozyme?

- A *ribozyme* is a catalytic RNA molecule (RNA enzyme).
- Ribozymes are seen in both prokaryotes and eukaryotes.
- Examples include the following:
  1. RNase P, a ribonucleoprotein that functions as an endonuclease and generates the mature 5' end of tRNA. The RNA component alone is sufficient to function as an endonuclease and the protein is required to stabilize the RNA or facilitate its functions.
  2. The self-splicing group I intron L-19 intervening sequence (IVS) in *Tetrahymena thermophila* L-19 RNA behaves as a *nucleotidyl transferase* and catalyzes transesterification reactions during rRNA splicing.
  3. *Peptidyltransferase* present in the larger subunit of the ribosome.

❑ Discuss the post-transcriptional modifications of eukaryotic mRNA.

- In eukaryotes, premRNA is synthesized as a very long precursor molecule called *heterogeneous nuclear RNA (hnRNA)* that complexes with proteins to form *heterogeneous nuclear ribonucleoproteins (hnRNPs)*.
- hnRNA contains both exons and introns.
- Mature mRNA molecules are generated after post-transcriptional modifications that include 5'-capping, 3'-cleavage and polyadenylation, splicing and methylation.
- ***5'-Capping***
  - Addition of the 5' cap of 7-methylguanosine (m^7G) is catalyzed by *guanyl transferase.*
  - The cap positions the mRNA on ribosomes and protects against 5'-exonucleases.
- ***3'-Cleavage and polyadenylation***

  A *poly A tail* of 100–200 residues is added to the 3' end generated by cleavage.
  - The *polyadenylation site* contains a *polyadenylation signal* 5' AAUAAA-3' followed by a 5'-YA-3', where Y = pyrimidine and a GU-rich sequence further downstream.
  - The polyadenylation signal is recognized by a specificity component CPSF and cleaved by the endonucleases CFI and CFII.
  - *Poly(A)polymerase (PAP)* adds the poly A tail to the 3' end of the cleaved premRNA: PAP has no requirement for template but requires RNA as a primer as well as CPSF and poly A binding protein (PABP).
- ***Splicing***
  - Involves removal of introns and joining of exons.
  - *Splice site consensus* (Fig. 30.9)
    a. Introns begin with 5' GU 3' and end with 5' AG 3'.
    b. A *polypyrimidine tract* is present before AG at the 3' end.
    c. About 10–40 nucleotides upstream of the polypyrimidine tract is a *branchpoint sequence.* In vertebrates, the branchpoint is 5' CURAY 3', where R = purine and Y = pyrimidine.

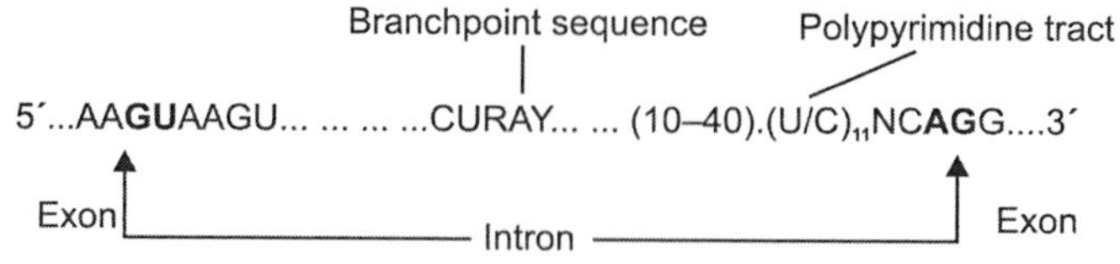

Fig 30.9 Splice site consensus

- Splicing is catalyzed by *spliceosome*, an RNA-protein complex that contains the primary transcript, the snRNAs U1, U2, U4, U5, and U6, and many proteins (Fig. 30.10).
- U1 snRNP makes a cut at the 5'-splice site. An unusual 5'-2'-phosphodiester bond is formed between the 5'-terminal G residue of the intron and the 2'-OH of an A residue within the intron. This generates a *lariat* structure with the A residue as the branch point.
- The 3'-splice site is cut to release the intron in lariat form.
- The lariat is linearized and degraded.
- The two exons are spliced by transfer of a phosphodiester bond to a different –OH group without loss of energy or ATP hydrolysis (*transesterification reaction).*

- **Methylation** of N6 position of A residues, especially when they occur in the sequence 5' RRACX 3', where X is rarely G.

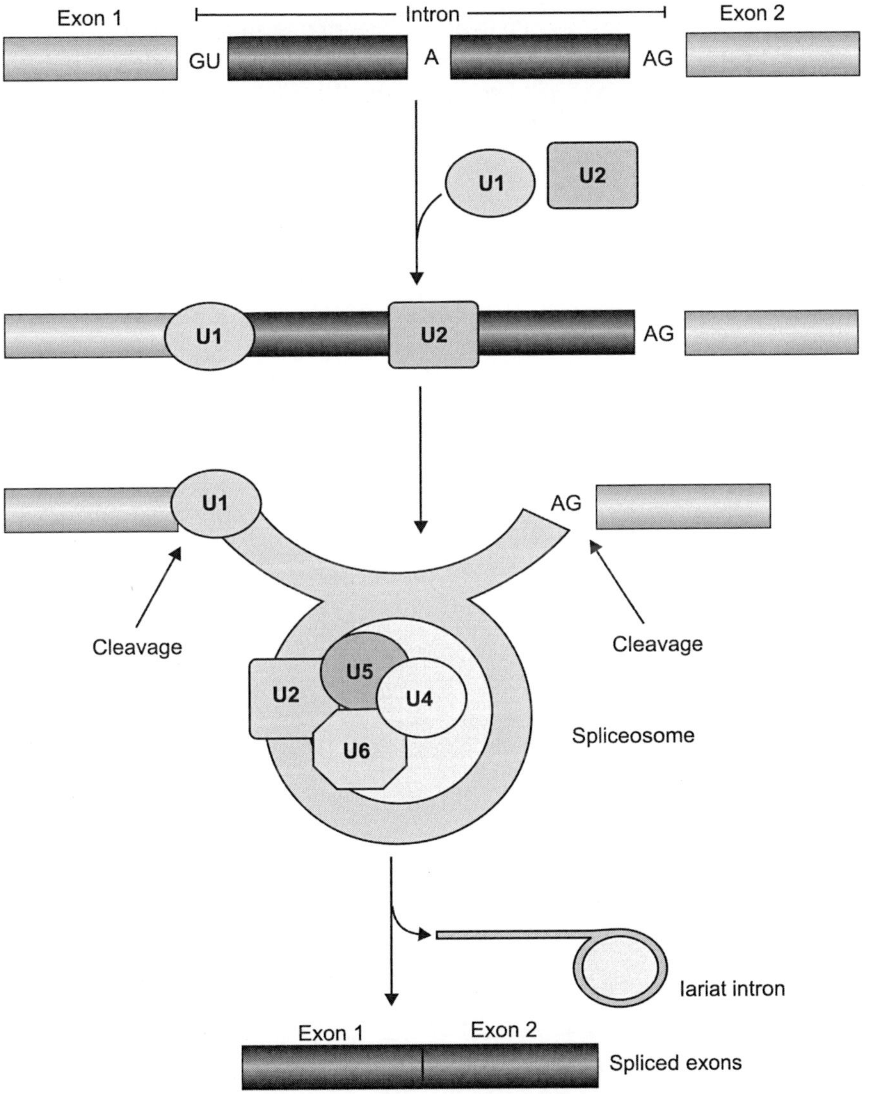

Fig. 30.10 Mechanism of eukaryotic mRNA splicing

❑ Write notes on RNA editing.

- *RNA editing* refers to a change in genetic information at the mRNA level.
- It occurs in organelle genomes as a co- or post-transcriptional event.
- RNA editing may involve insertion, deletion, or base modifications and results in the synthesis of a different polypeptide.
- The types of RNA editing include
  - *simple editing* involving single residue conversions. For example, apolipoprotein mRNA gives rise to a truncated protein in the small intestine because the codon for glutamine CAA is converted to the stop codon UAA by a C to T substitution.
  - *insertional editing* involves insertion of single nucleotide or small runs of nucleotides such as G insertions during transcription of the paramyxovirus P gene.
  - *pan editing* in which insertion/deletion of multiple uridine residues occurs.
  - *polyadenylation editing* in which premRNA lacking a stop codon is polyadenylated with the first one or two adenylate residues providing the missing information.

Chapter 31

Translation and Regulation of Gene Expression

❑ Outline the salient features of the genetic code.

The genetic code shows the following features:

1. It is *universal, i.e.*, it is similar in most organisms. However, the organelle genomes (mitochondria, chloroplasts) show slight deviations from the universal genetic code.
2. It is read in triplet nucleotide sequences termed *codons*, each codon representing one amino acid.
3. It is *unambiguous*. Each codon specifies a single amino acid.
4. It is *commaless, i.e.*, there is no punctuation to signal the end of one codon and the beginning of the next.
5. There are 64 codons in the genetic code, of which 61 code for the 20 amino acids (Table 31.1).
6. The initiating codon is AUG that encodes N-formylmethionine in prokaryotes and methionine in eukaryotes.
7. Three of the 64 codons do not code for any amino acid, and signal termination of the polypeptide chain. These are termed *stop codons, termination codons,* or *nonsense codons* and include UAA, UAG, and UGA.

Table 31.1 The genetic code

First position	Second position U	C	A	G	Third position
U	Phe	Ser	Tyr	Cys	U
	Phe	Ser	Tyr	Cys	C
	Leu	Ser	**Stop**	**Stop**	A
	Leu	Ser	**Stop**	Trp	G
C	Leu	Pro	His	Arg	U
	Leu	Pro	His	Arg	C
	Leu	Pro	Gln	Arg	A
	Leu	Pro	Gln	Arg	G
A	Ile	Thr	Asn	Ser	U
	Ile	Thr	Asn	Ser	C
	Ile	Thr	Lys	Arg	A
	Met	Thr	Lys	Arg	G
G	Val	Ala	Asp	Gly	U
	Val	Ala	Asp	Gly	C
	Val	Ala	Glu	Gly	A
	Val	Ala	Glu	Gly	G

8. The genetic code starts with the initiation codon AUG and ends with any one of the termination codons. The region of the mRNA that extends from the initiation codon to the termination codon is referred to as the *coding region* if it encodes a protein product and as an *open reading frame (ORF)* if the protein product is not known.
9. The genetic code is degenerate, *i.e.*, most of the amino acids are specified by more than one codon. For example, there are six codons for serine, four for glycine, and two for glutamate.
10. Degeneracy generally involves the third base in the codon, which is known as the *wobble base*. For example, the four codons for alanine GCU, GCC, GCA, and GCG differ only in the third base. The wobble base is responsible for the tRNAs to recognise more than one codon.

❑ What are the deviations shown by the mitochondrial genetic code from the universal genetic code?

In mitochondria,

- The codon for methionine is **AUA** and not AU**G**.
- UGA, a termination codon, encodes tryptophan.
- AGA signals termination instead of coding for arginine.

❑ Discuss in detail the steps in protein biosynthesis in prokaryotes.

The steps in protein biosynthesis include activation of amino acids, initiation of polypeptide chain synthesis, elongation, and termination.

◘ *Activation of amino acids*

- All the 20 amino acids are esterified to their corresponding tRNAs by *aminoacyl-tRNA synthetases* by the following reaction:

  Amino acid + tRNA + ATP ——> Aminoacyl-tRNA + AMP + PPi

- The process involves reaction of the amino acid with ATP to form an enzyme-bound aminoacyl adenylate intermediate, followed by transfer of the amino acyl group to the terminal adenine residue in the 3'CCA terminus of tRNA.
- The aminoacyl-tRNA synthetases have proofreading function.

◘ *Initiation of polypeptide chain synthesis*

- The initiating amino acid is N-formylmethionine encoded by the initiation codon AUG.
- The initiating tRNA is methionine $tRNA^{fmet}$ which accepts the formyl group. This differs from the tRNA species methionine-$tRNA^{met}$ that inserts methionine residues in internal positions of the polypeptide chain.
- *Requirements*
  i. the ribosome subunits—30S and 50S.
  ii. mRNA that contains codons specifying the amino acid sequence of the polypeptide chain.
  iii. initiating N-formylmethionyl $tRNA^{fmet}$.
  iv. initiation factors IF-1, IF-2, and IF-3.
  v. GTP.
- *Steps*
  i. Binding of IF-1, IF-2, and IF-3 to the 30S subunit together with GTP. This prevents association of the 30S subunit with the 50S subunit.
  ii. Binding of mRNA to the 30S subunit. mRNA is positioned for translation initiation by a polypurine stretch (5'AGGAGGU3') termed the *Shine-Dalgarno (SD) sequence* present approximately 7b upstream of the AUG codon that base pairs with a pyrimidine sequence in the 16S rRNA of the 30S subunit.

iii. Complexation of IF-2 with GTP which helps the initiator tRNA to bind to the 30S subunit and base pair with the initiation codon AUG on mRNA.

iv. Release of IF-3 from 30S to form the 30S *initiation complex*.

v. Binding of the 50S subunit to the 30S initiation complex to form the *70S initiation complex* accompanied by displacement of IF-1 and IF-2 and hydrolysis of GTP.

The 70S ribosome contains two sites for binding two tRNAs—the *peptidyl (P) site* for peptidyl tRNA, and the *aminoacyl (A) site* for the incoming aminoacyl-tRNA. Except the initiator tRNA which occupies the P site, all other tRNAs enter the A site (Fig. 31.1).

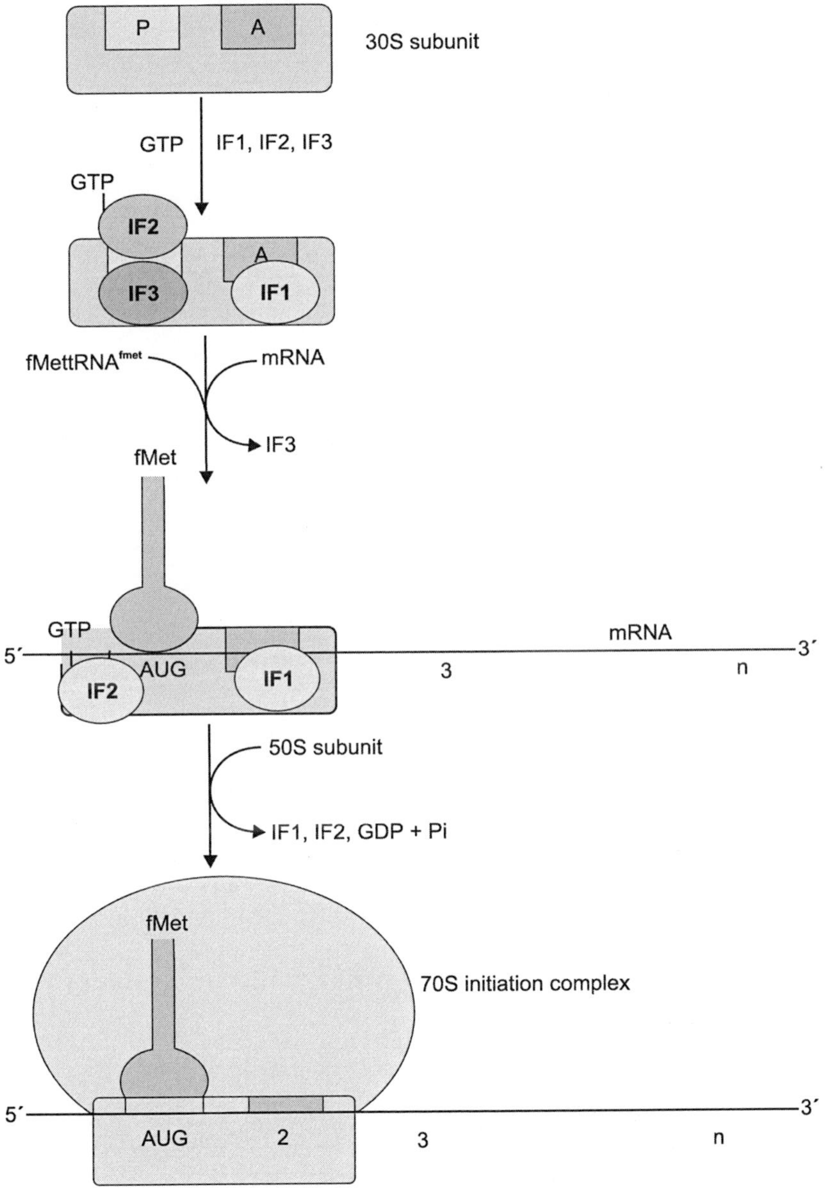

Fig. 31.1 Initiation of polypeptide chain synthesis

- ***Elongation***

  Requirements
  - 70S initiation complex
  - next incoming aminoacyl-tRNA
  - elongation factors EFTu, EFTs, and EFG
  - GTP.

  Stages
  1. *Binding of incoming aminoacyl-tRNA* to the A site of the initiation complex, mediated by EFTu that involves the following sequence of events:
     - The aminoacyl-tRNA binds as a ternary complex of EFTu-GTP-aminoacyl-tRNA.
     - GTP hydrolysis.
     - Release of EFTu-GDP.
     - Regeneration of the active EFTu-GTP by EFTs.

     This stage is blocked by tetracycline.
  2. *Peptide bond formation*
     - occurs between fmetRNA^{fmet} on P site and the newly bound aminoacyl-tRNA on the A site.
     - is catalyzed by *peptidyl transferase* present in the 50S subunit.
     - results in a dipeptidyltRNA bound to the A site leaving an uncharged initiator tRNA on the P site.

     This stage is inhibited by puromycin.
  3. *Translocation* catalyzed by EFG (translocase) and GTP involves
     - movement of the ribosome to the next codon on the mRNA
     - shifting of the dipeptidyl tRNA from the A site to the P site
     - removal of the uncharged tRNA from the P site.

The A site is now free for entry of the next aminoacyl-tRNA. EFG dissociates from the ribosome and GTP is hydrolyzed to GDP and Pi.

Stages 1–3 in elongation are repeated until a termination codon is encountered (Fig. 31.2).

- ***Termination***
  - occurs after the last amino acid is added.
  - is signaled by any one of the three termination codons—*UAG (amber)*, *UAA (ochre)*, and *UGA (opal)*.
  - is catalyzed by three release factors *RF-1, RF-2, and RF-3.*
  - involves release of the completed polypeptide chain, and dissociation of tRNAs and mRNA from the ribosome.
  - requires EF-G and a ribosome release factor for the complete dissociation of the subunits.

Enumerate the differences between prokaryotic and eukaryotic translation.

Although the basic mechanism of translation is similar in prokaryotes and eukaryotes, there are certain differences in the number of components and some of the steps.

- The sedimentation coefficient of ribosomes is 80S and subunits (larger 60S and smaller 40S).
- The mRNA has a 5' cap that has a role in translation initiation and in mRNA stability.
- The eukaryotic mRNA encodes a single polypeptide chain (*monocistronic*), unlike prokaryotic mRNA, which encodes several polypeptide chains (*polycistronic*).

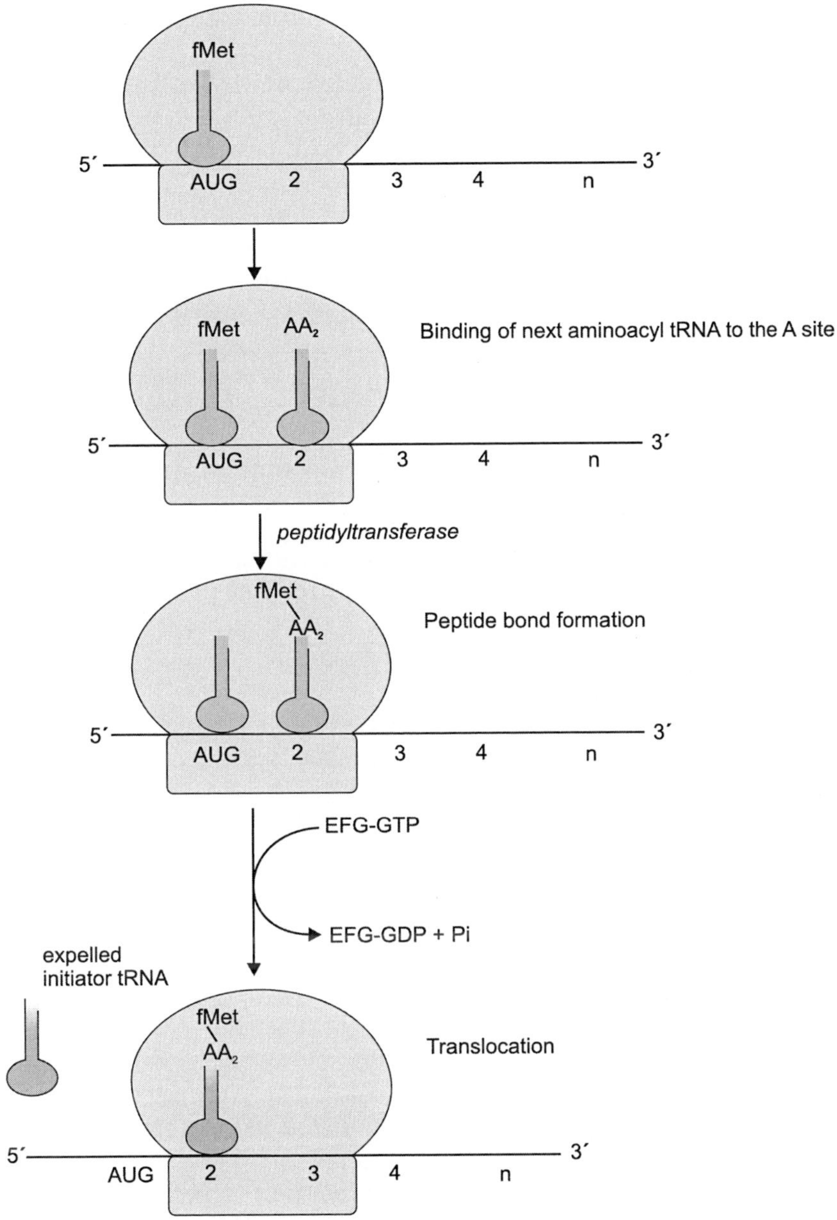

Fig. 31.2 Elongation of polypeptide chain

- The initiating amino acid is methionine and the initiator tRNA is $tRNA_i^{met}$ that is different from $tRNA_m$ recognized by internal methionine codons.
- SD sequence that directs initiation to AUG in prokaryotes is absent in eukaryotes. The start codon on the mRNA may be selected from within the *Kozak consensus sequence* 5'-CCRCCAUGG-3', where R = purine. This is achieved by scanning the mRNA by the 40S subunit containing the initiator tRNA.
- More number of factors are involved in eukaryotic initiation compared to prokaryotes. Nine initiation factors (labeled *eIF*) and a *cap binding protein* (*CBP*) have been identified.

- The order in which the initiation complex is assembled in eukaryotes is different from that seen in prokaryotes.
  - the initiator tRNA binds to the small subunit followed by mRNA binding.
  - binding of CBP to the 5'-cap promotes binding of mRNA as well as eIF-4A and eIF-4B.
  - ATP is hydrolyzed and bound by eIF-4A with unwinding of secondary structure in the 5'-untranslated region of the mRNA.
  - eIF-4C helps in association of 60S subunit with 40S.
  - eIF-5 displaces eIF-2 and eIF-3.
  - eIF-2B converts eIF-2-GDP released to eIF-2-GTP.
- The elongation factors eEF1α, eEF1$\beta\gamma$, and eEF2 are similar to prokaryotic EFTu, EFTs, and EFG, respectively, in their functions.
- A single release factor eRF that recognizes all the three stop codons is responsible for termination of polypeptide chain synthesis.

❑ Give an account of the inhibitors of protein biosynthesis.

The following agents inhibit protein biosynthesis:

- ***Antibiotics*** such as
  - *Tetracycline,* which blocks the A site on the ribosome and inhibits binding of aminoacyl-tRNAs.
  - *Streptomycin,* which binds to the 30S subunit and
    i. destabilizes the initiation complex
    ii. causes misreading of the genetic code
    iii. impairs dissociation of ribosomes into subunits
    iv. induces membrane damage.
  - *Erythromycin* binds to the 50S subunit inhibiting translocation.
  - *Chloramphenicol* blocks peptide bond formation by inhibiting *peptidyl transferase* on the 50S subunit.
  - *Cycloheximide* inhibits *peptidyl transferase* present in the 60S subunit of eukaryotic ribosomes.
  - *Puromycin,* a structural analog of the 3'-end of aminoacyl-tRNA, binds to the A site and forms peptidylpuromycin on the A site. It competes with aminoacyl-tRNA as an acceptor in the peptidyl transfer reaction leading to termination of protein synthesis.
- ***Toxins***
  - *Diphtheria toxin* secreted by *Corynebacterium diphtheriae* inhibits eukaryotic protein synthesis by transfer of ADP ribose from (NAD^+) to eEF-2 (*ADP ribosylation*). As a result, eEF-2 is inactivated leading to inhibition of translocation.
  - *Ricin* produced by castor beans inhibits eukaryotic protein synthesis by producing N-glycosidase, a toxin that removes an adenine residue from 28S rRNA.

❑ How are proteins targeted to different destinations in the cell after translation?

Proteins synthesized by ribosomes have several destinations. These include

- *Secretory proteins*
  - These contain *signal sequences (signal peptides)* on 13–36 amino acids with a stretch of hydrophobic amino acids in their N-terminal ends.
  - A *signal recognition particle (SRP)* recognizes the signal peptide and arrests translation.
  - The SRP–ribosome complex binds to a docking protein or SRP receptor on the endoplasmic reticulum (ER).

- Receptor proteins on the ER membrane bind tightly to the ribosome and release SRP allowing translation to continue.
- The nascent polypeptide enters the ER lumen where a signal peptidase cleaves the signal peptide.
- The protein is transported to the Golgi for glycosylation and then to different destinations based on the glycosylation pattern.

- *Plasma membrane proteins* are inserted into the ER membrane, transported to the Golgi and then to the cell surface.
- *Proteins of the endoplasmic reticulum* after synthesis on the rough endoplasmic reticulum pass into the lumen and are transported to the Golgi. Receptors in the Golgi bind to a C-terminal *retention signal* on these proteins with the amino acid sequence Lys-Asp-Glu-Leu (or KDEL using the one letter amino acid symbols). The protein is returned to the ER via vesicles.
- *Lysosomal proteins* are targeted to their destination by addition of a mannose-6-phosphate signal in the Golgi.
- *Nuclear proteins* which have a nuclear localization signal of 4–8 amino acids in the interior of the protein are taken up via nuclear pores by a process requiring ATP.
- *Mitochondria and chloroplast proteins* synthesized by free cytosolic ribosomes require a matrix targeting sequence, heat shock proteins, and ATP.

❑ Enumerate the various post-translational modifications.

A protein is biologically active only after it is folded in its native conformation and undergoes covalent modifications. The post-translational modifications include

1. *Modification of N-terminal amino acids* such as
   - removal of the formyl group of the initiating amino acid in prokaryotes.
   - removal of initiating methionine residue and one or more succeeding residues in eukaryotes.
   - acetylation of the N-terminal residue in some proteins.
2. Modification of C-terminal amino acids.
3. *Removal of signaling sequences* by specific peptidases.
4. *Phosphorylation* of the hydroxyl groups of ser, thr, and tyr residues of some proteins that confers a negative charge. This enables proteins such as casein in milk to bind Ca^{2+}.
5. *Hydroxylation* of proline and lysine residues in collagen.
6. *Carboxylation* of glutamate to form γ-carboxyglutamate (Gla) residues in prothrombin and other clotting factors by vitamin K-dependent carboxylases imparts a negative charge to Gla residues that bind Ca^{2+} during blood coagulation.
7. *Methylation* of lysine residues in muscle proteins to monomethyllysine and dimethyllysine.
8. *Attachment of carbohydrate side chains* to asn, ser, or thr residues of glycoproteins.
9. *Addition of prosthetic groups* such as biotin to acetyl CoA carboxylase.
10. *Formation of disulphide bonds* between cysteine residues.

❑ How are proteins degraded?

- Protein degradation occurs in both prokaryotes and eukaryotes.
- Proteins that are degraded include
  - defective or damaged proteins.
  - proteins with short half-lives such as those containing arg, his, ile, leu, lys, phe, trp, and tyr in their N-terminal ends. These amino acids have half-lives of 20–30 minutes.
  - proteins containing a destabilizing N-terminal amino acid such as asn, asp, gln, and glu.

- In bacteria, protein degradation is catalyzed by a protease called *Lon* (for "*long form*" of proteins), which requires hydrolysis of two ATPs for each peptide bond cleaved.
- In eukaryotes, a protein is marked for degradation by *ubiquitinylation.*
- This involves covalent linkage of free carboxyl group of a gly residue in ubiquitin to an ε-amino group of a lysine residue in the protein targeted for degradation.
- This is repeated many times to achieve polyubiquitinylation.
- The ubiquitinylated protein is digested by a *26S protease complex* (*proteasome*) in an ATP-dependent reaction and the intact ubiquitin released for reuse.

❑ What are the levels at which gene expression can be regulated?

The potential points at which the concentration of a protein can be regulated include

- transcription
- post-transcriptional processing
- mRNA degradation
- translation
- post-translational processing
- protein degradation.

Regulation at the level of transcription is the most common and most economical.

❑ Define an operon.

- an operon is a group of functionally related structural genes that map close to each other on the chromosome and are turned *on* or *off* through the same regulatory loci.
- these are mostly found in bacteria where clusters of genes are transcribed together.
- the *operon hypothesis* was proposed by *Francois Jacob* and *Jacques Monod* to explain regulation of gene expression in prokaryotes.

❑ Give an account of the structural genes and regulatory genes in the *lac* operon.

The *lac* operon functions to synthesize enzymes that metabolize lactose used as an energy source by *E.coli* when glucose is not available.

The *lac* operon contains the following genes (Fig. 31.3):

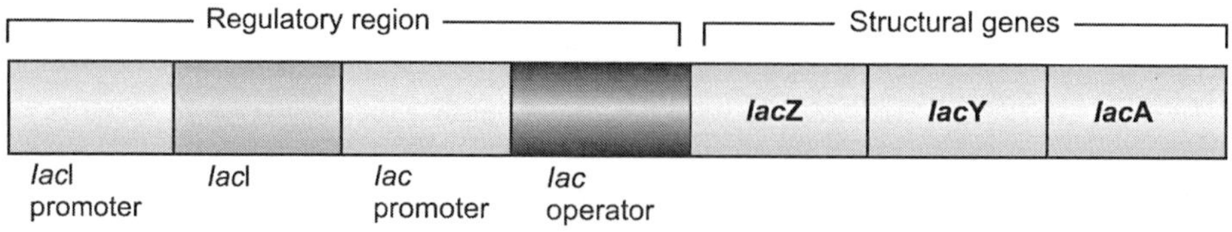

Fig. 31.3 Organization of the *lac* operon

- ***Structural genes***
  *lac*Z, *lac*Y, and *lac*A that encode *β-galactosidase, galactoside permease*, and *thiogalactoside transacetylase*, respectively.
- ***Regulatory genes*** that influence expression of the structural genes. These include
  - *i gene* present adjacent to the structural genes that encodes the *lac* repressor. *The repressor* binds to a specific DNA sequence termed the *operator* and prevents transcription initiation by RNA pol.

- *Operator* (O*lac*), a short, palindromic sequence extending from –5 to +21 that is bound by the repressor.
- *Promoter* located between the i gene and O*lac* contains binding sites for RNA polymerase and cyclic adenosine monophosphate receptor protein (CRP) or catabolite activator protein (CAP).

❑ Outline the mechanism of repression of the *lac* operon.

- The repressor is a heat labile protein
- In a wild type cell, there are approximately ten tetramers per wild type cell.
- In the absence of an inducer, one repressor binds to the operator while the rest bind randomly to random sites on DNA.
- Binding of the repressor to the operator prevents RNA pol from initiating transcription by converting the closed complex of DNA–RNA polymerase from a weak to a strong interaction.
- In the presence of an inducer, the repressor bound to operator is released and bound to random sites and the closed complex is converted to an open complex.

Figure 31.4 illustrates the mechanism of *lac* operon repression and induction.

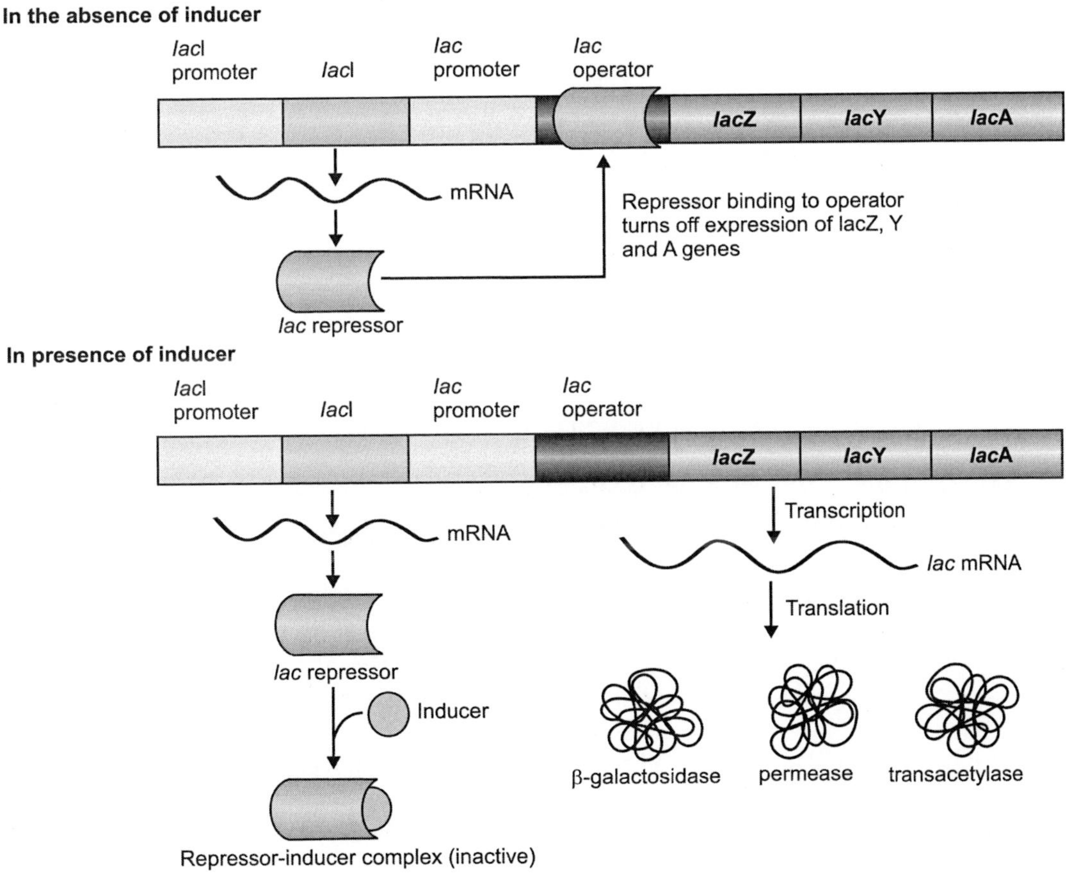

Fig. 31.4 Mechanism of *lac* operon repression and induction

❑ What is the effect of glucose on the *lac* operon?

- When glucose is available as energy source, it represses the *lac* operon *via* cAMP and CAP. This is termed *catabolite repression* (Fig. 31.5).
- When glucose concentration is low,
  - cAMP level is high.
  - cAMP binds CAP to form a cAMP–CAP complex that binds to the CAP site.
  - RNA pol binds to the promoter and initiates transcription of *lac* genes.
  - this is known as *positive transcriptional control.*
- When glucose concentration is high
  - cAMP is low.
  - cAMP–CAP complex is not formed.
  - *lac* operon expression is decreased.
  - this is known as *negative transcriptional control.*

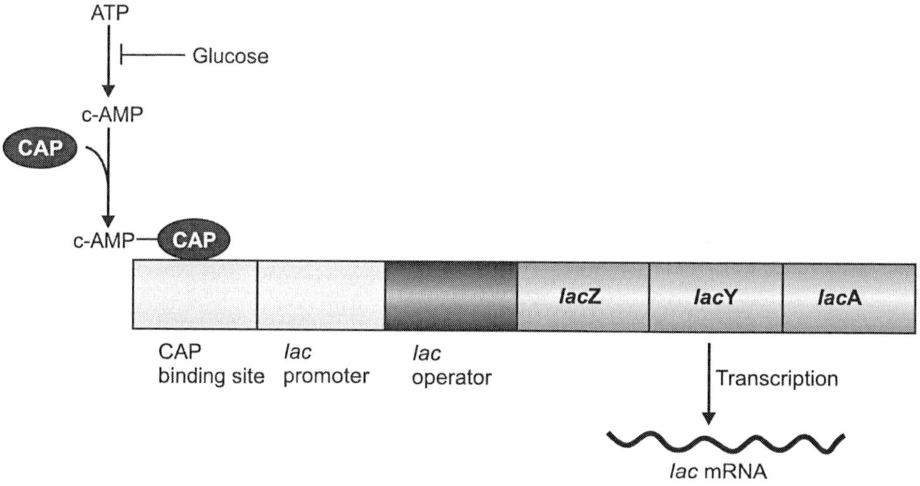

Fig. 31.5 Effect of glucose on *lac* operon expression

❑ Explain how gene expression is regulated in eukaryotes.

The eukaryotic genome is larger and more complex than the prokaryotic genome and is regulated at the following levels:

1. ***Alteration in gene content or position*** by
   - *Gene amplification* may be permanent as in the case of genes whose products are required in large amounts (*e.g.*, histones); or selective as in the case of the *dihydrofolate reductase (DHFR)* gene amplified in response to resistance to methotrexate, an antifolate cancer chemotherapeutic agent.
   - *Gene dimunition*, which involves inactivation and removal of genes from the genome such as loss of all genes in erythrocytes due to loss of the nucleus during development.
   - *Gene rearrangement* that is responsible for antibody diversity.
2. ***Transcriptional control*** manifested at the level of:
   - *Chromosome packaging* differs based on tissue specialization and differential gene expression. For example, the β-globin gene is in the active euchromatin in reticulocytes, and in the inactive heterochromatin in muscle cells. DNA methylation and histone acetylation also play a role in

gene expression. While methylation is associated with gene silencing, histone acetylation is involved in euchromatinization.

- *Individual gene regulation* by a large number of regulatory factors and sequence elements as well as by DNA bending.

3. ***Post-transcriptional regulation*** occurs by:
   - *Alternative splicing* by which the same gene produces different products in different cell types depending on which exons are spliced. For example, the same gene codes for calcitonin in the thyroid and calcitonin gene related peptide (CGRP) in the neurons. The two mRNAs contain the same first two exons but differ in the third exon and poly A sites.
   - *Regulation of RNA stability:* The half-life of mRNA increases with increasing length of the poly A tail and deletion of the 5'-noncoding sequence. Genes that encode an abundant product with a long duration of action, such as the β-globin gene, synthesize mRNA with long half-life; whereas genes whose products have a short duration of action, such as growth factors, generate mRNA with short half-life.
4. ***Translational Regulation*** is seen during the biosynthesis of hemoglobin. If heme required for hemoglobin synthesis is not available, globin synthesis is switched off by a protein synthesis inhibitor via the protein kinase cascade.

❑ How does an antisense RNA modulate gene expression? What are the applications of antisense technology?

- An *antisense RNA* is an RNA molecule with a sequence complementary to that of an mRNA.
- Antisense RNA modulates gene expression by
  - direct destabilization of a target mRNA by binding to it to form a duplex region that can be degraded by an endonuclease.
  - binding to mRNA and blocking its access to the ribosome, thereby preventing initiation of protein synthesis.
  - Causing premature termination of transcription by binding to a transcript and mimicking a terminator sequence.
- *Applications*
  - Analysis of the function of a regulatory gene.
  - Analysis of the timing of gene expression by turning the target gene off and on.
  - Development of new plant varieties.
  - Increasing the shelf-life of fruits by suppression of key enzymes in biosynthesis of ethylene essential for fruit ripening.
  - Prevention of viral infection.
  - Cancer treatment by targeting overexpressed oncogenes such as Bcl-2.

❑ Explain gene silencing by RNA interference.

- *RNA interference (RNAi)* refers to a post-transcriptional mechanism by which very small RNA molecules termed *short interfering RNAs (siRNAs)* regulate gene expression by silencing specific genes in higher eukaryotes.
- RNAi is activated when RNA molecules are present in the double-stranded state.
- The dsRNA is cleaved by *Dicer*, an endonuclease that generates short double-stranded siRNAs of 20–25 nucleotides.

- One of the strands of the RNA duplex is loaded into a ribonucleoprotein complex called *RISC (RNA-induced silencing complex)*, followed by base pairing between the RNA strand in RISC and a complementary mRNA strand. This leads to mRNA degradation or inhibition of protein synthesis.

Functions

- Regulation of developmental timing.
- Protection against viral infection in plants.
- Regulation of transposon activity.
- Formation of heterochromatin.

Chapter 32

Recombinant DNA Technology

❑ Define recombinant DNA.

A recombinant DNA (rDNA) is a DNA molecule constructed outside the living cell by joining natural or synthetic DNA segments to DNA molecules that can replicate to form multiple copies when introduced into a host cell. The rDNA molecule is also known as *chimeric DNA* or *chimera*.

❑ What is cloning?

Cloning is the process by which a segment of DNA is introduced into a host for replication to produce multiple copies of the DNA.

❑ What are the steps in a cloning experiment?

The steps involved in cloning include

1. Isolation of a specific gene or DNA sequence. The DNA sequence isolated for cloning is referred to as the '*foreign DNA*' or '*DNA insert*'. This step requires a donor DNA and restriction enzymes.
   Genomic DNA, which represents the complete genome of the organism is useful as source material if the purpose of cloning is to make multiple copies of a gene of interest or analyze gene expression. mRNA is useful as source material if production of a recombinant protein is desired.
2. Insertion of the foreign DNA into a carrier molecule known as a *vector*. Plasmids, phages, and cosmids are useful as cloning vectors.
3. Creation of the rDNA molecule by *splicing* (joining) the ends of the foreign DNA with the vector DNA.
4. Introduction into a suitable host for cloning (making multiple copies of the foreign DNA).
5. Multiplication and identification of clones containing the rDNA molecules. The rDNA molecule can be identified from a background of nonrecombinant molecules or from a population of other recombinants.

❑ Discuss the nomenclature, recognition sequence, types of cleavage, and mechanism of action of restriction endonucleases.

- Restriction enzymes are endonucleases found in bacteria. These enzymes cleave foreign invading DNA and protect bacteria.
- They are important tools for manipulation of DNA *in vitro*.
- There are three classes of restriction enzymes—type I, II and III. Only the type II enzymes are useful in gene cloning because they recognize a specific sequence of DNA and make double strand breaks within the sequence.

Nomenclature

1. The restriction enzymes are named after the bacterial species from which they are isolated. For example, *Eco*RI from *Escherichia coli.*
2. Each enzyme is designated by a symbol that consists of the first letter of the genus, the first two letters of the species, followed by a strain designation and a Roman numeral.
3. A large number of restriction enzymes have been isolated and characterized from a wide variety of organisms.

Examples of some restriction enzymes

1. *Eco*RI from *Escherichia coli* RY13 recognizes and cleaves within the sequence.

axis of symmetry

G A A T T C
C T T A A G

2. *Hind*III from *Hemophilus influenza* Rd recognizes and cleaves within the sequence.

A A G C T T
T T C G A A

3. *Pst*I from *Providencia stuartii* 164 recognizes and cleaves within the sequence.

C T G C A G
G A C G T C

Recognition sequence

- Each restriction enzyme recognizes a unique sequence of 4, 5, 6 or 8 nucleotides and makes double strand breaks within the sequence.
- The recognition sequences are *palindromes* (inverted repeats) that exhibit 2-fold rotational symmetry.
- The frequency with which a particular recognition sequence occurs on DNA is given by 4^n, where n is the length of the recognition sequence. This implies that theoretically, tetranucleotide sequences will occur once in every 256 bp and hexanucleotide sites every 4096 bp. Thus, a 4-cutter enzyme will cleave DNA more frequently than an 8-cutter.

Cleavage

- The type of fragments produced by a restriction enzyme depends on the recognition sequence and location of the cleavage site within the sequence.
- Cleavage can yield *blunt (flush) ends* or *sticky (staggered, cohesive) ends.*

Examples:

Blunt end cleavage

5′GTTAAC3′ —*Hpa*I→ 5′GTT + AAC3′
3′CAATTG5′ 3′CAA TTG5′

Sticky end cleavage

5′GAATTC3′ —*Eco*RI→ 5′G + AATTC3′
3′CTTAAG5′ 3′CTTAA G5′

Uses

- gene therapy
- restriction mapping
- DNA sequencing
- site-directed mutagenesis.

❑ What are the properties of an ideal vector?

An ideal cloning vector should have the following properties:

- Contain unique cleavage sites for restriction enzymes.
- Code for genetic information.
- Ability to replicate.
- Possess a non-essential region that can be removed for inserting the foreign DNA.

❑ Illustrate the use of plasmids as cloning vectors.

- Plasmids are small, circular, double-stranded, extrachromosomal DNA found in many bacterial species and in some eukaryotes.
- They replicate independently of chromosomal DNA and contain several or many genes.
- Generally, a plasmid that contains genes specifying resistance to two antibiotics is used. One of the genes helps to identify the bacteria carrying the plasmid, the other differentiates chimeric plasmids from parental vectors.
- The DNA insert size for plasmids ranges from 5–10 kb.
- ***Advantages***
  - Since the complete nucleic acid sequence of many plasmids is known, the cleavage sites of restriction enzymes for foreign DNA insertion is available.
  - They can be easily separated from bacterial chromosomes from which they differ in size, base composition, and density.
- ***Mechanism of cloning*:** Cloning a DNA insert into a plasmid involves the following steps (Fig. 32.1):

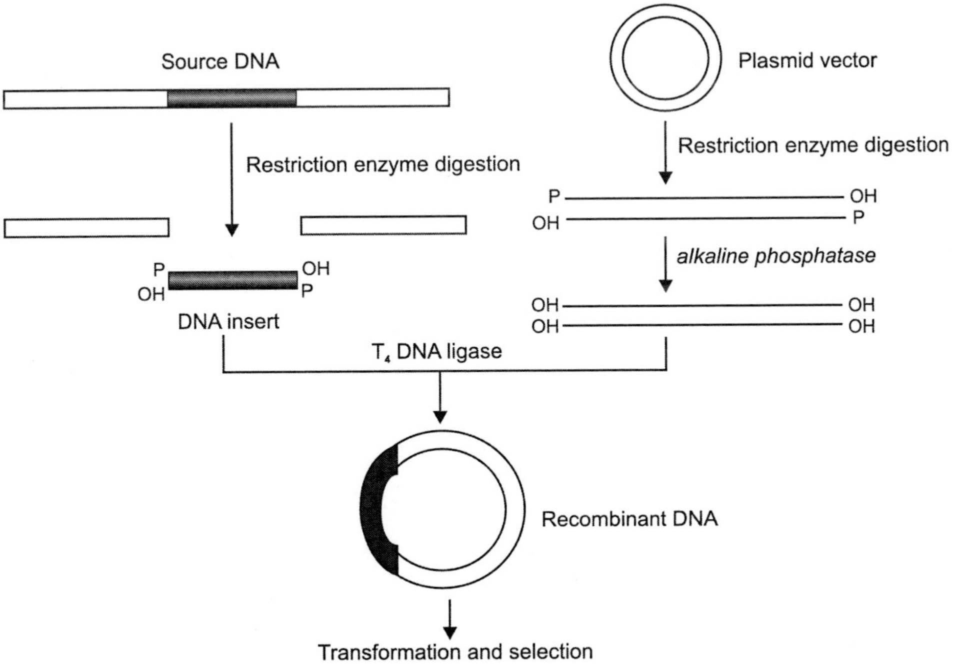

Fig. 32.1 Mechanism of cloning in plasmids

- The plasmid is cut with a restriction enzyme to generate a linear plasmid.
- The ends of the plasmid are joined to an insert DNA generated by cutting the source DNA with the same enzyme.
- The plasmid is prevented from recircularizing without gaining an insert by dephosphorylation of the 5' ends of the linear plasmid with *alkaline phosphatase*.
- The dephosphorylated plasmid DNA and insert DNA are joined by *T_4 DNA ligase*, which forms two phosphodiester bonds to generate chimeric plasmid.
- The two nicks in the chimeric plasmid are sealed by host cell *DNA ligase* after transformation.

❑ Describe the salient features of pBR322.

pBR322

- is the current standard, versatile, and widely used plasmid cloning vector.
- contains two antibiotic resistance genes—ampicillin (*amp^r*) and tetracycline (*tet^r*).
- has an origin of replication.
- contains recognition sites for several restriction enzymes. The foreign DNA is inserted into the *Pst*I site within the *amp^r* gene causing *insertional inactivation* of ampicillin resistance. This helps to identify recombinant plasmids with *tet^r amp^s* phenotype.

Figure 32.2 shows the genetic map of pBR 322.

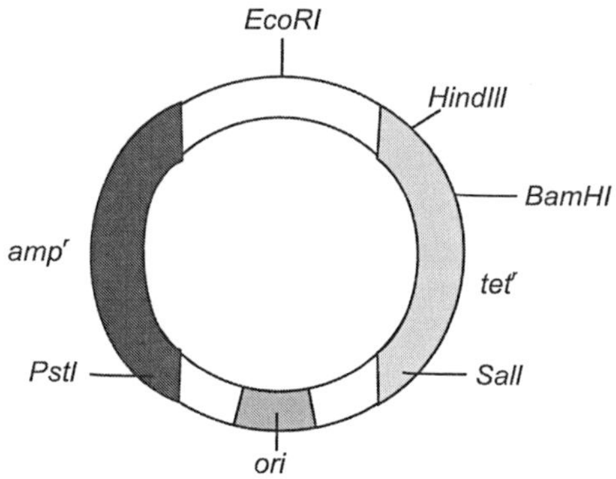

Fig. 32.2 Genetic map of pBR322

❑ Describe the use of bacteriophage λ as a cloning vector.

- Bacteriophage λ DNA is a linear duplex of ~48.5 kb containing short complementary cohesive termini of 12 nucleotides at its 5' ends termed *cos* sites.
- Since wild type phage λ contains restriction sites within essential regions, derivatives of phage λ have been developed as cloning vectors.
- *Mechanism of cloning* (Fig. 32.3)
  - Restriction enzyme cleavage of phage DNA produces two fragments that are joined to the ends of foreign DNA to form a *recombinant phage*.
  - Packaging of the recombinant phage into the phage head *in vitro* is essential for infection of host *E.coli* cells.
  - Phage infection causes formation of clear areas or *plaques* in the lawn of *E.coli* cells because lysis and reinfection prevent bacterial growth.
  - The λ DNA is purified from the plaques.
- *Advantages*
  - Several thousand plaques can be screened on a single Petri dish.
  - *In vitro* packaging into phage heads increases infectivity of rDNA molecules.
  - A *gene library* containing the complete genetic information of an organism can be stored in a few milliliters of broth.
  - The DNA insert size is 15–25 kb.

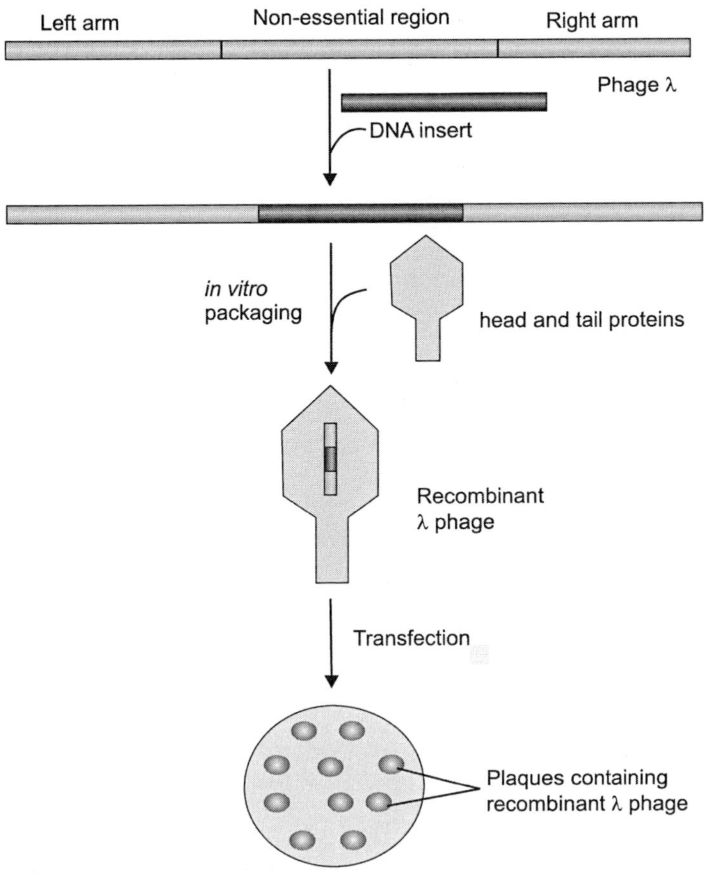

Fig. 32.3 Cloning in phage λ

❑ What are cosmids? How are they useful as cloning vectors?

- Cosmids are plasmids with λ phage *cos* sites.
- Restriction enzyme digestion linearises cosmid DNA, the ends of which can be ligated with the foreign DNA to form a recombinant cosmid.
- *Advantages*
  - Cosmids offer the best features of plasmids and phages such as propagation in bacteria in plasmid form and purification by packaging *in vitro* into phage heads.
  - They have higher infectivity than pure plasmids
  - The DNA insert size is 35–50 kb. The higher capacity is useful for generating genomic libraries.

Figure 32.4 shows the genetic map of a cosmid.

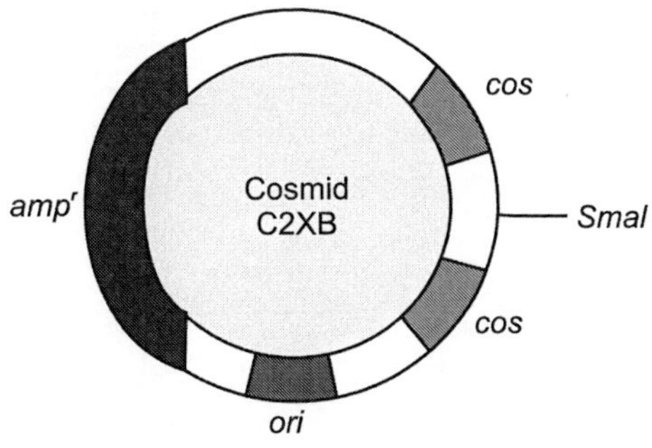

Fig. 32.4 Genetic map of cosmid C2XB

❑ Give an account of high-capacity cloning vectors.

- Vectors with high capacity are required for the analysis of large eukaryotic genomes and for genome mapping.
- The most widely used artificial chromosomes developed for this purpose include

Bacterial Artificial Chromosomes (BACs)

- These are based on single-copy sex factor F of *E.coli* and are useful for mapping and analysis of complex genomes.
- *Genetic map* (Fig. 32.5).
  BACs contain
  i. genes for replication (*ori*S and *rep*E)
  ii. genes for maintenance of copy number (*par*A and B)
  iii. chloramphenicol resistance marker (cm^r)
  iv. cleavage sites for several restriction enzymes.

Advantages

1. BACs can be introduced into *E.coli* efficiently.
2. They are capable of maintaining plant and human genomic fragments of greater than 300 kb for over 100 generations with a high degree of stability.

Disadvantages

1. It is difficult to identify foreign DNA inserts.
2. Because BACs cannot be amplified, large amounts of DNA cannot be isolated.

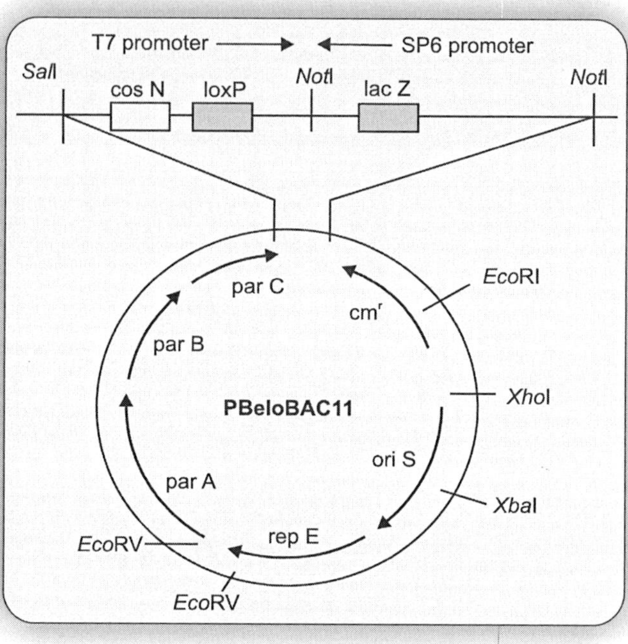

Fig. 32.5 Genetic map of BAC

Yeast Artificial Chromosomes (YACs)

Genetic map (Fig. 32.6)

i. ARS: Autonomously replicating sequence from yeast
ii. CEN4: Centromeres from yeast chromosome 4
iii. TEL: Telomeres from *Tetrahymena*
iv. URA3, TRP1: Yeast marker genes
v. Amp: amp resistance gene of pBR322
vi. *ori*: origin of replication of pBR322.

Advantages

1. DNA insert size is 50–2000 kb.
2. These are useful in constructing genomic libraries of higher organisms.

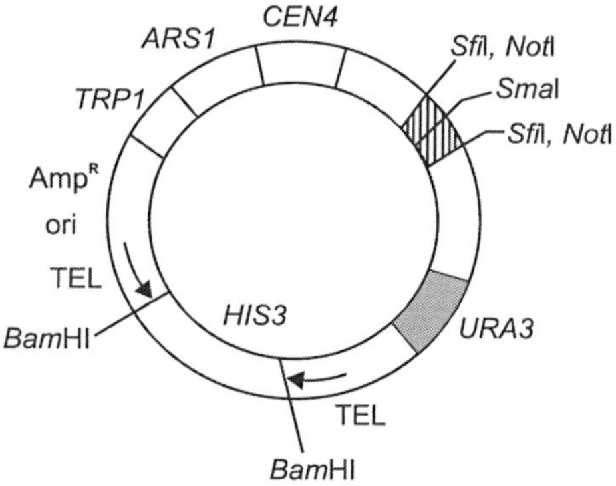

Fig. 32.6 Genetic map of YACs

Disadvantages

1. It is difficult to manipulate large DNA fragments and keep them intact.
2. Instability of the clones and deletion of regions from the inserts.
3. Since the YAC structure is similar to natural yeast chromosomes, separation of the two cannot be achieved using simple methods.

❑ Define the terms transformation and transfection.

- *Transformation* is the uptake of rDNA by the host cell. The host cell that has taken up the rDNA is said to be *transformed*.
- *Transfection* refers to the uptake of rDNA by eukaryotic cells.

❑ Elucidate the commonly used methods of gene transfer.

The commonly used methods of gene transfer include

◘ *Calcium phosphate co-precipitation*

- Bacterial cells when exposed to hypotonic calcium chloride at 0°C form spheroplasts that take up DNA after a mild heat shock.
- Transformation efficiency can be increased by co-precipitation of DNA with calcium phosphate.
- The basic protocol involves the following steps:
  i. DNA is mixed with buffered phosphate.
  ii. $CaCl_2$ is added to this to form a fine precipitate of calcium phosphate and DNA.
  iii. The precipitate is pipetted on to a monolayer of cells growing in a petridish and sufficient time allowed for the cells to take up DNA.
  iv. After a few hours, the precipitate is removed and the cells are incubated in fresh growth medium for several days.

◘ *Electroporation*

- This method involves electric field-mediated membrane permeabilization for DNA uptake.
- The cells are placed in a solution containing DNA and subjected to a brief electric pulse that causes formation of short-lived pores in the cell membrane.
- DNA enters the cell through these pores, which eventually close after 3–5 minutes.
- *Advantages*
  i. This method is highly versatile and can be used for a wide range of cell types. In addition to DNA, RNA, drugs, dyes, and antibodies can be introduced.
  ii. The method is rapid and does not require lengthy preparation of sample.

◘ *Microinjection*

- This is a highly versatile method of gene transfer.
- Although the technique is simple, it requires skill and practice.
- The foreign DNA can be targeted to specific sites in the cell such as cytoplasm or nucleus using finely constructed glass capillaries.
- The technique is performed with computer-assisted phase contrast microscope.
- A glass microcapillary filled with biological sample is introduced into the cell with the aid of a micromanipulator. An appropriate volume of the sample is injected by a syringe connected to the capillary.
- *Advantages*
  i. The transformation efficiency is approximately 100%.
  ii. Transformants are stable.
  iii. Carrier DNA is not required.
  iv. This can be used for any cell type.

❑ Explain how marker inactivation is used to identify rDNA.

Insertion of foreign DNA into the coding sequence of a gene inactivates it, leading to the expression of a new phenotype. This is known as *insertional inactivation* or *marker inactivation* that can be detected by antibiotic resistance or blue-white selection.

Antibiotic resistance

- Cloning into the *PstI* site in pBR322 changes the original phenotype from $amp^r tet^r$ to $amp^s tet^r$.
- Detection of the recombinant plasmids involves the following steps:
  i. Plating of the transformants on to a tetracycline-containing medium. This is the master plate. All the cells in the master plate survive and form colonies.
  ii. A replica of the master plate is grown in a medium containing ampicillin.
  iii. All the recombinants ($amp^s tet^r$) die, whereas nonrecombinants ($amp^r tet^r$) survive. The recombinants, identified by their absence in the replica plate are recovered from the master plate for analysis.

Figure 32.7 illustrates insertional inactivation by antibiotic resistance selection.

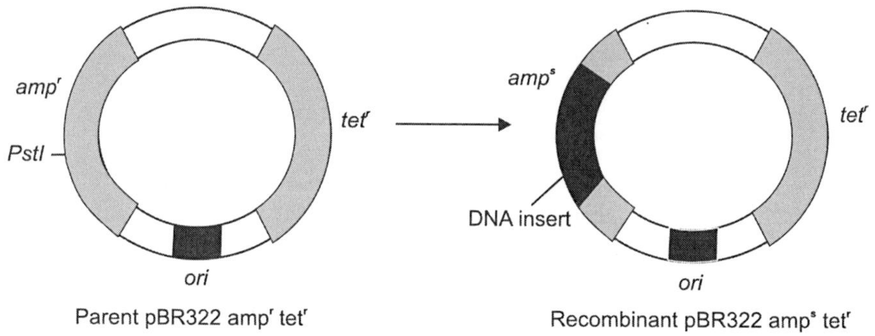

Fig. 32.7 Insertional inactivation by antibiotic resistance selection

Blue-white selection

- Some vectors have a *lac*Z gene incorporated, which is used as a marker.
- Insertion of the foreign DNA into the *lac*Z gene, which encodes *β-galactosidase*, can be detected by a color reaction.
- In presence of X-gal, the chromogenic substrate of *β-galactosidase*, plaques containing the nonrecombinants, which retain a functional *lac*Z gene, give rise to blue plaques.
- Recombinants, in which the *lac*Z gene has been inactivated by insertion, remain white.

Explain how colony hybridization allows detection of rDNA.

This technique involves the following steps:

- Transfer of colonies from agar plates to nitrocellulose filters (NCF) by replica plating. This is because NCF or nylon is more suitable for hybridization.
- Denaturation of the DNA fragments on NCF by alkali followed by several washes.
- Immobilization of the fragments to NCF by heating.
- Exposure to a specific radiolabeled DNA probe.
- Hybridization of the probe with the target DNA sequence homologous to the probe on the NCF.
- Washing to remove unbound probe, followed by drying and autoradiography.
- Positive colonies identified in the autoradiogram are picked from the master plate for further analysis.

Figure 32.8 depicts the protocol of colony hybridization.

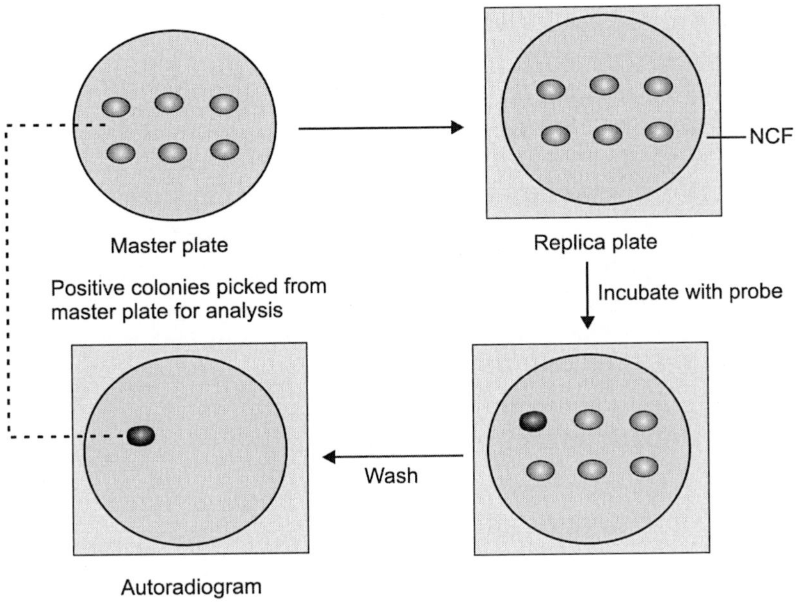

Fig. 32.8 Colony hybridization

❑ Describe the construction of a genomic library.

A genomic library is a collection of recombinant molecules that represent the entire genome of an organism. It contains a set of overlapping cloned fragments that are generated randomly and maintained in a stable form. Any gene of interest can be isolated from this library.

Steps in genomic library construction (Fig. 32.9)

1. Extraction of the total DNA from the organism, usually from embryos or sperm cells.

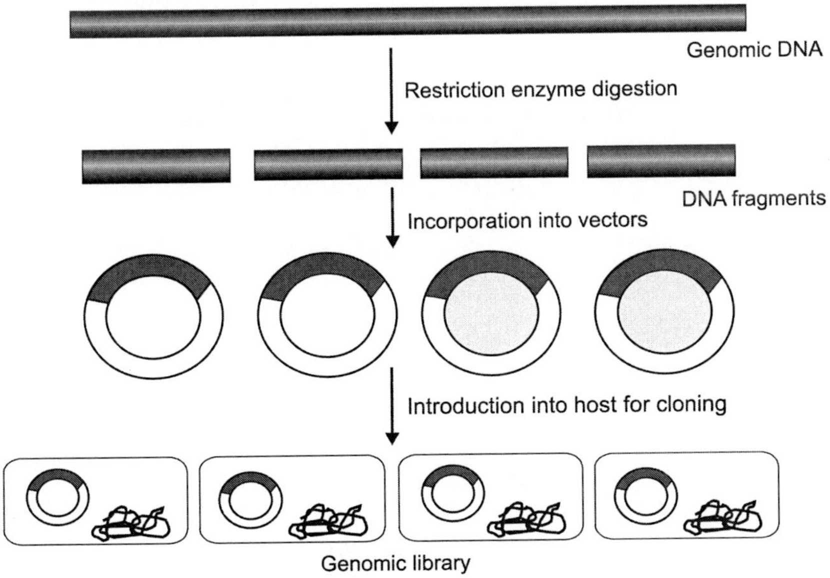

Fig. 32.9 Construction of a genomic library

2. Restriction enzyme digestion of the donor DNA and the bacteriophage λ vector. A restriction enzyme which yields sticky ends is used.
3. Annealing of the genomic and vector DNA by complementary base pairing.
4. Sealing of the ends of the donor DNA and vector DNA by ligase.
5. *In vitro* packaging of the recombinant DNA into phage heads and infection of host cells.
6. The viral DNA isolated from the cell lysates will contain the genes of the donor organism.

Factors that influence the construction of a genomic library include the size of the genome and the type of vector used. Small genomes such as that of *E.coli* (4.6×10^6 bp) will need fewer clones, and plasmid vectors can be used to maintain the genome library. Large, complex genomes such as the human genome (3×10^9 bp) require cloning vectors with higher capacity such as phages and cosmids for library construction.

❑ Elucidate the steps involved in the construction of a cDNA library.

Complementary DNA (cDNA) synthesized from mRNA by the enzyme reverse transcriptase (RTase) is a source of intron-free genetic information.

cDNA is useful for expressing eukaryotic genes in bacteria because bacteria cannot remove introns.

Steps in constructing a cDNA library (Fig. 32.10)

1. Extraction of total mRNA: Successful cDNA cloning depends on the purity of the mRNA template. Before isolating mRNA, endogenous RNase is inhibited by use of RNase inhibitors. The mRNA isolated should be free of protein and DNA. Eukaryotic mRNAs, which have a 3′-polyA tail, are separated from the other types of RNAs by affinity chromatography on oligo dT cellulose.

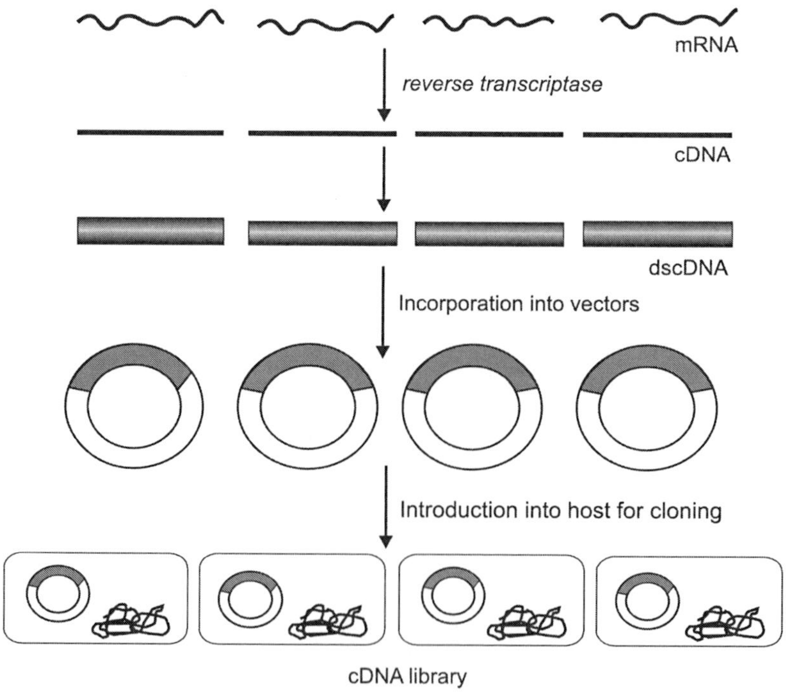

Fig. 32.10 Construction of a cDNA library

2. Reverse transcription of mRNA into cDNA: This is catalyzed by *reverse transcriptase*, an RNA-dependent DNA polymerase, which requires RNA template and a primer with a free 3'-OH group.
3. Formation of double-stranded cDNA and insertion into plasmid vectors.
4. Introduction of recombinant molecules into host cells.

❑ Enumerate the differences between genomic and cDNA libraries.

1. *Size and composition:* The cDNA library, which represents the mRNA population in a tissue, is much smaller than the genomic library which contains all the genes in an organism.
2. *Introns:* The smaller size of cDNA library is due to the absence of introns. The genomic library is large due to the presence of introns and regulatory sequences.
3. *Library construction:* While genomic library requires restriction enzyme digestion and phage vectors, cDNA library construction requires reverse transcriptase and plasmid vectors.
4. While genomic library represents the total genetic complement in the cell, a cDNA library indicates the specific state of cells or cell types. For example, while most of the mRNA in the β cells of pancreatic islets would be insulin mRNA, insulin DNA is only a small percentage of the total genome in a genomic library.

❑ Describe the procedure for cloning insulin gene in *E.coli.*

Bacteria, which normally do not produce insulin, can be manipulated by genetic engineering to produce multiple copies of the insulin gene.

Since bacteria cannot remove introns and carry out post-translational modifications, production of insulin protein in bacteria is difficult.

Insulin cDNA is more useful for cloning in *E.coli* than the native gene.

The protocol for cloning insulin involves the following steps:

1. Isolation of islet cells from pancreas.
2. Extraction of total RNA.
3. Separation of mRNA from rRNA and tRNA by passing through oligodT cellulose column.
4. Reverse transcription of mRNA to cDNA by RTase.
5. Hydrolysis of mRNA template by alkali.
6. Conversion of single-stranded cDNA to double-stranded cDNA. This can be accomplished by RTase, Klenow enzyme, or T4 DNA polymerase.
7. Removal of hairpin by S1 nuclease.
8. Attachment of a decanucleotide to both ends of cDNA by DNA ligase.
9. Cleavage of the decanucleotide by *Hind*III to generate sticky-ended cDNA .
10. Annealing of insulin cDNA with plasmid pMB9 also cut with *Hind*III.
11. Transformation of *E.coli* strain X 1776 by the recombinant plasmid.
12. Selection of transformed *E.coli* cells using tetracycline resistance.
13. Detection of insulin gene.
14. Inclusion of regulatory regions for transcription and translation in the vector ensures synthesis of insulin protein.

    In addition to cDNA, chemically synthesized DNA coding for each insulin chain can be used. The A and B chains produced separately are mixed together to generate the functional insulin protein.

❑ Define transgene and transgenesis.

A *transgene* is a gene from an organism that is introduced into the genome of another (usually multicellular) organism.

Transgenesis is the introduction of gene(s) into plant or animal cells with subsequent transmission of the transgene to successive generations.

❑ What are transgenic plants?

Transgenic plants are genetically transformed plants that express a particular trait or serve a specific need, and are produced by gene transfer methods.

❑ Elucidate the important delivery systems for gene transfer in plants.

Two important gene delivery systems for plants include

1. ***Agrobacterium-mediated transformation***
   - This is the most efficient method of gene transfer in plants.
   - *Agrobacterium tumefaciens* is a soil bacterium that causes *crown gall* tumors in dicotyledonous plants and contains *Ti (tumor inducing) plasmid*.
   - It has the following genes: *vir* for virulence, *onc* for tumor formation, and genes for synthesis and degradation of unusual amino acids termed opines (Fig. 32.11).
   - The gene for opine synthesis can be replaced by a selectable marker gene.

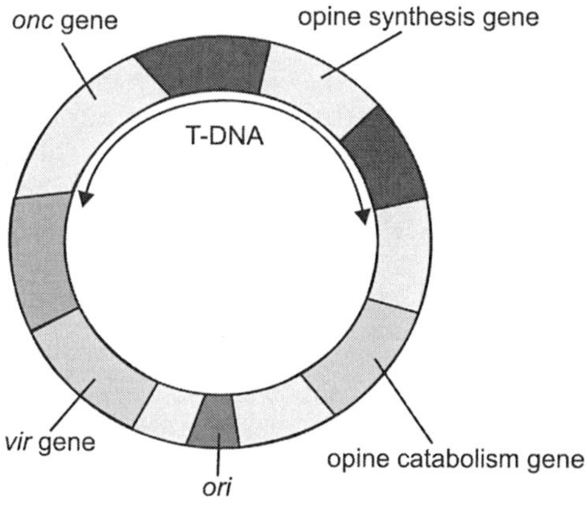

Fig. 32.11 Genetic map of Ti plasmid

2. ***Particle gun method***
   - this involves direct injection of DNA into whole cells.
   - minute tungsten beads coated with DNA are shot into plant cells at high velocity killing cells in the direct line of fire but penetrating into other cells.

❑ What are the applications of transgenic plants?

1. *Pest resistance* by introducing the toxin gene produced by the bacteria *Bacillus thuringiensis* (*B.t.*) into plants to kill a wide range of insect pests that attack plants.
2. *Herbicide tolerance* by producing plants that express a mutant enzyme not inhibited by glyphosate present in herbicides.
3. *Delay in fruit ripening* by
   a. insertion of the antisense version of the genes for ripening
   b. insertion of a gene that encodes an enzyme which prevents formation of ethylene required for ripening.
4. *Production of pharmaceutically important proteins* such as albumin, antibodies, and vaccines.

❑ Outline a method for the development of transgenic animals.

Microinjection is the method of choice for developing transgenic animals.

The steps include

1. Superovulation of donor females to induce a 3–5-fold increase in the number of fetilized eggs to be inoculated by microinjection.
2. After mating, the females are sacrificed and the fertilized eggs removed.
3. The transgene is injected into the male pronucleus.
4. Fusion of the two pronuclei forms the zygote.
5. Transfer of the embryos to the oviduct of a foster mother.
6. Testing of the offspring three weeks after birth for the presence of the transgene.
7. Breeding of the transgenic progeny to generate pure homozygous transgenic lines.

Limitations:

1. The efficiency of microinjection is only 3–5%.
2. The transgene may not be expressed in all the progeny.
3. The desired trait may not be present in all the transgenic pups.

❑ What are the applications of transgenic animals?

The applications of transgenic animals include:

1. *Development of disease and virus resistant transgenic animals* by immunological protection and transfer of genes coding for monoclonal antibodies.
2. *Expression of human growth hormone (hGH)* in cattle increases milk yield and produces leaner meat.
3. *Alteration in the properties of milk*:
   - Introduction of κ-casein transgene into animals leads to more consistent curd and enhanced cheese production.
   - Lactose-free milk produced by expression of *lactase* transgene in the mammary gland of cattle for consumption by people with lactose intolerance.
   - Introduction of lysozyme into milk can reduce bacterial content in milk and increase the consistency of curd.
4. *Production of pharmaceutical proteins*: Transgenic animals can be used as bioreactors to produce pharmaceutically important proteins such as hemoglobin and α_1-antitrypsin.
5. *Transgenic mice as models of human disease*: Mouse models for diseases like cancer and Alzheimer disease are useful to test potential therapeutic agents.

❑ Enumerate the applications of genetic engineering.

Genetic engineering has a wide range of applications in diverse fields, especially in medicine, agriculture and environment.

Medicine: The biomedical importance of rDNA technology include:

- understanding the molecular basis of disease.
- production of therapeutic drugs such as insulin in *E.coli*.
- vaccine production.
- *Gene therapy:* Clinical trials for gene therapy for cancer, AIDS, cardiovascular disease, and inherited disorders of metabolism have been undertaken.
- *DNA fingerprinting:* This is a technique based on the fact that each individual has a series of identical DNA sequences called repetitive DNA. The pattern, length, and number of these repeats are unique for each individual. This technique is of use in forensic sciences.

***Agriculture*:** rDNA technology has made possible the introduction of genes into plants that alter or produce a new trait. This has enabled:

- development of crop varieties that can fix nitrogen and resist salinity and drought.
- protection of crops against various types of pests either by use of microorganisms as pesticides or by creation of pest-resistant plants.

***Environment*:** rDNA technology has a role in tackling environmental problems.

- a recombinant *Pseudomonas* capable of clearing up oil spills in oceans has been produced.
- microorganisms can be engineered to degrade pesticides, heavy metals, and toxic wastes.
- agricultural waste can be used to produce biogas and fuels such as ethanol.

❑ What are the possible hazards of genetic engineering? Write a note on the safety aspects.

Hazards

1. rDNA technology involves the use of microorganisms that have been manipulated *in vitro*. These organisms are hazardous to the operator as well as the environment.
2. Environmental release of microorganisms can be deliberate or accidental. Deliberate release of microorganisms into the environment is useful in various ways including pest management, pollution control, oil recovery, and frost damage control. Chakraborthy has created a *Pseudomonas* species capable of cleaning oil spills in the ocean. These oil spills, which contain a mixture of hydrocarbons and other substances, can adversely affect marine organisms. Although oil-eating bacteria degrade the hydrocarbons, accidental release of these bacteria may result in loss of oil from oil wells.
3. Plasmids used as cloning vectors are capable of transferring antibiotic resistance to sensitive bacterial strains. This can cause problems in the management of infection.

Safety aspects

Environmental release of microorganisms can be prevented in two ways:

1. Physical containment, to prevent the escape of hazardous material from the laboratory.
2. Biological containment ensures that hazardous organisms can function only under laboratory conditions. Crippled microbes that are incapable of surviving outside the laboratory have been created.
   - All genetically manipulated organisms have to be subjected to review by Environmental Protection Agency (EPA) before approval.
   - Transgenic plants and animals used as bioreactors to produce pharmaceutically important proteins need to be evaluated for survival and fitness.
   - Before embarking on clinical trials for gene therapy in humans, it is essential to demonstrate the efficiency and absence of harmful side effects in animals.
   - The construction of harmful biological agents has been the most extreme hazard associated with rDNA technology. Since there is a possibility for a military use of this technology, approximately half the nations of the world are signatories to a treaty restricting the use of biological weapons.

Section 8

Analytical Techniques

- Analytical Techniques

Chapter 33

Analytical Techniques

❑ State Beer-Lambert law of absorption.

Beer-Lambert law states that the amount of light absorbed by a material is proportional to the concentration of the absorbing substance and to the thickness of the absorbing material.

❑ What is meant by absorption spectrum?

Absorption spectrum refers to the pattern of energy absorbed when light of varying wavelength passes through a substance. This is unique for each substance.

❑ Discuss the components of a spectrophotometer.

- A spectrophotometer is an instrument that enables analysis of absorption or emission of electromagnetic radiation as a function of wavelength.
- Components include the following (Fig. 33.1):
  - *Energy source*
    a. For ultraviolet radiation (UV): High pressure hydrogen or deuterium lamp that produces radiation in the 200–340 nm range.
    b. For visible radiation: A tungsten filament emitting radiation between 350 and 800 nm.
  - *Monochromator* is a wavelength selector that resolves polychromatic light into its individual wavelengths and isolates them to very narrow bands. This consists of
    a. an entrance slit for the polychromatic light.
    b. a lens or a mirror to collimate polychromatic light to the dispersion device.
    c. wavelength resolving device such as a prism or a diffraction grating.
    d. focussing lens or mirror.
    e. exit slit for the monochromator beam.
  - *Sample container*: While glass cuvettes are used for measurement in the visible region, quartz cuvettes are used for UV region.

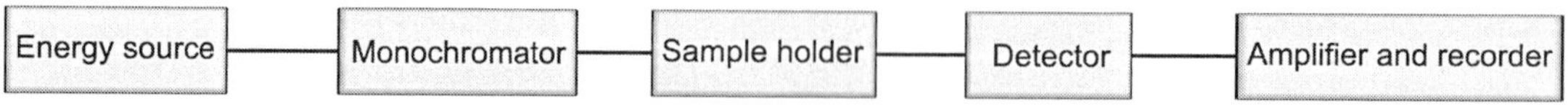

Fig. 33.1 Components of a spectrophotometer

- *Detectors* may be photocells, phototubes, or photomultiplier tubes. Photomultiplier tubes are the detectors of choice and consist of an evacuated glass tube containing the cathode and anode, and additional electrodes called dynodes. They amplify the initial photoelectric effect and are useful at very low light intensities.
- *Amplifier and recorder* help to translate the transmitted light into electrical signals that can be interpreted.

❑ What are the applications of UV-visible spectrophotometers?

UV-visible spectrophotometers have a wide range of applications that include:

1. Qualitative analysis by plotting absorbance spectrum
2. Quantitative analysis of nucleic acids and proteins
3. Assay of enzyme activity
4. Detection of impurities
5. Structural studies.

❑ Briefly explain turbidimetry and nephelometry.

- *Turbidimetry* is a technique that is used to measure the number of particles such as bacteria in a given suspension based on the *Tyndall effect*. When radiation passes through a suspension, the intensity of the transmitted light will be less than that of the incident light due to absorption by light scattering.
- *Nephelometry* is a technique that measures the intensity of light scattered by particles in suspension. The measurements are carried out at right angles to the direction of incident light. The method is more sensitive for low concentration and is used to estimate the concentration of microorganisms, analyse waste water, and evaluate the amount of haze present in beverages and pharmaceuticals.

❑ Define fluorescence.

Fluorescence is the phenomenon by which a molecule absorbs radiation and emits it at longer wavelength.

❑ Enumerate the components of a spectrofluorimeter.

The components of a spectrofluorimeter include the following (Fig. 33.2):

- *Radiant energy source* which may be mercury lamp or xenon arc.
- *Monochromators*: Two monochromators are used. The first is a prism which is placed before the sample, while the second is placed after the sample.
- *Detector* which is usually a photomultiplier
- *Amplifier*.

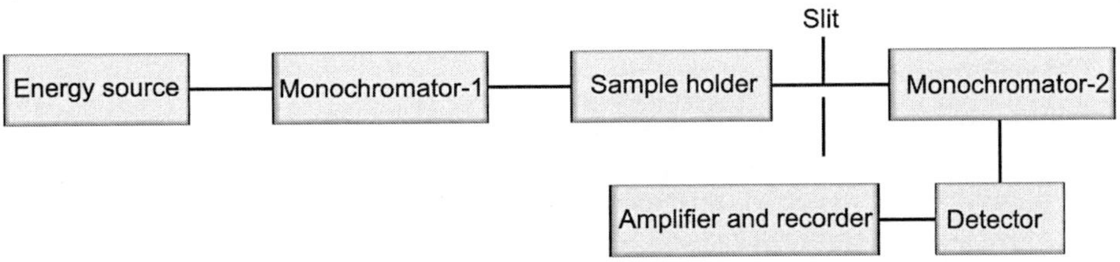

Fig. 33.2 Components of a spectrofluorimeter

❑ What are the applications of spectrofluorimetry?

1. Assay of vitamins (*e.g.*, riboflavin) and hormones (*e.g.*, cortisol).
2. Pesticide analysis.
3. Estimation of porphyrins.
4. Quantitative analysis of carcinogens and drugs.
5. Analysis of membrane structure using fluorescent probes such as anilinonaphthalene-8-sulphonate (ANS).
6. Subcellular localization of fluorescently labeled compounds by a combination of spectrofluorimetry with microscopy. This is useful in visualization of chromosomes.

❑ Define luminescence.

Luminescence refers to the phenomenon of light emission from a chemical reaction below the temperature of incandescence. This occurs as a result of oxidative reactions in solutions producing molecules in an excited state. These reactions may release energy as heat or as photons. Examples of luminescent compounds include *luciferin* and *luminol*.

❑ Differentiate chemiluminescence and bioluminescence.

- *Chemiluminescence* refers to luminescence produced by simple chemical means by oxidation of a luminescent compound such as *luminol* with hydrogen peroxide. Luminol is the most common chemiluminescent compound used as a marker in chemiluminescent assays. It is oxidized by hydrogen peroxide at pH 10–11 to release photons with maximal intensity at 430 nm.
- *Bioluminescence* is a sensitive technique in which luminescence is produced by the enzyme *luciferase.* Luciferase acts on *luciferin,* a chemiluminescent compound, to form *oxyluciferin* in the presence of ATP that can be measured by luminometry.

❑ What is flame photometry? What are the types of flame photometry?

- *Flame photometry* is a technique that is used to analyze the absorption or emission of light of specific wavelengths by excited atoms.
- The types of flame photometry include:
  - *emission flame photometry* in which molecules in a flame are volatilized to generate free atoms that are excited to higher energy levels. When these atoms return to the ground state, they produce a characteristic emission spectrum.
  - *atomic absorption flame photometry* which measures the absorption of a beam of monochromatic light by atoms in a flame.

❑ List the components of a flame spectrophotometer.

Emission flame photometer

- *Nebulisers or atomizers* convert the sample into a fine spray.
- *The flame*: Different gas mixtures varying in temperature are used to analyze different compounds. For example, acetylene at 2400°C is used for the assay of calcium, magnesium, and iron.
- *Monochromators* may be simple filters, prisms, or diffraction gratings.
- *Photocells:* A multichannel polychromator is used to measure six elements simultaneously.

Figure 33.3 gives a schematic representation of the components in a flame speactrophotometer.

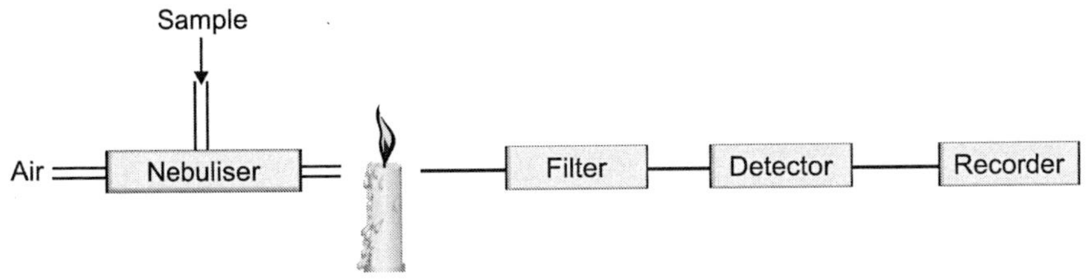

Fig. 33.3 Components of a flame spectrophotometer

Atomic absorption spectrophotometry

The instrumentation is similar to that of emission flame photometry except for the radiation source, which is a hollow cathode lined by the element to be analyzed.

❑ Enumerate the applications of flame photometry.

- Assay of elements in biological fluids, tissues, plants, and soils.
- Estimation of heavy metals.

❑ What is the unit of radioactivity?

Curie (Ci) is the unit of radioactivity. It is the quantity of radioactive material in which the number of nuclear disintegrations per second (d.p.s.) is the same as that in 1 g of radiation, *i.e.*, 3.7×10^{10} (d.p.s.). The curie refers to the number of disintegrations that actually occur in a sample rather than the number of disintegrations counted in a radiation chamber.

The microcurie (μCi) is Ci $\times 10^{-6}$ while the millicurie (mCi) is Ci $\times 10^{-3}$.

❑ What is specific activity?

- Specific activity refers to the quantity of radioisotope present per unit mass. It is expressed as disintegrations per second or disintegrations per minute, or in curies (mCi or μCi).
- Specific activity may be expressed as
  1. counts per minute/per second (c.p.m./c.p.s.) which is the recorded rate of radioactive decay.
  2. disintegrations per minute or second (d.p.m./d.p.s.) which is the actual rate of radioactive decay.
  3. curies (mCi or μCi) per unit of mass.

❑ Outline the principle of scintillation counting.

- Exposure of certain materials to ionizing radiation causes conversion of kinetic energy of particles or photons into flashes of light or scintillations. Substances that exhibit this property are termed *fluors* or *scintillators*.
- The light emitted by excitation of the fluor is directed to a photomultiplier, which converts light photons into an electrical pulse that is amplified, counted, and recorded.
- There are two types of scintillation counting—solid scintillation counting and liquid scintillation counting.

❑ Discuss the principle, instrumentation, and applications of solid scintillation counting.

Principle

- This technique involves interaction of radiation from a source physically external to the scintillator.
- The sample is placed next to a crystal of fluorescent material.
- The crystal used is
  - NaI for γ-isotopes
  - ZnS for α-particles
  - Organic scintillators of anthracene for β-particles.

Instrumentation

- A solid crystal of thallium activated NaI is optically coupled to the face of a photomultiplier. The γ-radiation passing into the fluor is dissipated by photomultiplier and pair production.
- *Preamplifier* shapes the pulse.
- *Line amplifier* strengthens the signal from the photomultiplier.
- *Discriminator* rejects pulses that are smaller than the predetermined size.
- *Scaler* records the counts.

Applications

- *In vitro* measurement of radioactivity of biological samples.
- *In vivo* measurement of radioactivity in specific organs.
- Radioimmunoassay for γ-emitting isotopes.

❑ Elucidate the principle and applications of liquid scintillation counting.

Principle

- This technique is based on the measurement of radiation from a source intimately in contact with the fluor, either dissolved or suspended in it.
- The radioactive sample is dissolved in a suitable solvent called the *scintillation cocktail* (toluene and xylene). The light emitted by the organic solvent when bombarded with radioactivity is of very short wavelength and cannot be detected.
- A compound that is capable of accepting energy from the solvent and fluorescing at a longer wavelength that can be detected is used. The compound is an organic liquid scintillator known as the *primary fluor* (*e.g.*, 2,5,diphenyloxazole, PPO).
- Since light detection by the primary fluor is not efficient, a *secondary fluor* (*e.g.*, 1,4,di [5-phenyloxazol-2-yl] benzene, POPOP) is added.
- Thus, the energy transfer in liquid scintillation counting is from the solvent via the primary fluor to the secondary fluor, which is then directed to the photomultiplier (Fig. 33.4).
- The light signals from the photomultiplier are transmitted via a preamplifier and a coincidence circuit to a scaler for counting.
- Interference in the energy transfer process occurs within the sample container during liquid scintillation counting. This phenomenon known as *quenching* decreases the number or intensity of light flashes produced and reduces counting efficiency. Quenching is of three types: optical quenching, color quenching, and chemical quenching.

- Quench correction can be done by using internal standards, external standards, or by the channels ratio method.

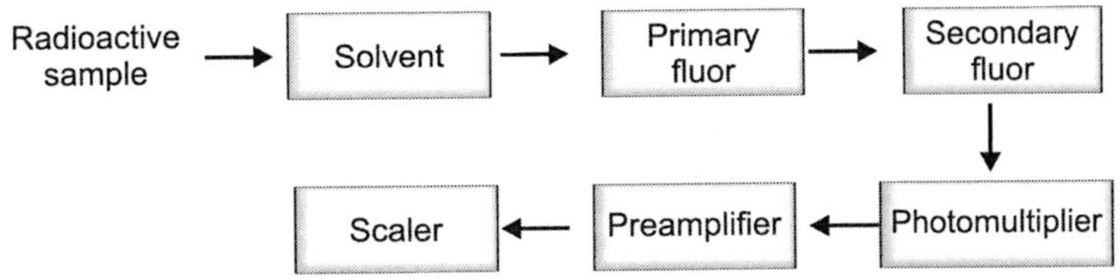

Fig. 33.4 Energy transfer process in liquid scintillation

Applications

Liquid scintillation counting is useful to quantitate low energy β-emitters such as ^{14}C, ^{3}H, and ^{35}S.

❑ What is autoradiography?

- *Autoradiography* is a technique used to locate an isotope in a tissue, cell, or molecule by analyzing images formed on X-ray film by emission of radioactive particles.
- The sample is placed in contact with an X-ray film. β Particles emitted from the sample activate silver halide grains in the emulsion, which is reduced to metallic silver when the film is developed.
- Autoradiography is used in genetic engineering to detect the hybridization of a radioactive DNA (probe) molecule to denatured DNA by the Southern transfer or colony hybridization procedures.

❑ Enumerate the applications of radioisotopes in biology.

Radioisotopes have the following applications:

1. Tracing metabolic pathways
2. Turnover of biomolecules
3. Analysis of absorption, accumulation, and translocation of compounds in plants and animals.
4. Drug development by determining site of accumulation, metabolism, and pharmacokinetics.
5. Analysis of enzyme and ligand binding.
6. Isotope dilution for trace element analysis.
7. Radiodating of rocks, fossils, and sediments.
8. Assessment of thyroid function using ^{131}I.
9. Assessment of renal function using ^{133}I iodohippuric acid.
10. Sterilization of food and laboratory equipment.

❑ What is chromatography?

- *Chromatography* is a technique used for the separation and identification of biomolecules based on partition and adsorption carried out on paper or in columns made of glass or metal.
- The technique uses a stationary phase and a mobile phase.
- The stationary phase may be a solid, a liquid, or a solid/liquid mixture, which is immobilized.
- The mobile phase may be liquid or gaseous, and flows over or through the stationary phase.

❑ What is partition chromatography?

Partition chromatography is the technique of separating substances based on partition between two immiscible solvents. It is used for the separation and identification of amino acids, carbohydrates, and fatty acids.

❑ What is adsorption chromatography?

- *Adsorption chromatography* is the separation of molecules using a stationary phase or support that contains binding sites for solute molecules.
- When the sample containing a mixture of substances is applied to the surface of an adsorbent, and a solvent allowed to flow across, some substances bind to the support, whereas others migrate at a rate that is dependent on the retardation by the column material.
- Generally, the stationary phase is solid and the mobile phase is a liquid except in gas liquid chromatography.
- Commonly used adsorbents include alumina, silica gel, activated carbon, and hydroxyapatite.

❑ Give a brief account of thin layer chromatography.

- *Thin layer chromatography* (*TLC*) is based on partitioning or distribution of components between two liquid phases.
- The distribution process may be based on adsorption, partition, ion-exchange, or molecular exclusion chromatography.
- The stationary phase is a uniform thin layer of adsorbent such as silica gel or alumina powder that is coated thinly on a glass, plastic, or metal foil plate.
- The mobile liquid phase passes through the thin layer plate and compounds distribute themselves between the two phases based on their distribution coefficients.
- Separation is carried out in a glass tank containing the developing solvent.
- Detection of separated compounds:
  - adsorbents containing a fluorescent dye allow examination under UV light.
  - spraying with 25–50% sulfuric acid in ethanol followed by heating results in appearance of the compounds as charred brown spots.
  - iodine vapor is used for organic compounds.
  - ninhydrin reagent gives a blue or purple color with amines and amino acids.
  - radioactive compounds are subjected to autoradiography and the spots detected as dark areas.
- *Applications*
  - separation of small molecules, amino acids, nucleic acids, lipids, and carbohydrates.
  - identification of drugs and contaminants.
  - phytochemical analysis of plant extracts.

❑ What are the types of gas chromatography?

- *Gas chromatography* is a separation technique in which the mobile phase is a gas and the stationary phase is a liquid or a solid.
- Gas chromatography is of two types:
  - Gas solid chromatography in which a gas functions as the mobile phase, and a solid adsorbent such as silica or alumina as the stationary phase. It has limited application and is not reproducible.
  - Gas liquid chromatography (GLC) in which the mobile phase is a gas and the stationary phase is a liquid adsorbed either to the inner surface of a column or to a solid support such as diatomaceous earth, Teflon powder, or fine glass beads.

❑ Give a brief account of the detectors used in GLC.

The commonly used detectors in GLC are

1. Flame ionization detector
   - Responds to all organic compounds.
   - Can detect as low as one nanogram of a compound.
   - One of the electrodes is the jet of a flame emanating from a mixture of hydrogen and air.
   - The other electrode is a platinum wire mounted near the tip of the flame.
   - Sample components are ionized in the flame and an increased signal is passed to the recorder.
2. Electron capture detector
   - Responds to halogen compounds which capture electrons.
   - Used to detect polychlorinated compounds and pesticides.
   - Has high sensitivity and can detect even picogram quantities.
   - A radioactive ^{63}Ni ionizes the column gas producing electrons which are captured by an electron capturing compound.
3. Thermionic detector
   - Works by ionic dissociation of a cesium bromide electrode.
   - Is useful in detecting organophosphorous pesticides.

❑ Enumerate the applications of GLC.

1. GLC is used to analyze volatile substances or substances that can be volatilized.
2. GLC is used for
   - separation of alcohols, fatty acids, and esters in biological samples
   - identification of components in flavoring agents and wines
   - detection of pesticides in biological samples.

❑ What is the principle of ion-exchange chromatography?

1. *Ion-exchange chromatography* is a technique by which molecules are separated based on their differences in acid–base behaviour.
2. The separations are carried out in columns packed with an ion-exchanger.

❑ What are the types of ion exchanger?

The types of ion exchanger include

1. Cation exchangers
   - contain negatively charged groups and attract positively charged ions.
   - are also called acidic ion exchangers.
   - examples include sulphonated polystyrene (Dowex 50), carboxymethyl dextran (CM-Sephadex).
   - strong cation exchangers are resins substituted with sulfonic acid groups (*e.g.*, Dowex 50).
   - weak cation exchangers are resins substituted with carboxylate groups (*e.g.*, CM-Sephadex).
2. Anion exchangers
   - contain positively charged groups and attract negatively charged anions.
   - are also called basic ion exchangers.
   - strong anion exchangers are resins containing tertiary amines (*e.g.*, diethylaminoethyl (DEAE) Sephadex).
   - weak anion exchangers are resins containing secondary amines (*e.g.*, DEAE cellulose).

❑ What are the applications of ion-exchange chromatography?

1. Separation of amino acids by a strong acid cation exchanger. This principle is used in the amino acid analyzer.
2. Determination of the base composition of nucleic acids.
3. Preparation of ultrapure, metal-ion free reagents.
4. Desalting.
5. Preparation of deionized water.
6. Separation of vitamins.

❑ Outline the principle of molecular sieve chromatography.

- *Molecular sieve chromatography,* also known as *gel filtration* or *molecular exclusion chromatography,* is a technique used to separate solutes based on their molecular size.
- The sample solution containing a mixture of proteins is passed through a column of polymeric material previously equilibrated with the buffer.
- The gels used to prepare the column may be cross-linked dextrans (*e.g.,* Sephadex), agarose (*e.g.,* Sepharose), polyacrylamide, or polystyrene.
- Proteins of different molecular size penetrate into the internal pores of the gel to varying degrees (Fig. 33.5).
- Large protein molecules that cannot enter the pores remain in the excluded volume, whereas small proteins enter the pores. Proteins of intermediate size are excluded based on their size.

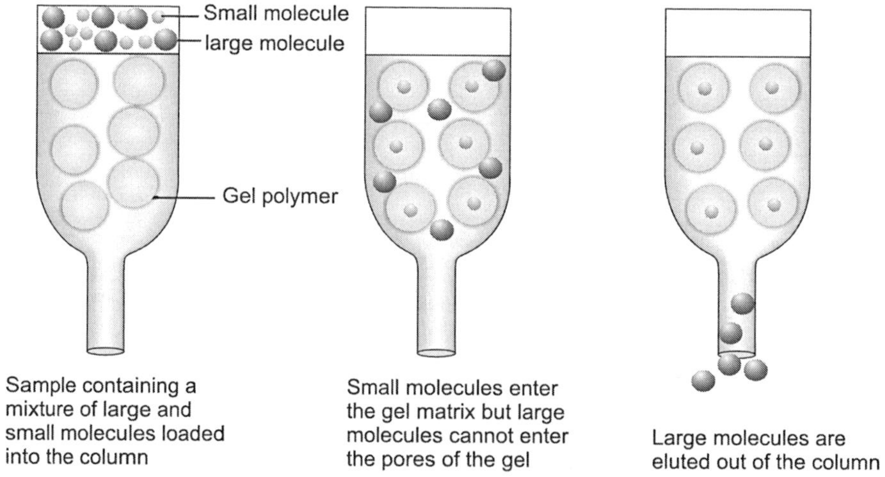

Fig. 33.5 Separation of molecules by molecular sieve chromatography

❑ What are the applications of molecular sieve chromatography?

The applications of molecular sieve chromatography include

1. Purification of proteins, viruses, enzymes, hormones, antibodies, nucleic acids, and polysaccharides.
2. Concentration of solutions containing high molecular weight substances. While water and low molecular weight compounds are absorbed by the gel, high molecular weight compounds that remain in solution become more concentrated without any change in their pH or ionic strength.
3. Desalting of low molecular weight compounds during protein purification.
4. Determination of molecular weight of macromolecules.

❑ Explain the principle of affinity chromatography.

- *Affinity chromatography* is a technique for separating molecules based on the specific affinity of a molecule for its ligand, such as an enzyme for its substrate.
- The sample to be separated is applied to a column of ligand bound to a matrix such as cross-linked dextran or agarose (Sepharose). The molecule that has high affinity for the ligand binds to it, while other molecules are eluted out. The bound molecule is eluted by changing the conditions so that binding does not occur.
- The types of affinity chromatography include
  i. *Immunoaffinity chromatography* in which a biological molecule is immobilized and a small molecule (ligand) bound to it, or the ligand is immobilized and the macromolecule bound to it. An antibody is used as the immobilized molecule to capture its antigen.
  ii. *Pseudoaffinity chromatography* in which a compound that resembles a biological ligand is immobilized on a solid material, and enzymes or other proteins are bound to it.
  iii. *Metal affinity chromatography* in which a metal ion is immobilized on a solid support and binds a specific macromolecule.

❑ Enumerate the applications of affinity chromatography.

1. Purification of enzymes, immunoglobulins, nucleic acids, and polysaccharides.
2. Separation of mixtures.
3. Isolation of glycoproteins by lectins.
4. Purification of IgG from serum or cell extracts by protein A-agarose.

❑ Discuss the principle, instrumentation, and applications of high-performance liquid chromatography.

Principle

High-performance liquid chromatography (HPLC) is a chromatographic technique carried out at high pressures using columns packed with tiny beads of matrix and automated equipment that enables highly reproducible profiles.

Instrumentation

The components in HPLC include (Fig. 33.6):

- Solvent reservoir made of steel to store the mobile phase.
- A high-pressure pump to drive the mobile phase through the column. The pump may be a holding coil, pneumatic amplifier, piston, or syringe.
- Sample injection device that may be a microsyringe or a small volume metal loop.

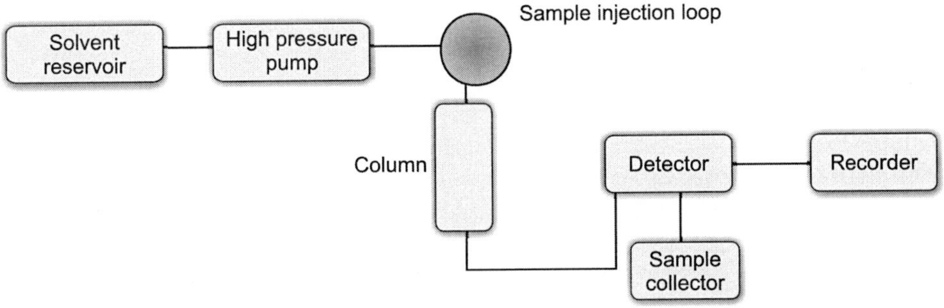

Fig. 33.6 Components of HPLC

- A column made of stainless steel in which separation takes place.
- Detector that may be of different types: Variable wavelength, scanning wavelength, fluorescence, or electrochemical.
- Recorder.

Applications

- Fractionation of proteins and peptides.
- Separation of monosaccharides, oligosaccharides, steroids.
- Analysis of vitamins.
- Analysis of pharmaceutical preparations.

❑ What is reverse phase chromatography?

Reverse phase chromatography is a chromatographic technique in which the stationary phase is hydrophobic and the mobile phase is polar.

❑ What is electrophoresis?

Electrophoresis refers to the migration of charged molecules such as proteins or nucleic acids under the influence of an electric field.

❑ Give a brief account of polyacrylamide gel electrophoresis.

- *Polyacrylamide gel electrophoresis* (PAGE) is a technique that separates molecules based on the molecular charge and size.
- *Gel preparation*
  - Polyacrylamide gels are prepared by polymerization of acrylamide monomers with *N, N'*-methylenebisacrylamide, a cross-linking agent, in the presence of ammonium persulphate which functions as an initiator and N, N, N', N'–tetramethylene diamine (TEMED), a free radical catlyst.
  - Low percentage gels have large pore size and are used to separate proteins and nucleic acids. High percentage gels with small pores are used to separate low molecular weight compounds.
- *Apparatus* for electrophoresis may be columns or slab gels.
- *Detection* is based on the type of molecule.
  - Coomassie brilliant blue is used for detection of proteins, Sudan black for lipoproteins, Periodic Acid Schiff (PAS) or lectins for glycoproteins.
- *Quantitation* can be done by scanning densitometry, elution and spectrophotometric measurement.
- *Applications*
  - Separation of nucleic acids and proteins.
  - DNA sequencing.
  - Restriction mapping.
  - Analysis of oligomeric proteins by using solubilizers such as urea, 2-mercaptoethanol, and sodium dodecylsulphate (SDS).
  - Determination of molecular weight of proteins by SDS-PAGE.

❑ Write notes on SDS-PAGE.

- Sodium dodecylsulphate (SDS) is an anionic detergent that disrupts the secondary, tertiary, and quaternary structures of proteins generating linear polypeptide chains.

- SDS binds to hydrophobic regions of denatured proteins and masks the native charge of the proteins.
- Under these conditions, the proteins have a constant charge to mass ratio and electrophoretic mobility is dependent only on the molecular size of the protein.
- This technique is used to determine the molecular weight of proteins.

❑ Differentiate native gels from gradient gels.

Native gels

- are polyacrylamide gels without SDS.
- are used to analyze proteins in their native state.
- separation of proteins occurs due to different electrophoretic mobilities and the sieving effects of the gel.

Gradient gels

- are polyacrylamide gels in which the acrylamide concentration ranges from 5% at the top of the gel to 25% at the bottom of the gel.
- are used to separate proteins of a wide range of molecular weights as well as proteins of very similar molecular weight.

❑ Outline the principle and applications of isoelectric focusing.

Principle

- *Isoelectric focusing* (IEF) is a technique that separates amphoteric molecules such as proteins based on the differences in their isoelectric pH (pI).
- Electrophoresis is carried out in a pH gradient created by introduction of ampholytes into the gel.
- Each protein moves in this gradient until it reaches a pH equal to its isoelectric point. At this point, the charge on the protein is zero and migration ceases and it forms a band.
- IEF is capable of separating proteins that differ in their isoelectric points by only 0.01 of a pH unit.

Applications

- Analysis of the microheterogeneity of a protein such as a protein which can exist in mono, di, and triphosphorylated forms.
- Identification of isoenzymes.
- Separation, purification, and characterization of proteins.

❑ What is cellulose acetate electrophoresis?

- *Cellulose acetate electrophoresis* is an electrophoretic technique that uses cellulose acetate paper as a supporting medium.
- It is used for the separation of glycoproteins, lipoproteins, and hemoglobin from blood.

❑ Outline the principle of agarose gel electrophoresis

- *Agarose gel electrophoresis* is an electrophoretic technique used to separate nucleic acids based on the resistance to movement of samples caused by a matrix composed of a highly purified form of agar.

- The concentration of agarose depends on the size of the molecules to be separated. While a 0.3% gel is suitable for separating 5–60 kb double- stranded DNA, a 0.8% gel is required for molecules between 0.5 and 10kb and a 2% gel for molecules in the size range 0.1–3 kb.
- For staining the bands, ethidium bromide, which gives an orange-red fluorescence under UV light, or SYBR Gold can be used.

❑ Elucidate the principle and applications of pulsed-field gel electrophoresis.

- *Pulsed-field gel electrophoresis* is an electrophoretic technique in which pulsed, alternating orthogonal fields are used to separate large DNA fragments in agarose gels.
- Two electric fields are alternately applied at different angles for a specified time period.
- While the first field causes stretching of coiled molecules in the horizontal plane and electrophoretic migration, application of the second field interrupts the first one driving the molecule to move in a new direction.
- Reorientation of the molecules to the second new field is length dependent; larger molecules take longer time period, whereas smaller molecules reorient rapidly. Molecules are separated based on size.

Applications

- Separation of large DNA molecules.
- Separation of whole chromosomes.
- Separation of major histocompatibility locus of several 1000 kb of DNA.

❑ Briefly explain the technique of Southern blotting.

Southern blotting refers to the transfer of DNA from agarose gel to nitrocellulose or nylon for hybridization.

Protocol

1. Restriction enzyme digestion of DNA to generate fragments.
2. Agarose gel electrophoresis of DNA fragments.
3. Denaturation of DNA by alkali to convert double-stranded DNA to single strands.
4. Transfer of separated fragments from agarose gel to a nitrocellulose or nylon filter by blotting, which involves the following steps:
   - Agarose gel is kept on a thick stack of filter paper placed in a tray filled with a buffer of high salt concentration.
   - A nitrocellulose filter is placed on top of the gel.
   - Thick stacks of dry filter paper are placed on top of the filter and pressed down by a weight.
   - The salt solution from the tray, which is attracted to the dry filter paper, passes through the gel and the nitrocellulose filter.
   - The DNA, which is carried with the salt solution, is trapped in the filter.
   - The filter paper is dried in vacuum at 80°C to immobilize DNA on the filter.
5. Hybridization of DNA on the filter with a radioactive probe (^{32}P DNA or RNA) at 68°C for 16–24 hours in a buffer of high ionic strength, which promotes hybridization.
6. Washing in stringent conditions to remove unbound probe. The filter is then dried.
7. Visualization by autoradiography (Fig. 33.7).

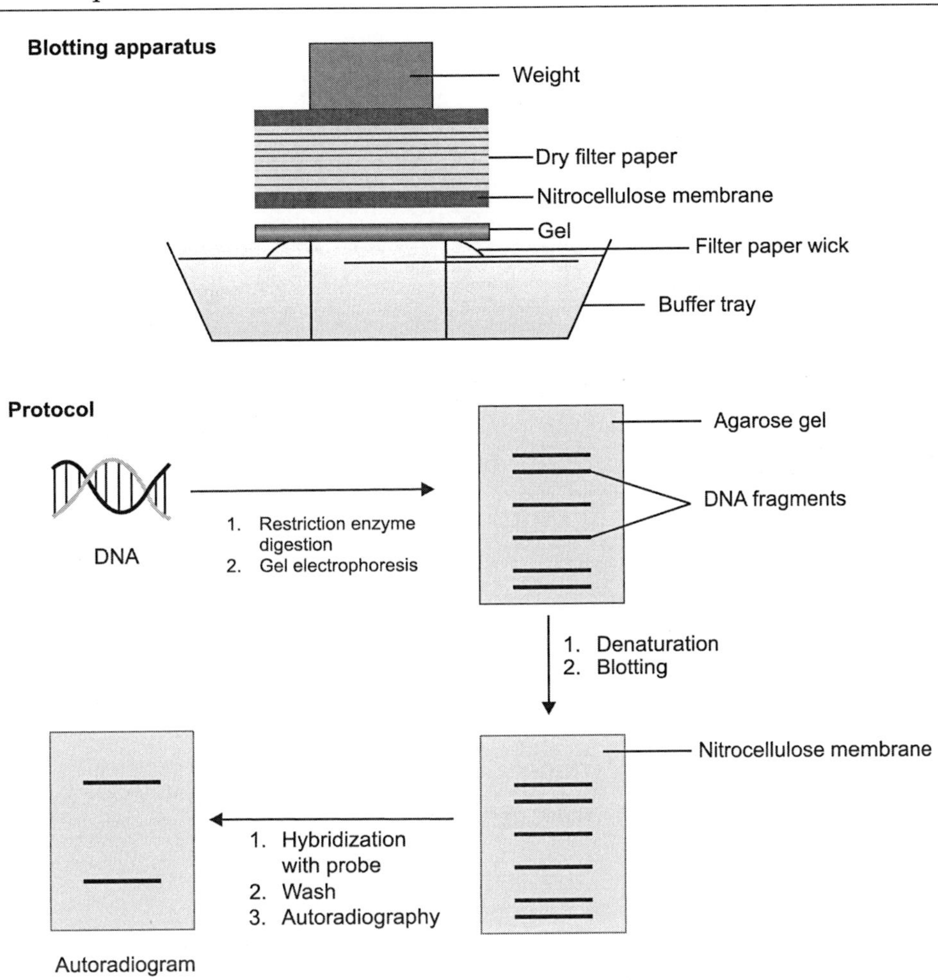

Fig. 33.7 Southern blotting technique

❑ Give a brief account of Northern blotting.

Northern blotting is a procedure by which RNA molecules are transferred from agarose gel to diazobenzyloxymethyl (DBM) paper or nylon membrane by capillary action for hybridization.

Protocol

- Digestion of RNA into fragments and size fractionation by agarose gel electrophoresis.
- Transfer of RNA from agarose to DBM paper by capillary action.
- Denaturation of RNA by formaldehyde.
- RNA immobilized on the membrane is hybridized to single-stranded DNA probes.
- Identification of the DNA–RNA hybrids by autoradiography.

Applications

- Determination of hybridization patterns in mRNA samples.
- Analysis of gene expression.
- Determination of the size of a transcript.

❑ Discuss the principle of Western blotting.

- Western blotting is a technique by which a protein is transferred from a polyacrylamide gel to a nitrocellulose or nylon membrane after electrophoresis.
- The protein molecule is identified by probing the blot with a specific antibody.
- Protein transfer can be done by capillary action or by electroblotting.
- Protein detection involves the following steps:
  - Treatment of the blot containing the proteins with 10% bovine serum albumin to block nonspecific binding.
  - Incubation of the blot with the antibody (primary antibody), which binds to the corresponding antigen on the blot.
  - Incubation of the blot with a second antibody (anti-antibody) that binds to the primary antibody to increase sensitivity and specificity.
  - The second antibody may be
    - i. radiolabeled. Exposure of the blot to a photographic film is used to demonstrate the presence of the antigen.
    - ii. tagged to an enzyme such as *horseradish peroxidase*. Incubation with the substrate gives a colored product, which indicates the presence of the antigen.
    - iii. labeled with fluorescein isothiocyanate and the antigen detected by measurement of fluorescence after UV irradiation of the blot.

❑ Describe the chemical method of DNA sequencing.

Chemical or *Maxam and Gilbert method* of DNA sequencing involves base-specific cleavage of DNA using chemical reagents.

Steps

1. Radioactive (^{32}P) labeling of the phosphate group of a DNA fragment.
2. Modification of the bases in DNA by specific chemicals. For example, methylation of N^7guanine by dimethylsulphate, hydrolysis of pyrimidines by hydrazine, high salt to inhibit thymine, and formic acid to cleave purines.
3. Cleavage of the modified base from the sugar-phosphate backbone by piperidine in such a way that only one or a few cleavages are generated per molecule to yield a ladder of progressively longer molecules.
4. Separation of the DNA fragments in a high-resolution sequencing gel containing 20% polyacylamide.
5. Detection of bands by autoradiography.

Advantage: The method is useful for both single- and double-stranded DNA.

Disadvantage: The method is not suitable for large-scale sequencing.

❑ Discuss the enzymatic method of DNA sequencing.

Enzymatic method or *Sanger (dideoxy) method* of DNA sequencing involves the use of four specific dideoxyribonucleotides (ddNTPs) to terminate DNA synthesis catalyzed by *DNA polymerase*. DNA polymerase extends the primer base paired to the template by the successive addition of dNTPs to the free 3'-OH of the primer. The ddNTPs, which lack the 3'-OH group, cause chain termination (Fig. 33.8).

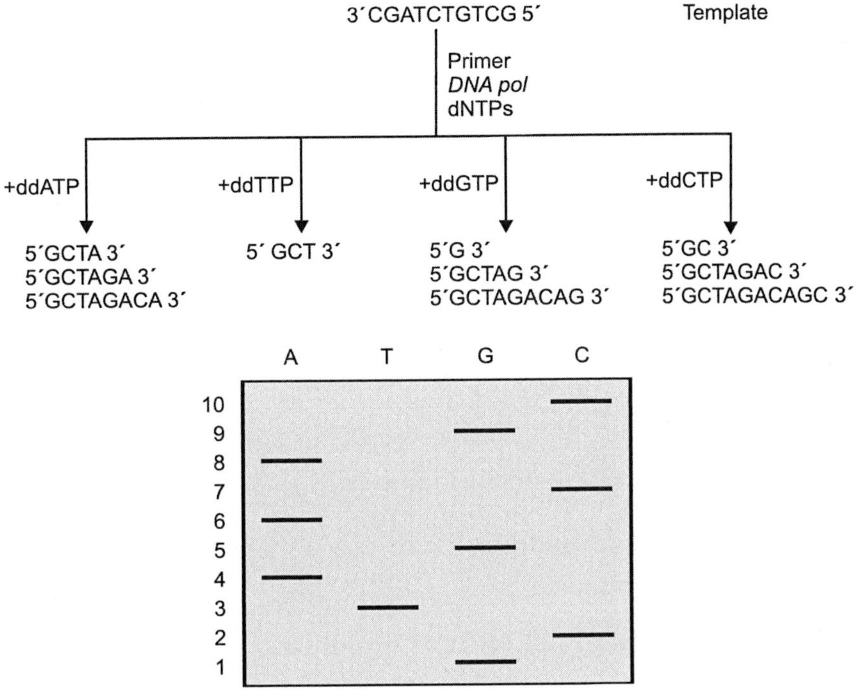

Fig. 33.8 Steps in dideoxy sequencing

Steps

1. Annealing of a short radiolabeled primer to the single-stranded DNA fragment to be sequenced.
2. Separation of the reaction mixture containing template-primer, DNA polymerase, and dNTPs into four sets.
3. Addition of one of the four specific ddNTPs (ddATP, ddGTP, ddCTP, ddTTP) to each set.
4. In a reaction set that contains ddGTP, DNA synthesis is terminated at positions where a G residue occurs. This results in several fragments of differing length, each of which terminates at a G residue.
5. The same protocol is followed for each of the other nucleotides.
6. Separation of the resulting DNA fragments by gel electrophoresis. The smaller DNA fragments that migrate faster and are found at the bottom of the gel represent nucleotides near the 5' end. The larger fragments found at the top represent nucleotides near the 3' end.
7. The sequence obtained is for the complementary strand from which the sequence of the template strand can be deduced.

❑ What is DNA fingerprinting?

- DNA fingerprinting is a technique developed by Alec Jeffreys for the identification of individuals based on the DNA profile.
- It is also known as *DNA profiling* or *molecular fingerprinting*.

- The technique is based on the fact that highly variable regions of the genome termed *variable number of tandem repeats (VNTRs)* in minisatellites and microsatellites are unique to each individual.
- The protocol involves
  - Digestion of DNA using a restriction enzyme such as *Hin*fI, which cuts outside the VNTR region.
  - Size fractionation of DNA fragments by gel electrophoresis.
  - Identification of the DNA pattern by:
    i. Southern transfer using a probe that binds to a polymorphic region of DNA.
    ii. polymerase chain reaction (PCR) using primers flanking the polymorphic region.
- Microsatellite tetranucleotide (CAGA) repeats are useful for forensic analysis.
- A single difference between sample and suspect rules out suspect.
- *Applications*
  - forensic analysis
  - paternity determination
  - pedigree analysis
  - migration studies
  - genetic analysis of agricultural crops and animals.

Discuss the principle and applications of the polymerase chain reaction.

Principle

- The polymerase chain reaction (PCR) is an *in vitro* technique for the selective amplification of specific sequences in DNA.
- *The reaction mixture* contains the following:
  i. Template DNA, is the target DNA to be amplified and can be from any source including blood, sperm, mouthwash, tissue sections, and paraffin blocks.
  ii. Two oligonucleotide primers complementary to sequences flanking the target DNA.
  iii. *Taq* DNA polymerase from the thermophilic microorganism *Thermus aquaticus.*
  iv. All the four dNTPS.
  v. Buffer.
- *The PCR cycle* consists of three stages (Fig. 33.9):
  i. *Denaturation* of template DNA by heating to 95°C for 1 min.
  ii. *Annealing* of the oligonucleotide primers to template DNA by cooling to approximately 55°C.
  iii. *Primer extension* by Taq polymerase by heating to 72°C.
- About 25–40 cycles are carried out to achieve 10^5 fold amplification of the original DNA target sequence.
- The PCR products after agarose gel electrophoresis are detected by ethidium bromide staining.

Applications

- Genomic cloning: PCR can amplify all the sequences in genomic DNA obviating the need for genomic libraries.
- Forensic science for analysis of DNA from trace amounts of biological sample obtained at the scene of a crime.
- Medical diagnostics.

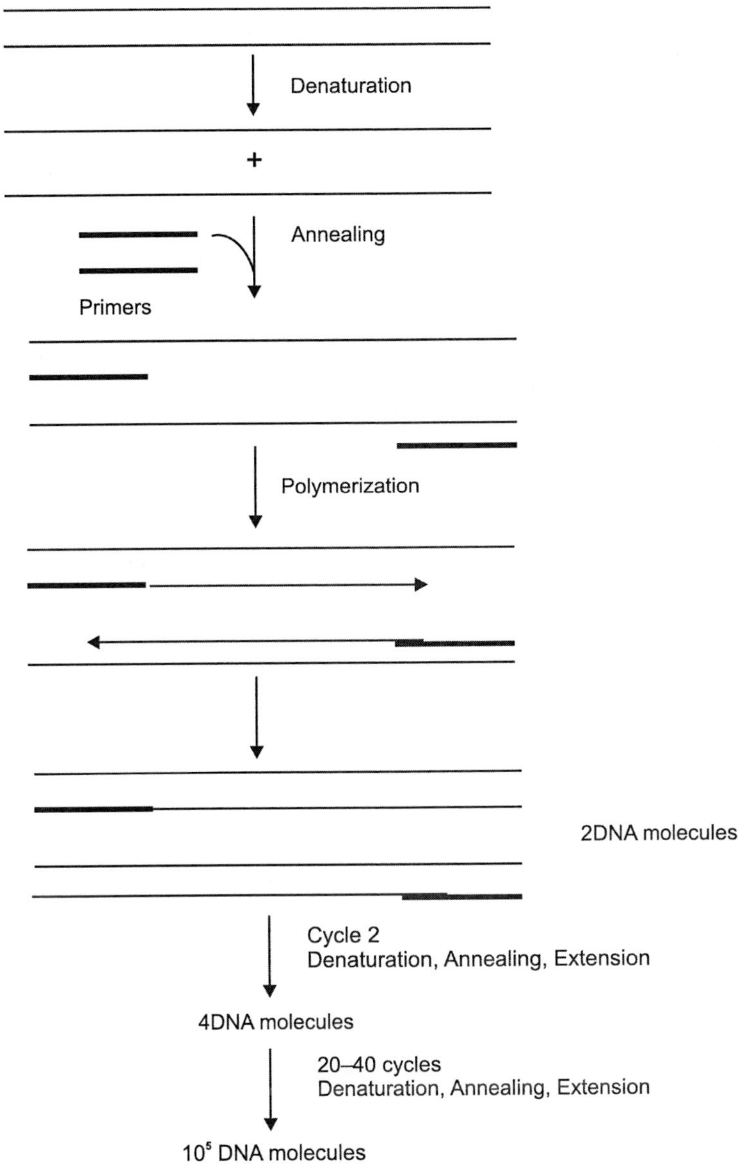

Fig. 33.9 The polymerase chain reaction

❑ What is RT-PCR?

Reverse transcriptase PCR (RT-PCR) is a technique involving reverse transcription of mRNA molecules to form cDNA, followed by amplification of the cDNA by PCR.

Applications

- RNA analysis.
- Construction of cDNA libraries.
- *Differential display* to understand diferentially expressed genes in two samples.

❑ What is real-time quantitative PCR?

- *Real-time PCR* is a technique for quantitation of PCR products using sensitive oligonucleotide probes, such as *Taqman* probe that contains a reporter fluorescent dye at the 5' end and a quencher at the 3' end.
- The probe anneals downstream to one of the primers.
- During primer extension, the probe is cleaved by the 5' → 3' exonuclease activity of Taq polymerase.
- This results in detachment of reporter from the quencher with increase in fluorescent signal.
- The fluorescence intensity increases with increase in the cycle number.

❑ Give a brief account of the importance of DNA microarrays in analyzing gene expression.

DNA microarrays consist of DNA fragments immobilized on to a solid surface. The microarray technique is useful for the analysis of genes and gene expression.

***Preparation of microarray*:** The microarrays are glass slides studded with thousands of DNA fragments, each of which functions as a probe for a specific gene. The DNA fragments after amplification by PCR are spotted on a solid surface, such as glass slides, silicon chips, or nitrocellulose sheets, with the help of a robotic device that deposits nanoliter quantities of DNA in organized patterns. Exposure to ultraviolet light cross-links DNA to the solid surface. The size of each spot in a microarray is typically 200 microns in diameter.

***Analysis*:** The microarray can be probed with fluorescently labeled cDNAs reverse transcribed from mRNAs. Hybridization occurs between the fluorescent cDNA probes and complementary sequences on the microarray. Fluorescent spots represent genes that are expressed in the sample. These can be identified by an automated scanning-laser microscope.

***Applications*:** Microarrays are useful for the following:

1. Gene discovery and gene mapping.
2. Analysis of gene expression: The RNA products of thousands of genes can be monitored simultaneously. This allows analysis of gene expression patterns during cell division and in response to hormones.
3. Disease diagnosis and mutational analysis.
4. Drug discovery and pharmacogenomics: Microarrays enable correlations between the genetic profile of patients and therapeutic response.

❑ What are stem cells?

- Stem cells are cells that are capable of producing different kinds of cells.
- Based on the ability to form different types of cells, stem cells may be divided into
  - i. *Totipotent stem cells* that can generate all cell types of the body. For example, a zygote gives rise to all cell types.
  - ii. *Pluripotent stem cells* that are capable of producing a number of different cell types but not all cell types. A classic example is the hematopoietic stem cell, which forms different types of blood cells but cannot form a muscle or a skin cell.
  - iii. *Unipotent stem cells* that form only one type of cell.

- Stem cells have two critical properties:
  i. The ability to reproduce indefinitely (*self-renewal*).
  ii. They are unspecialized (undifferentiated).
- Many stem cells also have the ability to differentiate to specialized cell types under certain conditions.

❑ How are stem cells divided based on the source/origin?

Stem cells are of two types based on the origin:
- Embryonic stem cells from embryos (<6 weeks).
- Adult stem cells present in tissues such as brain, spinal cord, and bone marrow.

❑ How are embryonic stem cells grown in the laboratory?

Embryonic stem cells can be grown in the laboratory by the following steps:
1. Transfer of the *inner cell mass* into a culture medium in a Petri dish coated with a *feeder layer* of mouse embryonic skin cells. The feeder layer provides a sticky surface for attachment of the inner cell mass and also provides nutrients.
2. Proliferation of the inner cell mass.
3. Subculturing of the cells for 6 months or more to generate millions of embryonic stem cells that are pluripotent.
4. The embryonic stem cell line so generated is undifferentiated.
5. The cells in the cell line can clump together to form *embryoid bodies* that can differentiate spontaneously into muscle or nerve cells.
6. The cell line can be directed to differentiate into specific cell types under special conditions. This is known as *directed differentiation*.

❑ What are the uses of stem cell technology?

The potential uses of stem cell technology include the following:
- To understand the process of human development and differentiation.
- To understand birth defects.
- Screening of drugs and toxins.
- Cell-based therapies of neurodegenerative disorders, heart disease, and cancer.

❑ What is nanotechnology?

- *Nanotechnology* refers to the study and manipulation of very small particles of nanometer scale.
- It involves design, characterization, production, and application of structures, devices, and systems that exploit the special properties at the nanoscale by manipulating shape and size.

❑ What is a nanometer?

A nanometer is 1 billionth (1,000,000,000) of a meter.

❑ What are the potential applications of nanotechnology?

Nanotechnology will impact a wide range of fields, including
- Biotechnology
- Solar energy

- Computers
- Stain resistant apparel
- Semiconductors
- Sensors that can detect a specific chemical or toxin
- Catalysts that improve efficiency of chemical processing, decreasing waste, and energy use.

Nanomedicine is expected to improve diagnostics and therapeutics. It is especially envisaged to be of immense benefit in the following areas:

- Drug development: Nanoparticles will be useful to increase the bioavailability of a drug.
- Cell-specific drugs.
- Cancer
  - i. Nanoparticles of cadmium selenide (quantum dots) that glow on exposure to UV light are useful in detecting tumors.
  - ii. Nanoshells coated with gold will be useful in killing tumors when targeted with antibodies or peptides.
  - iii. Visualization: Quantum dots attached to proteins can be used to penetrate cell walls.
  - iv. Nanorobots are speculated to repair or detect damage or infection.

Section 9

Special Topics

- Basic Principles of Oncology
- Genomics, Proteomics, and Bioinformatics

Chapter 34

Basic Principles of Oncology

❑ How does a malignant tumor differ from a benign tumor?

Tumors may be benign or malignant.

Benign tumors

- exhibit close resemblance to the tissue of origin.
- grow slowly by expansion and compress the surrounding tissues.
- are encapsulated and do not invade surrounding tissues.
- do not show any alteration in cell structure and function.
- are well differentiated.
- are generally localized and do not metastasize to other parts of the body.
- do not show recurrence.

Malignant tumors

- may sometimes lose resemblance to the tissue of origin.
- grow rapidly by infiltration and destroy adjacent tissues.
- are not encapsulated and invade surrounding tissues.
- the cell structure and functions are altered.
- may be well differentiated, moderately differentiated, or even poorly differentiated.
- metastasize to other parts of the body via blood and lymphatic channels.
- recurrence is common.

❑ What are the hallmark capabilities of cancer?

Cancer cells show the following hallmark capabilities that distinguish them from normal cells:

- ***Limitless replicative potential*:** Cancer results from genetic changes in a single cell that proliferates to form a clone of malignant cells. The cells in the clone are, however, not identical to each other and exhibit heterogeneity.
- ***Self-sufficiency in growth signals and insensitivity to antigrowth signals*:** Cancer cells are not subject to regulation by mechanisms that control normal cell growth.
- ***Evasion of apoptosis* (programmed cell death):** Cancer cells evade apoptotic mechanisms, thereby facilitating proliferation of genetically mutated cells.

- ***Angiogenesis*:** Cancer cells are capable of forming new blood vessels in order to nourish the tumor and to facilitate spread to other sites.
- ***Tissue invasion*:** This refers to the spread of cancer to other tissues. Invasion involves receptor-mediated attachment of tumor cells to matrix components, secretion of proteolytic enzymes (plasminogen activator, plasmin, cathepsins, type IV collagenase) that degrade the matrix, and migration of the tumor cells into the extracellular matrix.
- ***Metastasis*:** This refers to the ability of cancer cells to disseminate to other parts of the body via lymphatic or blood vessels, or by direct seeding into body cavities.

❑ Enumerate the growth characteristics exhibited by transformed cells in culture.

Cancer cells exhibit the following characteristics when grown in culture:

1. ***Alteration in morphology*:** Transformed cells have a round shape compared to flattened normal cells.
2. ***Anchorage independence*:** Transformed cells are capable of growing in suspension in semi-solid medium such as soft agar unlike normal cells, which require anchorage to a substratum.
3. ***Loss of contact inhibition of growth*:** Unlike normal cells which form only monolayers due to contact inhibition of growth, transformed cells form multilayers.
4. ***Decreased serum requirement*:** Transformed cells have diminished requirement for serum compared to normal cells due to *autocrine secretion* of growth factors.
5. ***Immortality*:** Malignant cells survive indefinitely in culture unlike normal cells, which survive only for 50–60 population doublings.

❑ Enumerate the morphological and ultrastructural changes that distinguish cancer cells from normal cells.

Nucleus

- Large relative to the cytoplasm.
- Hyperchromatic due to increased nucleoprotein.

Nucleolus

- Large.
- Pleomorphic.

Chromosomes

- *Aneuploid karyotype* (change from normal diploid state).
- Chromosomal abnormalities: Gain or loss of chromosomes, translocations, inversions, and duplications.

Cytoplasm

- Mitochondria: Decrease in size and number.
- Endoplasmic reticulum: Decrease in size and loss of function.
- Inclusions.

Cell surface

- Increased transport of amino acids and glucose into tumor cells.
- Increase in collagenase activity.
- Increased lectin agglutinability.

- Partial disorganization of cytoskeleton.
- Increased branching of glycan chains of glycoproteins.
- Decreased synthesis of glycosphingolipids.

❑ Elucidate the metabolic changes that take place in cancer cells.

The metabolic changes that occur in cancer provide a selective advantage to cancer cells for growth over normal cells. These include:

- selective activation of low K_m isoenzymes unresponsive to normal control mechanisms
- increased uptake and utilization of glucose
- increase in glycolysis with activation of key glycolytic enzymes
- increase in fatty acid oxidation, depletion of body fat
- negative nitrogen balance
- increase in amino acid uptake
- decreased protein synthesis with increase in protein catabolism
- protein energy malnutrition due to low intake, increased loss, or chemotherapy
- increase in polyamine synthesis
- increase in biosynthesis of purines, pyrimidines, and nucleic acids associated with decreased catabolism.

❑ What is a tumor marker?

- A tumor marker is a substance present in or produced by the tumor or by the host in response to the tumor. Tumor markers may be tumor-derived or tumor-associated.
- Tumor markers are useful in screening, diagnosis, prognosis, monitoring therapy, and in the detection of recurrence.

❑ Discuss the different classes of tumor markers.

Tumor markers are categorized into the following classes:

◘ *Oncofetal proteins*

- Present in high concentration in fetal life and low levels or traces in the adult.
- *Alpha fetoprotein (AFP)* is useful as a tumor marker in hepatocellular carcinoma and testicular teratomas.
- *Carcinoembryonic antigen (CEA)* is a marker for colorectal, gastrointestinal, lung, and breast cancer. However, CEA lacks specificity and sensitivity.

◘ *Hormones*

- Human chorionic gonadotrophin (hCG), a glycoprotein produced by syncytiotrophoblast cells of the placenta, which is increased in choriocarcinoma and testicular tumors.
- *Ectopic hormones* (secreted by tumors of non-endocrine organs) such as
  a. ACTH in small cell carcinoma of the lung.
  b. Calcitonin in medullary carcinoma of the thyroid.

◘ *Enzymes*

- Prostatic acid phosphatase (tartrate labile ACP) is increased in patients with disseminated prostatic carcinoma.

- Increase in plasma ALP in patients with ovarian cancer, testicular seminomas, and other tumors. A placental-like isoenzyme of ALP termed *'Reagan isoenzyme'* is seen in some patients with carcinoma of the bronchus.
- An increase in plasma lactate dehydrogenase (LD) is seen in lymphoma. LD-2 and LD-3 are increased in leukemia.
- Plasma gamma-glutamyl transpeptidase (GGT) is increased in hepatic metastasis.
- An increase in plasma AST activity with the presence of macro AST is seen in malignant infiltration of the liver.

◻ *Immunoglobulins*

Excess formation and urinary excretion of light chains as well as incomplete fragments of heavy chains or whole immunoglobulin molecules can be detected in urine or in circulation of patients with multiple myeloma and B-cell lymphomas. Serum electrophoresis reveals a monoclonal protein termed *M protein* as a sharp peak.

◻ *Tumor-associated antigens* such as

- *Carbohydrate antigen (CA) markers,* which are high molecular weight mucins or blood group specific antigens present on tumor cell surface or secreted by the tumor cells. The CA markers lack specificity except few such as CA125, which is specific for ovarian cancer.
- *Prostate specific antigen (PSA),* an organ-specific tumor marker and a sensitive tumor marker for advanced prostatic carcinoma together with digital rectal examination and transrectal ultrasonography.

❑ Give an account of agents that cause cancer.

Agents that cause cancer include radiation, viruses and chemicals.

Radiation

- Exposure to radiation occurs due to environment, occupation, or medical treatment.
- Both ultraviolet (UV) radiation and ionizing radiation (electromagnetic and particulate) can cause cancer by inducing
  a. chromosomal abnormalities such as point mutations, translocations, exchanges, and breaks.
  b. mutations in specific genes in target cells resulting in malignant transformation.
- UV radiation from sunlight induces skin cancer in fair skinned people and in people living close to the equator. Exposure to UV rays forms cyclobutane dimers between adjacent thymine residues in DNA that are normally repaired by the excision repair system. In *xeroderma pigmentosum,* in which the repair system is defective, UV-induced DNA damage is not repaired and eventually leads to mutations and skin cancer.
- Ionizing radiation causes cancer of the lung, thyroid, and breast.

Viruses

- Both DNA and RNA viruses cause cancer.
- Examples of viruses that induce cancer include Epstein-Barr virus that induces Burkitt's lymphoma and human T-cell leukemia virus that causes adult T-cell leukemia.
- The viral DNA integrates into the genome of the host cell and alters the structure and activity of the genes inducing neoplastic transformation.

Chemicals

- Environmental chemicals are responsible for more than 80% of cancer deaths.
- Exposure to carcinogenic chemicals occurs due to:
  a. *life style:* tobacco is known to cause cancers of a wide variety of organs such as lungs and oral cavity.
  b. *diet:* saturated fat is a risk factor for breast cancer.
  c. *occupation:* for example, workers exposed to aromatic amines in paint industry are susceptible to bladder cancer.
  d. drugs such as immunosuppressive agents can cause cancers of the reticuloendothelial system.
- Chemical carcinogens are structurally diverse and include polycyclic aromatic hydrocarbons (PAH), aromatic amines, azo dyes, alkylating and acylating agents, hydrazines, triazines, microbial agents (aflatoxin), plant products (cycasin), and inorganic chemicals (cobalt, chromium, nickel, etc.).

❑ Briefly explain the mechanism by which chemical carcinogens induce malignant transformation.

Carcinogens are classified into two categories:

- ***Direct acting carcinogens*** such as dimethylsulphate and β-propiolactone that are intrinsically electrophilic and bind to DNA to form adducts.
- ***Indirect carcinogens or procarcinogens*** that undergo metabolic activation to proximate and ultimate carcinogens catalyzed by phase I *cytochrome P450-dependent monoxygenases* and phase II detoxification enzymes. The ultimate carcinogens are electrophiles that bind to DNA to form adducts.

The critical target for both direct carcinogens and procarcinogens is DNA. If the damaged DNA undergoes replication before it is repaired, the damage becomes permanent and undergoes further mutations to induce malignant transformation.

❑ Discuss the multistep nature of carcinogenesis.

Carcinogenesis is a multistep process that can be divided into three stages:

- ***Initiation***
  - is an irreversible, rapid process.
  - involves carcinogen-induced damage to DNA.
  - damaged cells must undergo at least one cycle of replication to make the lesions permanent.
- ***Promotion***
  - is a slow and reversible process by which initiated cells develop into tumors.
  - does not involve direct DNA damage but causes indirect DNA damage *via* formation of reactive oxygen species.
  - induces pleiotropic effects such as changes in nucleic acid and polyamine synthesis, induction of enzymes, prostaglandin release, etc.
  - examples of promoting agents include the diterpene phorbol ester 12-O-tetradecanoylphorbol 13-acetate (TPA), the active component of croton oil, saccharin, alcohol, etc.
- ***Progression***

  Characterized by:
  - appearance of malignant neoplasms

- chromosomal translocations
- karyotypic instability
- deletions
- invasion and metastasis
- anaplasia (lack of differentiation).

❑ Explain one method for detecting the carcinogenicity of chemicals.

- The carcinogenicity of a chemical can be tested by long-term bioassays or by short-term tests.
- A number of short-term tests are available for testing the mutagenicity of potential carcinogens.
- The most widely used is the *Ames test* devised by Bruce Ames, which is based on the principle that most carcinogens are also mutagens.
- The test uses mutant strains of *Salmonella typhimurium* that are incapable of synthesizing histidine (his^-).
- When bacteria are exposed to a chemical mutagen, the mutation is corrected and the bacteria is transformed from his^- to his^+ (*back mutation*).
- For procarcinogens, the *S9 fraction* of the liver, which contains microsomes is used in conjunction with the Ames test.
- A large number of carcinogens give positive results with the Ames test.

❑ Define oncogenes, proto-oncogens, and tumor suppressor genes.

- An *oncogene* is a gene whose protein product is capable of causing cancer. Oncogenes are designated by a three-letter abbreviation. For example, the *abl* gene from *Abelson murine leukemia virus* causes chronic granulocytic leukemia. Oncogenes are dominant in nature and drive cells towards malignancy.
- *Proto-oncogenes* are normal cellular genes that are activated to form oncogenes.
- *Tumor suppressor genes* (*TSG*) are genes that can suppress tumor formation. These are also known as *anti-oncogenes*. These genes are recessive and drive cells to normality.
- Cancer is known to be caused by activation of oncogenes and/or inactivation of tumor suppressor genes.

❑ What are the functions of (proto) oncogene products?

- The products of proto-oncogenes regulate cell division and differentiation.
- They act at different levels:
  - on the cell surface as ligands and growth factors.
  - on the cell membrane as receptors and signal transducers.
  - in the cytoplasm linking the signals from cell membrane to the nucleus.
  - in the nucleus in transcription of specific genes.
- Oncogenes are mutated versions of proto-oncogenes that form abnormal products or cause gene overexpression.

Explain the mechanisms by which proto-oncogenes are converted to oncogenes.

The mechanisms by which proto-oncogenes can be activated to oncogenes include

- ***Promoter and enhancer insertion*:** Infection of chicken B-lymphocytes by the retrovirus avian leukemia virus (ALV) results in integration of a double- stranded cDNA copy of the virus termed the *provirus* into the host genome adjacent to the *myc* oncogene. The provirus is flanked at both ends by long terminal repeats (LTR) that function as promoters or enhancers and activates transcription and translation to form the *myc* mRNA and its protein product.
- ***Chromosomal translocations*:** A classic example is the reciprocal translocation of *abl* gene from chromosome 9 to 22 and *sis* gene from 22 to 9 in chronic myelogenous leukemia (CML). Translocation involves a small region of chromosome 22 (*Philadelphia chromosome)* termed *breakpoint cluster region* or *bcr*. This results in the formation of a new fusion protein with increased ability to phosphorylate tyrosine residues in proteins.
- ***Single point mutations*** especially of the *ras* gene occur in many tumors. These mutations reduce GTPase activity of p21 resulting in continuous stimulation of adenylate cyclase activity, and changes in cAMP dependent protein kinases leading to malignant transformation.
- ***Amplification*** of the proto-oncogenes— *c-myc, N-myc, H-ras, N-ras* and *K-ras* has been documented in several tumors. Amplification can be detected as *homogeneously staining regions (HSR)* on specific chromosomes or as extrachromosomal *double minutes*.

Enumerate the functions of tumor suppressor gene products.

- The tumor suppressor proteins function as negative regulators of cell growth and repress cell proliferation and induce differentiation.
- The products of TSGs function as
  1. transcription regulators (*p53*)
  2. cytoplasmic signal transducers (neurofibromin)
  3. cell adhesion molecules (*deleted in colon cancer* (DCC) gene).
- Mutational inactivation of TSGs causes uncontrolled cell proliferation and malignant transformation.

How does *p53* function as a TSG?

- The *p53* gene is located on the short arm of chromosome 17.
- It encodes a 53 kDa nuclear phosphoprotein essential for normal cell growth.
- *Functions of wild type p53 protein*
  - Transcription factor that regulates gene expression
  - Cell cycle arrest in G_1 phase following DNA damage
  - Stimulation of DNA repair
  - Induction of apoptosis by regulating Bcl-2/Bax ratio
  - Maintenance of genomic stability
- Mutations in the *p53* gene are found in a wide variety of cancers.

❑ Write a note on breast cancer susceptibility genes.

The breast cancer susceptibility genes include:

- *BRCA1*
  - gene location: chromosome 17q21.
  - protein: contains 1863 amino acids, functions as a transcription factor.
- *BRCA2*
  - gene location: chromosome 13q12-13.
  - protein: 3495 amino acids.
- The *BRCA* proteins regulate cell-cycle arrest, apoptosis, and DNA repair
- Mutations in *BRCA1* and *BRCA2* have been documented in breast and ovarian tumors.
- Mutation carriers are at high risk of developing breast cancer.

❑ Give an account of the major classes of chemotherapeutic drugs.

Chemotherapeutic drugs include the following major categories:

- ***Antimetabolites***
  - Analogs of nucleic acid precursors incorporated into nucleic acids, instead of the normal precursors.
  - 6-Mercaptopurine, a hypoxanthine analog is converted to 6-thioinosine monophosphate (6-thio-IMP) that inhibits purine biosynthesis.
  - 5-Fluorouracil (5-FU), an analog of uracil, blocks dTMP formation and DNA synthesis.
  - Methotrexate binds tightly to the enzyme dihydrofolate reductase (DHFR) that converts dihydrofolate to tetrahydrofolate required for purine and thymidylate synthesis.
- ***Antitumor antibiotics***
  - The anthracycline antibiotics daunorubicin and doxorubicin (adriamycin) intercalate into DNA causing DNA unwinding and strand breaks.
  - Bleomycin from *Streptomyces verticillus* causes DNA damage by generation of highly reactive oxygen species (ROS).
- ***Alkylating agents*** such as cyclophosphamide bind covalently to DNA.
- ***Platinum compounds*** such as cisplatin and carboplatin form interstrand cross-links with DNA.
- ***Plant derivatives*** such as:
  - epipodophyllotoxins etoposide (VP-16) and tenoposide (VM-26) from roots of May apple interrupt *DNA topoisomerase II*.
  - vinca alkaloids vincristine and vinblastine from periwinkle plant cause metaphase arrest by preventing polymerization of microtubes.
- ***Synthetic hormone analogs*** such as tamoxifen and cyproterone acetate compete with the hormone (estrogen and testosterone, respectively) for the cytosolic receptor but do not elicit the normal cellular response.
- ***Other treatment approaches*** include the following:
  - Monoclonal antibodies (MAbs) conjugated to drugs, toxins, and radionuclides.

- *Antibody-directed enzyme prodrug therapy (ADEPT),* in which MAbs target enzymes that activate drugs to tumor cells. This involves administration of a prodrug that is metabolized by the targeted enzyme to release a toxic metabolite into the tumor environment.
- *Genetic prodrug activation therapy (GPAT)* in which a gene encoding a prodrug metabolizing enzyme is linked to the c-erbB2 promoter and allowed to be overexpressed in tumor cells.
- Stimulation of natural antitumor response by cytokines produced by rDNA technology.
- Inhibition of lymphokine activated killer cells (LAKs) and cytotoxic T lymphocytes by interleukins.
- Upregulation of MHC Class I antigens by interferon γ.
- Induction of tumor cell apoptosis.
- Antisense therapy and RNA interference for gene silencing.

❑ What is multidrug resistance?

Cancer cells develop resistance to multiple chemotherapeutic drugs by
- decreased uptake
- inhibiting metabolic activation of the drug
- inducing changes in intracellular targets
- overexpression of *P-glycoprotein* encoded by the *mdr1* gene that causes efflux of drugs.

Chapter 35

Genomics, Proteomics, and Bioinformatics

❑ What is a gene map?

A *gene map* refers to the assignment of a gene to a locus on a chromosome.

❑ What is genomics?

Genomics is the discipline of mapping, analyzing, and sequencing genomes.

❑ What is the rationale for mapping genomes?

Mapping genes and genomes has the following uses:

1. Understanding the inheritance of particular traits and exploiting this for gene manipulation.
2. Selective breeding of plants and animals with desired characteristics.
3. Prenatal diagnosis of diseases.
4. Understanding the biological basis of the phenotype.
5. Gene tracking, which involves mapping genes for genetic diseases to particular loci on chromosomes and tracking them through pedigrees by linkage to a marker.
6. Drug development.
7. Comparative genome mapping that would help in understanding evolution.

❑ What are the types of gene map?

The types of gene map include

1. *Cytogenetic maps*
   - Provide visual appearance of chromosomes as light and dark bands when stained that can be analyzed for chromosomal aberrations and correlated with the phenotypes.
   - These maps however have low resolution.
2. *Genetic (linkage) maps*
   - are based on inheritance patterns.
   - assign specific DNA markers (genes or noncoding DNA) to linkage groups. The probability of two markers being inherited together decreases with increasing distance between them. The distance between two markers is measured in centimorgans (cM).
   - are useful to locate inherited diseases such as sickle cell anemia and locate genes whose functions are unknown.

3. *Physical (molecular) maps*
   - are created by ordering cloned fragments of genomic DNA.
   - are calibrated in units such as base pairs (bp), kilobase pairs (kb), and megabase pairs (Mb).
   - may be of low or high resolution.

❑ Give an account of low- and high-resolution physical maps.

- ***Low-resolution maps*** localize DNA markers to individual chromosomes or chromosome subregions and include
  - *chromosomal maps* in which DNA markers can be physically associated with particular chromosomal bands by *in situ* hybridization.
  - *cDNA maps* that pinpoint positions of exons to particular chromosomal regions.
- ***High-resolution physical mapping*** involves generation of DNA fragments by restriction enzymes, amplification of the fragments by cloning or PCR, electrophoretic separation of fragments, and visualization by direct DNA staining or by hybridization with probes. Chromosomes can be separated by flow sorting or by somatic cell hybridization, and single chromosomes mapped.

❑ Enumerate the molecular markers for mapping.

A marker is any inherited physical or molecular trait that differs among individuals. Markers must be polymorphic, *i.e.*, alternative forms must exist among individuals. In humans, naturally occurring DNA polymorphisms such as microsatellites and single nucleotide polymorphisms are used as markers.

Microsatellites are the markers of choice in human genome maps for the following reasons:
1. Ubiquity in the genome.
2. The marker density is 1 microsatellite for every 100 kb.
3. PCR typability.
4. Easy to clone and characterize.
5. Exhibit polymorphism in a block of tandem repeats, between organisms of the same species, and between cells of the same organism.
6. They are the loci for a number of inherited disorders.

Single Nucleotide Polymorphisms (SNPs) are the current preferred markers in human genome mapping because of their abundance and ease of automation. These are based on differences in DNA sequences between two individuals that occur once in every 500–1000 bases on an average. The SNPs are useful in identifying disease loci and individual variations in response to drugs.

❑ How is whole-genome sequencing done?

Since sequencers can only handle short segments of DNA, the genome is broken into fragments; the fragments are sequenced and then reassembled in the correct order to arrive at the sequence of the whole genome.

Approaches to whole-genome-sequencing

1. *Shotgun sequencing:* This method is useful for sequencing long fragments of DNA. The process involves breaking the genome into small fragments of ~1500 base pairs, each of which is cloned into an appropriate vector such as M13. From this library, clones are picked up at random for sequencing. The overlapping fragments can be reassembled into the full genome sequence using computer packages. The shotgun method is rapid but the reassembly of fragments is difficult. Although the same sequence must be done several times, the final sequence is highly reliable.

2. *Clone-by-clone approach*: This involves breaking the genome into fragments of 150,000 base pairs. Each fragment is cloned into an appropriate vector to generate an ordered library of clones. Each clone is then broken into smaller fragments of 500 bp sequences. Fragments from each small-insert clone library are sequenced and the overlaps used to reconstruct the original sequence. Although this method is reliable, it is slow.

❑ What is partial genome sequencing?

- Partial genome sequencing involves use of expressed sequence tags (ESTs), which are partial cDNA sequences specific for coding DNA (cDNAs are synthesized from mRNAs, which are the products of transcription of expressed genes).
- Partial sequencing of the genome is useful for
  - identifying genes in the genome.
  - determining chromosomal location of genes.
  - identifying coding regions in genomic sequences.
  - understanding gene expression in a particular tissue.

❑ What are the goals of the Human Genome Project (HGP)?

The HGP is a 13-year international research programme launched in 1990 to sequence the human genome. The coordinated effort involves 20 groups from six countries including the U.S., U.K., Japan, France, Germany, and China.

The goals of the HGP include the following:

1. To decode the entire 3 billion bases in human DNA.
2. To identify all the genes in human DNA.
3. Characterize the genomes of human and selected model organisms such as *E.coli* and *Drosophila* by complete mapping and sequencing of their DNA. Finding similarities between the human genome and the genomes of these organisms is valuable in understanding gene function.
4. Develop technologies for data analysis.
5. Train scientists to utilize the tools and resources for the development of human health.
6. Address the ethical, legal, and social issues (ELSI) that may arise from the project.

❑ Summarize the important results of the HGP.

1. The total size of the genome is 3.2 Gb, a large percentage of which is made up of different classes of repetitive sequences.
2. The pericentric and subtelomeric regions of the chromosomes contain large recent segmental duplications that constitute 5.3% of the genome.
3. The number of coding genes in the human genome has been estimated to be between 20,000 and 25,000 with the gene density being greater in GC-rich regions.
4. Introns are longer in human DNA compared to other organisms.
5. Only 94 of the 1278 protein families are vertebrate-specific. The complexity of the human genome may be due to:
   - the noncoding DNA
   - the number and combinations of protein domains
   - alternative splicing that generates a larger number of proteins.
6. Many human genes originate by horizontal transfer from bacteria.
7. The mutation rate in males is twice that in females.

❑ What are the potential benefits and risks of mapping the human genome?

Potential benefits

The HGP will have potential benefits in the following areas:

1. *Molecular Medicine and Pharmacogenomics*: The results of the HGP will help in
   - improved diagnosis of diseases
   - early detection of genetic predisposition to diseases
   - drug development based on SNPs (personalized medicine based on genetic make up of the individual)
   - gene therapy.
2. *Assesment of risk* from radiation, chemicals, toxins.
3. *DNA forensics*: Identification of potential suspects, victims of crime and catastrophe, establishing paternity, and identification of endangered species.
4. *Biology:* The results of the HGP will help in understanding evolution, migration of population groups, mutations, and provide solutions for environmental problems. It will also unravel biological processes such as development, differentiation, and coordination of cellular activities.

Potential risks

The results of the HGP will also have some potential risks such as discrimination based on gender, race, and genetic predisposition to diseases. The purpose of the ethical, legal, and social issues (*ELSI)* established as an integral part of the HGP is to anticipate and address the implications of mapping the human genome, prevent misinterpretation or misuse, and ensure that the information is used only for the benefit of individuals and society.

❑ What is proteomics?

Proteomics is the study of the protein complement of the genome. The proteome refers to the collection of all the proteins in a genome, cell, or tissue.

❑ Enumerate the tools of proteomics.

Identification and characterization of proteins requires the integration of four important tools:

1. Databases that provide a complete catalog of all the proteins expressed in an organism. These include protein, EST, and genome sequence databases. This will help to identify a protein from the sequence information or raw spectral data.
2. Mass spectrometry (MS): A mass spectrometer consists of a *high vacuum system, sample inlet* comprising a sample plate, *HPLC/GC/capillary electrophoresis system, solids probe, electron impact* or *direct chemical ionization chamber.*

   Two types of methods are used for peptide analysis:

   i. Matrix Assisted Laser Desorption Ionization-Time of Flight (MALDI-TOF) to measure the mass of peptides.
   ii. Electrospray Ionization (ESI)-tandem MS to obtain sequence data for peptides.

   MS data can provide
   - accurate molecular mass measurements of large proteins
   - accurate molecular mass measurements of peptides from proteolytic digests that can be directly searched against databases to identify target proteins.
   - sequence analysis of peptides obtained from proteolytic digests.

3. Software that can match MS data with specific protein sequences in databases.
4. Analytical protein separation technology that simplifies complex protein mixtures to individual proteins or small groups of proteins, and allows specific proteins to be targeted for analysis. The techniques used for this purpose include
   - two-dimensional sodium dodecylsulfate polyacrylamide gel electrophoresis (2D-SDS-PAGE)
   - 1D-SDS-PAGE
   - high-performance liquid chromatography (HPLC)
   - capillary electrophoresis
   - isoelectric focusing
   - affinity chromatography.

Integration of different techniques such as ion-exchange liquid chromatography (LC) in tandem with reverse-phase (RP)-HPLC is a more powerful approach for resolving complex peptide mixtures.

❑ Outline the principle, analysis, and applications of 2D-PAGE in proteomics.

Principle

- 2D PAGE is an important tool in protein separation and proteomic analysis.
- The technique involves separation of proteins in the first dimension based on isoelectric point (isoelectric focusing), followed by separation in the second dimension by molecular weight.
- About 5000 different protein spots can be identified on a large format 2D gel.

Analysis

- The gel patterns can be compared using 2D-gel analysis software such as the SWISS-2D PAGE software.
- Flicker program that superimposes two 2D patterns allows comparison of two gels. While spots that appear on both gels appear as fixed spots, a spot that appears on one gel and not on the other is seen as a flicker.
- Protein spots of interest can be identified by peptide mass-fingerprinting. This involves cutting the protein spot from the gel, digesting it with trypsin to form peptides, which are then analysed by MALDI-MS to obtain the mass of the peptides in the samples.
- The peptide mass-fingerprints can be compared with fingerprints from databases using programs such as *Mascot* or *Protein Prospector*.
- Partial amino acid sequence data from one of the peptides can be obtained by tandem mass spectrometry and used to search the protein sequence databases (SWISSPROT, TrEMBL) for sequence identity.

Applications

- Analysis of gene expression between two biological states such as between normal and diseased tissue.
- Analysis of the effect of a drug or toxin on cells.
- Detection and development of diagnostic markers of proteins.
- Comparison of pathogenic and nonpathogenic strains of bacteria.

❑ Briefly explain differential in-gel electrophoresis.

- Differential in-gel electrophoresis (DIGE) is an improvement of 2D PAGE technology in which two samples can be compared on a single gel.

- The two samples are labeled with two different fluorescent dyes (Cy3 and Cy5), mixed, and run on a single 2D gel.
- Proteins in the two samples can be identified by scanning the gel at two different wavelengths that excite the two dyes.
- If the protein spots fluoresce at both wavelengths, it indicates expression in both samples; whereas fluorescence at only one of the wavelengths reflects the presence of the protein only in one of the samples for which the wavelength is characteristic.

❑ Discuss the methods available to determine the function of a protein.

- The function of a protein may be determined by database search for sequence homology using programs such as BLAST and PSI-BLAST. If the protein shows homology with any protein of known function in the database, it provides information on the possible function of the protein.
- If the protein is *new* and does not show any homology with a known protein, the function of the protein may be determined by any one of the following methods:
  1. *Phylogenic profile method*: A functional link can be established if the unknown protein has the same phylogenic profile as other known proteins based on the assumption that only two proteins with similar functions would tend to be inherited together into a new species.
  2. *Method of correlated gene neighbors*: A functional link may be assumed if the genes for the two proteins are found to be neighbors in several different genomes. This is based on the fact that functionally related genes map close together in the genome. Although this method is highly applicable to microbial proteins, it also relevant to eukaryotic proteins to some extent.
  3. *Analysis of fusion*: If the protein products of two genes fused into a single gene in one organism are observed to interact, it may be inferred that protein interactions also occur in another organism where these two genes exist separately.
  4. *Protein–protein interactions*: Identification of protein–protein interactions by the yeast two-hybrid (Y2H) system that exploits the modular architecture of transcription factors gives a clue to the function of an unknown protein. Protein arrays are also useful in analyzing protein–protein interactions.

❑ Explain how the yeast two-hybrid (Y2H) system enables identification of protein–protein interactions.

- The *yeast two-hybrid (Y2H) system* is one of the most widely used methods for identifying protein–protein interactions.
- This technique exploits the modular architecture of transcription factors whose genes contain two domains—a *DNA-binding domain* and a *transactivation* domain. The transactivation domain is linked to a reporter gene that can be assayed.
- One of the domains is linked to the unknown protein and the other to a protein with which the unknown protein may interact. The two interacting proteins are termed *bait* and *prey*.
- The two domains are expressed in separate yeast cells and subsequently mated to generate diploid cells.
- If the bait and prey proteins bind to each other in the diploid cell, the two domains of the transcription factor are brought together resulting in transactivation of the reporter gene that can be detected by a *transcriptional reporter assay* based on expression of the reporter gene product.

❑ Write a note on protein arrays.

❑ Write a note on protein arrays.

1. The basic principle of protein arrays is similar to DNA microarrays (Chapter 34).
2. Proteins are immobilized on a solid support such as glass or nitrocellulose membrane in the form of small spots of 150–200 μm in size.
3. The protein whose function is unknown is incubated on the array, followed by washing and identification of the known protein on the array to which the unknown protein is bound.
4. Protein arrays are useful for studying protein–protein interactions.

❑ Elucidate the applications of proteomics.

The major applications of proteomics include the following:

1. *Mining*, which involves identifying most or all of the proteins in a sample. This helps to catalog the proteome directly rather than inferring it from gene expression data.
2. *Protein-expression profiling*, which refers to the identification of proteins in a sample as a function of:
   - a particular state of the cell or organism such as development, differentiation, disease, etc. For example, comparison of normal and diseased tissue will help in understanding differential protein expression in the disease state and aid in drug development.
   - exposure to a drug, physical or chemical stimulus.
3. *Protein-network mapping*, which is an approach to analyze protein–protein interactions in living systems. This helps in understanding metabolic and signal transduction pathways and in identification of the components of multisubunit proteins.
4. *Mapping of protein modifications* as a post-translational event as well as due to chemicals and drugs.

❑ What is bioinformatics?

Bioinformatics deals with computer-based analysis of large biological data sets. It has three components:

1. Creation of databases for the storage and management of large biological data.
2. Development of algorithms and statistics to determine relationships among members of large data sets.
3. Use of these tools for the analysis of various types of biological data, including DNA, RNA, and protein sequences, protein structures, gene expression profiles, and biochemical pathways.

❑ Give an account of the various biological databases.

Many biological databases are available for researchers, which include

- ***Primary sequence databases*** for the storage of raw experimental data.
  1. Primary databases that maintain DNA sequence data include
     - *GenBank*, maintained by the National Centre for Biotechnology Information (NCBI), USA.
     - *Nucleotide Sequence Database*, maintained by the European Molecular Biology Laboratory (EMBL).
     - *DNA Databank of Japan (DDBJ)*.
  2. There are two important primary databases for storing protein sequence data—*SWISSPROT* and *Translated EMBL (TrEMBL)* with cross-reference between them.

- ***Secondary databases***
  The primary databases contain a number of secondary databases for storing particular types of sequence data. For example, some of the secondary databases of GenBank such as:
  1. dbEST, to store expressed sequence tags (EST).
  2. dbGSS, to store single-pass genomic survey sequences (GSSs).
  3. dbSTS, to store sequence tagged sites (STS).
  4. high-throughput genomic database for storing unfinished genomic sequence data.

  The secondary databases for protein families and sequence patterns include PROSITE, PRINTS, and BLOCKS.

- ***Organism specific databases***
  These databases contain sequence data as well as information on gene expression, mutant phenotypes, genome maps, sequencing projects, and scientific literature. They also provide links for obtaining clones, mutants, and for contacting researchers. Information on any organism can be obtained by accessing *Genomes Online Database (GOLD)*.

- ***Miscellaneous databases*** include
  1. *Specialized databases* specific for rRNA, tRNA, introns, promoters, and other regulatory elements.
  2. *Online Mendelian Inheritance in Man (OMIM)*, a comprehensive database of human genes and genetic disorders maintained by NCBI.
  3. *Structural databases* that store information on nucleic acids, proteins, and other macromolecules. These include Protein Data Bank (PDB), Nucleic Acid Data Bank (NDB), Macromolecular Structure Database (MSD), and Molecular Modeling Database (MMDB).
  4. *Databases for higher-level functions* of a cell include Protein Information Resource (PIR) and Kyoto Encyclopaedia of Genes and Genomes (KEGG). KEGG contains secondary databases such as PATHWAY for data on metabolic pathways, GENES for data on genes and gene products, and LIGAND, with data for chemical compounds and cellular reactions.
  5. *Literature database*: The most comprehensive is *PubMed*, maintained by the NCBI.

What are the tools available for retrieval of biological data?

Although a large amount of biological information is available over the world wide web (WWW), the data are widely distributed. The efficient and most widely used mechanisms for *data retrieval* include *Entrez*, DBGET, and SRS (Sequence Retrieval System). Each of these tools allows text-based searching of a number of linked databases as well as sequence searching with utilities such as BLAST. They differ in the databases they cover and how retrieved information is accessed and presented.

What are orthologs and paralogs?

When two homologous genes in different organisms have the same function, they are known as *orthologs*. When two genes in the same organism have different functions, they are known as *paralogs*.

What is the rationale for doing sequence similarity searches?

- Sequence similarity searches extract sequences from a database that are similar to a query sequence.

- They are useful in the following ways:
  - to identify family members
  - to predict structure and function
  - to reveal evolutionary relationships between organisms.

❑ Discuss the software programmes for analyzing sequence data.

- Genes can be identified by searching for a known homolog in the database.
- If the putative protein encoded by the uncharacterized gene exhibits significant similarity to another protein of known function, it is evidence that the ORF is a *bona fide* new gene.
- Sequence similarity programs compare the query sequence to each sequence in the database followed by scoring the degree of homology.
- *FASTA* and *BLAST* are the most widely used software programmes for analyzing sequence data.
  - These are based on locating short stretches of identical or near identical matching letters termed *words*, followed by extension into longer alignments.
  - These programmes first translate a nucleic acid sequence in six reading frames to generate six possible protein sequences. This is followed by comparisons at the protein level.
  - Uses
    1. to search general purpose sequence databases as well as organism-specific databases.
    2. can be used to query nucleic acid or protein databases.
  - *Disadvantages*
    1. do not display optimal alignment between the query sequence and the target.
    2. only pairwise alignments are possible.
    3. multiple comparisons with a number of sequences cannot be made.

❑ What is the commonly used programme for multiple sequence alignments?

CLUSTAL is the most commonly used program for multiple sequence alignments. It generates a matrix of pairwise alignment scores after comparing all the sequences submitted. This is displayed as a *family tree* of sequences (*dendrogram*). The branching order of the dendrogram is used to achieve multiple alignment of all the proteins.

❑ Write notes on amino acid substitution matrices.

- Substitution score matrices allow detection of distant evolutionary relationships between protein sequences.
- Comparison of protein sequences can detect more distant relationships than comparison of nucleic acid sequences because protein-coding sequences evolve slowly in order to maintain protein structure and function.
- Changes in protein sequence during evolution tend to involve substitutions between amino acids with similar properties because these will maintain the structural stability of the protein. For example, changes between leucine and valine.
- The most widely used amino acid substitution matrices include *PAM 250 substitution matrix* and *BLOSUM 62*.

Index

B

D

G

H

I

N

O

P

Q

R

S

T

U